Astrophysical Radiation Hydrodynamics

NATO ASI Series

Advanced Science Institutes Series

A series presenting the results of activities sponsored by the NATO Science Committee, which aims at the dissemination of advanced scientific and technological knowledge, with a view to strengthening links between scientific communities.

The series is published by an international board of publishers in conjunction with the NATO Scientific Affairs Division

A Life Sciences B Physics	Plenum Publishing Corporation London and New York
C Mathematical and Physical Sciences	D. Reidel Publishing Company Dordrecht, Boston, Lancaster and Tokyo
D Behavioural and Social Sciences E Engineering and Materials Sciences	Martinus Nijhoff Publishers The Hague, Boston and Lancaster
F Computer and Systems Sciences G Ecological Sciences	Springer-Verlag Berlin, Heidelberg, New York and Tokyo

Series C: Mathematical and Physical Sciences Vol. 188

Astrophysical Radiation Hydrodynamics

edited by

Karl-Heinz A. Winkler

and

Michael L. Norman

Max-Planck-Institute for Physics and Astrophysics, Garching, F.R.G.
and
Los Alamos National Laboratory, Los Alamos, New Mexico, U.S.A.

D. Reidel Publishing Company

Dordrecht / Boston / Lancaster / Tokyo

Published in cooperation with NATO Scientific Affairs Division

Based on the Proceedings of the NATO Advanced Research Workshop on
Astrophysical Radiation Hydrodynamics
Garching, F.R.G.
August 2-13, 1982

Library of Congress Cataloging in Publication Data

NATO Advanced Research Workshop on Astrophysical Radiation Hydrodynamics
(1982: Garching bei München, Germany)
Astrophysical radiation hydrodynamics.

(NATO ASI series. Series C, Mathematical and physical sciences; vol. 188)
"Proceedings of the NATO Advanced Research Workshop on Astrophysical Radiation Hydrodynamics, Garching, F.R.G., August 2–13, 1982" – Verso t.p.
Includes index.
1. Gas dynamics–Congresses. 2. Radiative transfer–Congresses. 3. Hydrodynamics–Congresses. 4. Astrophysics–Comgresses. I. Winkler, K.-H. A. (Karl-Heinz A.), 1948- II. Norman, Michael L. III. Title. IV. Series: NATO ASI series. Series C, Mathematical and physical sciences, vol. 188.
QB466.G38N38 1982 523.01'92 86–22928
ISBN-13: 978-94-010-8612-7 e-ISBN-13: 978-94-009-4754-2
DOI: 10.1007/978-94-009-4754-2

Published by D. Reidel Publishing Company
P.O. Box 17, 3300 AA Dordrecht, Holland

Sold and distributed in the U.S.A. and Canada
by Kluwer Academic Publishers,
101 Philip Drive, Assinippi Park, Norwell, MA 02061, U.S.A.

In all other countries, sold and distributed
by Kluwer Academic Publishers Group,
P.O. Box 322, 3300 AH Dordrecht, Holland

D. Reidel Publishing Company is a member of the Kluwer Academic Publishers Group

CONTENTS

PREFACE

This NATO Advanced Research Workshop was devoted to the presentation, evaluation, and critical discussion of numerical methods in nonrelativistic and relativistic hydrodynamics, radiative transfer, and radiation-coupled hydrodynamics. The unifying theme of the lectures was the successful application of these methods to challenging problems in astrophysics. The workshop was subdivided into 3 somewhat independent topics, each with their own subtheme.

Under the heading <u>radiation hydrodynamics</u> were brought together context, theory, methodology, and application of radiative transfer and radiation hydrodynamics in astrophysics. The intimate coupling between astronomy and radiation physics was underscored by examples from past and present research. Frame-dependence of both the equation of transfer (plus moments) and the underlying radiation quantities was discussed and clarified. Limiting regimes in radiation-coupled flow were identified and described; the dynamic diffusion regime received special emphasis. Numerical methods for continuum and line transfer equations in a given background were presented. Two examples of methods for computing dynamically coupled radiation/matter fields were given. In 1-d and assuming LTE the complete equations of radiation hydrodynamics can be solved with current computers. Such is not the case in 2- or 3-d, which were identified as target areas for research. The use of flux-limiters was vigorously discussed in this connection, and enlivened the meeting.

Under the heading <u>gas dynamics</u> were presented 5 complementary computational approaches: finite difference, finite element, particle, spectral, and modal. Lectures focussed on similarities and differences in formulation and implementation, and identified flow regimes and structures that each method is best suited for. Two important innovations within the framework of finite difference methods emerged for the treatment of discontinuous flow: 1) implicit adaptive mesh, and 2) explicit nonlinearity. The type and use of basis functions in various methods became a consistent theme and a means of classification.

Thus, local (possibly constrained) basis functions are used in particle-mesh, finite element, and modern finite difference methods, and global basis functions are used in spectral and modal methods. Cross-fertilization and subsequent development of powerful hybrid methods was identified as a growth area.

Under the heading relativistic flows were examined methods for the evolution of space-times with and without matter present. Ultrarelativistic gas flows in flat space were shown to pose stringent requirements on methods of solution. An accurate method for such flows in 1-d was described. The 2 complementary approaches for the numerical solution of the Einstein equations were offered for comparison and evaluation. The well-established approach of solving the Cauchy problem was extensively reviewed, with an eye for future applications on next generation supercomputers. An application of these techniques to collapsing Zel'dovich pancakes in the early universe was given. The second approach of solving the characteristic initial value problem was introduced, and novel techniques for integration through caustics and up to singularities were described.

This NATO Advanced Research Workshop was financially sponsored by the NATO Scientific Affairs Division and the Max-Planck-Institut fuer Physik und Astrophysik. The organizers wish to express their gratitude for this support.

Karl-Heinz A. Winkler
Michael L. Norman

PROBLEMS IN ASTROPHYSICAL RADIATION HYDRODYNAMICS

John I. Castor

Lawrence Livermore National Laboratory

1. ABSTRACT

The basic equations of radiation hydrodynamics are discussed in the regime that the radiation is dynamically as well as thermally important. Particular attention is paid to the question of what constitutes an acceptable approximate non-relativistic system of dynamical equations for matter and radiation in this regime. Further discussion is devoted to two classes of application of these ideas. The first class consists of problems dominated by line radiation, which is sensitive to the velocity field through the Doppler effect. The second class is of problems in which the advection of radiation by moving matter dominates radiation diffusion.

2. INTRODUCTION

The objective of my lectures in this workshop is to try to illuminate some of the phenomena that arise in the coupling of radiation with matter in the presence of a material velocity field. My discussion will be entirely analytic, and often very approximate, so perhaps a few words are in order about the relationship of this kind of discussion to the intensely computational subject addressed by the majority of the speakers.

The first reason I would advance why analytic theory is helpful in a computational discipline is that it can be very difficult to calculate phenomena correctly if one does not have a fairly good idea of what to expect ahead of time. That is,

K.-H. A. Winkler and M. L. Norman (eds.), Astrophysical Radiation Hydrodynamics, 1–43.

the choice of zoning, time steps, frequency bin structure, discrete angles, and also of the dominant physical processes, depends on the nature of the solution of the problem, especially if the result is to be obtained with reasonable economy. Thus the availability of an analytic caricature of the problem that displays the main phenomena may be a great aid in setting up the numerical calculation. The second reason is that in many cases the rough analytic answer may be good enough. An answer that is correct in the order of magnitude may be the most refined theory that is warranted when the data are at least that inaccurate; when the data are indeed very good, a rough calculation may suffice to show that a certain theory is hopeless. A somewhat extreme statement of this point of view is: Computational methods exist to provide accurate answers, when these are needed, to problems whose approximate solutions are already known from other means. While this is certainly arguable, it is a viewpoint that represents a counterweight to the computing-nature-as-it-is philosophy, which can result in the waste of alarming amounts of computer time on ill-conceived problems. Of course, an intelligent combination of analytic estimation and economical computation is best.

My first lecture will be devoted to a general discussion of the equations of radiation hydrodynamics, and in particular of the photon transport equation for a moving medium. These remarks are intended to be complementary to the thorough discussion by D. Mihalas, also in this workshop. Even at this late date there appears to remain some confusion about the choices the theoretician can make of the best form of transport equation for his own problem, and even some confusion about the fact that choices do exist, and must be made, either explicitly or inadvertently. So the first lecture will begin with a discussion of the most general form of the transport equation, then proceed to its $O(v/c)$ reduction, and a further simplification of that to a practical system for use in cases of modest velocity. The limiting forms of the radiation hydrodynamics equations in optical thick and optically thin regimes will be discussed, with further approximations that can be made in these cases.

The second lecture will emphasize transport of spectral line radiation, for which the effects of a velocity field can easily be very large. The methods of approximation breaks into categories for subsonic, transonic and hypersonic problems. The second category is naturally the most difficult, and I will call attention to the lack of a generally useful computational method for such problems. The hypersonic class is the one for which Sobolev's approximation is designed, and I will discuss some interesting phenomena that arise when radiation treated in the Sobolev approximation is coupled to matter. As pointed out by

Karp, Lasher, Chan and Salpeter, there are many astrophysical flow problems in which the medium is optically thick, but the velocity gradient is large enough to smear the lines until they cover a substantial part of the spectrum and thereby affect the radiation diffusion rate substantially. The method they devised to deal with this situation is a presently underutilized tool.

The third lecture is devoted to a discussion of two examples of "dynamic diffusion" of radiation. This is the nomenclature introduced by Mihalas to distinguish two types of diffusion. In the diffusion (optically thick) limit, the dynamical equation for the radiation energy density as a function of inertial coordinates has the form of an advection-diffusion equation with a source term and a sink term. The relative importance of advection versus diffusion gives two cases: "static diffusion" (diffusion dominates) and "dynamic diffusion" (advection dominates). The dynamic diffusion regime with decoupled matter and radiation is relatively unfamiliar to astrophysicists, so a discussion of these two examples may be a useful illustration of the phenomena that can occur. The examples, drawn from the winds-and-accretion literature of high-energy astrophysics, are of the influence of cyclotron line radiation on neutron star accretion, specifically the work of Langer and Rappaport, and the wind associated with supercritical disc accretion, studied by Meier.

3. FIRST LECTURE -- EQUATIONS OF RADIATION HYDRODYNAMICS

There are two kinds of radiation hydrodynamics: the ordinary kind and the subtle kind. The ordinary kind is represented by problems for which the Boltzmann number $B_o = F_m/F_R$ is of order unity, in which F_m is a material heat flux $\rho C_p T v$, and F_R is the radiative flux σT^4. The typical ratio of radiation energy density to matter energy density is $(F_R/c)/(\rho C_p T) = (v/c)B_o^{-1}$, so in this kind of problem radiation energy and radiation pressure are $O(v/c)$ and play only minor roles. This has the consequence that every term in the transport equation involving the velocity or a time derivative can be dropped with no great effect. This leads to the simple static transfer equation with which we are all familiar. The subtle kind of radiation hydrodynamics is represented by problems for which the Boltzmann number is small, of order v/c, so the radiation energy density and matter energy density are comparable, while the radiative energy flux overwhelms the other fluxes. In this case most of the terms in the transport equation that could be dropped before must now be retained, since otherwise a completely wrong answer would be obtained. In ordinary radiation hydrodynamics the thermal effect of radiation is considerable, but its mechanical effect is not. In subtle radiation hydrodynamics the mechanical effect of radiation is also important. The subtle kind of radiation hydrodynamics is both more trouble for the theorist and more interesting, and it is that I want to discuss.

3.1. Relativistic photon transport

I would like to begin the discussion of the equations of radiation hydrodynamics with the photon Boltzmann equation in its relativistic form. Please note: the use of some of the methods of general relativity (GR) in the development does not mean that this work is tied to typical GR problems. In fact, I am not going to say anything about gravity, field equations, etc., etc. The point is that the covariant language of relativity is a handy way of getting answers to otherwise awful problems and avoiding some conceptual pitfalls along the way. This language also incorporates special relativity, which of course is essential in a proper formulation.

Definitions: The intensity function is the product of the occupation number per photon state and a phase-space factor involving the frequency that changes from one observer's frame to another. The occupation number itself is a Lorentz invariant, so this is taken as the invariant intensity,

$$\mathscr{I} = \frac{I_\nu}{\nu^3} . \tag{1}$$

Likewise, frequency-dependent factors appear in the quantum expressions for the absorption coefficient k (length^{-1}) and the emission coefficient j_ν (energy/(volume time frequency·solid angle)). Dividing out these factors puts them into invariant form:

$$\sigma = \nu k_\nu \tag{2}$$

and

$$\eta = j_\nu/\nu^2. \tag{3}$$

Now, what does it mean to say that the variables $\mathscr{I}$, σ, and η are invariant? It does not mean that they are independent of any of the variables space, time, frequency or direction, although they may be. What it does mean is that different observers in relative motion will agree on their values, after they have properly worked out the correspondences of frequency and direction.

The properly covariant transport equation is the following:

$$\frac{d\mathcal{I}}{d\lambda} = -\sigma\mathcal{I} + \eta \ , \tag{4}$$

in which λ is a scalar parameter measuring displacement along the photon's world line, that is,

$$\frac{dx^\alpha}{d\lambda} = M^\alpha \ , \tag{5}$$

in which $(x^\alpha) = (x^1, x^2, x^3, t)$ is the position four-vector and $(M^\alpha) = \nu(n^1, n^2, n^3, 1)$ is the photon momentum four-vector. I will use the conventions that α runs from 1 to 4, time being the fourth coordinate, and that the Minkowski metric is $ds^2 = -(dx^1)^2 - (dx^2)^2 - (dx^3)^2 + dt^2$. The 3-vector $\underset{\sim}{n} = (n_1, n_2, n_3)$ is the ordinary unit direction vector of the photon. Now although M^α is a 4-vector, it is a null vector and so it has only 3 functionally independent components, for example M^1, M^2, M^3. Not to prejudice the formalism, suppose that the four components of M^α are functions to be specified later of three variables y^1, y^2 and y^3. Thus the invariant intensity, as well as the invariant opacity and emission coefficient, depend on the seven variables x^1, x^2, x^3, t, y^1, y^2, and y^3. Doing out the derivative in equation (4) using the chain rule, and making use of equation (5), leads to

$$\frac{d\mathcal{I}}{d\lambda} = M^\alpha \frac{\partial\mathcal{I}}{\partial x^\alpha} + \frac{dy^a}{d\lambda}\frac{\partial\mathcal{I}}{\partial y^a} \ , \tag{6}$$

where the <u>roman</u> index a runs from 1 to 3.

The hard part of obtaining the right transport equation lies in finding the derivatives $dy^a/d\lambda$. The physical basis for this task is the assumption that the photon momentum is unchanged, in some sense, as the photon moves along its path. Now, it is not literally unchanged unless space-time is flat and the metric is constant. More general cases invoke the concept of <u>parallel transport</u>. What does this mean? It is a rule for carrying a 4-vector such as M^α along any curve in space-time. In order to do the carrying from the point on the curve labelled by λ to the point labelled $\lambda + d\lambda$, a transformation is made locally to a coordinate system for which the metric is constant up to terms of $(\text{displacement})^1$. With respect to this special coordinate system, M^α should have the same components at λ and at $\lambda + d\lambda$. When this condition is expressed in terms of the components in the general coordinate system, it is

$$\frac{\delta M^\alpha}{\delta\lambda} = \frac{dM^\alpha}{d\lambda} + \Gamma^\alpha_{\beta\gamma} M^\beta \frac{dx^\gamma}{d\lambda} = 0 \ , \tag{7}$$

in which the Γ symbols are the Christoffel symbols derived from the metric tensor.

Now we can say how y^a should change with λ. Let the 4-momentum $M^\alpha(x^\alpha, y^a)$ at λ be parallel-transported to $\lambda + d\lambda$. Scalar products and lengths of 4-vectors are unchanged by being parallel-transported, so if M^α was a null vector at λ, its transported self is a null vector at $\lambda + d\lambda$. As such, it must agree with $M^\alpha(x^\alpha + M^\alpha d\lambda, y^a + dy^a)$ for some choice of the perturbations dy^a to the null-cone coordinates y^a. This leads to the equation

$$M^\beta \frac{\partial M^\alpha}{\partial x^\beta} + \Gamma^\alpha_{\beta\gamma} M^\beta M^\gamma + \frac{dy^a}{d\lambda} \frac{\partial M^\alpha}{\partial y^a} = 0 \ , \tag{8}$$

or, in more compact form,

$$\frac{dy^a}{d\lambda} = y^a_\beta M^\beta \ , \tag{9}$$

where

$$M^\alpha_{;\beta} + \frac{\partial M^\alpha}{\partial y^a} y^a_\beta = 0 \ , \tag{10}$$

and $M^\alpha_{;\beta}$ is the usual covariant derivative of the contravariant vector M^α with the parameters y^a held fixed. For each of the four values of β equation (10) represents a set of four linear equations for the three unknowns y^a_β. That these have a solution is ensured by the conditions that M^α be a null vector at all x^α, whatever the choice of y^a. Finally, we find the transport equation

$$M^\alpha \frac{\partial \mathscr{I}}{\partial x^\alpha} + M^\beta y^a_\beta \frac{\partial \mathscr{I}}{\partial y^a} = -\sigma\mathscr{I} + \eta \ . \tag{11}$$

A summary of where we are: We use the invariant intensity, and absorption and emission coefficients, to free us from reference to any particular frame. We can choose any coordinate system x^α that we find to be convenient; it certainly does not have to be the rest frame of the fluid (not a practical choice in multidimensional problems anyway). Then we can choose a functional representation of the photon 4-momentum in terms of 3 felicitously chosen variables y^a. The choice of these variables is in no way tied to the choice of our coordinate frame. In fact, for many problems the most convenient choice of frame is the Eulerian one of a fixed inertial frame, and the most convenient choice of y^a would be the 3-momentum of the photon in the co-moving frame, expressed in polar coordinates. The functional form of M^α in term of space-time and null-cone coordinates, plus some algebra, leads to the desired transport equation. If it is desired, the factors of ν can be reinstated in the transport equation, as known functions of y^a, to give an equation for the intensity as seen by an observer in some frame. If y^a is based on the comoving-frame 3-momentum, it would be natural to use the intensity in the comoving frame as well.

In numerical computations there is a further choice that must be made. This is of the computational mesh. One choice is the Eulerian one of a mesh that is fixed at discrete values of the chosen coordinates x^α. However, the mesh may also move in time, and if the nodes of the mesh are required to move with the local fluid velocity the Lagrangian system of equations results. It must be emphasized that one is using a Lagrangian mesh in this case, not Lagrangian coordinates. Lagrangian coordinates, in the sense of an orthogonal coordinate system with respect to which the fluid is everywhere and at all times at rest, do not exist for fluid flows unless the vorticity vanishes. This is true even if orthogonality only between space and time is desired.

The Lagrangian approach really consists of a formulation of dynamical equations with respect to inertial coordinates, followed by the appropriate transformation of the time derivative to give the evolution equations with respect to the moving mesh. Thus a perfectly reasonable way of doing radiation hydrodynamics is to solve on a Lagrangian mesh the equations with respect to inertial coordinates for the radiation field in the comoving frame! And of course other combinations are equally possible.

We can proceed further in reducing the transport equation is we assume a linear relation between y^a and the x^α-frame components of M:

$$y^a = A^a_\alpha M^\alpha , \tag{12}$$

which is to say that we have selected 3 covariant vectors A^a_α, a = 1 to 3, that must be space-like, and define y^a as the projection of the 4-momentum onto these. If we add a fourth time-like vector A^4_α, then the four covariant vectors $A^{\alpha'}_\alpha$, with $\alpha' = 1$ to 4, form a covariant basis set at the point x. This basis set replaces the natural covariant basis $\partial/\partial x^\alpha$, and presumably we can choose the new basis in such a way as to make the transport equation nicer. If the components of the four covariant vectors are arranged as the rows of a 4x4 matrix, this matrix gives the transformation from the natural basis to the new one, for contravariant vectors. The matrix inverse of $(A^{\alpha'}_{\alpha})$ is another matrix $(A^{\alpha}_{\alpha'})$ of which the columns comprise four contravariant vectors that form a contravariant basis at x^α. The inverse relation to equation (12) is then

$$M^\alpha = A^\alpha_{\alpha'}\, y^{\alpha'} \ , \tag{13}$$

in which a term in y^4 also enters, but y^4 is determined by the requirement that M^α be null:

$$g_{\alpha\beta}\, A^\alpha_{\alpha'}\, A^\beta_{\beta'}\, y^{\alpha'}\, y^{\beta'} = 0 \ , \tag{14}$$

a quadratic equation for y^4, given values of y^a, with one positive root.

Equation (13) is the representation of M^α in terms of y^a that is needed for the transport equation. Carrying out the algebraic steps needed to find y^a_α gives the result

$$y^a_\alpha = -A^a_\beta\, A^\beta_{\alpha';\alpha}\, y^{\alpha'} = A^a_{\beta;\alpha}\, A^\beta_{\alpha'}\, y^{\alpha'} = M^\beta\, A^a_{\beta;\alpha} \ , \tag{15}$$

where the second form follows from the first since the product of inverse matrices is the constant identity, of which the derivative vanishes. Substitution into equation (11) gives this form for the transport equation:

$$M^\alpha \frac{\partial \mathscr{I}}{\partial x^\alpha} + M^\alpha\, M^\beta\, A^a_{\beta;\alpha} \frac{\partial \mathscr{I}}{\partial y^a} = -\sigma \mathscr{I} + \eta \ . \tag{16}$$

This gives a physical interpretation of the coefficient $dy^a/d\lambda$ of $\partial\mathscr{I}/\partial y^a$: it is the <u>strain</u> in the M direction associated with a field of displacements along the ath basis vector.

For the further specialization of the transport equation we assume (1) that spacetime is flat and that the x^α frame is Minkowskian; (2) that transformation (13) is a Lorentz transformation, so the y basis is orthonormal, i.e., the y^a are <u>tetrad components</u>; (3) the Lorentz transformation is a simple boost by the fluid velocity, of which the 3-space components are v^i. The tetrad components then consist of a time-like one, which is just the projection on the fluid 4-velocity (that was the object of the transformation), and space-like ones that are the projections on the original space-like unit vectors boosted by v^i. The transformation of the time-like component is the Doppler effect, and the transformation of the space-like ones is the abberation effect. Working out the partial derivatives of the usual Lorentz transformation formulae leads to the following result

$$\frac{d\mathscr{I}}{d\lambda} + \frac{dy^a}{d\lambda}\frac{\partial\mathscr{I}}{\partial y^a} = -\sigma\mathscr{I} + \eta \ , \qquad (17)$$

where

$$\frac{dy^a}{d\lambda} = \frac{\gamma^2}{\gamma+1}\, y^i\left(v^a\frac{dv^i}{d\lambda} - v^i\frac{dv^a}{d\lambda}\right) - y^4\left(\gamma\frac{dv^a}{d\lambda} + \frac{\gamma^3}{\gamma+1}\, v^i\frac{dv^i}{d\lambda}\, v^a\right) \qquad (18)$$

(with $\gamma = (1 - v^2)^{-1/2}$), and the operator $d/d\lambda$ can have any of the following forms

$$\frac{df}{d\lambda} \equiv y^{\alpha'} f_{,\alpha'} = A^{\alpha}_{\alpha'} f_{,\alpha}\, y^{\alpha'} = M^{\alpha} f_{,\alpha} \ , \qquad (19)$$

in which $A^{\alpha}_{\alpha'}$ is the Lorentz transformation matrix. $f_{,\alpha'}$ represents derivatives in the tetrad directions, while $f_{,\alpha} = \partial f/\partial x^{\alpha}$. When the A's are substituted equation (17) becomes quite long, so I will not write it here.

The terms in equation (18) arise from the Lorentz transformation from the tetrad defined by v^i to the one defined by $v^i + (dv^i/d\lambda)d\lambda$. This transformation consists of a boost by roughly dv^i plus a rotation by a small angle around the axis $\underset{\sim}{v} \times d\underset{\sim}{v}$. The latter effect is essentially the Thomas precession. Specifically, the y^i term in equation (18) is the Thomas precession term, and the y^4 term represents the Doppler and aberration effects associated with the boost by $\approx dv^i$. The boost is not precisely by dv^i, as we see. In fact, the coefficient of y^4 in equation (19) consists of the space-like tetrad components of the 4-vector $du^{\alpha}/d\lambda$, where u^{α} is the fluid 4-velocity. If the velocity referred to a local almost comoving frame is $\underset{\sim}{v}'$, these components do form the vector $d\underset{\sim}{v}'/d\lambda$.

3.2. O(v/c) transport equation

When the terms in the fluid velocity higher than the first are discarded, the non-relativistic (NR) limit, the equations simplify greatly. The Thomas precession vanishes, and the infinitesimal boost is precisely by dv^i. Furthermore, the transformation to the tetrad components is linear in v. The transport equation becomes

$$(y^4 + y^i v^i)\mathscr{I}_{,4} + (y^i + y^4 v^i)\mathscr{I}_{,i} - y^4\left(y^4 v^i_{,4} + y^j v^i_{,j}\right)\frac{\partial \mathscr{I}}{\partial y^i} = -\sigma\mathscr{I} + \eta \; . \quad (20)$$

This equation is equivalent to the one given by Buchler (1982). Restoring the factors of c required for dimensional consistency and using a compact notation gives Buchler's form:

$$\left(1 + \frac{\underset{\sim}{n} \cdot \underset{\sim}{v}}{c}\right)\frac{1}{c}\,\frac{\partial I_\nu}{\partial t} + \left(\underset{\sim}{n} + \frac{\underset{\sim}{v}}{c}\right)\cdot \underset{\sim}{\nabla} I_\nu - \left(\frac{\underset{\sim}{a} \cdot \underset{\sim}{n}}{c} + \underset{\sim}{n} \cdot \underset{\sim}{\nabla}\underset{\sim}{v} \cdot \underset{\sim}{n}\right)\frac{\nu}{c}\,\frac{\partial I_\nu}{\partial \nu}$$
$$- \frac{\partial}{\partial \underset{\sim}{n}} \cdot \left(P\left(\frac{\underset{\sim}{a}}{c^2} + \frac{\underset{\sim}{n} \cdot \underset{\sim}{\nabla}\underset{\sim}{v}}{c}\right) I_\nu\right) + \left(\frac{\underset{\sim}{a} \cdot \underset{\sim}{n}}{c^2} + \frac{\underset{\sim}{\nabla} \cdot \underset{\sim}{v}}{c}\right) I_\nu = -k_\nu I_\nu + j_\nu \; , \quad (21)$$

where the new symbols are the operator P that projects a vector perpendicular to $\underset{\sim}{n}$, the differential operator $\partial/\partial\underset{\sim}{n}\cdot$, which is the divergence operator on the unit sphere, and the acceleration vector $\underset{\sim}{a} = d\underset{\sim}{v}/dt$. The notations $\underset{\sim}{n}$ and ν are now used for the direction and frequency in the comoving frame. This is equivalent to the following conservation-law form:

$$\frac{1}{c}\,\frac{\partial}{\partial t}\left[\left(1 + \frac{\underset{\sim}{n} \cdot \underset{\sim}{v}}{c}\right)\frac{I_\nu}{\nu}\right] + \underset{\sim}{\nabla} \cdot \left[\left(\underset{\sim}{n} + \frac{\underset{\sim}{v}}{c}\right)\frac{I_\nu}{\nu}\right] + \frac{\partial}{\partial \nu}\left[-\left(\frac{\underset{\sim}{a} \cdot \underset{\sim}{n}}{c^2} + \frac{\underset{\sim}{n} \cdot \underset{\sim}{\nabla}\underset{\sim}{v} \cdot \underset{\sim}{n}}{c}\right)\nu\,\frac{I_\nu}{\nu}\right]$$
$$+ \frac{\partial}{\partial \underset{\sim}{n}} \cdot \left[-\,P\left(\frac{\underset{\sim}{a}}{c^2} + \frac{\underset{\sim}{n} \cdot \underset{\sim}{\nabla}\underset{\sim}{v}}{c}\right)\frac{I_\nu}{\nu}\right] = -k_\nu \frac{I_\nu}{\nu} + \frac{j_\nu}{\nu} \; . \quad (22)$$

The quantity that is being conserved is $(1 + \underset{\sim}{n} \cdot \underset{\sim}{v}/c)$ times the photon number density as measured in the comoving frame. It is not hard to show that the Doppler factor in this expression accounts for the difference in number density when the volume element is fixed in the inertial frame rather than fixed in the comoving frame. Thus the conserved quantity is indeed the inertial frame number density of photons in a unit comoving frame momentum volume. Equation (22) is a multidimensional continuity equation, with the advection "velocities"

$$\frac{d\underset{\sim}{r}}{dt} = \frac{\underset{\sim}{n}c + \underset{\sim}{v}}{1 + \frac{\underset{\sim}{n} \cdot \underset{\sim}{v}}{c}} , \tag{23}$$

$$\frac{d\nu}{dt} = -\left(\frac{\underset{\sim}{a} \cdot \underset{\sim}{n}}{c} + \underset{\sim}{n} \cdot \nabla\underset{\sim}{v} \cdot \underset{\sim}{n}\right)\nu , \tag{24}$$

and

$$\frac{d\underset{\sim}{n}}{dt} = -P\left(\frac{\underset{\sim}{a}}{c} + \underset{\sim}{n} \cdot \nabla\underset{\sim}{v}\right), \tag{25}$$

to first order in $\underset{\sim}{v}$.

An analysis of equation (21) by Buchler, and of the similar one dimensional equation by Mihalas and Weaver (1982), shows that certain of the terms are never important in any low-velocity problem, regardless of the time scales and optical thickness. These are all the terms with c^2 in the denominator in equation (22): the acceleration terms and the Doppler factor in the time derivative. Dropping them gives this equation

$$\frac{1}{c}\frac{\partial}{\partial t}\left(\frac{I_\nu}{\nu}\right) + \frac{1}{c}\underset{\sim}{\nabla} \cdot \left(\underset{\sim}{v}\frac{I_\nu}{\nu}\right) + \underset{\sim}{\nabla} \cdot \left(\underset{\sim}{n}\frac{I_\nu}{\nu}\right)$$

$$+ \frac{\partial}{\partial\nu}\left(-\frac{\underset{\sim}{n} \cdot \nabla\underset{\sim}{v} \cdot \underset{\sim}{n}}{c} I_\nu\right) + \frac{\partial}{\partial\underset{\sim}{n}} \cdot \left(-P\frac{\underset{\sim}{n} \cdot \nabla\underset{\sim}{v}}{c}\frac{I_\nu}{\nu}\right)$$

$$= -k_\nu\frac{I_\nu}{\nu} + \frac{j_\nu}{\nu} . \tag{26}$$

There are four terms in equation (26) with c in the denominator. These are responsible for the dynamical effects of radiation. They are the time derivative term, which we may call the retardation effect; the divergence term with the fluid velocity, which we may call the advection effect (for the tendency of the radiation to be carried with the fluid, owing to this term); the frequency derivative, which we may call the Doppler effect; and the angular divergence, which we may call the abberation effect.

3.3. Radiation moments and coupling to matter

The first two moments of equation (26) with respect to angle are very important, since they form the basis of many computational methods, as well as leading to the momentum and energy conservation laws. In the first, unweighted, moment, the source-sink terms and the spatial transport term are subject to severe cancellation, and for this reason the 1/c terms can play a major role; this is not true of the $\underset{\sim}{n}$-weighted moment, for which only the time derivative term among the 1/c terms can be important. This leads to this reduced set of moment equations:

$$\frac{\partial E_\nu}{\partial t} + \underset{\sim}{\nabla} \cdot (\underset{\sim}{v} E_\nu) + \underset{\sim}{\nabla} \cdot \underset{\sim}{F}_\nu - \nu \frac{\partial \underset{\approx}{P}_\nu}{\partial \nu} : \underset{\sim\sim}{\nabla v} = -k_\nu \, cE_\nu + 4\pi j_\nu \quad (27)$$

$$\frac{1}{c} \frac{\partial \underset{\sim}{F}_\nu}{\partial t} + c \, \underset{\sim}{\nabla} \cdot \underset{\approx}{P}_\nu = -k_\nu \underset{\sim}{F}_\nu \, , \quad (28)$$

where $\underset{\approx}{P}_\nu$ is the monochromatic pressure tensor. Their frequency-integrated form is

$$\frac{\partial E}{\partial t} + \underset{\sim}{\nabla} \cdot (\underset{\sim}{v} E) + \underset{\sim}{\nabla} \cdot \underset{\sim}{F} + \underset{\approx}{P} : \underset{\sim\sim}{\nabla v} = \int_0^\infty (-k_\nu c E_\nu + 4\pi j_\nu) \, d\nu \quad (29)$$

$$\frac{1}{c} \frac{\partial \underset{\sim}{F}}{\partial t} + c \, \underset{\sim}{\nabla} \cdot \underset{\approx}{P} = - \int_0^\infty k_\nu \underset{\sim}{F}_\nu \, d\nu \, . \quad (30)$$

The equations of radiation hydrodynamics are completed by adding to the foregoing dynamical equations for the radiation field the fluid equations for the material. In the form compatible with our v/c approximation of the radiation equations these are

$$\frac{\partial \rho_0}{\partial t} + \underset{\sim}{\nabla} \cdot (\rho_0 \underset{\sim}{v}) = 0 \quad (31)$$

$$\rho_0\left(\frac{\partial E_m}{\partial t} + \underset{\sim}{v} \cdot \underset{\sim}{\nabla} E_m\right) + p_m \underset{\sim}{\nabla} \cdot \underset{\sim}{v}$$

$$= - \int_0^\infty (4\pi j_\nu - k_\nu cE_\nu)\, d\nu \qquad (32)$$

and

$$\left(\rho_0 + \frac{\rho_0 E_m}{c^2} + \frac{p_m}{c^2}\right)\left(\frac{\partial \underset{\sim}{v}}{\partial t} + \underset{\sim}{v} \cdot \underset{\sim}{\nabla}\underset{\sim}{v}\right) + \underset{\sim}{\nabla}\, p_m$$

$$+ \frac{\underset{\sim}{v}}{c^2}\frac{\partial p_m}{\partial t} = \frac{1}{c}\int_0^\infty k_\nu \underset{\sim}{F}_\nu\, d\nu\,, \qquad (33)$$

where the new variables are ρ_0, the density of proper mass, not including excitation energy, in the fluid frame, E_m, the material internal energy per unit proper mass, and p_m, the material pressure. To allow for the possibility that the material may be thermally relativistic (the sound speed approaches c even though $v << c$) the enthalpy contribution to the inertial mass is retained, as is the correct form for the fluid-frame pressure gradient. The reason for this is seen when $(1 + E_m/c^2)\underset{\sim}{v}$ times equation (31) and $\underset{\sim}{v}/c^2$ times equation (32) are added to equation (33). The result, neglecting a v^2/c^2 term, is

$$\frac{\partial}{\partial t}\left[\left(\rho_0 + \frac{\rho_0 E_m}{c^2} + \frac{p_m}{c^2}\right)\underset{\sim}{v}\right]$$

$$+ \underset{\sim}{\nabla} \cdot \left[\left(\rho_0 + \rho_0\frac{E_m}{c^2} + \frac{p_m}{c^2}\right)\underset{\sim}{v}\,\underset{\sim}{v}\right] + \underset{\sim}{\nabla}\, p_m$$

$$= \frac{1}{c}\int_0^\infty k_\nu \underset{\sim}{F}_\nu\, d\nu - \frac{\underset{\sim}{v}}{c^2}\int_0^\infty (4\pi j_\nu - k_\nu cE_\nu)\, d\nu\,. \qquad (34)$$

We see that the momentum coupling term between radiation and matter has different forms in equations (33) and (34). In both cases the force that appears on the right-hand side consists of the space part of the 4-force coupling radiation and matter, but in the acceleration equation (33) the fluid-frame space part appears, while in the momentum conservation equation (34) the inertial frame space part appears. The additional term in the force in equation (34) can be thought of as the reaction force due to isotropic absorption and emission in the fluid frame, which is anisotropic in the inertial frame owing to the abberation effect. This correction can dominate the one due to the fluid-frame flux if the medium is transparent and emits strongly.

The form of radiation momentum conservation comparable with equation (34) is

$$\frac{1}{c^2}\frac{\partial}{\partial t}\left(\underset{\sim}{F} + \underset{\sim}{v}E + \underset{\approx}{P}\cdot\underset{\sim}{v}\right) + \underset{\sim}{\nabla}\cdot\left(\underset{\approx}{P} + \frac{\underset{\sim\sim}{vF} + \underset{\sim\sim}{Fv}}{c^2}\right) = -\frac{1}{c}\int_0^\infty k_\nu \underset{\sim}{F}_\nu \, d\nu + \frac{\underset{\sim}{v}}{c^2}\int_0^\infty \left(4\pi j_\nu - k_\nu cE_\nu\right) d\nu \,, \qquad (35)$$

and the corresponding radiation energy equation in conservation-law form is

$$\frac{\partial}{\partial t}\left(E + \frac{2\underset{\sim}{v}}{c^2}\cdot\underset{\sim}{F}\right) + \underset{\sim}{\nabla}\cdot\left(\underset{\sim}{F} + \underset{\sim}{v}E + \underset{\approx}{P}\cdot\underset{\sim}{v}\right) = \int_0^\infty \left(4\pi j_\nu - k_\nu cE_\nu\right) d\nu - \frac{\underset{\sim}{v}}{c}\cdot\int_0^\infty k_\nu \underset{\sim}{F}_\nu \, d\nu \,. \qquad (36)$$

Some problems arise with these equations, however. First of all, in order to put the radiation moments into conservative form, *all the c^{-1} and c^{-2} terms must be kept in the transport equation.* The c^{-2} terms are needed to get the correct velocity terms under the time derivatives, and all the c^{-1} terms are needed to get the velocity terms under the space derivatives. For this reason equations (35) and (36) cannot be obtained from equations (29) and (30). Also, while the right-hand sides of equations (35) and (34) cancel, thus ensuring total momentum conservation, the material total energy equation, the right-hand side of which cancels the right-hand side of equation (36), can only be obtained relativistically. I have not been able to find a *conservative* system of dynamically non-relativistic but thermally relativistic hydro equations. The problem is that the relativistic variation of mass with velocity contributes a term to the total energy just equal to the change in the kinetic energy density due to the variation of mass density with the internal energy. The correct total energy equation cannot be obtained unless both effects are included.

It is not hard to construct an example for which these difficulties arise. Consider a steady flow with a uniform high velocity, which is everywhere transparent but which emits uniformly and isotropically (in the fluid frame) within a fixed slab perpendicular to the flow. The velocity is large enough and the slab is transparent enough that v/c is large compared with the optical thickness of the slab. In this case, the reaction or recoil force, proportional to the velocity times the emissivity, dominates the usual force proportional to the opacity times the flux. In this situation the force is dynamically significant only if the matter is very hot: the equilibrium radiation energy density should be larger than the rest-energy density. If the simplified equations (27) and (28) are used to solve this problem, the correct leading terms are found for the expansions of E and F in v/c. However the subsequent terms are in error, and as a result the leading terms in ∇E and $\nabla\cdot P$ are wrong. For the correct solution, $c\nabla\cdot P$ does not balance $-\int k_\nu F_\nu d\nu$, it balances $-v\cdot\nabla F/c$. The body force in the comoving frame, the quantity on the right-hand side of equation (33), does come out right, since the leading term in F is right, but this agrees with the left side of equation (30) only if the incorrect P is used there. If the incorrect P is used, the left side of equation (35) will not agree with the right side. Thus consistent use of the simplified equations (27) and (28) for the radiation, together with the material mass, acceleration and internal energy equations, (31), (32), and (33), will give essentially the right radiation field, and essentially the right motion and temperature of the material, at the cost of conserving momentum and energy exactly. This example is one in which the material energy and momentum densities are small compared with those of the radiation field, so the total conservation equations are not appropriate for finding the small material parts.

With cold -- thermally non-relativistic -- material, finding a conservative system is easier. The c^{-2} terms in all the material equations can be dropped. It is then immaterial whether the radiation force in the comoving frame or the fluid frame is used. The simple forms (29) and (30) of the radiation conservation laws can be used, which allows momentum to be conserved exactly, and energy to be conserved except for a negligible error $(v\cdot\partial F/\partial t)/c^2$. If the time derivative term is dropped in equation (31), which then becomes the static flux equation, both momentum and energy can be conserved. An inspection of the equations to see what has been lost shows that the radiation momentum density has disappeared. Indeed, radiation momentum density can never be important in a thermally non-relativistic problem.

3.4. The right approximation for the problem

A primary question to be answered when the equations of radiation hydrodynamics are being formulated is: what complexities should be retained in the theory when it is applied in different situations? The different situations we have in mind are specified by different strengths of coupling between matter and radiation, or otherwise put, different optical thicknesses. Radiation hydrodynamics is a theory of a two-component system. One component consists of all the material particles, which are lumped together and called matter for our purpose, and the other component consists of the photons. The three coupling regimes are: strong coupling, which produces radiation-matter equilibration, weak coupling, for which the radiation field can be specified _a priori_, and the in-between case of moderate coupling.

I will call the strong coupling regime the _stellar interior_ limit. The two conditions that must be met in order for this regime to apply are (1) that the distance radiation diffuses before exchanging appreciable energy with the matter is small compared with the characteristic length of the system, and (2) that the time required for this to occur is small compared with the characteristic time. Diffusion theory can then be applied in the local comoving frame. This leads to the result that the radiation field is the equilibrium blackbody field plus small corrections of order the mean free path times the Planck function gradient, the mean free time times the time derivative of the Planck function, and the mean free path and mean free time times respectively the velocity gradient divided by c and the acceleration divided by c, both multiplied by the Planck function. The velocity and acceleration corrections represent the effects of radiative viscosity. Since the radiation stress-energy tensor is known explicitly in terms of the state of the matter, it can be added to the matter stress-energy tensor, so it is no longer necessary to carry the radiation-matter coupling terms. In the fluid frame the radiation corrections consist of adding $aT^4/3$ to the pressure, adding aT^4/ρ to the internal energy, accounting for the diffusion energy flux, and including the viscosity-like velocity gradient and acceleration terms in the internal energy and pressure tensor. In stating this, which seems to be obvious and represents _the way stellar interiors theory has been done_ since the time of Eddington, I am making the point that this approach _is fully consistent with even a relativistic treatment of radiation transport_. The only question would be whether additional v^2/c^2 terms should enter if the flow is dynamically relativistic. But since the velocity enters the transport equation in the comoving frame only through gradient terms and time derivatives, it must enter the diffusion theory in combination with the mean free path, thus v^2/c^2 terms come in only when the diffusion theory is taken to second order in the mean free path. This assumes that relativistic kinematics is used to evaluate the velocity gradient and acceleration in the comoving frame in terms of the inertial frame velocity.

The weak and moderated coupling regimes I will call the atmospheres limit. Weak coupling results when the system is transparent to essentially all radiation. Then there are two cases for how the radiation is produced. One case is that the radiation is externally imposed from presumably known sources. Then the radiation transport equation can be discarded, since the radiation is known, and the radiation hydrodynamics problem becomes a pure hydrodynamics problem, but with source terms of momentum and energy given by the radiation-matter coupling. Optical thinness ensures that the coupling terms have a negligible effect on the radiation field. The coupling terms can still be significant if the Boltzmann number for the problem is small enough, but since the coupling is weakened by a factor τ, the optical thickness of the system, the Boltzmann number must be less than τ for radiation to be energetically important, and less than $\tau v/c$ for radiation to be dynamically important. These conditions are easily realized in a tenuous gas next to a bright radiation source, such as a stellar atmosphere.

The second case of weak coupling is the one in which the radiation is internally generated in the medium. The magnitude of this radiation contains a factor τ, so the upper limits on the Boltzmann number for radiative heating and momentum coupling to be significant are further reduced by a factor τ. In fact, radiative heating is moot, since radiative cooling is sure to overwhelm it in this case. In the rare case that the material is so hot that momentum coupling to the internally-generated radiation is significant, the problem must be treated in the same way as for moderate coupling.

In the strong coupling and weak coupling regimes the two-component problem reduces to a one-component problem, either by equilibration of the two components (strong coupling) or by having one component -- radiation -- fixed by boundary conditions (weak coupling). The moderate coupling regime is the one of true radiation hydrodynamics, since the full system of material and radiation equations must be solved simultaneously. Of course any problem that spans a large range of density will likely have regions that are in the strong coupling limit, and other regions that are in the weak coupling limit, with interfaces between them that fall in the moderate coupling regime. It may be that the entire problem must be treated with the general methods in this case. More interesting in the sense of creating new effects are problems that contain extended regions of moderate coupling. This can arise in a variety of ways, but I will focus on two: an atmosphere in which the spectral lines are opaque, but the continuum between the lines is transparent; and a continuum scattering problem in which the system is opaque to scattering, but rare enough that the radiation is not thermalized.

In the thick lines/thin continuum problem, the large majority of radiation energy and momentum resides in the continuum, and does not couple to the matter. For this reason the total radiation conservation laws are not useful, and a simpler procedure is to take the continuum radiation as given and solve only for the radiation field at line frequencies. The transport equation can then be simplified by omitting all c^{-2} terms and all c^{-1} terms except for the Doppler term, which is enhanced in importance owing to the rapid frequency variation of the radiation field in the lines.

In the continuum scattering problem the full radiation equations as discussed above (viz., equations [27] and [28]) must be used, but owing to the optical thickness of the problem, the diffusion approximation is successful. The frequency spectrum of the radiation can be expected to depart significantly from Planckian, so a frequency-dependent treatment is needed.

A discussion of these two problems will occupy the balance of these lectures.

4. SECOND LECTURE -- METHODS FOR LINE-DOMINATED PROBLEMS

At frequencies that lie within a spectral line the absorption coefficient may be orders of magnitude greater than at continuum frequencies, with the result that regions of even modest optical depth in the continuum become so opaque in the lines that the flux at line frequencies, given by the diffusion approximation, is quite negligible, and the contribution made by these frequencies to the momentum and energy coupling terms (viz., equations [29] and [30]) is to the continuum contribution as the fraction of the spectrum they occupy. If the lines are sparse in the spectrum, their effect in this limit is negligible. However, as the region becomes partially transparent in the lines, the flux is no longer inversely related to the absorption coefficient, but tends to be comparable in and out of the lines, and the lines have an importance in the coupling terms far greater than their relative bandwidth. This is the regime that needs to be investigated more carefully.

The effect of lines is also greatly enhanced by a velocity gradient. Such a gradient acts like a source of line broadening, increasing the line width and reducing the peak opacity. The fraction of the spectrum influenced by lines is increased, as is their effect on the coupling terms since their saturation is reduced with the opacity. At typical temperatures and densities lines normally have widths $\Delta\nu/\nu$ that are 10^{-4} or less, so the effect of a velocity of only 30 km s^{-1} will be major for transport in the lines. The sensitivity to advection, abberation and retardation is no greater at line frequencies than at any others, so the Doppler term overwhelms the other c^{-1} effects in the co-moving frame (CMF) transport equation.

The transport equation (26), simplified by the omission of retardation, advection and abberation, becomes

$$\underset{\sim}{n} \cdot \underset{\sim}{\nabla} I_\nu - \frac{1}{c} \underset{\sim}{n} \cdot \underset{\sim}{\nabla} \underset{\sim}{v} \cdot \underset{\sim}{n}\, \nu \frac{\partial I_\nu}{\partial \nu} = j_\nu - k_\nu I_\nu \,. \tag{37}$$

The comparison of the transport terms on the left side -- spatial transport and frequency transport -- leads us to define three velocity regimes since different calculational methods are used in each regime. These are (1) static regime, $v \ll v_{th}$; (2) transition or, roughly, transonic regime, $v \approx v_{th}$; and (3) Sobolev or hypersonic regime, $v \gg v_{th}$. Here v_{th} stands for the intrinsic line width in velocity units. In the most common case, the principal mechanism of line broadening is the thermal Doppler effect, and v_{th} is $(2/3)^{1/2}$ times the rms speed of a line-emitting atom or ion. This is a few times less than the speed of sound, depending on the atomic weight of the species. The calculational methods differ in that in the static case the Doppler term is negligible, and in the Sobolev case the spatial transport term is negligible. These are relatively easy to treat, since equation (37) becomes an ODE. In the transition case the two transport terms are comparable and the PDE character of equation (37) is essential.

4.1. Line transport in the static regime

In the static regime the transport equation is the familiar equation of transfer, and methods for solving it in spectral lines are discussed by, for example, Mihalas (1978). Analytic approximations for the momentum coupling term -- the force -- due to a line are given by Castor (1974). The behavior of the force due to a line in a plane-parallel stellar atmosphere can be understood from the following rough formula:

$$f_{line} \approx -\Delta\nu \frac{4\pi}{3c} \frac{dJ_{\nu_0}}{dz} , \tag{38}$$

in which $\Delta\nu$ is the equivalent width of the line (essentially the width of the frequency interval in which line opacity exceeds the continuous opacity) and J_{ν_0} is the angle-averaged intensity at the center of the line. We see that it is the depth dependence of the mean intensity that determines the line force. Deep in the atmosphere the mean intensity will be closely equal to the Planck function at the local temperature, both at line center and at other frequencies. Then we see that the line force is very small in comparison with the force due to the continuum opacity, since the equivalent width of the line is a small part of the whole spectrum. However, near the edge of the atmosphere, where the radiation escapes freely, the line-center mean intensity can be substantially less than the mean intensity in the continuum or the Planck function. Just where the line-center intensity begins to depart from the continuum intensity, and how much less it is at the edge of the atmosphere, depends on the details of the formation mechanism of the line. One general rule is that the stronger the line is, i.e., the greater its opacity is, the closer to the edge of the atmosphere the drop in line-center intensity is. Thus strong lines and weak lines all have about the same pressure jump across the whole atmosphere, but for a strong line the pressure gradient occurs in low density material and therefore produces a larger force per unit mass than for weaker lines. The line force has its maximum at the edge where all the radiation can escape, and the maximum value can approach the multiple of the star's gravity given by

$$\frac{f_{line}}{\rho g} \simeq \frac{\kappa_{line}}{\kappa_{es}} \frac{v_{th}}{c} \frac{L}{L_{ED}} \tag{39}$$

in terms of the ratio of line opacity to the electron scattering opacity and the ratio of the stellar luminosity L to the Eddington limit luminosity $L_{ED} = 4\pi G\mathcal{M}c/\kappa_{es}$. This limit for the force is approached if the emergent flux at line center is not too much less than the continuum flux. The point to be noted about this formula is that the opacity ratio can be 10^7 for the resonance line of an abundant ion, and v_{th}/c is about 10^{-4}, so the force due just to a single line could be 10^3 times greater than gravity for a star not far from the Eddington limit! This idea was the departure point for thinking about radiatively-driven stellar winds (Castor, Abbott and Klein 1975; referred to below as CAK). What this estimate shows is that for some stars a static state is impossible. The wind itself has a velocity much larger than v_{th}, and the Sobolev method or the transitional methods are needed to treat it.

4.2. Transonic line transport

In the transitional case the space and frequency derivative terms are comparable, and the transport equation must be solved as a PDE in those variables. In 1-D slab symmetry the intensity is a function of z, the comoving-frame frequency ν, and μ, the z component of the direction vector. It obeys this equation

$$\mu \frac{\partial I_\nu}{\partial z} - \frac{\mu^2}{c} \frac{dv}{dz} \nu \frac{\partial I_\nu}{\partial \nu} = k_L \phi(\nu) (S - I_\nu) + k_c (S_c - I_\nu) , \qquad (40)$$

in which k_L is the integrated line absorption coefficient, k_c is the continuous absorption coefficient, $\phi(\nu)$ is the normalized line profile, S_c is the known continuum source function and S is the line source function, which in the usual case of complete frequency redistribution has the form

$$S = (1 - \varepsilon) \int_0^\infty d\nu \, \phi(\nu) \, \frac{1}{2} \int_{-1}^{1} I_\nu(\mu) \, d\mu + \varepsilon B_\nu \qquad (41)$$

in terms of the intensity, the Planck function and a thermalization parameter ε.

Methods to solve equations (40) and (41) have been presented by Noerdlinger and Rybicki (1974) and Mihalas, Kunasz and Hummer (1975, 1976). The latter authors considered the case of spherical symmetry. All used the Feautrier transformation, that is, they separated the intensity into the sum of an even function of μ and an odd function of μ, which obey coupled equations. The equations are spatially differenced in a staggered fashion: the even components are centered in zones, and the odd components at zonal interfaces. This differencing of $\partial/\partial z$ is second-order accurate. Noerdlinger and Rybicki used centered implicit frequency differencing, while Mihalas, Kunasz and Hummer used a donor-cell type frequency differencing. The donor-cell scheme is only first-order accurate, but is more robust for optically thick lines, when equation (40) becomes stiff. There is some question about the accuracy of the centered frequency differencing in this case.

Equation (40) is to be solved between certain limits in z, the boundaries of the problem in physical space, and between certain limits in ν outside of which the opacity contribution of the line in question is negligible. In other words, the equation is to be solved within a rectangle in the z,ν plane. The boundary conditions that may be applied to make this problem properly posed are to specify the intensity at one point on each characteristic of the equation, the point where that characteristic, traced in the direction of radiation flow, first crosses the perimeter of the rectangle. Therein lie the difficulties that afflict these methods for complex problems. While it is easy to see that the boundary data can always be given at the inflow boundary in z, i.e., at z_{min} for $\mu > 0$ and at z_{max} for $\mu < 0$, the inflow boundary in ν is defined only when dv/dz has constant sign. If dv/dz is positive everywhere, then ν_{max} is the inflow boundary; if dv/dz is negative everywhere, then ν_{min} is the inflow boundary. If v(z) is not monotonic, i.e., if dv/dz changes sign, then data may be given at ν_{max} for some range of z, and at ν_{min} for another range of z, but for still other values of z the characteristics that enter the rectangle at ν_{max} or ν_{min} are the same ones that left the rectangle at some other z, so the incident intensity at the first z is equal to the emergent intensity at the second z. This kind of mapping function for the boundary data would be very cumbersome to handle. In practice, only for the monotonic velocity case have results been given in the literature.

All the previous authors have used direct solution methods for the large linear system that results from the differencing of equations (40) and (41). Noerdlinger and Rybicki (1974) and Mihalas, Kunasz and Hummer (1976) solved the system in block tri-diagonal form for which the blocks are indexed by z and the internal structure of each block refers to the frequencies and directions. Mihalas, Kunasz and Hummer (1975) solved the system by formally expressing the intensity in terms of the line source function at each frequency and direction, then at the end solving a full system for the source function vector. The first method of solution is called the Feautrier elimination scheme, from the paper by Feautrier (1964) in which was proposed, and the second method is called the Rybicki elimination scheme, from Rybicki (1971).

The problem of the proper frequency boundary conditions can be circumvented by making the frequency range computed, from ν_{min} to ν_{max}, large enough -- more than twice the full range of the flow velocity. Then the frequency boundary data are irrelevant, and can be set to zero at the inflow frequency for each z. To avoid an excessive total number of frequencies the size of the frequency increment can be graded from a fraction of a Doppler width at the center of the line to a fraction of the flow velocity Doppler shift at ν_{max} and ν_{min}. This would only roughly double the number of frequency points from the number required for the line proper. There still remains the problem that the stable donor-cell frequency differencing requires the correct choice of donor frequency, and in the Mihalas, Kunasz and Hummer (1975) method this must be the same at all z, since their method consists of a forward and reverse sweep in z within a unidirectional sweep in ν.

If we return to the first-order form of the transport equation, that is, abandon the Feautrier even-odd parts of the intensity in favor of the intensity itself at positive and negative angles, the calculation can be organized as a unidirectional sweep in ν within a unidirectional sweep in z, for each angle. This allows the correct direction of the ν sweep to be chosen for each z. The Rybicki elimination scheme can still be used, with the same operation count, and the only price that must be paid is either some loss of accuracy of the spatial differencing or the necessity of using exponentials, for example, in the difference formulae.

Another approach to the problem is to make it an iterative one. The PDE solution as just described can be used as a method for evaluating the intensity given the source function. This can be used within an iteration at each step of which the difference between an accurately calculated intensity and an intensity calculated using a crude intensity-source function relation is used, with the crude relation, to improve the current source function estimate and make another accurate intensity calculation. Iterative methods of this kind have been presented by Cannon (1973a,b) and Scharmer (1981). Convergence is obtained, in favorable cases, in a few iterations, and the total time required is largely spent in the accurate intensity calculations. This can result in a severalfold improvement in speed compared with even the Rybicki elimination scheme, which is usually the fastest of the direct solution methods.

In the preceding paragraphs I have discussed the limitations of the presently available methods for solving line transport problems in the transitional -- transonic -- regime, and have made some suggestions of directions for further work. This remains an area that is ripe for development.

4.3. Line transport in the Sobolev regime

The Sobolev or hypersonic regime is the one in which the Doppler term in the transport equation overwhelms the spatial transport term. The Sobolev approximation (Sobolev 1947; Castor 1970, 1974) consists of simply neglecting spatial transport. The picture one has is that spatial transport occurs only in the continuum, and can be treated trivially if the continuum is optically thin, as we assume here. The Doppler shift due to the fluid flow then feeds radiation from the continuum into the line. The simplified form of equation (37) for the Sobolev approximation is

$$-\frac{\underset{\sim}{n} \cdot \underset{\sim\sim}{\nabla v} \cdot \underset{\sim}{n}}{c} \, \nu \frac{\partial I_\nu}{\partial \nu} = k_L \, \phi(\nu) \, (S - I_\nu) \, . \tag{42}$$

The line width is assumed negligible compared with ν, so the factor ν is treated as a constant. The dependence on the line profile function is absorbed by defining a new frequency variable y:

$$y = \begin{cases} \int_{\nu}^{\infty} \phi(\nu')\, d\nu' & \underset{\sim}{n} \cdot \underset{\sim}{\nabla} \underset{\sim}{v} \cdot \underset{\sim}{n} > 0 \\ \int_{0}^{\nu} \phi(\nu')\, d\nu' & \underset{\sim}{n} \cdot \underset{\sim}{\nabla} \underset{\sim}{v} \cdot \underset{\sim}{n} < 0 \end{cases} \tag{43}$$

which runs from 0 at the "inflow" side of the line to 1 at the "outflow" side. The intensity thus satisfies the following ODE:

$$\frac{\partial I_\nu}{\partial y} = \tau_S(\underset{\sim}{n})\,(S - I_\nu) \tag{44}$$

in which $\tau_S(\underset{\sim}{n})$ is the Sobolev optical depth defined by

$$\tau_S(\underset{\sim}{n}) = \frac{k_L\, c}{\nu |\underset{\sim}{n} \cdot \underset{\sim}{\nabla} \underset{\sim}{v} \cdot \underset{\sim}{n}|}\ . \tag{45}$$

The solution of equation (44) is

$$I_\nu = I_c\, e^{-\tau_S y} + S\left(1 - e^{-\tau_S y}\right) \tag{46}$$

where I_c is the intensity in the continuum on the inflow side of the line. The dependence of I_c, I_ν and τ_S on $\underset{\sim}{n}$ has not been shown.

The results for the frequency-averaged angle moments of the intensity are expressed in terms of escape probabilities. The angle dependent escape probability $\beta(\underset{\sim}{n})$ is defined as the probability that a photon emitted in the direction $\underset{\sim}{n}$ with an initial frequency ν in the comoving frame sampled from the distribution $\phi(\nu)$ will escape to infinity, or at least to a region where the velocity is different by many line widths, without being reabsorbed. The expression for $\beta(\underset{\sim}{n})$ is

$$\beta(\underset{\sim}{n}) = \frac{1 - e^{-\tau_S(\underset{\sim}{n})}}{\tau_S(\underset{\sim}{n})}. \tag{47}$$

The average of $\beta(\underset{\sim}{n})$ over direction is the angle-mean escape probability, or simply the escape probability, $\bar{\beta}$ given by

$$\bar{\beta} = \int \beta(\underset{\sim}{n}) \frac{d\Omega}{4\pi} = \int \frac{d\Omega}{4\pi} \frac{1 - e^{-\tau_S(\underset{\sim}{n})}}{\tau_S(\underset{\sim}{n})} . \tag{48}$$

The key results of the Sobolev theory are expressions for $\bar{J}$, the angle and frequency-averaged intensity which enters the non-LTE kinetic equations for atomic level populations, and the line force. These are

$$\bar{J} = S(1 - \bar{\beta}) + \int \frac{d\Omega}{4\pi} \beta(\underset{\sim}{n}) I_c(\underset{\sim}{n}) \tag{49}$$

and

$$\underset{\sim}{f}_{line} = \frac{k_L}{c} \int d\Omega \, \beta(\underset{\sim}{n}) \, \underset{\sim}{n} \, I_c(\underset{\sim}{n}) . \tag{50}$$

When equation (49) for $\bar{J}$ is combined with equation (41) for S in terms of $\bar{J}$ the result is an explicit formula for S in terms of the temperature, thermalization parameter ε, and the locally-defined optical depth. There is no coupling to other depths, and there are no differential equations to solve. This simplicity makes the Sobolev method very attractive. Expression (50) for the line force is even simpler, since the force does not even depend on the local source function (owing to the near isotropy of the locally-emitted radiation) and can be computed directly from the continuous spectrum, given the local optical depth.

The order of magnitude of the force per unit volume due to a single line is that given in equation (39), if the Sobolev optical depth τ_S is less than unity, and smaller by a factor τ_S when τ_S is larger than unity. In the latter case the force contributed by the line is independent of the strength of the line, but proportional to the velocity gradient. When the force is summed for all the lines in the spectrum, a total is obtained which is essentially the force due to one strong line multiplied by the number of lines for which $\tau_S > 1$. It is a non-linear function of the velocity gradient. In the case that the distribution function of line strengths is a power law, which is often a reasonable approximation, the total force is also a power law in the velocity gradient, with an exponent between 0 and 1 (cf., CAK). The force due to the lines can be non-dimensionalized by expressing it as a multiple of the force due to electron scattering of the continuum radiation:

$$\underset{\sim}{f}_{line} = \left(\frac{\kappa_{es}\rho F}{c}\right) \frac{1}{F} \int d\Omega \, M\left(\frac{\kappa_{es}\rho v_{th}}{|\underset{\sim}{n} \cdot \nabla \underset{\sim}{v} \cdot \underset{\sim}{n}|}\right) \underset{\sim}{n} I_c(\underset{\sim}{n}) \tag{51}$$

where M(t) is a dimensionless function of t, which is not the time but a measure of the continuum optical thickness of the layer across which the velocity changes by the intrinsic line width. The function M depends implicitly on the ionization and excitation state of the material, since these determine the opacities of the various lines. The most complete and accurate calculations of M(t) have been made by Abbott (1977, 1982), and the results are presented graphically in that paper; they can indeed be fitted to a power law: $M \propto t^{-0.6}$ approximately. There is an additional weak dependence on ρ that arises from the variation of the ionization state.

4.4. Dynamical effect of line radiation

In order to illustrate the dynamical effects that can occur when the force due to an ensemble of lines is substantial, let's consider the following problem. Let there be a flow in the radial direction in a possibly stratified atmosphere with a collimated beam of radiation parallel to the flow, for example a stellar wind. The energy coupling between matter and radiation we suppose to be strong enough to hold the material at a fixed temperature. We want to consider what kinds of waves are produced when this flow is perturbed. The following discussion is similar to that of Abbott (1980). The equations of motion are written

$$\frac{\partial \rho}{\partial t} + \underset{\sim}{\nabla} \cdot (\rho \underset{\sim}{v}) = 0 \tag{52}$$

and

$$\frac{\partial \underset{\sim}{v}}{\partial t} + \underset{\sim}{v} \cdot \underset{\sim}{\nabla}\, \underset{\sim}{v} + \frac{a^2}{\rho} \underset{\sim}{\nabla} \rho = \frac{f_{line}}{\rho} \underset{\sim}{e}_r \tag{53}$$

where a is the isothermal sound speed and $\underset{\sim}{e}_r$ is the unit radial vector. We suppose that f_{line}/ρ is a function of $(dv_r/dr)/\rho$, as suggested by equation (51). Equations (52) and (53) can be linearized around the hypothesized flow, and the dispersion relation obtained for short wavelength perturbations (i.e., wavelength small compared with the scale of the stratification) is found to be

$$\frac{\omega}{k} = v_{or} \cos\theta + \Delta V \tag{54}$$

where θ is the angle between the propagation vector k and the radius, and the velocity shift ΔV satisfies the cubic equation

$$(\Delta V)^3 + A \cos\theta\, (\Delta V)^2 - a^2\, \Delta V - a^2 A \cos\theta \sin^2\theta = 0\ , \tag{55}$$

in which $A = [d(f_{line}/\rho)/d(\rho^{-1} dv_r/dr)]/\rho_0$.

If there were no line force, the wave modes found from equations (54) and (55) would be just the usual acoustic wave modes, and a root at $\Delta V = 0$ representing the degenerate internal waves which cannot be supported by the assumed isothermal variations. If instead f_{line}/ρ is comparable to the acceleration vdv/dr and v is hypersonic, then the factor A is of order $v >> a$, and the roots of the cubic give the following for the wave modes

(radiation wave mode)

$$\Delta V \simeq -A \cos\theta \tag{56}$$

$$\frac{\delta\rho}{\rho_0} \simeq \frac{\delta v_r}{\Delta V} \tag{57}$$

$$\underset{\sim}{u} \simeq \underset{\sim}{v}_0 - A\,\underset{\sim}{e}_r \tag{58}$$

and

(acoustic modes)

$$\Delta V \simeq \pm\, a \sin\theta \tag{59}$$

$$\delta\underset{\sim}{v} \simeq \frac{\delta\rho}{\rho_0}\, a\, \hat{k}_\perp \tag{60}$$

$$\underset{\sim}{u} \simeq \underset{\sim}{v}_0 \pm a\, \hat{k}_\perp \tag{61}$$

In these formulae the quantities prefixed by δ are the wave amplitudes, the ones subscripted by o are the unperturbed values, $\hat{k}_\perp$ is the unit vector in the direction of the projection of $\underset{\sim}{k}$ perpendicular to the radius. $\underset{\sim}{u}$ is the group velocity of the wave in each case. What we see is that the acoustic waves can propagate only transverse to the radius, while the radiation wave propagates radially, and with a velocity that is inward with respect to the flow for the case we are considering.

4.5. Unstable radiation waves

The wave modes that have been described above are the ones that are obtained when the Sobolev approximation is valid, so that the force is a function of the local velocity gradient. This approximation is valid only so long as the spatial wavelength of the fluctuations is large compared with the distance over which the velocity varies by the line width, v_{th}. Let's call this distance ℓ. When the wavelength is smaller than ℓ the fluctuation in the force depends not on the velocity gradient, but on the velocity fluctuation itself. Both cases are obtained by letting the fluctuating force be proportional to the fluctuation of the velocity gradient averaged over a region of size about equal to ℓ located just interior to the radius of interest. The dispersion relation for the radiation mode for the case $a = \theta = 0$ is then

$$\omega_{real} \simeq k \left[v_o - \frac{A}{1 + (k\ell)^2} \right] \tag{62}$$

$$\omega_{imag} \simeq \frac{A}{\ell} \; \frac{(k\ell)^2}{1 + (k\ell)^2} \; . \tag{63}$$

Indeed, equation (63) shows that when the wavelength is not very large compared with ℓ the wave has exponential growth in time. Even for small k, large wavelengths, there is exponential growth at a rate proportional to k^2, which is characteristic of negative viscosity. The corresponding negative kinematic viscosity coefficient is $A\ell$. Since A is typically the same order as the flow velocity, the negative Reynolds number for the problem is of order L/ℓ, where L is the relevant length scale. This ratio varies from about one, near the sonic point of the wind, to about 100 when the wind is approaching full velocity. We can conclude that the negative viscosity should be important, at least near the sonic point, and that instabilities will grow and presumably produce shock waves. These are essentially the conclusions that have been reached by MacGregor, Hartmann and Raymond (1979), Lucy and White (1980) and Owocki and Rybicki (1983).

4.6. Radiatively driven steady winds

A steady stellar wind is a flow that satisfies equation (53), or its equivalent, without the time derivative term. This becomes an ordinary differential equation for the radial component of velocity, and, as discussed by Parker (1958) and many others, a smooth flow can be obtained only if a regularity condition is obeyed at every point where the velocity of one of the linearized wave modes vanishes. For the case considered by Parker, this means the sonic point in the flow, and a complete analogy exists with the deLaval nozzle for which the velocity vanishes at the throat of the nozzle. Furthermore, in Parker's theory the only physically acceptable wind solution _is_ the one which passes through such a sonic point. In general the flow also contains a shock outside the sonic point, which provides the correct pressure match of the wind to the interstellar medium.

When the equation of motion includes the force due to the momentum transfer in line scattering, as in equation (53), the situation becomes more complex. That is because the line force depends globally on the velocity field, so equation (53) is no longer a simple ODE for the velocity. Certain idealizations of this dependence can be made, treating the line force in different ways, and each of these leads to a Parker-type ODE with a singular point like Parker's sonic point, but very different ones for the different idealizations. If f_{line} is treated as a known function of either the radius or the local velocity, then it has no effect on the wave speeds (only on their growth), so the critical point is just the sonic point, and it turns out that the regularity condition says in effect that f_{line} balances the gravity at that point. If f_{line} is treated as a known function of the local velocity gradient, as in the Sobolev approximation, then the appropriate wave speed becomes that of the radiation wave described by equations (56)-(58), and the regularity condition is more complicated. In each case, the singular point is the outermost one that can send signals both inward and outward.

The paradoxical situation that arises is that if indeed there is a smooth steady flow that obeys equation (53) with an accurately calculated f_{line}, it should be described well by all the idealizations. It should obey the regularity condition at the sonic point expected if $f_{line}(r)$ is fixed, and also the CAK regularity condition at the Sobolev approximation singular point, where v is nearly equal to A, defined above. Indeed, this seems to be true for the models that have been made in the two ways. For the Sobolev CAK model f_{line} does in fact nearly balance the gravity at the sonic point (guaranteed by the ODE). Weber's (1981) wind models, which use the CMF method in place of the Sobolev approximation and which also differ from CAK in other respects (discrete line groups, accurate angle integration), have the same characteristics as matched Sobolev models. If we accept that the same flow can be modelled in quite different ways, what about the critical point? We may ask, what really is the outermost point that can send signals upstream? The answer is: what kind of signals? If the radiation field is frozen, then indeed, an ordinary sound wave is the signal that can be sent, and the sonic point is the singular point. If the radiation field is affected, then it becomes the medium that can carry a signal upstream much more quickly than sound can, and it is not surprising that the wave speed and singular point are changed. We are led once again to consider the wavelength of the wave to decide what kind of role the radiation plays, and come again to the dispersion relations of equations (62) and 63).
Equation (62) shows that the wave speed is A, directed inward relative to the fluid, for wavelengths large compared with ℓ, and small, i.e., of order a, for wavelengths small compared with ℓ. So the conclusion is that an observer situated outside the

sonic point and inside the Sobolev critical point of a wind can send medium and long pulses inward, but not short pulses.

The rather complex picture that emerges from a consideration of the stellar wind driven by line radiation, as indicated above, is that the relative stability of the Sobolev wind model to long-wavelength perturbations may mean that this model gives a valid picture of the wind structure in the large, meaning smoothed over moderate increments in space and time, but that the instability to short wavelengths will necessarily produce small-scale fluctuations around that model. It is virtually impossible to quantify this statement, owing to the lack of a time-dependent hydrodynamic calculation of the wind that can treat the nonlinear development of the short-wavelength instabilities. This is an area that calls for the application of better numerical techniques for radiation hydrodynamics in line-dominated problems, such as an improved comoving-frame method, as discussed above.

4.7. Velocity-modified diffusion

The final subject in the area of line-dominated problems that I want to discuss is what I will call velocity-modified diffusion (VMD). This is a version of the Sobolev theory that is adapted to solve the following kind of problem: a region filled with material in motion such that the region is optically thick in the continuum, that is, the continuum mean free path is small compared with the size L of the region, but with a large enough velocity gradient that the continuum optical thickness associated with the thickness ℓ of the local constant-velocity surface is small, that is, the continuum mean free path is larger than ℓ. The Sobolev approximation discussed earlier assumed free-streaming radiation in the continuum; velocity-modified diffusion extends that to the case of opaque continua. If the continuum mean free path diminishes to the point that it no longer exceeds ℓ, then the effect of the velocity gradient on the diffusion process becomes negligible and the ordinary diffusion limit is recovered. The intermediate case exists only if the velocity span V across the region is hypersonic, i.e., large compared with the intrinsic line width. This is actually the case in many astrophysical situations, and in fact conditions in moving stellar envelopes often fall into the VMD regime, as pointed out by Karp, Lasher, Chan and Salpeter (1977), to whom this theory is due.

The heart of the VMD approximation is the calculation of the mean free path of a photon emitted at a particular fluid-frame frequency ν_0. For this calculation the approximation is taken of an infinite homogeneous medium with a uniform velocity gradient. Then there is a constant ratio between displacement in space and displacement in the logarithm of the fluid-frame frequency for the photon, so the mean free path can be expressed as

$$\lambda^{eff}(\nu_o) = \frac{dx}{d\,\ell n\,\nu}\left\langle \ell n\,\frac{\nu_{final}}{\nu_o}\right\rangle . \tag{64}$$

in which x is the distance along the photon path. The mean can be calculated assuming an exponential distribution for the running optical depth $\tau(\nu)$ along the path:

$$\lambda^{eff}(\nu_o) = \int_0^\infty \frac{1}{\kappa_\nu \rho}\, e^{-\tau(\nu)}\, d\tau(\nu) \tag{65}$$

and

$$\tau(\nu) = s \int_\nu^{\nu_o} \frac{\kappa_{\nu'}}{\kappa_c} \frac{d\nu'}{\nu'} , \tag{66}$$

in which the constant s, the same as c/v_{th} times the variable t of equation (51), is

$$s = \frac{\kappa_c\ \rho c}{\underset{\sim}{n} \cdot \underset{\sim}{\nabla}\,\underset{\sim}{v} \cdot \underset{\sim}{n}} . \tag{67}$$

It should be noted that these relations hold for either sign of s. If the velocity gradient is one of expansion for the direction in question, then s is positive, and the frequency ν steadily decreases from the initial frequency ν_o. If the velocity gradient is compressive, s is negative and the frequency increases from ν_o. The range in $|s|$ for which the VMD approximation is useful is from the lower limit c/V, below which the continuum radiation is free-streaming, to the upper limit c/v_{th}, above which the ordinary diffusion theory is applicable.

Karp, Lasher, Chan and Salpeter evaluate the integral in equation (65) by assuming a negligible intrinsic line width, so that $\tau(\nu)$ has the form of a linear ramp with a jump discontinuity at each line frequency. Their result is

$$\lambda^{eff}(\nu_o) = \lambda_c \left(1 - \varepsilon_{\nu_o}\right) , \tag{68}$$

in which λ_c is the continuum mean free path and the correction factor ε_{ν_o} is given by

$$\varepsilon_{\nu_o} = \sum_{j=J}^{\infty} \left(\frac{\nu_j}{\nu_o}\right)^s \exp\left(-\sum_{i=J}^{j-1} \tau_i\right) \left(1 - e^{-\tau_j}\right) . \tag{69}$$

The indices i and j in equation (69) run over the list of absorbing lines, and the ordering is such that for positive (negative) values of s, the lines are ordered in decreasing (increasing) order of frequency, and the term J in the summations stands for the line with frequency closest to but less (greater) than ν_o. The variable τ_i is the Sobolev optical depth of the ith line, given by equation (45).

Karp, Lasher, Chan and Salpeter supply a table of values of the Rosseland mean of ε_{ν_o}, which demonstrates that the lines can reduce the mean free path by a factor 2 for s of order 1000, and a factor 4 or more for s of order 100. (The effect is inversely related to s.)

The conclusion to be drawn from this work is that a person who is contemplating doing a diffusion calculation of a moving medium should have a look at the value of s for the problem. If this is significantly smaller than the limit c/v_{th}, then the VMD correction factor $1 - \varepsilon_{\nu_o}$ should be applied. This can be in the form of a Rosseland average for the whole spectrum (a result given by Karp et al.) or separate Rosseland averages for individual coarse frequency groups. Of course, if one chooses to use so many frequency groups that all the lines can be resolved, the CMF approach is applicable. The use of diffusion coefficients based on the VMD approximation requires one further comment. Since all the quantities entering equation (69) are angle-dependent, the effective mean free path is also, which means that the diffusion is anisotropic and is described by a diffusion coefficient tensor. The easiest way of obtaining this tensor is by a low-order angle quadrature of the (probably tabular) angle-dependent mean free path. The quadrature procedure will require special care in the event that the three principal values of the strain tensor include both positive and negative values.

5. THIRD LECTURE -- DYNAMIC DIFFUSION PROBLEMS

A second class of radiation hydrodynamics problem that requires a full two-fluid treatment is of problems for which the continuum is optically thick but effectively thin, and such that advection of radiation competes with radiation diffusion. The distinction between optical and effective thickness arises if the continuous opacity is dominated by scattering. Then being optically thick means that the size of the region is many scattering mean free paths, while being effectively thin means that the size of the region is less than the distance a photon would diffuse by scattering before suffering an absorption or other thermalizing collision. In this kind of problem the lines are often saturated, as discussed above, to the degree that their role is minor in energy and momentum coupling to the matter. It is the effective thinness of these problems which prevents assuming equilibration of radiation and matter, as in the stellar interior regime. However, the optical thickness of the problem allows using the Eddington approximation, so equations (27) and (28), closed by neglecting $\partial \underset{\sim}{F}_\nu/\partial t$ and replacing $\underset{\approx}{P}_\nu$ by $E_\nu/3$, give an adequate description.

In these scattering-dominated problems the coupling of matter and radiation is relatively weak, and is provided by absorption, such as inverse bremsstrahlung, and by the Compton effect. For the latter, the energy exchanges between photons and electrons can be described using a Fokker-Planck approximation, which leads to the Kompaneets (1957) equation. If these are taken for the coupling terms in equation (27), we find the following dynamical equation for the radiation field:

$$\frac{\partial E_\nu}{\partial t} + \left[\underset{\sim}{\nabla} \cdot (\underset{\sim}{v} E_\nu) - \frac{1}{3} \underset{\sim}{\nabla} \cdot \underset{\sim}{v}\, \nu \frac{\partial E_\nu}{\partial \nu}\right]$$

$$- \underset{\sim}{\nabla} \cdot \left(\frac{\lambda_\nu c}{3} \underset{\sim}{\nabla} E_\nu\right) = \kappa_{abs}(\nu)\, \rho (4\pi B_\nu - cE_\nu)$$

$$+ \kappa_{es}\, \rho c \frac{kT}{mc^2}\, \nu \frac{\partial}{\partial \nu}\left[\nu \frac{\partial E_\nu}{\partial \nu} + \left(\frac{h\nu}{kT} - 3\right) E_\nu + \frac{c^3}{8\pi \nu^2 kT} E_\nu^2\right] \qquad (70)$$

The terms in equation (70) have been divided so that the ones on the left side of the equation represent transport in the most general sense, while the ones on the right describe coupling to the matter. With the exception of the Doppler term (with the frequency derivative), the operators on the left are independent of frequency in the large part of the spectrum where Compton scattering is the dominant opacity. We may think of this left-side operator as governing the dependence on space and time of the field of photons, created through the action of a source term on the right side, that have suffered a variable number N of scatterings. This part of the problem is geometry dependent, and the important processes are advection and diffusion of the photons. But this description does not require any knowledge of the frequency spectrum of the photons. Conversely, the right side operator governs the evolution of the spectrum of the photons as a function of the number of scatterings that have occurred. This description does not depend on the geometry except insofar as the temperature, etc., are inhomogeneous. (The Doppler term, though part of the transport operator, must be included with the right-side terms in this dichotomy.)

Whether the problem in question is one of temporal evolution, spatial diffusion or spatial advection is established by the competition of the left-side terms. For each of these cases there will be the appropriate distribution function of the number of scatterings for the photons seen by an observer at the selected observation point. The terms on the right side that compete are the absorption and Comptonization terms, and the comparison of these introduces a transitional frequency where the two effects are comparable. For a particular number of scatterings, the photon spectrum will be equilibrated to a Bose or Planck distribution to a degree that depends on the frequency as well as the number of scatterings. Folding the distribution of the number of scatterings, determined by the left-side operator, with the spectrum vs. number of scatterings determined by the right-side operator, gives the observed spectrum.

5.1. Comptonization

The Comptonization problem is the problem of the right side of equation (70). It has been discussed thoroughly by Chapline and Stevens (1973) and Rybicki and Lightman (1979) and in several other references cited by Rybicki and Lightman. A recent refined analytic treatment is the one by Caflisch and Levermore (1983). Since this subject can become rather involved, I will describe only the main features here.

The strength of the Comptonization operator is measured by the so-called Compton parameter y defined by

$$y = \frac{4kT}{mc^2} N . \qquad (71)$$

When $y \ll 1$, Comptonization is unimportant with a bremsstrahlung source spectrum, and the spectrum evolved through N scatterings is simply described by saying that it equals the equilibrium Planck spectrum at frequencies below ν_{cr} defined by

$$\frac{\kappa_{es}}{\kappa_{abs}(\nu_{cr})} = N , \qquad (72)$$

and equals the accumulated source of photons at frequencies above ν_{cr}. When $y = 1$, ν_{cr} coincides with the frequency ν_{coh} (cf., Rybicki and Lightman) defined by

$$\frac{\kappa_{es}}{\kappa_{abs}(\nu_{coh})} = \frac{mc^2}{4kT} \, . \tag{73}$$

ν_{coh} is the dividing frequency for the importance of absorption vs. Comptonization. At frequencies below ν_{coh} absorption dominates, and at frequencies above ν_{coh} Comptonization dominates. So when $y >> 1$, the photon spectrum is equilibrated at frequencies below ν_{coh}. Above ν_{coh} the spectrum is more complicated. If N is not sufficiently large for the entire spectrum to be equilibrated, i.e., $N < \kappa_{es}/\kappa_{abs}(4kT/h)$, then the spectrum above ν_{coh} consists of a rather flat energy distribution on which is superimposed a Wien peak. The amplitude of the Wien peak is reduced from the blackbody peak by about the factor $N\kappa_{abs}(4kT/h)/\kappa_{es}$. For still larger values of N the entire spectrum is blackbody.

If the advection term on the left side is dominant, then the effect of the Doppler term on the evolution of the spectrum should also be considered. For large y the Doppler term cannot compete with the Comptonization term, and so has little effect. For small y the effect of the Doppler term is that for frequencies above ν_{cr} the spectrum is the accumulated photon source Doppler shifted by the amount corresponding to the bulk strain experienced by the matter between the points of photon creation and observation.

5.2. Static vs. dynamic diffusion

The three terms on the left side of equation (70) apart from the Doppler term comprise an advection-diffusion operator. The question that immediately arises is whether advection or diffusion dominates. Assuming that Compton scattering is the dominant opacity, that question is answered as follows:

$$\text{diffusion (static diffusion)} \qquad \tau_{es} \frac{v}{c} << 1 \tag{74}$$

$$\text{advection (dynamic diffusion)} \qquad \tau_{es} \frac{v}{c} >> 1 \, , \tag{75}$$

where τ_{es} is the typical Compton optical thickness of the system.

The radiation field that evolves according to equation (70) has an initial transient that lasts a time

$$t_{diff} = \frac{\tau_{es}^2}{\kappa_{es}\rho c} \tag{76}$$

in the static diffusion regime, and a time

$$t_{flow} = \frac{\tau_{es}}{\kappa_{es}\rho v} \tag{77}$$

in the dynamic diffusion regime, and in general the smaller of the two times. Apart from the transient behavior, the temporal evolution proceeds on the time scale t_{evol} of the source and boundary data. The comparison of the three time scales leads to the three regimes

temporal evolution (infinite medium)	$t_{evol} << t_{flow}, t_{diff}$
diffusion (static)	$t_{diff} << t_{evol}, t_{flow}$
advection (static)	$t_{flow} << t_{evol}, t_{diff}$.

The representative value of N for each of these cases is just the number of scatterings that occur in the appropriate length of time:

temporal evolution	$N = \kappa_{es} \rho c t_{evol}$
diffusion	$N = \kappa_{es} \rho c t_{diff} = \tau_{es}^2$
advection	$N = \kappa_{es} \rho c t_{flow} = \tau_{es} \frac{c}{v}$. (78)

These are the values of N to use in equations (71) and (72) to estimate the observed spectrum. The spectrum in fact depends not only on the typical N but also on the distribution of N. For example, in the diffusion case the probability distribution is of the form $p(N) \sim N^{-1/2}$ up to the typical N listed above. This power-law behavior leads to the modified blackbody spectrum $F_\nu \sim (\kappa_{abs}(\nu))^{1/2} B_\nu$ discussed by Rybicki and Lightman. This power-law part of the distribution function of N is absent for the temporal and dynamic cases.

The production of the observed spectrum can be summarized by saying that the spectrum can be calculated in a geometry- and kinematics-independent way in terms of N (except for the effect of the Doppler term in a limited domain), as described in the references. The geometry and kinematics enter in finding the appropriate distribution of N, which is not a difficult problem in general.

For some problems the mechanical effect of radiation dominates that of the matter, i.e., $B_o << v/c$, and the dynamical part of the radiation hydrodynamics problem can be divorced from the question of the formation of the spectrum. This is true to a greater or lesser degree of the two examples that I will discuss next.

5.3. First example -- supercritical wind

Circumstances exist in which a more or less spherical star has a luminosity that exceeds the Eddington limit $L_{ED} = 4\pi G\mathcal{M}c/\kappa_{es}$. These include, for example, a compact object that is accreting matter from a disk at a rate such that the release of gravitational energy exceeds L_{ED}. Another example is an X-ray burst source, a similar compact object that undergoes a thermonuclear outburst after a critical amount of nuclear fuel has accumulated on its surface. It is impossible for a steady structure to exist for such stars in which this super-Eddington luminosity is removed by radiation alone, since the super-Eddington flux would produce a force on the matter in excess of gravity. If there is a steady structure with a total energy loss $L_T > L_{ED}$ then it must incorporate a wind so that the sum of the wind energy loss and the sub-Eddington radiative energy loss equals L_T. (I am putting aside the possibility of energy transport by other mechanisms such as convection and waves.) For objects that exist only a short time in the high-power state, such as the X-ray burst sources alluded to above, a wind may or may not be produced, depending on the comparison of the lifetime of the high-power state with the energy storage time scale for the portion of the star through which the large luminosity must flow. But for the present purpose I will assume that the lifetime of the high-power state is indeed very long, and that a wind results.

The discussion of supercritical winds by Meier (1982a) provides a nice illustration of the ideas of dynamic vs. static diffusion and of decoupling between matter and radiation, so I will use this paper for my example.

The model to be considered is a compact star into a certain outer layer of which energy is injected at a rate L_T by either release of gravitational energy of an accreting wind (with subsequent spreading over the whole surface) or by thermonuclear burning; it is not critical which energy source applies. The hypothesis is that in and near the region of the energy injection the radiative flux in the comoving frame rises to a value close to L_{ED}. The excess of L_T above L_{ED} is carried by the wind, so

$$L_T - L_{ED} \simeq \dot{\mathcal{M}}\left[\frac{v^2}{2} - \frac{G\mathcal{M}}{r} + E_T + \frac{P_T}{\rho}\right], \qquad (79)$$

where E_T and P_T are the total energy per unit mass and the total pressure. Another assumption is that the radiative energy and pressure are overwhelming, so E_T can be replaced by E/ρ and P_T by P, where E and $P \approx E/3$ are the radiation energy density and pressure. A consequence of the dominance of radiation is that the net energy exchange between matter and radiation must be small, since the matter is unable to retain an appreciable amount of energy. Thus the radiation energy density obeys the sourceless equation

$$\underset{\sim}{v} \cdot \underset{\sim}{\nabla} E + \frac{4}{3} (\underset{\sim}{\nabla} \cdot \underset{\sim}{v}) E - \underset{\sim}{\nabla} \cdot \left(\frac{c}{3\kappa_{es}\rho} \underset{\sim}{\nabla} E \right) = 0 , \tag{80}$$

and the only other equations needed to describe the problem are mass conservation

$$4\pi\rho v r^2 = \dot{\mathcal{M}} \tag{81}$$

and the material momentum equation

$$\underset{\sim}{v} \cdot \underset{\sim\sim}{\nabla v} + \frac{1}{3\rho} \underset{\sim}{\nabla} E = \underset{\sim}{g} . \tag{82}$$

The fact that energy exchange with the matter is negligible in the <u>radiation</u> (<u>not</u> matter) energy balance allows us to put aside the question of the radiation spectrum in considering the dynamics.

The first two terms in equation (80) are the same order, namely, the order of the advection operator, while the third term is the diffusion operator. The relative magnitudes of these terms gives the comparisons of equations (74) and (75), with τ_{es} estimated as $\kappa_{es}\rho r$. We notice that since $\rho v r^2$ is a constant, from equation (81), the quantity $\tau_{es} v/c$ decreases with r. Thus the dynamic diffusion regime exists at small r, while the static diffusion regime exists at large r. The crossover radius is called by Meier the <u>trapping radius</u>, r_t. (Trapping is another name for the situation in which the radiation is advected with the flow faster than it diffuses.) Interior to the trapping radius the first two terms of equation (80) dominate the third, therefore they must cancel each other, and the radiation obeys the $\gamma = 4/3$ adiabatic law

$$E \propto \rho^{4/3} . \tag{83}$$

Exterior to the trapping radius the third term should dominate. This can occur only if the flux is nearly divergence-free,

$$\underset{\sim}{\nabla} \cdot \underset{\sim}{F} = - \underset{\sim}{\nabla} \cdot \left(\frac{c}{3\kappa_{es}\rho} \underset{\sim}{\nabla} E \right) \simeq 0 , \tag{84}$$

which is the flux-freezing condition. But Meier notes that the momentum equation itself ensures the vanishing of the flux divergence if the velocity is very much subsonic, in which case the first two terms of equation (80) must balance and equation (83) is still valid. Thus the adiabatic radius, defined as the outermost radius where equation (83) applies, cannot be inside the trapping radius, but it can be outside the trapping radius if the latter occurs in the subsonic region.

The density of the wind determines the ordering of the sonic and trapping radii, and the actual photosphere or radius of last scattering, r_{sc} defined by $\kappa_{es}\rho r_{sc} = 1$. The various orderings give the cases called A, B, C1, C2, and D by Meier. The densest wind is for the largest ratio of L_T to L_{ED}, and this is Meier's case D. This is the case for which the trapping radius is in the supersonic region, and therefore the adiabatic radius coincides with the trapping radius.

For the whole structure of the wind out to and beyond the sonic radius in Meier's case D, the adiabatic relation of E to ρ may be used, which turns equations (81) and (82) into Parker's standard adiabatic wind model, with $\gamma = 4/3$. All the details of the wind are determined once two parameters are given, which can be taken to be the total wind energy flux, fixed by equation (79), and the constant radiation entropy per unit mass. The further condition that the wind must be in pressure balance with the compact star's envelope across the injection region fixes the entropy.

The structure of the wind derived from Parker's model is the following for the subsonic region:

$$r < r_s \qquad \begin{aligned} v &\propto r \\ E &\propto r^{-4} \\ L &= L_{ED} \end{aligned} \tag{85}$$

(L is the comoving-frame radiative luminosity), and for the supersonic region:

$$r_s < r < r_{ad} \qquad \begin{aligned} v &= \text{constant} \\ E &\propto r^{-8/3} \\ L &\propto r^{1/3} \ . \end{aligned} \tag{86}$$

For the structure of wind outside the adiabatic radius, we see that the velocity remains constant, while E changes from the variation given by equation (83) to that given by equation (84). Thus L remains constant at the value it has at the adiabatic radius, and E(r) adjusts itself accordingly. The structure is

$$r_{ad} < r < r_{sc} \qquad \begin{aligned} v &= \text{constant} \\ E &\propto r^{-3} \\ L &= \text{constant} \end{aligned} \qquad (87)$$

between the adiabatic radius and the photosphere, and

$$r > r_{sc} \qquad \begin{aligned} v &= \text{constant} \\ E &\propto r^{-2} \\ L &= \text{constant} \end{aligned} \qquad (88)$$

exterior to the photosphere.

One interesting feature of the energetics of this model is that the dominant energy loss mechanism changes from the advection of radiation at the base of the wind to kinetic energy far out in the wind, and at no point is the comoving-frame radiative luminosity dominant. Another feature is that the emergent radiative luminosity is in fact larger than L_{ED}. The factor by which it is larger, from equation (86), is the cube root of r_{ad}/r_s. The ratio of the final kinetic energy loss to L_{ED} is just r_{ad}/r_s, so a super-Eddington luminosity can be obtained from a supercritical wind provided the mechanical power is even larger, as the cube (Rees 1977).

Obtaining the emergent spectrum from the wind is a good bit more complicated than finding the dynamical structure (Meier 1982b). In rough outline the spectrum is that described by Rybicki and Lightman and others, with regions of Planck shape, modified Planck shape, and a Wien peak normalized in this case to the known value of the emergent flux. What is difficult is that the relevant temperature for the different parts of the spectrum should be the material temperature at some appropriate depth. (Perhaps the depth where N, the number of scatterings to escape, is $\kappa_{es}/\kappa_{abs}(\nu)$ for the Planck and modified Planck regions, and $mc^2/4kT$ for the Wien peak.) The run of material temperature can only be found by solving for it self-consistently with the run of the radiation spectrum; a discussion of this is too involved to give here, and the reader is referred to Meier (1982b).

5.4. Second example -- cyclotron line emission in X-ray pulsars

The mechanism by which X-ray pulsars produce their radiation has been investigated intensively, and several competing models have been advanced. It is not my intent to review this subject, but rather to select a particular model with which I am familiar and which illustrates, I think, the concept of dynamic diffusion. This is the model of Langer and Rappaport (1982), which is similar in many respects to the model of Klein, Arons and Lea (1983).

A brief description of the model is the following. Matter accreting onto the neutron star is channeled by the 10^{12} G field onto a small region near the magnetic polar cap. The models differ in whether the accretion column is filled with matter or hollow, but in either case a collisionless shock is assumed to arrest the matter at a few tenths of a neutron-star radius above its surface. The temperature of the ions is elevated to the order of 40 MeV, but the electrons, which are heated by electron-ion collisions, attain a temperature of only 20 keV or so, a factor m_e/m_p less than the ion temperature. The precise value of the electron temperature depends on the electron-ion energy exchange rate. At this temperature and the appropriate density the principal energy loss by the electrons is the collisional excitation of the Landau states of the electrons, and bremsstrahlung is secondary. This is due to the fact that $kT_e = 20$ keV is comparable with the cyclotron energy, which is assumed to be of order 40 keV to explain the observations of Her X-1. The excited Landau states quickly decay, producing cyclotron line emission. The very large electron scattering cross section at the cyclotron resonance ensures that the line photons are trapped within the accretion column, since typical values of the Compton optical radius of the column are of order unity, and the resonant cross section is some 10^4 times larger than the Compton cross section. Since the post-shock flow speed is still of order $v/c = 10^{-1}$, $\tau v/c$ is large and the condition for dynamic diffusion applies.

The picture that then emerges is that the subsonic post-shock part of the accretion column is supported against gravity by the pressure of cyclotron line radiation which is created in a radiation zone just behind the shock and then trapped in the gas as it flows downstream. It is very difficult for this radiation to be destroyed or leave the column, since the destruction mechanisms of inverse bremsstrahlung and collisional de-excitation of the excited Landau levels, and the lateral diffusion of line radiation in the column, are all weak. It turns out that the most important loss mechanism is the scattering of the line photons in the non-resonant polarization mode (Ventura 1979), which can scatter the photons into a region of frequency-angle space in which the resonant cross section is small and therefore the escape of the photon from the column is likely.

If there is no loss of line radiation whatever, then the structure of the column follows equations (82) and (83) discussed in connection with the supercritical wind. Since the flow is subsonic, the structure approximates that of an n = 3 polytropic static envelope model. Of course, this would fail to properly match the conditions at the surface of the neutron star. The height of the shock above the neutron star surface must be sufficient for the line radiation to have time to escape, which produces a much more rapid increase of density with decreasing radius as required to match the neutron star. Thus the steady-state height of the shock is determined by the time necessary before a sufficient number of non-resonant scatterings has occurred to lead to the escape of the photon.

The gravitational energy of the accreting material is finally lost to the system as radiation from the side of the column (or possibly through its hollow core) near its base, as the photons that have been non-resonantly scattered escape from the line. The spectral distribution that would be expected is roughly that of an emission line with perhaps a central dip at the resonant frequency. At this time it is not clear whether this spectrum is consistent with that observed for X-ray sources such as Her X-1 in which line features have been seen. Another question that must be considered is whether this model gives a fair representation of the pulse profile.

6. ACKNOWLEDGMENT

This work was performed under the auspices of the U.S. Department of Energy by Lawrence Livermore National Laboratory under contract No. W-7405-Eng-48.

7. REFERENCES

Abbott, D.C. 1977, unpublished Ph. D. thesis, University of Colorado.
Abbott, D.C. 1980, Astrophys. J., 242, p. 1183.
Abbott, D.C. 1982, Astrophys. J., 259, p. 282.
Buchler, J.-R. 1982, Los Alamos National Laboratory report, No. LA-UR-82-1336.
Caflisch, R.E. and Levermore, C.D. 1983, in preparation.
Cannon, C.J. 1973a, J. Quant. Spectr. Rad. Transf., 13, p. 627.
Cannon, C.J. 1973b, Astrophys. J., 185, p. 621.
Castor, J.I. 1970, Mon. Not. Roy. Astr. Soc., 149, p. 111.
Castor, J.I. 1974, Mon. Not. Roy. Astr. Soc., 169, p. 279.
Castor, J.I., Abbott, D.C. and Klein, R.I. 1975, Astrophys. J., 195, p. 157 (CAK).
Chapline, F.G. and Stevens, J. 1973, Astrophys. J., 184, p. 1041.

Feautrier, P. 1964, C.R. Acad. Sci. Paris, 258, p. 3189.
Karp, A.H., Lasher, G., Chan, K.L. and Salpeter, E.E. 1977, Astrophys. J., 214, p. 161.
Klein, R.I., Arons, J. and Lea, S.M. 1983, presented at the Santa Cruz workshop on High Energy Transients.
Kompaneets, A.S. 1957, Sov. Phys. JETP, 4, p. 730.
Langer, S.H. and Rappaport, S. 1982, Astrophys. J., 257, p. 733.
Lucy, L.B. and White, R.L. 1980, Astrophys. J., 241, p. 300.
MacGregor, K.B., Hartman, L. and Raymond, J.C. 1979, Astrophys. J., 231, p. 514.
Meier, D.L. 1982a, Astrophys. J., 256, p. 681.
Meier, D.L. 1982b, Astrophys. J., 256, p. 693.
Mihalas, D. 1978, "Stellar Atmospheres", second edition (San Francisco: Freeman).
Mihalas, D., Kunasz, P.B. and Hummer, D.G. 1975, Astrophys. J., 202, p. 465.
Mihalas, D., Kunasz, P.B. and Hummer, D.G. 1976, Astrophys. J., 210, p. 419.
Mihalas, D. and Weaver, R. 1982, Los Alamos National Laboratory report No. LA-UR-82-606.
Noerdlinger, P. and Rybicki, G. 1974, Astrophys. J., 193, p. 651.
Owocki, S.P. and Rybicki, G. 1982, Bull. Amer. Astr. Soc., 14, p. 920
Parker, E.N. 1958, Astrophys. J., 128, p. 664.
Rees, M.J. 1977, presented at the NATO Advanced Study Institute "Active Galactic Nuclei", Cambridge, England.
Rybicki, G.B. 1971, J. Quant. Spectr. Rad. Transf., 11, p. 589.
Rybicki, G.B. and Lightman, A.P. 1979, "Radiative Processes in Astrophysics" (New York: Wiley-Interscience).
Scharmer, G.B. 1981, Astrophys. J., 249, p. 720.
Sobolev, V.V. 1947, English edition 1960, "Moving Envelopes of Stars" (Cambridge, Mass.: Harvard University Press).
Ventura, J. 1979, Phys. Rev., D19, p. 1684.
Weber, S.V. 1981, Astrophys. J., 243, p. 954.

THE EQUATIONS OF RADIATION HYDRODYNAMICS

Dimitri Mihalas

X-Division Consultant
Los Alamos National Laboratory

ABSTRACT: The purpose of this paper is to give an overview of the role of radiation in the transport of energy and momentum in a combined matter-radiation fluid. The transport equation for a moving radiating fluid is presented in both a fully Eulerian and a fully Lagrangean formulation, along with conservation equations describing the dynamics of the fluid. Special attention is paid to the problem of deriving equations that are mutually consistent in each frame, and between frames, to O(v/c). A detailed analysis is made to show that in situations of broad interest, terms that are formally of O(v/c) actually dominate the solution, demonstrating that it is essential 1) to pay scrupulous attention to the question of the frame dependence in formulating the equations, and 2) to solve the equations to O(v/c) in quite general circumstances.

These points will be illustrated in the context of the non-equilibrium radiation diffusion limit, and a sketch of how the Lagrangean equations are to be solved will be presented.

I. INTRODUCTION

Radiation plays an important role in a wide variety of laboratory and astrophysical flows. Usually radiative exchange will be the most effective energy transport mechanism at temperatures above an eV. At higher temperatures, radiation will make a large, perhaps dominant, contribution to the total energy density, energy flux, and stress in the radiating fluid. To describe the dynamics of such flow it is thus essential to have conservation equations that account accurately for the radiative contributions to flow dynamics.

K.-H. A. Winkler and M. L. Norman (eds.), Astrophysical Radiation Hydrodynamics, 45–69.

An estimate of the local importance of radiation in the fluid is given by the ratio

$$R \equiv \frac{\text{material internal energy density}}{\text{radiation energy density}} = \frac{3kN}{2aT^3} = \frac{2.8\times10^{-2}N}{T^3} \qquad (1.1)$$

where N is the particle density and T is the temperature. One easily finds that $R \lesssim 1$ when $T_{keV} \gtrsim 1.75\,\rho^{1/3}$ where ρ [gm/cm^{-3}] is the density of the material. It is then immediately obvious that when temperatures rise above a few keV (e.g., in stellar interiors, astrophysical X-ray sources, high-temperature laboratory plasmas), radiation dominates the energy density (and pressure) of the radiating fluid.

The situation for energy transport is even more dramatic. A measure of the efficiency of radiative transport is given by the Boltzmann number

$$Bo \equiv \frac{\text{material enthalpy flux}}{\text{radiative flux from free surface}} = \frac{\rho C_p T v}{\sigma T^4} \,. \qquad (1.2)$$

One finds that Bo is approximately (v/c) times R as defined in Eq. (1.1), where v is a typical fluid velocity. In most astrophysical and laboratory flows one has $(v/c) \ll 1$ [rarely does (v/c) approach 10^{-2}], which shows that radiation normally will be, by far, the most effective vehicle for energy transport. Another way of understanding this conclusion is to note that in opaque material both radiation diffusion and heat conduction have the mathematical form $\underset{\sim}{F} = -K\underset{\sim}{\nabla}T$ where K is an appropriate conduction coefficient. For particles (in this case photons and material particles) of equal energy, K for each species is proportional to the particle velocity and mean-free-path. In both respects the photons win because $c \gg v$, and usually the photon mean-free-path set by material opacity is much greater than a typical molecular mean-free-path.

One of the goals of our discussion is to formulate dynamical equations that provide an accurate description of radiative effects in the flow. As we shall see, one achieves such a description only by paying scrupulous attention to the frame in which the equations are posed, and the quantities they contain are measured. The differences between frames are formally O(v/c), and at first sight one might expect the effects mentioned to be of negligible importance. Deeper analysis shows, however, that in many regimes of great interest and practical importance these seemingly small terms can actually dominate all others in the equations, and therefore must inevitably affect significantly the dynamics of a flow. Indeed, the failure to account carefully for O(v/c) terms has often led to serious misconceptions in past treatments of radiation hydrodynamics, and basic inconsistencies between various formulations. We believe that these issues can now be settled once and for all, and that we can agree, finally, on the correct set of equations to be solved. (Solving them is, of course, another matter!).

For brevity we shall state equations without detailed derivations. (For further details see References 1-3). For simplicity we shall focus on one-dimensional flows, though many of our results hold in general geometries. Further, we shall not discuss numerical methods in detail.

II. LORENTZ TRANSFORMATION OF RADIATION TRANSPORT QUANTITIES

In order to be able to write the transfer equation in different frames, we must know the Lorentz transformation properties of its constituents. In what follows the suffix "o" will denote quantities measured in the comoving fluid frame (i.e., the Lagrangean frame), in which material properties are isotropic; unadorned symbols denote quantities in the fixed lab frame.

A. The Photon Four-Momentum

By an application of a Lorentz transformation to the photon four-moment

$$M^{\alpha} = (h\nu/c)\ (1, \underset{\sim}{n}) \tag{2.1}$$

one finds

$$\nu_o = \gamma(1 - \beta\mu)\nu \tag{2.2}$$

and

$$\mu_o = (\mu - \beta)/(1 - \beta\mu) \tag{2.3}$$

where $\beta \equiv v/c, \gamma \equiv (1 - \beta^2)^{-1/2}$, and μ is the cosine of the angle between $\underset{\sim}{n}$, the direction of photon propagation, and the velocity $\underset{\sim}{v}$ (which, for one-dimensional flow we assume lies along the z axis or in the radial direction for planar and spherical geometry respectively).

B. Specific Intensity, Opacity, Emissivity

For convenience we characterize the radiation field macroscopically in terms of the <u>specific intenstiy</u> $I(\underset{\sim}{x}, t; \underset{\sim}{n}, \nu)$, which, for one-dimensional flows is equivalent to $I(\mu, \nu)$ (suppressing reference to x and t for brevity). L.H. Thomas[4] showed that

$$I(\mu,\nu) = (\nu/\nu_o)^3 \quad I_o(\mu_o,\nu_o). \tag{2.4}$$

We characterize the opacity of the material by a macroscopic absorption coefficient $\chi(\underset{\sim}{x},t;\underset{\sim}{n},\nu)$, which will, in general, be anisotropic because of velocity-induced Doppler-shift and aberration effects. Thomas showed that

$$\chi(\mu,\nu) = (\nu_o/\nu)\ \chi_o\ (\nu_o) \tag{2.5}$$

where χ_o (measured in the fluid frame) is isotropic. Similarly, the emissivity transforms as

$$\eta(\nu,\nu) = (\nu/\nu_o)^2\ \eta_o\ (\nu_o)\ . \tag{2.6}$$

C. The Radiation Stress-Energy Tensor

The stress-energy tensor for the radiation field is given by

$$R^{\alpha\beta} = \frac{1}{c}\int_0^\infty d\nu \oint d\omega\ I(\underset{\sim}{n},\nu)n^\alpha n^\beta,\quad (\alpha,\beta = 0,\ldots 3)\ , \tag{2.7}$$

where n^i is the ith component of the photon propagation vector $\underset{\sim}{n}$, and $n^o \equiv 1$. An equivalent form of Eq. (2.7) is

$$\underset{\approx}{R} = \begin{pmatrix} E & c^{-1}\underset{\sim}{F} \\ c^{-1}\underset{\sim}{F} & \underset{\approx}{P} \end{pmatrix}\ , \tag{2.8}$$

where

$$E \equiv \frac{1}{c}\int_0^\infty d\nu \oint d\omega\ I(\underset{\sim}{n},\nu) \tag{2.9}$$

is the radiation energy density,

$$F \equiv \int_0^\infty d\nu \oint d\omega\ I(\underset{\sim}{n},\nu)\underset{\sim}{n} \tag{2.10}$$

is the radiation flux, and

$$P \equiv \frac{1}{c}\int_0^\infty d\nu \oint d\omega\ I(\underset{\sim}{n},\nu)\underset{\sim}{n}\underset{\sim}{n} \tag{2.11}$$

is the radiation stress (pressure) tensor. Applying a Lorentz transformation to Eq. (2.8), and retaining terms only to O(v/c) in a spherical flow we find that lab-frame radiation quantities are related to their fluid-frame counterparts by

$$E = E_o + 2(\beta/c)F_o \quad , \tag{2.12}$$

$$F = F_o + v(E_o + P_o), \tag{2.13}$$

and

$$P = P_o + 2(\beta/c)F_o \quad . \tag{2.14}$$

III. INERTIAL-FRAME (EULERIAN) EQUATIONS OF RADIATION HYDRODYNAMICS

A. The Inertial-Frame Transfer Equation and Its Moments

The time-dependent transfer equation in the lab frame is

$$\left(\frac{1}{c}\frac{\partial}{\partial t} + \underset{\sim}{n}\cdot\underset{\sim}{\nabla}\right)I(\underset{\sim}{x},t;\underset{\sim}{n},\nu) = \eta(\underset{\sim}{x},t;\underset{\sim}{n},\nu) - \chi(\underset{\sim}{x},t;\underset{\sim}{n},\nu)I(\underset{\sim}{x},t;\underset{\sim}{n},\nu) \quad . \tag{3.1}$$

Taking the zeroth and first moment of Eq. (3.1) over angle we obtain the monochromatic moment equations

$$\frac{\partial E_\nu}{\partial t} + \underset{\sim}{\nabla}\cdot\underset{\sim}{F}_\nu = \oint\left[\eta(\underset{\sim}{x},t;\underset{\sim}{n},\nu) - \chi(\underset{\sim}{x},t;\underset{\sim}{n},\nu)I(\underset{\sim}{x},t;\underset{\sim}{n},\nu)\right]d\omega \tag{3.2}$$

and

$$\frac{1}{c^2}\frac{\partial \underset{\sim}{F}_\nu}{\partial t} + \underset{\sim}{\nabla}\cdot\underset{\approx}{P}_\nu = \frac{1}{c}\oint\left[\eta(\underset{\sim}{x},t;\underset{\sim}{n},\nu) - \chi(\underset{\sim}{x},t;\underset{\sim}{n},\nu)I(\underset{\sim}{x},t;\underset{\sim}{n},\nu)\right]\underset{\sim}{n}\,d\omega \quad , \tag{3.3}$$

which, when integrated over frequency, yield the radiation energy equation (Eulerian)

$$\frac{\partial E}{\partial t} + \underset{\sim}{\nabla}\cdot\underset{\sim}{F} = \int_0^\infty d\nu \oint d\omega\left[\eta(\underset{\sim}{x},t;\underset{\sim}{n},\nu) - \chi(\underset{\sim}{x},t;\underset{\sim}{n},\nu)I(\underset{\sim}{x},t;\underset{\sim}{n},\nu)\right] \tag{3.4}$$

and the radiation momentum equation (Eulerian)

$$\frac{1}{c^2}\frac{\partial \underset{\sim}{F}}{\partial t} + \underset{\sim}{\nabla}\cdot \underset{\approx}{P}$$

$$= \int_0^\infty d\nu \oint d\omega \left[\eta(\underset{\sim}{x},t;\underset{\sim}{n},\nu) - \chi(\underset{\sim}{x},t;\underset{\sim}{n},\nu) I(\underset{\sim}{x},t;\underset{\sim}{n},\nu)\right] \underset{\sim}{n} \ . \qquad (3.5)$$

Equations (3.4) and (3.5) are, of course, consistent with the general dynamical equation for the radiation field

$$R^{\alpha\beta}_{\ ;\beta} = -G^{\alpha} \qquad (3.6)$$

where G^{α} is the density of radiation four-force density acting on the material. In particular

$$G^{o} = c^{-1}\int_0^\infty d\nu \oint d\omega \left[\chi(\underset{\sim}{n},\nu) I(\underset{\sim}{n},\nu) - \eta(\underset{\sim}{n},\nu)\right] \qquad (3.7)$$

is easily seen to be c^{-1} times the rate of radiative energy input, per unit volume, into the material, while

$$G^{i} = c^{-1}\int_0^\infty d\nu \oint d\omega \left[\chi(\underset{\sim}{n},\nu) I(\underset{\sim}{n},\nu) - (\underset{\sim}{n},\nu)\right] n^{i} \qquad (3.8)$$

is the net rate of radiative momentum input.

B. Mixed-Frame Equations

The essential problem with equations (3.2)-(3.5), (3.7) and (3.8) is that, in accordance with Eq. (2.2), the material properties must be evaluated at fluid frame frequencies

$$\nu_o = \gamma(1 - \underset{\sim}{n}\cdot\underset{\sim}{v}/c)\nu \ , \qquad (3.9)$$

which implies that the lab-frame absorption-emission coefficients are effectively anisotropic. Hence, the radiation energy and momentum absorption-emission terms must be computed from cumbersome souble integrals over both angle and frequency.

In principle it is possible to evaluate these integrals by brute-force computations. However, to do so: a) is costly, and b) deprives us useful insight. We can get a deeper picture of what is going on by arguing that because $(v/c) \ll 1$ we can make a first-order expansion of χ and η around the observer-frame (lab-frame) frequency ν. With the aid of Eqs. (2.2) and (2.4)-(2.6) we then obtain the transfer equation

$$\left(\frac{1}{c}\frac{\partial}{\partial t} + \underset{\sim}{n}\cdot\underset{\sim}{\nabla}\right)I(\underset{\sim}{n}, \ ) = \eta_o(\nu) - \chi_o(\nu)I(\underset{\sim}{n},\nu)$$

$$+ \left(\frac{\underset{\sim}{n}\cdot\underset{\sim}{v}}{c}\right)\{2\eta_o(\nu) - \nu\frac{\partial\eta_o}{\partial\nu} + \left[\chi_o(\nu) + \nu\frac{\partial\chi_o}{\partial\nu}\right]I(\underset{\sim}{n},\nu)\} \ . \tag{3.10}$$

Here χ_o and η_o are now fluid-frame quantities and are therefore isotropic. One expects this kind of an expansion procedure to be reasonable for continua, which are relatively smooth, because the opacity will change only slightly over a Doppler width $\Delta\nu_D \sim \nu\ v/c$. The procedure clearly will not work for spectral lines where $\Delta\nu_D$ will be of the same order of the line-width over which the change in opacity is substantial.

Integrating over angle and frequency we obtain the radiation energy equation

$$\frac{\partial E}{\partial t} + \underset{\sim}{\nabla}\cdot\underset{\sim}{F} = \int_0^\infty [4\pi\eta_o(\nu) - \chi_o(\nu)cE_\nu]\,d\nu$$

$$+ \frac{\underset{\sim}{v}}{c}\cdot\int_0^\infty [\chi_o(\nu) + \nu\frac{\partial\chi_o}{\partial\nu}]\underset{\sim}{F}_\nu d\nu \tag{3.11}$$

and the radiation momentum equation

$$\frac{1}{c^2}\frac{\partial\underset{\sim}{F}}{\partial t} + \underset{\sim}{\nabla}\cdot\underset{\approx}{P} = -\frac{1}{c}\int_0^\infty \chi_o(\nu)\underset{\sim}{F}_\nu d\nu + \frac{4\pi\underset{\sim}{v}}{c^2}\int_0^\infty \eta_o(\nu)d\nu$$

$$+ \frac{\underset{\sim}{v}}{c}\cdot\int_0^\infty\left[\chi_o(\nu) + \nu\frac{\partial\chi_o}{\partial\nu}\right]\underset{\approx}{P}_\nu d\nu \ . \tag{3.12}$$

Equations (3.11) and (3.12) are <u>mixed-frame</u> radiation equations because the material properties (χ and η) are measured in the fluid frame, whereas the space-time, angles, frequencies, and radiation quantities are all measured in the lab frame.

Aside from the annoying $O(v/c)$ terms on the right-hand side Eqs. (3.11) and (3.12) strongly resemble the inertial-frame equations [Eqs. (3.4) and (3.5)], if we agree to ignore the anisotropy of the material coefficients in the latter. Inasmuch as $(v/c) \ll 1$, it is natural to ask whether we can simply drop the extra terms in Eqs. (3.11) and (3.12) and treat the radiation as if the material were static (even though time-dependent).

We now show that the answer to this question is "no" because although the velocity-dependent terms are formally $O(v/c)$

can actually dominate the other terms in these equations in various regimes of importance. To extract the essential physical flavor, suppose the material is grey; then

$$\frac{\partial E}{\partial t} + \nabla\cdot\mathbf{F} = \chi(4\pi B - cE) + \frac{\chi}{c}\mathbf{v}\cdot\mathbf{F} \tag{3.13}$$

and

$$\frac{1}{c^2}\frac{\partial \mathbf{F}}{\partial t} + \nabla\cdot\mathbf{P} = -\frac{\chi}{c}\mathbf{F} + \frac{\chi}{c}\left(\frac{4\pi B}{c}\mathbf{v} + \mathbf{v}\cdot\mathbf{P}\right) . \tag{3.14}$$

Consider first the radiation energy equation, [Eq. (3.13)]. If we ignore the velocity-dependent term we ignore a term equal to the rate of work done by radiation on the fluid. Clearly this is a serious error when the radiation field is strong and the radiation force is large. Indeed, in the diffusion limit where $(\lambda_p/\ell) \ll 1$, the fluid-frame energy-density equilibrates to $E_o = 4\pi B/c$, hence from the Eq. (2.12) we have $4\pi B - cE = -2\mathbf{v}\cdot\mathbf{F}/c + O(v^2/c^2)$. Equation (3.13) then reduces to

$$\frac{\partial E}{\partial t} + \nabla\cdot\mathbf{F} = -\frac{\chi}{c}\mathbf{v}\cdot\mathbf{F} , \tag{3.15}$$

which makes the physically sensible statement that the rate of change of the radiation energy density equals the flow of radiation out of that region of space minus the rate of work done by the radiation field on the material. Two points can be made here: 1) Omission of the v F term in Eq. (3.13) would produce an error equal in size to omitting the net absorption-emission term in the diffusion regime [which is unacceptable for non equilibrium diffusion; cf. Eq. (5.5)]. 2) The ratio of $(\chi\mathbf{v}\cdot\mathbf{F}/c)$ to $\nabla\cdot\mathbf{F}$ is $O(\ell v/\lambda_p c)$, so in the dynamic diffusion regime where $(v/c) \gtrsim (\lambda_p/\ell)$ the velocity-dependent term dominates the energy balance.

Consider now the momentum equation, [Eq. (3.14)]. On a fluid-flow timescale $t_f \sim \ell/v$ the time derivative term is $O(\lambda_p v/\ell c)$ relative to $(\chi F/c)$, and therefore is negligible in opaque material. Equation (3.14) can then be written

$$\mathbf{F} = -(c/\chi)\nabla\cdot\mathbf{P} + (4\pi B/c)\mathbf{v} + \mathbf{v}\cdot\mathbf{P} . \tag{3.16}$$

But to $O(v/c)$, $(4\pi B/c) = E_o$, and $\mathbf{P}_o \rightarrow \frac{1}{3}E_o\mathbf{I}$; furthermore, $\mathbf{F}_o = -(c/\chi)\nabla\cdot\mathbf{P}_o$. Using these results in Eq. (3.16) we find

$$\mathbf{F} = \mathbf{F}_o + E_o\mathbf{v} + \mathbf{v}\cdot\mathbf{P}_o + O(v^2/c^2) \rightarrow \mathbf{F}_o + \frac{4}{3}E_o\mathbf{v} , \tag{3.17}$$

which is nothing more than the Lorentz transformation connecting the inertial-frame (Eulerian) flux $\underset{\sim}{F}$ and the Lagrangian flux $\underset{\sim}{F}_o$. In other words, if we omit the O(v/c) terms in Eq. (3.14), we fail to discriminate between the Eulerian and Lagrangean fluxes. To do so may lead to serious errors! To illustrate, consider the interior of a star. It is easy to show that in a star $(F_o/cE_o) \sim (T_{eff}/T)^4$, where T_{eff} is the effective temperature defined by astronomers. For the Sun, $T_{eff} \approx 6 \times 10^3$K, whereas at the Sun's center, $T \sim 6 \times 10^6$K. Thus $(vE_o/F_o) \approx (v/c)(T/T_{eff})^4 \sim 10^{12}(v/c)$. It is then obvious that even a miniscule velocity implies an enormous difference between $\underset{\sim}{F}_o$ and $\underset{\sim}{F}$ (i.e., dynamic diffusion, see below), a point often overlooked in formulating radiation energy and momentum equations!

The main thrust of the preceding few paragraphs is that terms that are formally O(v/c) can actually dominate the flow of radiation energy and momentum. One therefore must make a careful analysis of these terms. In doing so we consider both the streaming limit (optically thin material) and the diffusion limit (opaque material). In the streaming limit we just analyze two characteristic timescales: the radiation-flow timescale $t_r \sim \ell/c$, and the fluid-flow timescale $t_f \sim \ell/v$, where ℓ is a characteristic length in the flow. In the diffusion limit we have two regimes: We have static diffusion when t_f is larger than the radiation diffusion time $t_d \sim (\ell^2/c\lambda_p)$ where λ_p is a photon meanfree-path. In this case we have $(v/c) < (\lambda_p/\ell)$. We have dynamic diffusion when $t_f < t_d$ [or $(v/c) > (\lambda_p/\ell)$]. In static diffusion, the diffusion process itself sets the rate of radiant energy flow through the medium; in dynamic diffusion, flow-induced changes in local properties of the medium drive changes in the radiation field faster than they would have occured if only diffusion were operative.

The guidline is that a term that is always (i.e., in all regimes) O(v/c), or less, relative to the dominant terms in the equations can, in fact, be dropped because $(v/c) \ll 1$. However, if any term is found to be of the same order (or larger than) the dominant terms (e.g., the spatial operators) in any regime, then we must retain that term in the equation from the outset because in general our flows will span both optically thick and optically thin regions, and in general fluid velocities will be small in some regions, but large in others.

The analysis is straightforward but tedious, so we state only the results.[2,3] One finds that:

(a) to obtain a correct solution of the mixed-frame transfer equation [Eq. (3.10)] on a fluid-flow timescale t_f, it is necessary to retain the differential operator on the left-hand side, and all terms on the right-hand side; but the $(\partial/\partial t)$ term can be dropped;

(b) to solve the mixed-frame radiation energy and momentum equations [Eqs. (3.11) and (3.12)] correctly on a fluid-flow time-scale we must retain all terms in both equations except $(\partial/\partial t)$ in the momentum equation, which may be dropped;

(c) if we wish to follow radiation flow on a timescale t_r, then we must retain the $(\partial/\partial t)$ terms in the transfer equation and momentum equations as well.

Unfortunately, these requirements make the equations cumbersome to solve! However, we emphasize that all the terms mentioned above must be included, not to achieve accuracy to $O(v/c)$ (as is sometimes asserted), but rather to assure that we have taken into account all dominant terms in the various important physical regimes we encounter. There are no shortcuts! We shall see below that the situation is a little more favorable in the Lagrangian frame.

C. Inertial-Frame Dynamical Equations for a Radiation Fluid

Let $M^{\alpha\beta}$ and $R^{\alpha\beta}$ be the stress-energy tensors for the material and radiation components, respectively, of the radiating fluid. Let F and G be, respectively, the four-force density exerted on the fluid from external sources (e.g., gravity) and on the material by the radiation. We can then write dynamical equations for the radiating fluid either by considering both F^α and G^α to act on the matter, so that

$$M^{\alpha\beta}_{;\beta} = F^\alpha + G^\alpha \tag{3.18}$$

or we can consider the external forces to act on the radiating fluid comprising both matter and radiation, in which case

$$(M^{\alpha\beta} + R^{\alpha\beta})_{;\beta} = F^\alpha . \tag{3.19}$$

Equations (3.18) and (3.19) are, of course, equivalent by virtue of Eq. (3.6); Equation (3.18) is perhaps the more natural physical picture in optically thin material, whereas Eq. (3.19) seems most natural in opaque material (especially if it is in thermal equilibrium).

Writing out Eqs. (3.18) or (3.19) in Cartesian coordinates one readily finds the relativistically correct momentum equation for a radiating fluid

$$\rho_* \left(\frac{D\mathbf{v}}{Dt}\right) = \mathbf{f} - \nabla p - \frac{\mathbf{v}}{c^2}\left(\frac{\partial p}{\partial t} + \mathbf{v}\cdot\mathbf{f}\right) - \left(\nabla\cdot\mathsf{P} + \frac{1}{c^2}\frac{\partial \mathbf{F}}{\partial t}\right)$$
$$+ \frac{\mathbf{v}}{c^2}\left(\frac{\partial E}{\partial t} + \nabla\cdot\mathbf{F}\right) . \tag{3.20}$$

Here $\rho_* \equiv \gamma\rho_{ooo}$, where, following L.H. Thomas,[4] $\rho_{ooo} \equiv \rho_o(1 + e/c^2) + (p/c^2)$. Here ρ_o is the proper density, p and e are the gas pressure and material internal energy, and $\underset{\sim}{f}$ is the external force density. Despite the appearence of the operator (D/Dt), Eq. (3.20) is not fully Lagrangean because all radiation quantities are measured in the lab-frame instead of the fluid frame; we shall call such equations quasi-Lagrangean. The fully Lagrangean form is given in Eq. (4.12).

The distinction between ρ_* and ρ is $O(v^2/c^2)$, as is the term containing $(p_{,t} + \underset{\sim}{v}\cdot\underset{\sim}{f})$ on a fluid-flow timescale; the same is true for $(\partial E/\partial t)$. Thus the momentum equation correct to O(v/c) is

$$\rho(D\underset{\sim}{v}/Dt) = \underset{\sim}{f} - \underset{\sim}{\nabla}p - [\underset{\sim}{\nabla}\cdot\underset{\approx}{P} + c^{-2}(\partial\underset{\sim}{F}/\partial t)] + c^{-2}\underset{\sim}{v}(\underset{\sim}{\nabla}\cdot\underset{\sim}{F}). \qquad (3.21)$$

The O(v/c) terms are of logical importance when we want to demonstrate consistency among various forms of the momentum equations[2,3] (e.g., between the Eulerian and Lagrangean frames); they can be ignored for practical computations. Thus the momentum equation suited to practical work is

$$\rho(D\underset{\sim}{v}/Dt) = \underset{\sim}{f} - \underset{\sim}{\nabla}p - [\underset{\sim}{\nabla}\cdot\underset{\approx}{P} + c^{-2}(\partial\underset{\sim}{F}/\partial t)] \,. \qquad (3.22)$$

In Eq. (3.22) we can also drop $(\partial F/\partial t)$ unless a) we are in the streaming regime, and b) we must work on a radiation-flow timescale t_R (e.g., in the outer layers of a laser-fusion pellet).

We can write several forms of the energy equation; we restrict attention to the gas energy equation for a radiating fluid (i.e., the first law of thermodynamics for the material). By suitable manipulation of Eq. (3.18) one can derive the relativistically correct equation

$$\rho_o\left[\frac{De}{D\tau} + p\,\frac{D}{D\tau}\left(\frac{1}{\rho_o}\right)\right] = -\,V_\alpha F^\alpha - V_\alpha G^\alpha \qquad (3.23)$$

where τ is proper time and V^α is the four velocity of the flow. For ordinary body forces (e.g., gravity) that do not affect the proper density of the material, $V_\alpha F^\alpha \equiv 0$; this is not true for radiation, which can change the internal excitation and ionization state of the material. Reducing Eq. (3.23) to O(v/c) we have

$$\rho\left[\frac{De}{Dt} + p\,\frac{D}{Dt}\left(\frac{1}{\rho}\right)\right] = -\left(\frac{\partial E}{\partial t} + \underset{\sim}{\nabla}\cdot\underset{\sim}{F}\right) + \underset{\sim}{v}\cdot\left(\frac{1}{c^2}\frac{\partial \underset{\sim}{F}}{\partial t} + \underset{\sim}{\nabla}\cdot\underset{\approx}{P}\right) \,. \qquad (3.24)$$

In the presence of thermonuclear energy release, proper mass is not conserved, and $-V^\alpha F^\alpha = -c\,(d\rho_o/dt) = \rho_o\varepsilon_o$ where ε_o is the rate of energy-release, per gram, measured in the fluid frame; this term must be added to the right-hand side of Eq. (3.24).[4] For practical computations we can drop terms that are never larger than O(v/c); the only term we can omit is $(\partial \underset{\sim}{F}/\partial t)$. We can then rewrite Eq. (3.24) in the strictly Eulerian form

$$(\rho e + E)_{,t} + \underset{\sim}{\nabla}\cdot(\rho e\underset{\sim}{v} + \underset{\sim}{F}) = \underset{\sim}{v}\cdot\underset{\sim}{\nabla}\cdot\underset{\approx}{P} - p\underset{\sim}{\nabla}\cdot\underset{\sim}{v} + \rho\varepsilon, \tag{3.25}$$

or in the quasi-Lagrangean form

$$\rho\left[\frac{D}{Dt}\left(e + \frac{E}{\rho}\right) + p\,\frac{D}{Dt}\left(\frac{1}{\rho}\right)\right] + \underset{\sim}{\nabla}\cdot(\underset{\sim}{F} - \underset{\sim}{v}E) = \underset{\sim}{v}\cdot\underset{\sim}{\nabla}\cdot\underset{\approx}{P} + \rho\varepsilon. \tag{3.26}$$

Thus to calculate the flow of a radiating fluid in the lab-frame (i.e., an Eulerian solution) we must solve Eqs. (3.22) and (3.25), or Eq. (3.26), along with the equation of continuity. The radiation quantities appearing in these equations are all measured in the lab-frame and must be computed from Eqs. (3.11) and (3.12), or Eqs. (3.13) and (3.14). There are many problems to be faced in solving Eqs. (3.11) and (3.12) including the usual ones of obtaining a closure relation between $\underset{\approx}{P}$ and E (variable Eddington factor), and of flux-limiting if we drop $c^{-2}(\partial \underset{\sim}{F}/\partial t)$ from Eq. (3.12) in optically thin material and then attempt to follow radiation flows. Here we merely wish to emphasize that in addition to the problems just mentioned, one must also face the problem of including all the velocity-dependent terms on the right-hand side of Eqs. (3.11) and (3.12), for otherwise one will not: a) obtain the correct Eulerian flux $\underset{\sim}{F}$, and b) obtain correct radiation energy balance. This conclusion, however unpleasant, seems inescapable.

IV. COMOVING-FRAME (LAGRANGEAN) EQUATIONS OF RADIATION HYDRODYNAMICS

A. The Lagrangean Transfer Equation

In the Lagrangean transfer equation we express all quantities (material properties, radiation quantities, angles, frequencies) in the comoving frame of the fluid, and with respect to a comoving coordinate mesh (e.g., a radial mass-variable). The complication to be faced is that this is not an inertial frame (fluid elements accelerate relative to one another) so photon trajectories are no longer straight lines at constant frequency (i.e., the photon four-momentum is not invariant in the curved spacetime of the acceleration fluid frame). Deep and beautiful discussions of the Lagrangean equation have been given by Castor[5] and Buchler;[6] a simpler derivation can be found in a paper by Mihalas.[7]

Assume one-dimensional spherical symmetry. Then using Eqs. (2.4) - (2.6) in the transfer equation

$$\left[\frac{1}{c}\frac{\partial}{\partial t} + \mu\frac{\partial}{\partial r} + \frac{(1-\mu^2)}{r}\frac{\partial}{\partial \mu}\right] I(\mu,\nu) = (\eta,\nu) - \chi(\mu,\nu)I(\mu,\nu) \tag{4.1}$$

one obtains

$$\begin{aligned}
&\left(\frac{\nu}{\nu_o}\right)\left[\frac{1}{c}\frac{\partial}{\partial t} + \mu\frac{\partial}{\partial r} + \frac{(1-\mu^2)}{r}\frac{\partial}{\partial\mu}\right] I_o(\mu_o,\nu_o) \\
&- 3\left(\frac{\nu}{\nu_o^2}\right)\left[\frac{1}{c}\frac{\mu\nu_o}{\partial t} + \mu\frac{\partial\nu_o}{\partial r} + \frac{(1-\mu^2)}{r}\frac{\partial\nu_o}{\partial\mu}\right] I_o(\mu_o,\nu_o) \\
&= \eta_o(\nu_o) - \chi_o(\nu_o)I_o(\mu_o,\nu_o)
\end{aligned} \tag{4.2}$$

where now all quantities with an affix "o" are measured in the Lagrangean frame. By expanding the derivatives in a chain rule, for example

$$\left.\frac{\partial}{\partial r}\right|_{t\mu\nu} \rightarrow \left.\frac{\partial}{\partial r}\right|_{t\mu_o\nu_o} + \left.\frac{\partial\mu_o}{\partial r}\right|_{t\mu\nu}\frac{\partial}{\partial\mu_o} + \left.\frac{\partial\nu_o}{\partial r}\right|_{t\mu\nu}\frac{\partial}{\partial\nu_o}, \tag{4.3}$$

etc., and by making repeated use of Eqs. (2.2) and (2.3) one can eliminate all reference to the inertial-frame (μ,ν) in favor of (μ_o,ν_o). Reducing the results to $O(v/c)$, i.e., letting $\gamma \rightarrow 1$, one finds, after a fair amount of algebra, the Lagrangean transfer equation

$$\begin{aligned}
&\frac{\rho}{c}\frac{D}{Dt}\left[\frac{I_o(\mu_o,\nu_o)}{\rho}\right] + 4\pi r^2\rho\mu_o\frac{\partial I_o(\mu_o,\nu_o)}{\partial M_r} + \frac{(1-\mu_o^2)}{r}\frac{\partial I_o(\mu_o,\nu_o)}{\partial\mu_o} \\
&+ \left[(1-3\mu_o^2)\frac{v}{cr} - \frac{\mu_o^2}{c}\frac{D\ell n\rho}{Dt} + \frac{2\mu_o a}{c^2}\right] I_o(\mu_o,\nu_o) \\
&+ \frac{1}{c}\left\{\left(\frac{3v}{r} + \frac{D\ell n\rho}{Dt}\right)\frac{\partial}{\partial\mu_o}\left[\mu_o(1-\mu_o^2)I_o(\mu_o,\nu_o)\right]\right. \\
&- \frac{a}{c^2}\frac{\partial}{\partial\mu_o}\left[(1-\mu_o^2)I_o(\mu_o,\nu_o)\right] \\
&- \left.\left[(1-3\mu_o^2)\frac{v}{r} - \mu_o^2\frac{D\ell n\rho}{Dt} + \frac{\mu_o a}{c^2}\right]\frac{\partial}{\partial\nu_o}\left[\nu_o I_o(\mu_o,\nu_o)\right]\right\} \\
&= \eta_o(\nu_o) - \chi_o(\nu_o)I_o(\mu_o,\nu_o) .
\end{aligned} \tag{4.4}$$

Here $dM_r \equiv 4\pi r^2\rho dr$, a Lagrangean mass variable and $a \equiv (\partial v/\partial t)$ is the fluid acceleration. Equation (4.4) is impressively complicated! The complications arise because this equation accounts for the change in photon angle and frequency along its trajectory through fluid elements moving differentially with respect to one another.

Taking the zeroth the first moments of Eq. (4.4) against μ_o we obtain the frequency-dependent moment equations

$$\rho \frac{D}{Dt}\left[\frac{E_o(\nu_o)}{\rho}\right] + 4\pi\rho \frac{\partial}{\partial M_r}\left[r^2 F_o(\nu_o)\right] - \frac{D\ell n\rho}{Dt} P_o(\nu_o)$$

$$- \frac{v}{r}\left[3P_o(\nu_o) - E_o(\nu_o)\right] + \frac{2a}{c^2} F_o(\nu_o)$$

$$+ \frac{\partial}{\partial\nu_o}\left[\nu_o \left\{\frac{v}{r}\left[3P_o(\nu_o) - E_o(\nu_o)\right] + \frac{D\ell n\rho}{Dt} P_o(\nu_o) - \frac{a}{c^2} F_o(\nu_o)\right\}\right]$$

$$= 4\pi\ \eta_o(\nu_o) - c\chi_o(\nu_o)E_o(\nu_o)\ , \qquad (4.5)$$

and

$$\frac{\rho}{c^2}\frac{D}{Dt}\left[\frac{F_o(\nu_o)}{\rho}\right] + 4\pi r^2\rho \frac{\partial P_o(\nu_o)}{\partial M_r} + \frac{1}{r}\left[3P_o(\nu_o) - E_o(\nu_o)\right]$$

$$+ \frac{1}{c^2}\left(\frac{\partial v}{\partial r}\right)F_o(\nu_o) + \frac{a}{c^2}\left[E_o(\nu_o) + P_o(\nu_o)\right]$$

$$+ \frac{\partial}{\partial\nu_o}\left[\nu_o \left\{\frac{v}{c^2 r}\left[3Q_o(\nu_o) - F_o(\nu_o)\right] + \frac{1}{c^2}\frac{D\ell n\rho}{Dt} Q_o(\nu_o) - \frac{a}{c^2} P_o(\nu_o)\right\}\right]$$

$$= \frac{-\chi_o(\nu_o)}{c} F_o(\nu_o)\ . \qquad (4.6)$$

Here $Q_o(\nu_o) \equiv 2\pi \int_{-1}^{1} I_o(\mu_o, \nu_o)\mu_o^3 d\mu_o$.

Finally, integrating over frequency we obtain the Lagrangean radiation energy equation

$$\rho \frac{D}{Dt}\left(\frac{E_o}{\rho}\right) + 4\pi\rho \frac{\partial(r^2 F_o)}{\partial M_r} - \frac{D\ell n\rho}{Dt} P_o - \frac{v}{r}(3P_o - E_o) + \frac{2aF_o}{c^2}$$

$$= \int_0^\infty \left[4\pi\rho_o(\nu_o) - c\chi_o(\nu_o)E_o(\nu_o)\right] d\nu_o \qquad (4.7)$$

and the Lagrangean radiation momentum equation

$$\frac{\rho}{c^2}\frac{D}{Dt}\left(\frac{F_o}{\rho}\right) + 4\pi r^2 \rho \frac{\partial P_o}{\partial M_r} + \frac{3P_o - E_o}{r} + \frac{1}{c^2}\left(\frac{\partial v}{\partial r}\right)F_o$$

$$+ \frac{a}{c^2}(E_o + P_o) = -\frac{1}{c}\int_0^\infty \chi_o(\nu_o)F_o(\nu_o)\,d\nu_o \quad . \qquad (4.8)$$

We see that the matter radiation interaction integrals are now simple, single integrals over frequency against isotropic material coefficients evaluated in the fluid frame.

Equations (4.7) and (4.8) apply in spherical symmetry. Buchler[6] has written tensorial forms of these equations [and also of the full transfer equation, Eq. (4.4) and the frequency-dependent moment equations, Eqs. (4.5) and (4.6)] applicable in any geometry:

$$\rho \frac{D}{Dt}\left(\frac{E_o}{\rho}\right) + \underset{\sim}{\nabla}\cdot\underset{\sim}{F}_o + \underset{\approx}{P}_o : \underset{\sim}{\nabla}\underset{\sim}{v} + \frac{2}{c^2}\,\underset{\sim}{a}\cdot\underset{\sim}{F}_o + cG^o_o = 0 \qquad (4.9)$$

and

$$\frac{1}{c^2}\frac{D\underset{\sim}{F}_o}{Dt} + \underset{\sim}{\nabla}\cdot\underset{\approx}{P}_o + \frac{1}{c^2}\left[\underset{\sim}{F}_o\cdot(\underset{\sim}{\nabla}\underset{\sim}{v}) + (\underset{\sim}{\nabla}\cdot\underset{\sim}{v})\underset{\sim}{F}_o\right]$$

$$+ \frac{1}{c^2}\left(E_o\underset{\sim}{a} + \underset{\sim}{a}\cdot\underset{\approx}{P}_o\right) + \underset{\sim}{G}_o = 0 \quad . \qquad (4.10)$$

Recently Buchler[8] has given explicit expressions for all of these equations in the case of cylindrical symmetry.

We emphasize that all radiation quantities appearing in Eqs. (4.4) - (4.10) are measured in the comoving frame.

Equations (4.7) - (4.10) are correct to $O(v/c)$. We now analyze them, as we did the inertial-frame equations, to find which terms must be retained because they are competitive with, or even dominate, the "usual" terms in one or another regime. One finds that the acceleration-dependent terms are alsways at most $O(v/c)$ and hence can be dropped. Aside from this term, all other terms must be retained in the radiation energy equation [Eq. (4.7)]. In stark contrast one finds that all of the velocity-dependent terms in the radiation momentum equation are $O(v/c)$, as is D/Dt on a fluid-flow timescale. Thus for fluid-flow timescales we can simplify Eq. (4.8) to

$$4\pi r^2 \rho \frac{\partial P_o}{\partial M_r} + \frac{3P_o - E_o}{r} = -\frac{1}{c}\int_0^\infty \chi_o(\nu)F_o(\nu_o)\,d\nu_o \quad . \qquad (4.11)$$

If we must work on a radiation-flow timescale then we should add a term c^{-2} (DF_o/Dt) to the left-hand side of Eq. (4.11). In most respects, Eqs. (4.7) and (4.11) are a good deal simpler to work with than their Eulerian counterparts, Eqs. (3.11) and (3.12).

B. Lagrangean Dynamical Equations for a Radiating Fluid

One derives a relativistically correct Lagrangean momentum equation by reducing Eq. (3.20) to the comoving frame, in which $\underset{\sim}{v} = 0$ instantaneously, which yields

$$\rho_{ooo}(D\underset{\sim}{v}/Dt) = \underset{\sim}{f} - \underset{\sim}{\nabla} p + \underset{\sim}{G}_o = \underset{\sim}{f} - \underset{\sim}{\nabla} p + c^{-1}\int_0^\infty \chi_o(\nu_o)\underset{\sim}{F}_o(\nu_o)d\nu_o . \tag{4.12}$$

To $O(v/c)$ we can replace ρ_{ooo} in Eq. (4.12) with ρ_o (or ρ). In view of Eq. (4.11) we can also write

$$\rho(D\underset{\sim}{v}/Dt) = \underset{\sim}{f} - \underset{\sim}{\nabla} p - \underset{\sim}{\nabla}\cdot\underset{\approx}{P}_o ; \tag{4.13}$$

on a radiation-flow timescale in the streaming limit one should also add a term $-(\rho/c^2)[D(F_o/\rho)/Dt]$ to the right-hand side of Eq. (4.13).

Similarly, the relativistically correct Lagrangean gas-energy equation is

$$\rho_o\left[\frac{De}{Dt} + p\frac{D}{Dt}\left(\frac{1}{\rho_o}\right)\right] = \int_0^\infty [c\chi_o(\nu_o)E_o(\nu_o) - 4\pi\eta_o(\nu_o)]d\nu_o + \rho_o\varepsilon , \tag{4.14}$$

which says that the rate of change of material internal energy plus the work done by the material in expansion equals the rate of energy input to the material, per gram, from radiation and thermonuclear sources.

The radiation energy equation can be written in an entirely analogous form:

$$\rho_o\left[\frac{D}{Dt}\left(\frac{E_o}{\rho_o}\right) + P_o\frac{D}{Dt}\left(\frac{1}{\rho_o}\right) - (3P_o - E_o)\frac{v}{\rho_o r}\right] = -\frac{1}{r^2}\frac{\partial(r^2F_o)}{\partial r} + \int_0^\infty [4\pi\rho_o(\nu_o) - c\chi_o(\nu_o)E_o(\nu_o)]\,d\nu_o . \tag{4.15}$$

Recognizing the second and third term on the left-hand side as $\underset{\approx}{P}_o:\underset{\sim}{\nabla}\underset{\sim}{v}$ in spherical geometry, we see that Eq. (4.15) states that the rate of change of radiant energy density plus the rate of work done by radiation equals the net rate of radiant energy release by the material minus the net outflow of radiation out of the material element. It is worth noting in passing that had we neglected terms that are formally $O(v/c)$, the terms accounting for the rate of work done by the radiation field would not appear in Eq. (4.15) (these terms have often been added by after-the-fact ad hoc arguments), and the distinction between $\rho[D(E_0/\rho)/Dt]$ and (DE_0/Dt) would have been lost. Finally, adding Eqs. (4.14) and (4.15) we obtain the first law of thermodynamics for the radiating fluid

$$\frac{D}{Dt}\left(e + \frac{E_0}{\rho}\right) + p\frac{D}{Dt}\left(\frac{1}{\rho}\right) + \left[P_0 \frac{D}{Dt}\left(\frac{1}{\rho}\right) - (3P_0 - E_0)\frac{v}{\rho r}\right]$$
$$= -\frac{\partial}{\partial M_r}(4\pi r^2 F_0) + \varepsilon. \tag{4.16}$$

Taking the dot product of $\underset{\sim}{v}$ with Eq. (4.12) we obtain a mechanical energy equation, which when added to Eq. (4.16), gives the total energy equation

$$\frac{D}{Dt}\left(e + \frac{E_0}{\rho} + \frac{1}{2}v^2 - \frac{GM_r}{r}\right)$$
$$+ \frac{\partial}{\partial M_r}\{4\pi r^2[v(p + P_0) + F_0]\} = \varepsilon. \tag{4.17}$$

To calculate the flow in the Lagrangean frame we must solve the momentum equation [Eq. (4.13)] simultaneously with an energy equation. For the energy equation we may choose Eq. (4.14), (4.16), or (4.17). Of these, Eq. (4.14) is a natural form for optically thin material, inasmuch as it displays explicitly the energy exchange between the material and radiation in terms of direct gains and losses. In opaque material the net emission-absorption term of Eq. (4.14) vanishes to high order, the radiation energy and pressure approach equilibrium, and the pressure becomes isotropic; Equation (4.17) then provides the most natural form of the energy equation.

These material equations are to be solved simultaneously with the radiation energy and momentum equations [Eqs. (4.7) and (4.8)]. If (and only if!) the $O(v/c)$ terms in Eq. (4.7) have been retained, the choice of the material energy equation is irrelevant because Eq. (4.7) assures exact consistency among the various forms. But if any of the terms are omitted from Eq. (4.7) this exact reduction is not possible, and Eq. (4.14) can then lead to large errors in the material temperature in opaque material, whereas Eq. (4.16) can lead to large errors in the material temperature in transparent regions. According to Castor[5] the error in the temperature is $O(P/p)$! As all real flows of interest span both the optically thick and thin regimes, it is important to treat Eq. (4.7) correctly.

V. THE DIFFUSION REGIME

The conceptually most informative way to derive the diffusion-limit radiation equations is to specialize the Lagrang an equations to opaque material in which $(\lambda_p/\ell) \ll 1$. In this limit the radiation field in the Lagrangean frame isotropizes to a high degree, so $P_o(\nu_o) \to 1/3\, E_o(\nu_o)$. Then discarding terms of $O(\lambda_p v/\ell c)$, which are truly negligible, Eq. (4.5) and (4.6) become

$$\underset{\sim}{F}_o(\nu_o) = - \frac{c}{3\chi_o(\nu_o)} \underset{\sim}{\nabla} E_o(\nu_o) \tag{5.1}$$

and

$$\rho\left[\frac{D}{dt}\left[\frac{E_o(\nu_o)}{\rho_o}\right] + \frac{1}{3}\{E_o(\nu_o) - \frac{\partial}{\partial\nu_o}\left[\nu_o E_o(\nu_o)\right]\}\frac{D}{Dt}\left(\frac{1}{\rho}\right)\right]$$

$$= \underset{\sim}{\nabla}\cdot\left[\frac{c}{3\chi_o(\nu_o)} \underset{\sim}{\nabla} E_o(\nu_o)\right] + \kappa_o(\nu_o)\left[4\pi B(\nu_o,T) - cE_o(\nu_o)\right], \tag{5.2}$$

where we have assumed $\chi_o(\nu_o) = \kappa_o(\nu_o) + \sigma_o$, and $\eta_o = \kappa_o B + \sigma(cE_o/4\pi)$ (i.e., LTE). Integrating over all frequencies we obtain the total radiation flux

$$\underset{\sim}{F}_o = - \frac{c}{3\bar{\chi}} \underset{\sim}{\nabla} E_o \tag{5.3}$$

and the radiation energy equation (the <u>nonequilibrium diffusion equation</u>)

$$\rho\left[\frac{D}{Dt}\left(\frac{E_o}{\rho}\right) + \frac{1}{3} E_o \frac{D}{Dt}\left(\frac{1}{\rho}\right)\right] = \underset{\sim}{\nabla}\cdot\left(\frac{c}{3\bar{\chi}} \underset{\sim}{\nabla} E_o\right)$$

$$+ c(\kappa_P a T^4 - \kappa_E E_o) . \tag{5.4}$$

Here κ_P is the usual Planck mean opacity, while κ_E and $\bar{\chi}$ are the absorption mean and flux mean (or some approximation to those means). It is often convenient to parametrize the radiation energy density as $E_o = aT_R^4$, in which case we get the <u>two-temperature diffusion equation</u>

$$\rho\left[\frac{D}{Dt}\left(\frac{aT_R^4}{\rho}\right) + \frac{1}{3} aT_R^4 \frac{D}{Dt}\left(\frac{1}{\rho}\right)\right] = \underset{\sim}{\nabla}\cdot\left(\frac{4acT_R^3}{3\chi} \nabla T_R\right)$$

$$+ ac(\kappa_P T^4 - \kappa_E T_R^4). \tag{5.5}$$

To solve the fluid dynamics we adjoin the momentum equation and the material energy equation, which may be written as

$$\rho \left[\frac{De}{Dt} + p \frac{D}{Dt} \left(\frac{1}{\rho}\right)\right] = ac(\kappa_E T_R^4 - \kappa_P T^4) + \rho\varepsilon . \qquad (5.6)$$

We obtain the standard equilibrium diffusion energy equation by setting $T_R \equiv T$, and adding Eq. (5.5) to Eq. (5.6) which gives the first law of thermodynamics for the radiating fluid.

These equations are so important that they merit a couple of comments. First, it should be stressed very strongly that all variables here are in the Lagrangean frame. This comment is most relevant to the flux: it cannot be overemphasized that Eq. (5.3) should not be used as an expression for the diffusion-limit flux in Eulerian codes. Rather, in the diffusion regime

$$\begin{aligned} \underset{\sim}{F}_{\text{Eulerian}} &= \underset{\sim}{F}_{\text{Lagrangean}} + \frac{4}{3} E_o \underset{\sim}{v} \\ &= \frac{4acT_R^3}{3\bar{\chi}} \underset{\sim}{\nabla} T_R + \frac{4}{3} aT_R^4 \underset{\sim}{v} . \end{aligned} \qquad (5.7)$$

As remarked earlier, at high temperatures for dynamic diffusion (i.e., $\frac{v}{c} \frac{\ell}{\lambda} \gtrsim 1$) the second term can vastly outweigh the first. Likewise, it should be noted that the parameter T_R is defined in terms of the energy density in the Lagrangean frame, not the Eulerian frame. From Eq. (2.12) the Eulerian radiation energy density is

$$E = aT_R^4 - (8aT_R^3/3c\bar{\chi}) \underset{\sim}{v} \cdot \underset{\sim}{\nabla} T ; \qquad (5.8)$$

the distinction becomes crucial when we calculate the net absorption-emission term as in Eq. (3.11). Indeed, only if we use Eqs. (5.7) and (5.8) is it possible to recover Eqs. (5.5) and (5.6) consequently from the Eulerian equations [Eqs. (3.11) and (3.24)].

The significance of these remarks becomes clearer in the context of earlier formulations of diffusion theory, e.g., Campbell and Nelson[9] or Freeman.[10] In essence these analyses use Eqs. (3.13) and (3.14), drop the $\underset{\sim}{v}$-dependent terms, and assume $P^{ij} = \frac{1}{3} E\delta^{ij}$ (appropriate for diffusion), obtaining finally an energy equation of the form

$$(\partial E/\partial t) = \chi(4\pi B - cE) + \underset{\sim}{\nabla} \cdot [(c/3\chi)\underset{\sim}{\nabla} E] . \qquad (5.9)$$

The problem is that Eq. (5.9) is not correct in either the Lagrangian or Eulerian frame. Thus, while the right-hand side of Eq. (5.9) looks identical to the right-hand side of the Lagrangean equation [Eq. (5.4)], the equation is not Lagrangean because on the left-hand side the distinction between $(\partial E/\partial t)$ and

$\rho D(E/\rho)/Dt$ is not made, and the term $\frac{1}{3}\rho E\, D(1/\rho)/Dt$ corresponding to the rate of work done by radiation pressure is missing. Likewise, comparing Eq. (5.9) with the Eulerian equation Eq. 3.13) we see that the right-hand side lacks a term $(\chi/c)\underset{\sim}{v}\cdot\underset{\sim}{F}$ corresponding to the rate of work done by radiation forces on the material, and further that in the flux divergence, the comoving-frame flux is used instead of the lab-frame flux, thus omitting the dominant term that discriminates $\underset{\sim}{F}_E$ from $\underset{\sim}{F}_L$.

If we wish to treat multigroup diffusion we should solve Eq. (5.2). The new term that appears is the frequency derivative $\partial/\partial\nu_o$; inasmuch as this term is conservative (vanishes when integrated over all frequency) it cannot affect total energy balance and it is natural to ask if we can't simply drop it and avoid the mathematical complication. The answer is "no" because if we assume an "infinitely opaque" (i.e. $1/\chi\to 0$) pure scattering ($\kappa\to 0$) medium, then Eq. (5.2) without the frequency derivative becomes

$$\frac{D}{Dt}\left[\ln E_o(\nu_o)\right] - \frac{4}{3}\frac{D\ln\rho}{Dt} = 0 \ , \tag{5.10}$$

which implies $E_o(\nu_o) \propto \rho^{4/3}$. This result is correct for the total energy density in an adiabatic enclosure, but not E_ν. Indeed the very fact that $E \propto \rho^{4/3}$, implies that $T_R \propto \rho^{1/3}$, which in turn implies that the spectral distribution must change when the enclosure expands or contracts. Put another way, it is the frequency derivative that accounts for the progressive redshift (blueshift) of photons in a space undergoing expansion (compression); further, it allows redistribution of radiation to produce a non-Planckian spectrum in the general nonequilibrium case (e.g., a cascade of photons from the spectral peak into a high-energy or low-energy tail). Note that this effect is always present (i.e., it does not depend on a particular physical mechanism such as Compton scattering), and must be accounted for if one wishes to obtain the correct spectral distribution (which, presumably, is the whole point of doing the multigroup calculation). Numerically one would handle the ν-derivative with upstream differencing, ie., coupling group g with group g+1 (g-1) when the material expands (contracts), where $E_{g-1} < E_g < E_{g+1}$.

VI. THE TRANSPORT REGIME

The diffusion approximation inevitably breaks down near boundaries where the material becomes transparent, and use of the diffusion equation in the optically thin limit can lead to nonphysical results. In particular one may find that the magnitude of the energy flux $|\underset{\sim}{F}|$ exceeds cE, which is nonsensical. In practice this has led to the use of ad hoc flux limiters, which provide a fixup to the diffusion equation in the streaming regime. As is shown elsewhere[11] the root of the problem is that the diffusion equation doesn't limit correctly to the wave equation in optically thin material, and to avoid having to introduce flux limiters one must solve the full time-dependent transport equation. Flux limiters have other disadvantages as well, particularly in multidimensional geometries. For example, even if we suppose that the flux obtained in a flux-limited diffusion calculation has the correct magnitude, we cannot guarantee that it points in the correct direction because $\underset{\sim}{F}$ is actually proportional to $\nabla \cdot \underset{\approx}{P}$, not $\underset{\sim}{\nabla} E$; the prediction of flux-limited diffusion will be correct only if $\underset{\approx}{P}$ is isotropic. Similarly, the radiation pressure work term is actually $\underset{\approx}{P}:\underset{\sim}{\nabla}\underset{\sim}{v}$, which will not, in general, equal $\frac{1}{3} E \frac{D}{Dt} (\frac{1}{\rho})$ as is assumed in flux-limited diffusion. The difference between these two may be quite significant for shear flows in complex geometries with highly anisotropic radiation fields. In short, we require a full transfer calculation not only to obtain the correct time-development of the radiation field, but also the correct geometrical distribution of the radiation field.

Needless to say, this presents a problem of truly formidable difficulty, and no fully satisfactory solution has yet been obtained, certainly not to O(v/c). Given the magnitude of the effort that will be required to develop a fully satisfactory code, it is perhaps helpful to consider the relative advantages and disadvantages of the Eulerian and Lagrangean formulations. Consider first the Eulerian equations [Eqs. (3.11) and (3.12)]. There are three important disadvantages of these equations: 1) Derivatives of the opacity appear on the right-hand side. In most materials the frequency-spectrum of the opacity is extremely jagged, and it is not at all clear how one can represent these derivatives. 2) We are required to represent velocity-dependent terms explicitly on the right-hand side. This may be difficult to do accurately because of mis-centering of variables on the mesh, yet the solution, particularly in the diffusion limit, literally hangs on these terms particularly the flux - see Eq. (3.7) and related discussion! . Furthermore, subtleties of interpretation enter [see Eq. (5.8)] in obtaining physically sensible results in the diffusion limit. 3) Scattering terms (not discussed here) are extremely clumsy and difficult to handle correctly[12] in this frame. The main advantage of the Eulerian frame is that one is free to choose a regular mesh (although even here one is faced with the problem of having sufficient resolution in opaque materials and having mixed cells!). In our opinion these disadvantages are quite severe.

The situation for the Lagrangean equations is rather different. There are three basic advantages to these equations: 1) From a fundamental physical standpoint the Lagrangian frame is the proper frame, and it is the frame in which we can most naturally compute material properties (e.g., opacities), do details of physics (e.g., non LTE rate equations), and handle complex radiation-material interactions (e.g., scattering with partial redistribution) trivially. 2) As we saw in Sec. V, the Lagrangean equations limit naturally and unambiguously to the correct diffusion-regime results. Indeed the Lagrangean frame is the preferred frame of diffusion theory. 3) Although the multigroup equations contain a frequency derivative, it is the derivative of the spectral distribution of the radiation field, which in general (particularly in the diffusion regime!) will be much smoother (perhaps Planckian) than the opacity spectrum. Moreover, the derivative terms are missing entirely from the frequency-integrated equations, which most directly describe the dynamics of the flow. There are two basic disadvantages of the Lagrangean frame. 1) The full angle-frequency-dependent transfer equation is extremely complicated; too complicated to treat in general. But we shall show below that it is possible to use a simpler equation that retains the essential physics and omits needless complications. 2) While the Lagrang an equations are straightforward to apply in 1D geometry, they are more difficult to formulate in 2D and 3D (see Buchler[8] for the case of 2D cylindrical symmetry), and present substantial difficulties when they must be solved on an irregular mesh in, say, two dimensions. However, this difficulty is fundamentally a numerical/technical problem, which we believe can be solved. In our opinion the great physical advantages offered by the Lagrangean frame are compellingly attractive, and should strongly motivate even a major effort to overcome the technical difficulties.

It is instructive to sketch how a fully Lagrangean solution proceeds in 1D, because this illustrates nicely some of the basic physics. Defining the Eddington factor

$$f_\nu \equiv P_\nu/E_\nu = \int_{-1}^{1} I_\nu(\mu)\mu^2 d\mu \Big/ \int_{-1}^{1} I_\nu)\mu)d\mu \tag{6.1}$$

and the sphericity factor

$$\ln q_\nu \equiv \int_{r_0}^{r} \frac{3f_\nu - 1}{f_\nu r'}\, dr' , \tag{6.2}$$

the Lagrangean frequency-dependent moment equations become

$$\frac{D}{Dt}\left(\frac{E_\nu}{\rho}\right) + \left[f_\nu \frac{D}{Dt}\left(\frac{1}{\rho}\right) - (3f_\nu - 1)\frac{v}{\rho r}\right] E_\nu$$

$$- \frac{\partial}{\partial\nu}\left\{\left[f_\nu \frac{D}{Dt}\left(\frac{1}{\rho}\right) - (3f_\nu - 1)\frac{v}{\rho r}\right] \nu E_\nu\right\}$$

$$= \frac{\kappa_\nu}{\rho}(4\pi B_\nu - cE_\nu) - \frac{\partial}{\partial M_r}(4\pi r^2 F_\nu) \tag{6.3}$$

and

$$\frac{1}{c^2}\frac{DF_\nu}{Dt} + \frac{4\pi r^2 \rho}{q_\nu}\frac{\partial(f_\nu q_\nu E_\nu)}{\partial M_r} = -\frac{\chi_\nu F_\nu}{c} . \tag{6.4}$$

Here we have omitted terms found to be inessential according to the analysis described in Sec. IV. Integrating Eqs. (6.3) and (6.4) over frequency we obtain the radiation energy equation

$$\frac{D}{Dt}\left(\frac{E}{\rho}\right) + \left[f\frac{D}{Dt}\left(\frac{1}{\rho}\right) + (3f-1)\frac{v}{\rho r}\right]E$$

$$= \frac{1}{\rho}\left(4\pi\kappa_P B - c\kappa_E E\right) - \frac{\partial(4\pi r^2 F)}{\partial M_r} , \tag{6.5}$$

and the radiation momentum equation

$$\frac{1}{c^2}\frac{DF}{Dt} + \frac{4\pi r^2\rho}{q}\frac{\partial(fqE)}{\partial M_r} = -\frac{\chi_F F}{c} , \tag{6.6}$$

where

$$f \equiv \int_0^\infty f_\nu e_\nu d\nu \tag{6.7}$$

$$\kappa_E = \int_0^\infty \kappa_\nu e_\nu d\nu \tag{6.8}$$

and

$$\chi_F \equiv \int_0^\infty \chi_\nu \phi_\nu d\nu , \tag{6.9}$$

where $e_\nu \equiv (E_\nu/E)$ and $\phi_\nu \equiv (F_\nu/F)$.

The basic procedure for solving this system invokes the multigroup/grey approach first introduced in the VERA code.[13] Thus if we assume that f, q, (κ_E/κ_p), and (χ_F/χ_R) are given, where χ_R is the usual Rosseland mean opacity, we can write difference equations representing Eqs. (6.5) and (6.6) on a staggered mesh (E at cell centers, F at all surfaces). For stability one may use a fully implicit time-differencing (backward Euler scheme). These equations, coupled with the material energy equation [Eq. (4.14)] pose a nonlinear system for the temperature distribution at the advanced time-level t^{n+1}. They can be linearized and solved by a Newton-Raphson procedure.

The results can be no better than our knowledge of the geometric factors f and q, and the profile functions e_ν and ϕ_ν needed to construct κ_E and χ_F. Thus given $T^{n+1}(M_r)$, we must update these parameters. To obtain f_ν (hence q_ν) we must in principle solve the full angle-frequency dependent transfer equation (which accounts fully for the curved photon paths and differential Doppler shifts a photon experiences in the differentially moving fluid frame) shows that is hopelessly complicated. But noticing that all the terms in the curly brackets vanish when integrated over angle and frequency, we use instead the model Lagrangean transfer equation

$$\frac{\rho}{c}\frac{D}{Dt}\left(\frac{I_\nu}{\rho}\right) + 4\pi r^2\rho\mu\frac{\partial I_\nu}{\partial M_r} + \frac{(1-\mu^2)}{r}\frac{\partial I_\nu}{\partial\mu}$$
$$= \eta_\nu - \left[\chi_\nu + (1 - 3\mu^2)\frac{v}{cr} - \frac{\mu^2}{c}\frac{D\ln\rho}{Dt}\right] I_\nu . \qquad (6.10)$$

We solve Eq. (6.10) for $I^{n+1}(\mu,\nu)$ assuming $T(M_r)$ is given (i.e., we accept the temperature distribution obtained above). While omitting the complications caused by aberration and Doppler shifts, which are not essential in the computation of the ratio f_ν, this equation has the following virtues:

(a) It accounts for the time-dependence of the radiation field.
(b) The rays are now straight (no aberration).
(c) It is consistent with the integrated moment equations. In particular the two velocity-dependent terms retained on the right-hand side yield the radiation work term in Eq. (6.5).
(d) It is no more difficult to solve than the standard equation for a static medium.

Equation (6.10), while not perfect, thus accounts for the dominant effects of geometry and time-dependence (as well as the one crucial v-dependent term), and should yield fairly accurate values for f_ν (which, in principle, is dominated mainly by geometric effects and gradients in source terms). On the other hand, Eq. (6.10) is probably not adequate to determine spectral profiles because it omits the important frequency-derivative term.

To obtain e_ν and ϕ_ν, we next solve Eqs. (6.3) and (6.4) assuming now that f_ν, q_ν, and T^{n+1} are given. For economy the frequency-coupling terms can be moved to the right-hand side and iterated. From these calculations we accept only the profiles $e_g \equiv E_g/\Sigma E_g$ and $\phi_g \equiv F_g/\Sigma F_g$ for each group. We can then update κ_E, χ_F, and f and q.

Using these improved values for the geometric and spectrum-dependent quantities we can re-solve Eqs. (6.5) and (6.6) for an improved T^{n+1}, and iterate the whole procedure to convergence if desired. Experience with astrophysical calculations indicates the procedure converges quickly. In media with material discontinuities there may be problems crossing zones, but these are not peculiar to the Lagrangean equations, and can be handled by standard fixups.

REFERENCES

1. Pomraning, G.C. 1973, The Equations of Radiation Hydrodynamics, Oxford: Pergamon).

2. Mihalas, D. and Weaver, R.P. 1982, Los Alamos National Laboratory Report No. LA-UR-82-606.

3. Mihalas, D. and Mihalas, B.R.W. 1983, Foundations of Radiation Hydrodynamics, (Oxford: Oxford University Press), in preparation.

4. Thomas, L.H. 1930, Quant. J. of Math. (Oxford), 1, 239.

5. Castor, J.I. 1972, Astrophys. J., 178, 779.

6. Buchler, J.R. 1979, J. Quant. Spectrosc. Rad. Transf., 22, 293.

7. Mihalas, D. 1980, Astrophys. J., 237, 574.

8. Buchler, J.R. 1982, preprint: "Radiation Transfer in the Fluid Frame."

9. Campbell, P.M. and Nelson, R.G. 1964, Lawrence Radiation Laboratory Report No. UCRL-12411.

10. Freeman, B.E. 1965, Los Alamos National Laboratory Report No. LA-3377.

11. Mihalas, D. and Weaver, R.P. 1982, J. Quant. Spectrosc. Rad. Transf., in press: "Time-Dependent Radiative Transfer with Automatic Flux Limiting."

12. Fraser, A.R. 1966, U.K. Atomic Weapons Research Establishment Report No. O-82/65.

13. Freeman, B.E. and Hauser, L.E. 1969, Systems, Science and Software, P.O. Box 1620, La Jolla, CA, Report No. 3SIR-38.

WH80s: NUMERICAL RADIATION HYDRODYNAMICS

Karl-Heinz A. Winkler and Michael L. Norman

Los Alamos National Laboratory and
Max-Planck-Institut fuer Physik und Astrophysik

ABSTRACT

An implicit, adaptive-mesh technique for radiation hydrodynamics computations in one spatial dimension is described in detail. The equation of radiative transfer and its moments are formulated in a fully adaptive coordinate system in space, angle and frequency. Integral conservation laws for the radiation quantities, which form the basis for our method, are then given. A hierarchy of splitting schemes for reducing this 3-dimensional problem to ones of lower dimensionality are discussed, and our variable Eddington factor approach is described and put into this context. The embodiment of this method is the WH80s radiation hydrodynamics code, which further incorporates self-gravity, realistic equation-of-state and opacity data, nuclear energy generation and convection. Spatial and temporal differencing techniques, the adaptive mesh equation, and our tensor artificial viscosity formulation implemented in WH80s are described. A section is devoted to "Principles of Coding", in which we illustrate a structured programming style that we have developed which greatly aids writing and debugging large coupled-implicit systems. Finally, several applications of WH80s are given which show the power and versatility of adaptive mesh techniques to problems with a variety of spatial and temporal scales such as arise in astrophysics.

K.-H. A. Winkler and M. L. Norman (eds.), Astrophysical Radiation Hydrodynamics, 71–139.

I. INTRODUCTION

In this paper we shall attempt to collect into one place and describe the physical and numerical principles which underly the WH80s radiation hydrodynamics code. In so doing, we hope that this paper will be useful to students and researchers alike who are developing radiation hydro codes for a variety of applications in astrophysics where the energy and momentum in the radiation field contribute to, if not dominate, the structure and dynamics of the flow. Some examples where this is the case are supernovae, stellar winds, star formation, accreting neutron stars and black holes, and pulsating stars.

WH80s solves the comoving-frame equations of Newtonian radiation hydrodynamics in one spatial dimension correct to first order in the ratio of the fluid velocity to the speed of light (an application of WH80s to relativistic, nonradiating flow is discussed by Norman and Winkler, this book.) The hierarchy of moment equations of the radiative transfer equation is closed via a variable Eddington factor which is evaluated at each point in space-time by a separate solution of the tranfer equation. WH80s is an outgrowth of a computer code developed by Winkler (1976) and Tscharnuter and Winkler (1979) for studying the gravitational collapse of interstellar clouds to form protostars (Winkler and Newman 1980a,b). A novel feature of WH80s and its forerunner is that the physical equations are cast on an adaptive mesh which is neither Eulerian nor Lagrangian. Unlike numerous "rezoning" schemes, the distribution of computational zones is governed by an adaptive mesh equation which is solved implicitly coupled to the set of dynamic equations. As discussed in Winkler, Norman and Newman (1984), an adaptive mesh is absolutely essential for correctly modeling thin radiative ionization zones which are non-stationary with respect to both space and mass coordinates. These occur in abundance in astrophysics, as a brief consideration of the list of objects above will attest.

In this paper we descibe the current state of the art of our adaptive mesh technology. Since its first implementation in Winkler (1976), the adaptive mesh equation has undergone numerous extentions and reformulations so that it can selectively and stably resolve and track with an unprecedented amount of grid refinement ($>10^6$) multiple and interacting features (e.g., shocks, contact discontinuities, compositional discontinuities) which arise within a time-dependent flow. This development is chronicled in Winkler, Norman and Newman (1984) and Winkler, Mihalas and Norman (1985).

In Sec. 2 we present the adaptive mesh radiation transfer equation from which the moment equations we solve are derived. In Sec. 3 we derive monochromatic and gray adaptive-mesh radiahydrodynamics equations and discuss computational strategies for their solutions. WH80s solves the gray equations at present, incorporating self-gravity, tabular equations-of-state and opacity data, nuclear energy generation and convection, which is also described. In Sec. 4 the adaptive mesh, tensor artificial viscosity, and other numerical tools for stable and accurate computations are discussed. In Sec. 5 we present the finite difference equations solved in WH80s and describe their method of solution. Section 6 is devoted to "Principles of Coding", in which we illustrate a structured programming style that we have developed that greatly aids writing and debugging large coupled-implicit systems. Finally, several applications of WH80s are given in Sec. 7 which show the capability and versatility of adaptive mesh techniques to problems with interacting nonlinear waves and a range of spatial scales as arise in radiating shocks.

II. ADAPTIVE MESH RADIATION TRANSFER

The Eulerian radiation transfer equation can be written, Mihalas (this book)

$$\left[\frac{1}{c}\frac{\partial}{\partial t} + (\underline{n}\cdot\underline{\nabla})\right] I(\underline{x},t;\underline{n},\nu) = \eta(\underline{x},t;\underline{n},\nu) - \chi(\underline{x},t;\underline{n},\nu) I(\underline{x},t;\underline{n},\nu) \quad (2.1)$$

In this equation, all radiation quantities are measured with respect to an inertial frame. Here I denotes the specific intensity, η the emissivity, χ the macroscopic absorption coefficient, $(\partial/\partial t)$ the Eulerian time derivative, $\underline{\nabla}$ the gradient operator, and c the speed of light. I, η and χ are functions of position, of time t, of the direction of photon propagation $\underline{n}$, and of the frequency ν of the radiation. From a computational point of view the problem with Eqn. (2.1) is that η and χ are essentially anisotropic because of velocity-induced Doppler-shift and aberration effects.

Therefore, we choose the Lagrangean radiation transfer equation in spherical symmetry as a starting point for constructing a numerical scheme for 1-D radiation hydrodynamic problems. In this equation

$$\frac{\rho}{c}\frac{D}{Dt} \qquad \left[\frac{I_0(\mu_0,\nu_0)}{\rho}\right]$$

$$+ \frac{\partial}{\partial \mathrm{Vol}} \left[\mu_o r^2 \qquad I_o(\mu_o,\nu_o) \right]$$

$$+ \frac{\partial}{\partial \mu_o} \left[(1-\mu_o^2) \left\{ \frac{1}{r} + \frac{\mu_o}{c}\left(\frac{3u}{r} + \frac{D\ln\rho}{Dt}\right) - \frac{a}{c^2} \right\} I_o(\mu_o,\nu_o) \right]$$

$$+ \frac{\partial}{\partial \nu_o} \left[-\nu_o \left\{ (1-3\mu_o^2)\frac{u}{cr} - \frac{\mu_o^2}{c}\frac{D\ln\rho}{Dt} + \frac{\mu_o a}{c^2} \right\} I_o(\mu_o,\nu_o) \right]$$

$$= \eta_o(\nu_o) - \left[\chi_o(\nu_o) + \left\{ (1-3\mu_o^2)\frac{u}{cr} - \frac{\mu_o^2}{c}\frac{D\ln\rho}{Dt} + 2\frac{\mu_o a}{c^2} \right\} \right] I_o(\mu_o,\nu_o), \tag{2.2}$$

first derived by Castor (1972) and written in almost conservative form by Mihalas and Mihalas (1984), the isotropic absorption and emission coefficients η_o and χ_o are a function of position r, of time t, and of frequency ν_o alone. For brevity, we shall henceforth suppress reference to r and t. Equation (2.2) is correct to O(u/c), and all radiation quantities are defined with respect to a frame comoving with the fluid. In particular, all quantities with an affix "o" are measured in this Lagrangean coordinate system. Thus I_o denotes the specific intensity, ν_o the frequency of the radiation, η_o the emissivity, and χ_o the macroscopic absorption coefficient in the comoving frame. μ_o is the cosine of the angle θ between the direction of photon propagation and velocity $\underline{u} \equiv (u,0,0)$, (u,0,0 being the vector components in the r, θ, ϕ - direction of the polar coordinates we use). ρ is the gas density, r is the radius of a sphere with volume V(r) and mass (the Lagrangean variable) $m \equiv \int_0^{V(r)} \rho \, d\mathrm{Vol}$, and $d\mathrm{Vol} \equiv (1/3)d(r^3)$. (D/Dt) is the Lagrangean time derivative and $a \equiv (\partial u/\partial t)$ is the Eulerian fluid acceleration.

In the following, radiation quantities are still defined in and measured with respect to the Lagrangean frame, but the governing equations themselves will be evaluated in completely different coordinate systems, chosen according to the numerical requirements of our method of solution, Winkler (1976), Tscharnuter and Winkler (1979). This new set of "adaptive-mesh radiation equations", still determining the familiar Lagrangean radiation quantities, can be derived by applying a Galilean transformation to Eqn. (2.2), Tscharnuter and Winkler (1979), Mihalas and Mihalas (1984). Introducing an adaptive mesh in radius, angle, and frequency let us define

$$\underline{u}_{gr} = \left(\frac{r}{dt}\right) \quad \text{and} \quad \underline{u}_{rel} = \underline{u} - \underline{u}_{gr}. \tag{2.3}$$

Furthermore we set

$$\underline{\tilde{u}} \equiv (u;0), \quad \underline{\tilde{u}}_{gr} \equiv \left(\underline{u}_{gr}, \frac{d\mu_o}{dt}, \frac{d\nu_o}{dt}\right)$$

$$\underline{\tilde{u}}_{rel} = \tilde{u} - \tilde{u}_{gr} = \left(u - \frac{dr}{dt}, -\frac{d\mu_o}{dt}, -\frac{d\nu_o}{dt}\right) \tag{2.4}$$

$$\underline{\tilde{\nabla}} = \left(\frac{\partial}{\partial r}, \frac{\partial}{\partial\mu_o}, \frac{\partial}{\nu_o}\right)$$

Then Eqn. (2.2) can be rewritten in the following way (Winkler, Norman, and Mihalas (1984)) and we obtain an adaptive mesh radiation transfer equation.

$$\frac{1}{c}\left[\frac{d}{dt}\left(I_o(\mu_o,\nu_o) + (\underline{\tilde{u}}_{rel}\cdot\underline{\tilde{\nabla}}) I_o(\mu_o,\nu_o)\right) + I_o(\mu_o,\nu_o)(\underline{\nabla}\cdot\underline{u})\right]$$

$$+ \frac{\partial}{\partial Vol} \left[\mu_o r^2 \; I_o(\mu_o,\nu_o)\right]$$

$$+ \frac{\partial}{\partial\mu_o} \left[(1-\mu_o^2)\left\{\frac{1}{r} + \frac{\mu_o}{c}\left(\frac{3u}{r} - \frac{\partial(r^2u)}{\partial Vol}\right) - \frac{a}{c^2}\right\} I_o(\mu_o,\nu_o)\right]$$

$$+ \frac{\partial}{\partial\nu_o} \left[-\nu_o \left\{(1-3\mu_o^2)\frac{u}{cr} + \frac{\mu_o^2}{c}\frac{\partial(r^2u)}{\partial Vol} + \frac{\mu_o a}{c^2}\right\} I_o(\mu_o,\nu_o)\right]$$

$$= \eta_o(\nu_o) - \left[\chi_o(\nu_o) + \left\{(1-3\mu_o^2)\frac{u}{cr} + \frac{\mu_o^2}{c}\frac{\partial(r^2u)}{\partial Vol} + 2\frac{\mu_o a}{c^2}\right\}\right] I_o(\mu_o,\nu_o)]$$

$$\tag{2.5}$$

Depending on the velocity field $\underline{u}_{gr}$ for the grid motion, we can evaluate the Lagrangean radiation quantities either in a Lagrangean frame ($\underline{u}_{rel} \equiv 0$, $\underline{u} \equiv \underline{u}_{gr}$), an Eulerian frame ($\underline{u}_{gr} \equiv 0$, $\underline{u}_{rel} \equiv \underline{u}$), or an arbritrary coordinate system, neither fixed in space or mass. It should be emphasized that in this way we transform the coordinates with respect to which we express the equations, but not the frame with respect to which we measure the physical variables. In fact, because the radiation quantities are not transformed to and measured with respect to the moving mesh, grid velocities can be arbitrarily large, e.g. exceeding by far the speed of light, without violating any laws of physics or leading to erroneous results. Typical values for the grid motion in the collapse phases of star formation are of order $u_{gr} \simeq 10^3\,c$.

Integrating Eqn. (2.5) against $d\widetilde{Vol}$, we obtain finally the integral adaptive mesh radiation transfer equation in conservation form

$$\frac{d}{dt}\Big[\int_{\tilde{V}(t)} \frac{I_o(\mu_o,\nu_o)}{c}\, d\widetilde{Vol}\Big] + \int_{\partial\tilde{V}} \frac{I_o(\mu_o,\nu_o)}{c}\, (\underline{\tilde{u}}_{rel}\cdot \underline{d\tilde{S}})$$

$$+ \int_{\tilde{V}} \frac{\partial}{\partial\, Vol}\quad [\mu_o r^2 \qquad I_o(\mu_o,\nu_o)]\, d\widetilde{Vol}$$

$$+ \int_{\tilde{V}} \frac{\partial}{\partial \mu_o}\quad [(1-\mu_o^2)\,\{\frac{1}{r} + \frac{\mu_o}{c}(\frac{3u}{r} - \frac{\partial(r^2u)}{\partial\, Vol}) - \frac{a}{c^2}\}\, I_o(\mu_o,\nu_o)]\, d\widetilde{Vol}$$

$$+ \int_{\tilde{V}} \frac{\partial}{\partial \nu_o}\quad [-\nu_o \quad \{(1-3\mu_o^2)\frac{u}{cr} + \frac{\mu_o^2}{c}\frac{\partial(r^2u)}{\partial\, Vol} + \frac{\mu_o a}{c^2}\}\, I_o(\mu_o,\nu_o)]\, d\widetilde{Vol}$$

$$= \int_{\tilde{V}} (\eta_o(\nu_o) - [\chi_o(\nu_o) + \{(1-3\mu_o^2)\frac{u}{cr} + \frac{\mu_o^2}{c}\frac{\partial(r^2u)}{\partial\, Vol} + 2\frac{\mu_o a}{c^2}\}]\, I_o(\mu_o,\nu_o))\, d\widetilde{Vol}, \tag{2.6}$$

where $d\widetilde{Vol} \equiv dVol\cdot d\mu_o \cdot d\nu_o$, $\underline{d\tilde{S}}$ is the corresponding surface element with outwards-pointing normal, and $\partial\tilde{V}$ is the boundary of the volume $\tilde{V}$ under consideration. In this equation the time derivative of the volume integral is expressed entirely by the difference of surface terms, except for the source term on the RHS of the equation. Consequently, we can arrange the numerical integration in such a way that no global conservation errors are introduced by it. Henceforth, in actual numerical simulations we will use the radiation transfer equation audits moments only in the form of Eqn. (2.6).

For a thorough and detailed discussion of the advantages and disadvantages of the Eulerian, Lagrangean, and mixed Eulerian-Lagrangean radiation transfer equation, the reader is refered to Mihalas (this book). Our goal is now to derive an approximate, but nevertheless accurate, description of the matter-radiation-interaction, accessible to numerical solution by an adaptive-mesh finite-difference method.

III. ADAPTIVE MESH RADIATION HYDRODYNAMICS

The 0th, 1st, 2nd, and 3rd moments of intensity J_o, H_o, K_o, and N_o are defined by

$$J_o(\nu_o) \equiv \frac{1}{2}\int_{-1}^{+1} I_o(\mu_o,\nu_o) \quad d\mu_o \tag{3.1}$$

$$H_o(\nu_o) \equiv \frac{1}{2}\int_{-1}^{+1} I_o(\mu_o,\nu_o)\, \mu_o \, d\mu_o \tag{3.2}$$

$$K_o(\nu_o) \equiv \frac{1}{2}\int_{-1}^{+1} I_o(\mu_o,\nu_o)\, \mu_o^2 \, d\mu_o \tag{3.3}$$

$$N_o(\nu_o) \equiv \frac{1}{2}\int_{-1}^{+1} I_o(\mu_o,\nu_o)\, \mu_o^3 \, d\mu_o \tag{3.4}$$

Integrating against ν_o we obtain

$$J_o \equiv \int_0^\infty J_o(\nu_o)\, d\nu_o \tag{3.5}$$

$$H_o \equiv \int_0^\infty H_o(\nu_o)\, d\nu_o \tag{3.6}$$

$$K_o \equiv \int_0^\infty K_o(\nu_o)\, d\nu_o \tag{3.7}$$

The radiation energy density E_o, radiation flux F_o, radiation pressure P_o, the radiation quantity Q_o, and their corresponding frequency-dependent counterparts $E_o(\nu_o)$, $F_o(\nu_o)$, $P_o(\nu_o)$, and $Q_o(\nu_o)$ are related to the moments of intensity by

$$E_0(\nu_0) = \frac{4\pi}{c} J_0(\nu_0) \tag{3.8}$$

$$F_0(\nu_0) = 4\pi \; H_0(\nu_0) \tag{3.9}$$

$$P_0(\nu_0) = \frac{4\pi}{c} K_0(\nu_0) \tag{3.10}$$

$$Q_0(\nu_0) \equiv 4\pi \; N_0(\nu_0) \tag{3.11}$$

$$E_0 = \frac{4\pi}{c} J_0 \tag{3.12}$$

$$F_0 = 4\pi \; H_0 \tag{3.13}$$

$$P_0 = \frac{4\pi}{c} K_0 \tag{3.14}$$

The Eddington factors $g_0(\nu_0)$, g_0, $f_0(\nu_0)$, f_0, and $h_0(\nu_0)$ are given by

$$g_0(\nu_0) \equiv \frac{H_0(\nu_0)}{J_0(\nu_0)} = \frac{c\ F_0(\nu_0)}{E_0(\nu_0)} \tag{3.15}$$

$$f_0(\nu_0) \equiv \frac{K_0(\nu_0)}{J_0(\nu_0)} = \frac{P_0(\nu_0)}{E_0(\nu_0)} \tag{3.16}$$

$$h_0(\nu_0) \equiv \frac{N_0(\nu_0)}{H_0(\nu_0)} = \frac{Q_0(\nu_0)}{F_0(\nu_0)} \tag{3.17}$$

$$g_0 \equiv \frac{H_0}{J_0} = \frac{c\ F_0}{E_0} \tag{3.18}$$

$$f_0 \equiv \frac{K_0}{J_0} = \frac{P_0}{E_0} \tag{3.19}$$

Taking the 0th and 1st moments of Eqn. (2.6) against μ_0 we obtain the frequency-dependent adaptive mesh 0th moment equation

$$\frac{d}{dt}\left[\int_{V(t)} E_0(\nu_0)\, dVol\right] + \int_{\partial V} E_0(\nu_0)\,(\underline{u}_{rel}\cdot \underline{dS}) + \int_V \frac{\partial}{\partial Vol}\left[r^2 F_0(\nu_0)\right] dVol$$

$$+ \int_V \frac{\partial}{\partial \nu_0}\left[-\nu_0\left\{(E_0(\nu_0)-3P_0(\nu_0))\frac{u}{cr} + P_0(\nu_0)\frac{\partial(r^2u)}{\partial Vol} + \frac{aF_0(\nu_0)}{c^2}\right\}\right] dVol$$

$$= \int_V \left[\quad -\left\{(E_0(\nu_0)-3P_0(\nu_0))\frac{u}{cr} + P_0(\nu_0)\frac{\partial(r^2u)}{\partial Vol} + \frac{2aF_0(\nu_0)}{c^2}\right\}\right] dVol$$

$$+ \int_V \left[4\pi\eta_0(\nu_0) - c\chi_0(\nu_0)E_0(\nu_0)\right] dVol \tag{3.20}$$

and the frequency-dependent adaptive mesh 1st moment equation

$$\frac{d}{dt}\left[\int_{V(t)} \frac{F_0(\nu_0)}{c^2}\, dVol\right] + \int_{\partial V} \frac{F_0(\nu_0)}{c^2}\,(\underline{u}_{rel}\cdot \underline{dS})$$

$$+ \int_V \left[\frac{\partial P_0(\nu_0)}{\partial r} + \frac{(3P_0(\nu_0)-E_0(\nu_0))}{r} + \frac{F_0(\nu_0)}{c^2}\frac{\partial u}{\partial r} + \frac{a}{c^2}(E_0(\nu_0)+P_0(\nu_0))\right] dVol$$

$$+ \int_V \frac{\partial}{\partial \nu_0}\left[-\nu_0\left\{(F_0(\nu_0)-3Q_0(\nu_0))\frac{u}{cr} + \frac{Q_0(\nu_0)}{c^2}\frac{\partial(r^2u)}{\partial Vol} + \frac{a}{c^2}P_0(\nu_0)\right\}\right] dVol$$

$$= -\int_V \left[\frac{\chi_0(\nu_0)}{c} F_0(\nu_0)\right] dVol \tag{3.21}$$

Integration of equation (3.20) and (3.21) over frequency results in the adaptive mesh radiation energy equation

$$\frac{d}{dt}\left[\int_{V(t)} E_o \, dVol\right] + \int_{\partial V} E_o \, (\underline{u}_{rel} \cdot \underline{dS}) + \int_V \frac{\partial}{\partial Vol}\left[r^2 F_o\right] dVol$$

$$+ \int_V \left[(E_o - 3P_o)\frac{u}{cr} + P_o \frac{\partial r^2 u}{\partial Vol} + \frac{2aF_o}{c^2}\right] dVol$$

$$= + \int_V \left\{ \int_0^\infty \left[4\pi\eta_o(\nu_o) - c\chi_o(\nu_o)E_o(\nu_o)\right] d\nu_o \right\} dVol \qquad (3.22)$$

and the adaptive mesh radiation momentum equation

$$\frac{d}{dt}\left[\int_{V(t)} \frac{F_o}{c^2} \, dVol\right] + \int_{\partial V} \frac{F_o}{c^2} (\underline{u}_{rel} \cdot \underline{dS})$$

$$+ \int_V \left[\frac{\partial P_o}{\partial r} + \frac{(3P_o - E_o)}{r} + \frac{F_o}{c^2}\frac{\partial u}{\partial r} + \frac{a}{c^2}(E_o + P_o)\right] dVol$$

$$= - \int_V \left\{\frac{1}{c}\int_0^\infty \left[\chi_o(\nu_o) \, F_o(\nu_o) d\nu_o\right\} dVol. \qquad (3.23)$$

In Eqns. (3.20) and (3.21), or (3.22) and (3.23) $\underline{u}_{rel} \equiv (u_{rel}, 0, -d\nu_o/dt)$, or $\underline{u}_{rel} \equiv (u_{rel}, 0, 0)$, respectively, $dVol \equiv (1/3)d(r^3)d\nu_o$ or $dVol \equiv (1/3)d(r^3)$, respectively, and $\underline{dS}$ is the corresponding surface element with outwards pointing normal. Notice that in this form of the equations the adaptive mesh does not produce extra nonconservative terms.

The evolution of selfgravitating gas flows is determined by (in addition to the radiation transfer equation) Poisson's equation and Euler's equations of gas dynamics. We use these equations in the following form.

Poisson equation

$$dm = \rho \, dVol \qquad (3.24)$$

continuity equation

$$\frac{d}{dt}\Big[\int_{V(t)} \rho \; dVol\Big] + \int_{\partial V} \rho \; (\underline{u}_{rel} \cdot \underline{dS}) = 0. \qquad (3.25)$$

internal energy equation

$$\frac{d}{dt}\Big[\int_{V(t)} \rho e \; dVol\Big] + \int_{\partial V} \rho e \; (\underline{u}_{rel} \cdot \underline{dS}) + \int_{V} \Big[P \frac{\partial(r^2 u)}{\partial Vol} + \frac{\partial(r^2 F_{conv})}{\partial Vol}\Big] dVol$$

$$= - \int_{V} \Big\{ \int_0^{\infty} [4\pi\eta_0(\nu_0) - c\chi_0(\nu_0)E_0(\nu_0)] d\nu_0 \Big\} dVol + \int_{V} \varepsilon_N \rho \; dVol \qquad (3.26)$$

equation of motion

$$\frac{d}{dt}\Big[\int_{V(t)} \rho u \; dVol\Big] + \int_{\partial V} \rho u \; (\underline{u}_{rel} \cdot \underline{dS}) + \int_{V} \Big[\frac{\partial P}{\partial r} + \frac{4\pi Gm}{r^2} \rho \Big] dVol$$

$$= + \int_{V} \Big\{ \frac{1}{c} \int_0^{\infty} \chi_0(\nu_0) \; F_0(\nu_0) \; d\nu_0 \Big\} dVol \qquad (3.27)$$

Here e is the internal energy per gram of the gas, P the gas pressure, F_{conv} the convective flux (considered to be negligible in the momentum balance), ε_N the nuclear energy generation per gram and unit time, G the gravitational constant. In case we want to compute normal gas flow without the influence of gravity we set G = 0. For numerical reasons we solve the internal energy equation (3.26) and use the total energy Eqn. (3.28) to check the accuracy of the numerical solution.

$$\frac{d}{dt}\Big[\int_{V(t)} E_t \; dVol\Big] + \int_{\partial V} E_t \; (\underline{u}_{rel} \cdot \underline{dS})$$

$$+ \int_{\partial V} [(P + P_0)\underline{u} + \underline{F}_0 + \underline{F}_{conv}] \cdot \underline{dS} = \int_{V} \varepsilon_N \rho \; dVol \qquad (3.28)$$

$$E_t \equiv \rho e + E_0 + \frac{1}{2} \rho u^2 - \frac{4\pi Gm}{r} \rho \qquad (3.29)$$

Here, E_t is the total energy per unit volume, $\underline{F}_0 \equiv (F_0, 0, 0)$, $F_{conv} \equiv (F_{conv}, 0, 0)$. Equation (3.28) can be derived from

Eqns. (3.22), (3.23), (3.26), (3.27), and (3.29). Integrating Eqn. (3.28) against time we find

$$\int_V E_t \, dVol = Const. - \Big\{ \int_{time} \Big(\int_{\partial V} E_t(\underline{u}_{rel} \cdot \underline{dS}) \Big) dt + \int_{time} \Big(\int_{\partial V} ([(P + P_o)\underline{u} + \underline{F}_o + \underline{F}_{conv}] \cdot \underline{dS}) \Big) dt - \int_{time} \Big(\int_V \varepsilon_N \rho \, dVol \Big) \, dt \Big\} \quad (3.30)$$

Here $\int_{time} \ldots dt$ is the time integral over the entire flow evolution, $\int_V \ldots dVol$ the volume integral of the entire computational space, and in the surface integral $\int_{\partial V} (\underline{\ldots \cdot dS})$, the only nonvanishing terms are at the outer and inner surface of our computational mesh. Const. is a constant of the time integration.

Note in passing that if we set $\underline{u}_{gr} \equiv 0$, i.e. $\underline{u}_{rel} \equiv \underline{u}$, in any of the above integral equations we obtain the general rule for the transformation of Eqns. (not the variables) from Eulerian coordinates to the adaptive mesh

$$\frac{\partial}{\partial t} \Big[\int_V \ldots dVol \Big] \equiv \frac{d}{dt} \Big[\int_V \ldots dVol \Big] - \int_{\partial V} \ldots (\underline{u}_{gr} \cdot \underline{dS}) \quad (3.31)$$

Similarly we find for the transformation from Langrangean coordinates ($\underline{u}_{rel} \equiv 0$, $u \equiv u_{gr}$) to the adaptive mesh

$$\frac{D}{Dt} \Big[\int_V \ldots dVol \Big] \equiv \frac{d}{dt} \Big[\int_V \ldots dVol \Big] + \int_{\partial V} \ldots (\underline{u}_{rel} \cdot \underline{dS}) \quad (3.32)$$

Equations (3.31) and (3.32) are merely gereralized versions of the Reynolds Transport Theorem, see e.g. Mihalas and Mihalas (1984) or Winkler, Norman, and Mihalas (1984).

It is instructive to point out that the matter-radiation coupling in the energy and momentum Eqns. (3.22), (3.23), (3.26), and (3.27) is described by frequency-averaged radiation quantities only. The energy-coupling term C_E and the momentum-coupling term C_M are

$$C_E \equiv \int_V \{\int_0^\infty [4\pi\eta_o(\nu_o) - c\chi_o(\nu_o)E_o(\nu_o)]\ d\nu_o\}\ dVol \tag{3.33}$$

$$C_M \equiv \int_V \{\frac{1}{c}\int_0^\infty \chi_o(\nu_o)\ F_o(\nu_o)d\nu_o\}\ dVol \tag{3.34}$$

The macroscopic mass absorption coefficient $\kappa_o(\nu_o,T,\rho)$ [$cm^2\,g^{-1}$], related to the macroscopic absorption coefficient $\chi_o(\nu_o,T,\rho)$ [cm^{-1}]

$$\chi_o(\nu_o) = \rho\ \kappa_o(\nu_o), \tag{3.35}$$

consists of 2 parts, the true absorption coefficient $\kappa_o^a(\nu_o,T,\rho)$ and the scattering coefficient $\kappa_o^s(\nu_o,T,\rho)$

$$\kappa_o(\nu_o) = \kappa_o^a(\nu_o) + \kappa_o^s(\nu_o). \tag{3.36}$$

In these relations we have suppressed the T and ρ dependence for brevity. Assuming LTE, the emissivity η_o can be written

$$\eta_o(\nu_o) = \rho\kappa_o(\nu_o)\ S_o(\nu_o), \tag{3.37}$$

where

$$S_o(\nu_o) \equiv \frac{\kappa_o^s(\nu_o)}{\kappa_o(\nu_o)}\ J_o(\nu_o) + \frac{\kappa_o^a(\nu_o)}{\kappa_o(\nu_o)}\ B_o(T,\nu_o) \tag{3.38}$$

or

$$S_o(\nu_o) = \frac{\kappa_o^s(\nu_o)}{\kappa_o(\nu_o)} \frac{cE_o(\nu_o)}{4\pi} + \frac{\kappa_o^a(\nu_o)}{\kappa_o(\nu_o)} B_o(T,\nu_o) \tag{3.39}$$

$$B_o(T,\nu_o) \equiv \frac{2h\nu_o^3}{c^2} \frac{1}{\left(e^{\frac{h\nu_o}{kT}} - 1\right)} \tag{3.40}$$

Here $S_o(\nu_o)$ is the source function, $B_o(T,\nu_o)$ the Planck function, T the gas temperature, h Planck's constant, k Boltzmann's constant. Notice that we have constructed the radiation hydrodynamics equations in such a way that T enters explicitly only through this source function (the constitutive relations are assumed to be functions of e and ρ). Integrating $B_o(T,\nu_o)$ against ν_o we get

$$B_o(T) \equiv \int_0^\infty B_o(T,\nu_o)\, d\nu_o = \frac{\sigma_R}{\pi} T^4 \tag{3.41}$$

where σ_R is the Stefan-Boltzmann constant. For convenience we also define a radiation temperature T_R

$$E_o = \frac{4\pi}{c} J_o = \frac{4\pi}{c} \frac{\sigma_R}{\pi} T_R^4 = \frac{4\sigma_R}{c} T_R^4 = a_R T_R^4 \tag{3.42}$$

where a_R is the radiation constant.

For later use we now introduce the Planck mean, the Rosseland mean, the absorption mean, and the flux mean opacity κ_P, κ_R, κ_E, κ_F

$$\kappa_P \equiv \int_0^\infty \kappa_o^a(\nu_o) \frac{B_o(T,\nu_o)}{B_o(T)} d\nu_o \tag{3.43}$$

$$\kappa_E \equiv \int_0^\infty \kappa_0^a(\nu_0) \frac{E_0(\nu_0)}{E_0} d\nu_0 \tag{3.44}$$

$$\kappa_F \equiv \int_0^\infty \left(\kappa_0^a(\nu_0) + \kappa_0^s(\nu_0)\right) \frac{F_0(\nu_0)}{F_0} d\nu_0 \tag{3.45}$$

$$\frac{1}{\kappa_R} = \int_0^\infty \frac{1}{\left(\kappa_0^a(\nu_0) + \kappa_0^s(\nu_0)\right)} \frac{\dfrac{\partial B_0(T,\nu_0)}{\partial T}}{\dfrac{4\sigma_R}{\pi} T^3} d\nu_0 \tag{3.46}$$

where κ_R is derived from the definition

$$\frac{1}{\kappa_R} \equiv \int_0^\infty \frac{1}{\left(\kappa_0^a(\nu_0) + \kappa_0^s(\nu_0)\right)} \frac{\dfrac{\partial B_0(T,\nu_0)}{\partial T}}{\displaystyle\int_0^\infty \frac{\partial B_0(T,\nu_0)}{\partial T} d\nu_0} d\nu_0 \tag{3.47}$$

Also recall from Eqns. (3.16) and (3.19)

$$f_0 = \int_0^\infty f_0(\nu_0) \frac{E_0(\nu_0)}{E_0} d\nu_0 \tag{3.48}$$

We define

$$k_E \equiv \frac{\kappa_E}{\kappa_P} \tag{3.49}$$

$$k_F \equiv \frac{\kappa_F}{\kappa_R} \tag{3.50}$$

If we accept that the gas temperature T can always be expressed as a function of e and ρ, i.e. $T(e,\rho)$, then we find the following dependences of the mean opacities. Whereas

$$\kappa_P[T(e,\rho),\rho] \tag{3.51}$$

$$\kappa_R[T(e,\rho),\rho] \tag{3.52}$$

are a function of e and ρ alone, the other in addition depend also on the spectral profiles

$$\kappa_E[T(e,\rho),\ \rho,\ \frac{E_0(\nu_0)}{E_0}] \tag{3.53}$$

$$\kappa_F[T(e,\rho),\ \rho,\ \frac{F_0(\nu_0)}{F_0}] \tag{3.54}$$

which are problem-dependent, and therefore not known beforehand in general. By building the ratios k_E and k_F we are trying to separate the e-ρ-dependence of κ_E and κ_F from the dependence on the spectral profiles. In other words, κ_E and k_F describe the difference in the mean opacities due to deviations from the Planckian spectrum.

$$\kappa_E = k_E[\frac{E_0(\nu_0)}{E_0}]\ \kappa_P[T(e,\rho),\ \rho] \tag{3.55}$$

$$\kappa_F = k_F[\frac{F_0(\nu_0)}{F_0}]\ \kappa_R[T(e,\rho),\ \rho] \tag{3.56}$$

For a large collection of frequency dependent and mean opacities, as well as other constitutive relations, the reader is refered to the "Astrophysical Opacity Library" of Huebner, Merts, Magee, and Argo (1977).

Finally we now can rewrite the matter-radiation-coupling terms C_E and C_M

$$C_E = \int_V \left[4\pi\kappa_P B_o(T) - c\kappa_E E_o \right] \rho \, dVol \tag{3.57}$$

or

$$C_E = \int_V c\left[\kappa_P a_R T^4 - \kappa_E E_o\right] \rho \, dVol \tag{3.58}$$

and

$$C_M = \int_V \frac{\kappa_F}{c} F_o \, \rho \, dVol \tag{3.59}$$

Before we outline different numerical approaches for solving the radiation hydrodynamics equations let us investigate some aspects of the other constitutive relations.

For a given set of nuclear abundances, T and P are related to e and ρ via the equation of state (EOS) relations

$$T = T_{EOS}(e,\rho) \tag{3.60}$$

$$P = P_{EOS}\left[T_{EOS}(e,\rho),\ \rho\right], \tag{3.61}$$

typically given in form of monotonic spline interpolated tables. For an ideal gas the EOS relations can be given explicitly

$$T = (\Gamma - 1)\ \frac{\mu}{R}\ e \tag{3.62}$$

$$P = (\Gamma - 1)\quad e\ \rho, \tag{3.63}$$

where Γ is the ratio of the specific heats, μ the mean molecular weight, and R the universal gas constant.

To compute ε_N(T, ρ, nuclear abundance) properly a network of nuclear abundances and reactions has to be solved. This network determines the evolution of the nuclear abundances, and hence the nuclear energy generation per gram and unit time. As in many problems the typical time scale for the dynamical evolution is short compared to the nuclear time scale, hence the numerical solution of the nuclear reaction network can be decoupled from the numerical solution procedure for the radiation hydro equations. We assume the nuclear abundances to be constant during one timestep, and update the abundances in between timesteps. However, the density dependence and the (extremely sensitive!) temperature dependence of ε_N is explicitly taken into account. Therefore, within the radiation hydrodynamics equations we have the following dependence of ε_N

$$\varepsilon_N = \varepsilon[T_{EOS}(e,\rho),\ \rho] \tag{3.64}$$

Besides the radiation flux, the convective flux can be an important physical source for the transport of energy. According to the Schwarzschild criterion, convective energy transport can be dominant in regions where the radiative temperature gradient ∇_{rad} is steeper than the adiabatic temperature gradient ∇_{ad}

$$\nabla_{rad} \equiv \left(\frac{\partial \ln T}{\partial \ln P}\right)_{rad} > \left(\frac{\partial \ln T}{\partial \ln P}\right)_{ad} \equiv \nabla_{ad}, \tag{3.65}$$

where ∇, ∂ denote a spatial difference. In this case, the convective energy transport can be computed e.g. from the mixing length theory of convection, Böhm-Vitense, (1958). As a discussion of this topic is outside the scope of this paper we refer to the literature (see e.g. Cox and Giuli, (1968)) for the details of the computation of the convective flux F_{conv}.

Finally, having described all the relevant equations for the matter-radiation interaction, we are now in a position to make a reasonable and justifiable choice for the principal numerical solution-procedure. Assuming LTE (local thermodynamic equilibrium), the truly 3-dimensional system 1 of radiation transfer and hydrodynamic equations (besides various relations given throughout Sections II and III)

system 1: equations $(r,\mu_0,\nu_0$ + time)
(2.6) + (3.24),(3.25)
(3.26),(3.27) (3.66)

governs the evolution of time-dependent, spherically symmetric, (allegedly 1-D) radiation-hydro flows. In principle one could solve this coupled system of equations with straightforward, upwind-difference techniques. However, the 3-D nature of system 1 prohibits us from doing so. Such a direct procedure would require excessive execution times on present state-of-the-art computers and, therefore, has to wait for advances in technology for realization.

One alternative is to split the 3-dimensional system 1 into two 2-dimensional systems

system 2a: simplified version of equation (2.6) — $(r,\mu_0$ + time parameterized by $\nu_0)$
$\downarrow$
$f_0(\nu_0), h_0(\nu_0)$
$\uparrow$
system 2b: equations (3.20),(3.21),(3.24), (3.25),(3.26),(3.27). — $(r,\nu_0$ + time) (3.67)

In this so-called multigroup approach, the frequency derivative has been ignored (or is at least not fully, implicitly coupled) in the transfer Eqn. (2.6) but retained in the other equations. System 2a is then coupled into system 2b via the Eddington factors $f_0(\nu_0)$ and $h_0(\nu_0)$. One modification of system 2b is to couple the frequency dependent radiation moment Eqn. (3.20), (3.21) with the internal energy Eqn. (3.26) implicitly, but solve the remaining equations explicitly.

The other alternative, which we adopt, is the so-called multigroup/grey approach, first realized in the VERA code, Freeman and Hauser (1968), described and analyzed in detail by Mihalas and Mihalas (1984). In this method, system 1 is split into one 2-dimensional and two 1-dimensional systems

system 3a: simplified version of equation (2.6) — $(r,\mu_0$ + time parameterized by $\nu_0)$
$\downarrow$
$f_0(\nu_0), h_0(\nu_0)$
$\uparrow$
system 3b: equations (3.20),(3.21) — (r + time parameterized by $\nu_0)$
$\downarrow$
f_0, k_E, k_F
$\uparrow$
system 3c: equations (3.22),(3.23),(3.24), (3.25),(3.26),(3.27). — (r + time) (3.68)

The coupling of system 3a into the system 3b is again done via the Eddington factors $f_0(\nu_0)$ and $h_0(\nu_0)$, whereas system 3b is coupled with system 3c via f_0, k_E, and k_F. A solution of system 3a can be obtained with standard numerical techniques, see e.g. Yorke this book. Preferably, the transfer Eqn. (2.6) is solved in such a way that the numerical integration over μ_0 results exactly in the moment equations, Mihalas (private communication). The frequency derivatives in system 3b are kept but treated explicitly, i.e. iterated, whereas the other parts of the equations are solved implicitly. This procedure requires the solution of a number of systems with two coupled equations. Generally, the entire system (3a, 3b, and 3c) is iterated to convergence. For grey radiative transfer problems, step 2 of the method (system 3b) is superflous.

In the remainder of this paper we will exclusively be concerned with the numerical solution procedure for the system 3c. Assuming f_0, k_E, and k_F to be known from the previous steps of the method, and considering T, P, ε_N, κ_P, and κ_R to be known functions of e and ρ, we finally obtain our system of partial differential equations for the following 8 dependent variables.

$$F_{conv}, E_0, F_0, e, u, \rho, m, r. \tag{3.69}$$

Together with appropriate boundary conditions, an adaptive mesh equation for the grid motion (see Section IV), and an equation for the convective flux, system 3c of radiation energy and radiation momentum, Poisson, continuity, internal energy, and momentum equation determines the evolution of the 8 dependent variables. In addition to many other reasons, see Tscharnuter and Winkler (1979), our system of 8 partial differential equations will be solved with implicit techniques because the numerical stability criterion (Courant condition) for explicit methods puts a severe (prohibitive) upper limit on the timestep of the numerical computation for astrophysical radiation flows. Our 2 independent variables are (as we will see in Section V on the numerical equations)

time-index n
grid-index j.

Of course, we can ask whether such a huge computational effort in obtaining the correct Eddington factor f_0 is justified? Or should we use instead e.g. radiation diffusion, nonequlibrium diffusion, multigroup diffusion, flux limiters (for a critical review see Mihalas and Mihalas, 1984)?

In the grey case, f is directly obtained from the 2-dimensional system 3a and used only to close the radiation moment Eqns. (3.22) and (3.23). The answer to our first question is: absolutely yes. To our second question: in general, no (approximations are only valid in special restricted cases). Consider, for example, the outer parts of extended envelopes around stars, which are typical for star formation processes going on in interstellar clouds. In this case the radiation momentum Eqn. (3.23) can be used in the streaming limit on a fluid-flow time scale. It reduces to

$$\int_V \left[\frac{\partial P_o}{\partial r} + \frac{3P_o - E_o}{r}\right] dVol = -\frac{1}{c}\int_V \kappa_F F_o \rho \, dVol \qquad (3.70)$$

With the help of the sphericity factor q_o, introduced by Auer (1971)

$$\ln q_o \equiv \int_{r_{center}}^{r} \frac{3f_o - 1}{f_o \; r'} dr' \qquad (3.71)$$

Eqn. (3.70) can be rewritten

$$F_o = -\frac{c}{q_o} \frac{\partial (q_o \; f_o \; E_o)}{\rho \; \kappa_F \; \partial r} \qquad (3.72)$$

Back to our problem; f_o first increases strongly with r, approaching 1, then drops due to the fact that the whole cloud is embedded in a diluted isotropic background stellar radiation field. Assuming the wrong behaviour of f_o (e.g. $f_o \equiv (1/3)$) in this circumstance will lead to catastrophic consequences for the evaluation of F_o and E_o. By virtue of Eqn. (3.72), wrongly estimated gradients of $q_o f_o$ can lead to grossly over- or underestimated radiation fluxes. This in turn couples back onto E_o through the $\int_V \frac{\partial}{\partial Vol}\left[r^2 F_o\right] dVol$ - term in the radiation energy Eqn. (3.22). As a consequence, energy is first artificially taken out of the radiation energy density, supporting an artificially high radiation flux. Further outside, in a reverse situation, the radiation flux is artificially funneled into the radiation energy density. This numerical instability, caused by a physically incorrect estimate of f_o, feeds on itself and leads to faster and faster changes in E_o and F_o, finally determining completely the time scale for the numerical evolution. In the end, one obtains nonsensical results $E_0 < 0$, $|g_0| >> 1$ (the physically correct limits are $E \geqslant 0$, $|g_0| \leqslant 1$). This, of course, is unacceptable.

Therefore, we will always compute f_0 from the solution of an admittedly simplified version of the radiation transfer Eqn. (2.6). This way, we have a much better chance to obtain physically correct and accurate numerical solutions of the radiation hydrodynamics equations, avoiding the numerical instability just described. It is also a good practice to constantly monitor

$$|g_0| \equiv \left|\frac{cF_0}{E_0}\right| \leqslant 1 \, , \tag{3.73}$$

where E_0 and F_0 have been obtained from the solution of the radiation moment Eqns. (3.22) and (3.23), coupled with the hydrodynamical equations.

In the literature, the radiation momentum equation is usually written in terms of the sphericity factor q_0. We refrain from this practice so as to keep the number of dependent variables of our system of equations small, in order to minimize computational costs. Using q_0 explicitly in our implicit, adaptive mesh method increases the number of dependent variables from 8 to 9. This leads to an increase in the computational costs of about 42%, as the costs scale roughly to the cube of the number of dependent variables per gridpoint for direct linear algebra methods. In this case, q_0 itself could be determined from the ordinary differential equation

$$d(\ln q_0) = \frac{3f_0 - 1}{f_0 r} \, dr \tag{3.74}$$

where f_0 at gridpoint j is considered to be constant during a time step of our system of partial differential equations. In the following, we present a set of numerical equations which closely approximates the just derived analytical equations.

IV. COMPUTATIONAL TOOLS

The physical equations derived in the previous two sections must be modified with additional terms and equations in order to make them numerically tractable. In the following, we will first describe the evolution equations for the adaptive mesh on which we want to solve the physical equations. Next, we describe artificial viscosity and artificial diffusion techniques.

A. Adaptive Mesh

The grid distribution $\{r_j\}$ as a function of grid-index j is determined from the grid Eqn. (4.1), which is solved implicitly with the radiation hydrodynamics equations. For a definition of how our variables are defined on the mesh, i.e. interface or zone centered, we refer to Section V. As shown by Eqn. (4.1), the grid equation contains three key elements, each of which plays a distinctive role. First, the zone-centered structure function f^s provides an objective measure of the "structure" of the solution at a given time-level. The term Δf^s thus acts as an interface-centered driving term to the rest of the equation, which is globally elliptic, but may be locally parabolic (Winkler, et. al. 1985). Next, the radial function f^r contains the terms that define mesh intervals, stabilize the mesh, and set both a ceiling and a floor on the mesh spacing. As discussed by Winkler, Norman, and Newman (1984) these powerful operators allow meshpoints to move instantly to those parts of the computational domain where new structures are developing and better grid resolution is needed, and have proven successful in a wide variety of applications.

In some flows structure may temporarily disappear locally in some part of the flow, only to reappear a few timesteps later, e.g. when a shock reflects from a wall. In this event an instantaneous response of the mesh in that part of the flow is neither necessary nor desirable. Therefore to avoid unnecessary (and perhaps catastrophic) double rearrangement of the mesh in such circumstances we introduce a parabolic term $(\delta r_j/\delta t)$ which is allowed to be operative only in certain parts of the flow. The coefficient of this term is constructed in such a way as to contain "memory" about the history of the structure in the flow, and is used to provide an asymmetric time-filtering of the grid motion in the sense that it is set to zero in those parts of the flow where the structure function is increasing (hence instantaneous grid response is needed), and is set to a positive number, with a fadeout over a specified number of timesteps in those parts of the flow where the structure function is decreasing. The effect of this operation is to allow gridpoints to diffuse out of newly structureless regions only slowly, hence to keep grid points available temporarily should they suddenly be needed a few timesteps later. This method has proven very effective in controlling transient runaways of the grid distribution, and has improved overall performance of the code. (It was developed after the NATO ARW in collaboration with D. Mihalas and is described fully in Winkler et. al. 1985).

Details of the asymmetric time-filtering algorithm are as follows:

$$- \mathrm{CVMGP}(0.,1.,\delta f)\left(\frac{\delta f}{f^n}\right)^2 \frac{T_{scale}}{R_{scale}} \frac{\delta r}{\delta t} + \Delta f^r + \Delta f^s = 0. \tag{4.1a}$$

$$- \mathrm{CVMGP}(0.,1.,\delta f)\left(\frac{\delta f}{f^n}\right)^2 \frac{T_{scale}}{r} \frac{\delta r}{\delta t} + \Delta f^{r\ell} + \Delta f^s = 0. \tag{4.1b}$$

where

$$\delta f = f^s - f^n \tag{4.2}$$

$$f^{n+1} = f^s + (1-\varepsilon)\,(f^n-f^s)\ \mathrm{CVMGP}\ (0.,1.,\delta f) \tag{4.3}$$

$$\mathrm{CVMGP}(X1,X2,FLAG) = \begin{cases} X1 & \text{if } FLAG > 0. \\ X2 & \text{otherwise} \end{cases}$$

Here CVMGP is the vector merge operator defined as CVMGP(X,Y,Z) = X, if $Z > 0$; = Y, if $Z < 0$. Eqn. (4.1a) is used when the range of the spatial domain is small, and Eqn. (4.1b) is used when the mesh has to cover many orders of magnitude in radius. Both f^s and f^r as well as the operator Δ are dimensionless; hence from dimensional arguments one sees that the coefficient of $(\delta r/\delta t)$ must have units [T/L]. This ratio provides a measure of (the reciprocal of) the characteristic diffusion speed of the mesh relative to grid-index. The parameter ε specifies the number of timesteps over with old structure is to be "remembered". The quadratic factor $(\delta f/f)^2$ varies between 0 and 1, and provides a smooth (differentiable) off-on switch for the time-derivative term. The spatial operators defining, conditioning, and stabilizing the radial distribution of the mesh are shown in detail in Eqns. (4.4a) - (4.5b)

$$
\begin{aligned}
f_j^r = \; & W_{r0} \, \Delta r_j && \text{S0} \\
& + W_{r2} \, \Delta^2 r_j \, / \, (\Delta r_{j-1} \, \Delta r_{j+1}) && \text{S2} \\
& + W_{r4} \, (\Delta r_{j-2} \Delta^6 r_j \Delta r_{j+2}) \, / \, (\Delta^4 r_{j-1} \, \Delta^4 r_{j+1}) && \text{S4} \\
& + W_{r_{max}} \, (\Delta r_j \, / \, \Delta r_{max})^n && \text{Max} \\
& - W_{r_{min}} \, (\Delta r_{min} \, / \, \Delta r_j \,)^n && \text{Min} \\
& + W_{\ell r} \, (\Delta r_{j-1} - \Delta r_{j+1} \, / \, (\Delta r_j \,)^2 && \text{Ratio} \\
& + W_{\ell r_{max}} \, [(\Delta r_{j-1} - \Delta r_{j+1} \, / \, (\Delta r_j \, \Delta \ell r_{max})]^n && \text{Max Ratio}
\end{aligned}
\qquad (4.4a)
$$

$$f_j^{r\ell} = \text{(Same operators as in Eq. (4.4a), but applied to } \Delta r\ell_j) \qquad (4.4b)$$

where

$$\Delta r_j = (r_j - r_{j+1}) \, / \, R_{scale} = \left(\frac{1}{R_{scale}} \Delta r\right)_j \qquad (4.5a)$$

$$\Delta r\ell_j = (r_j - r_{j+1}) \, / \, (r_j + r_{j+1}) = (\Delta \ln r)_j \qquad (4.5b)$$

In Eqns. (4.4) the W´s represent weights assigned to each term (see Winkler, Norman, and Newman (1984) for examples and discussion), and R_{scale}, Δr_{max}, $\Delta \ell r_{max}$, and Δr_{min} are scale factors. The term S0 sets the spatial scale, the spatially symmetric operators in terms S2 and S4 stabilize the mesh and also guarantee positivity of our metric, terms Max and Min set, respectively, upper and lower limits on the mesh spacing. Whereas the term Max Ratio sets an upper limit on the spatial change in grid size. The exponent n is a free parameter, and typically is set to 4. The higher order stabilizing operator S4 was developed after the NATO ARW in collaboration with B. Scheurer and D. Mihalas.

Thus far we have discussed only the "defensive" measures (existence, positivity, stability) which we build into the grid equation. In the absence of any driving terms the grid equation with appropriate boundary conditions would result in a grid distribution equidistant in either r of in ln r. Therefore the success of the ability of the grid to detect and resolve arbitrary structures in the flow depends heavily on the quality of the structure function f^s, which drives the grid into the desired distribution. This function is designed to resolve all important flow variables equally well in all parts of the mesh. Terms included in the structure function to date are

$$
\begin{aligned}
f^s_j = \; & W_m \, \Delta m_j && \text{T 1}\\
+ & W_{m\ell} \, \Delta m\ell_j && \text{T 2}\\
+ & W_\mu \, \Delta \mu_j && \text{T 3}\\
+ & W_{\rho\ell} \, \Delta \rho\ell_j && \text{T 4}\\
+ & W_{e\ell} \, \Delta e\ell_j && \text{T 5}\\
+ & W_{P\ell} \, \Delta P\ell_j && \text{T 6}\\
+ & W_{\kappa\rho\ell} \, \Delta \kappa\rho\ell_j && \text{T 7}\\
+ & W_{E_0\ell} \, \Delta E_0\ell_j && \text{T 8}\\
+ & W_{\varepsilon_N\ell} \, \Delta \varepsilon_N\ell_j && \text{T 9}\\
+ & W_{q\ell} \, \Delta q\ell_j && \text{T10}\\
+ & W_u \, \Delta u_j && \text{T11}\\
+ & W_{u\ell} \, \Delta u\ell_j && \text{T12} \quad (4.6)
\end{aligned}
$$

where

$$\Delta m_j = (m_j - m_{j+1}) \; / \; M_{scale} \qquad (4.7a)$$

$$\Delta m\ell_j = (m_j - m_{j+1}) \; / \; (m_j + m_{j+1}) \qquad (4.7b)$$

$$\Delta \mu_j = (\mu_{j-1} - \mu_{j+1})^2 \qquad (4.8)$$

$$\Delta \rho\ell_j = [(\rho_{j-1} - \rho_{j+1}) \; / \; \rho_j]^2 \qquad (4.9)$$

$$\Delta e\ell_j = [(e_{j-1} - e_{j+1}) \; / \; e_j]^2 \qquad (4.10)$$

$$\Delta P\ell_j = [(P_{j-1} - P_{j+1}) \; / \; P_j]^2 \qquad (4.11)$$

$$\Delta\kappa\rho\ell_j = [(\kappa_{j-1}\rho_{j-1} - \kappa_{j+1}\rho_{j+1}) / \kappa_j\rho_j]^2 \tag{4.12}$$

$$\Delta E_o\ell_j = [(E_{o_{j-1}} - E_{o_{j+1}}) / E_{o_j}]^2 \tag{4.13}$$

$$\Delta\varepsilon_N\ell_j = [(\varepsilon_{N_{j-1}} - \varepsilon_{N_{j+1}}) / (\varepsilon_{N_j} + \varepsilon_N \text{ floor})]^2 \tag{4.14}$$

$$\Delta q\ell_j = [(q_{j-1} - q_{j+1}) / q_j]^2 \tag{4.15}$$

$$q_j = \left(\frac{P_Q}{P + \rho u^2}\right) + q \text{ floor} \tag{4.16}$$

$$\Delta u_k = [\, u_j - u_{j+1}]^2 \quad \text{CVMGP}(1.,0.,u_j - u_{j+1}) \tag{4.17a}$$

$$\Delta u\ell_j = \frac{[u_j - u_{j+1}]^2}{(u_j^2 + u_{j+1}^2 + cu\ell)} \text{CVMGP}(1.,0.,u_j - u_{j+1}) \tag{4.17b}$$

The term T1 (or T2) tries to resolve m (or ln m); T3 the mean molecular weight μ; T4, T5, and T6 the quantities ln ρ, ln e, and ln P; T7 the logarithm of the opacity; T8 the logarithm of the radiation energy density ln E_o; and T9 the logarithm of the nuclear energy generation rate. T10 pulls grid points into shocks by resolving the ratio of viscous pressure P_Q to gas pressure P and ram pressure ρu^2. T11 focuses only on rarefaction fans in planar geometry, and T12 is used as a special purpose tool to resolve the head of the rarefaction fan in a shocktube simulation.

In order to guarantee that indeed all variables are smoothly resolved everywhere in the flow, we may apply the maximum operator T4 of Eqn. (4.4) to any of the variables used in the structure function f_j^s of Eqn.(4.6). That way we can enforce that none of the variables change by say more than 20% from gridpoint to gridpoint. The operators doing this are as follows

$$f_j^s \rightarrow f_j^s + \ldots + W\, V_{max} \left(\frac{\Delta V_j}{\Delta V_{max}}\right)^n$$

$$+ W\, V\ell_{max} \left(\frac{\Delta V\ell_j}{\Delta V\ell_{max}}\right)^n \tag{4.6a}$$

where V stands for any of the variables considered, $V\ell$ = ln V, and ΔV_{max}, $\Delta V\ell_{max}$ are constants.

The evolution of our thinking concerning the structure function f^s can be traced in Winkler (1976), Tscharnuter and Winkler (1979), Winkler and Newman (1980 a,b), and Winkler, Norman, and Newman (1984). The last reference contains an extensive discussion of the spatial stabilization term T2 of Eqn (4.4). Further details and a thorough investigation of the basic concept of an adaptive mesh, including the analysis of a variety of possible operators, as well as space stabilization and time filtering, can be found in Winkler et. al. (1985). It should also be noted that a diffusion equation for dynamical rezoning of a Lagrangean mesh has been used by Castor, et. al. (1977). Our algorithm differs significantly from theirs in two important respects: (1) Our elliptic operators have a precise and specifiable meaning and function, in contrast to their diffusion term which was constructed heuristically. (2) We use our diffusive term in a highly discriminating manner, whereas it is always present in their method.

B. Tensor Artificial Viscosity

Shock waves that arise in the flow are not treated as fluid discontinuities, but rather are spread out over a desired length by artificial viscosity. Our formulation differs significantly from the familiar artificial viscosity of von Neumann and Richtmyer (1950) in order to satisfy additional requirements imposed by the use of an adaptive grid and the particular nature of the protostar problem, see Tscharnuter and Winkler (1979). Our starting point is to define a viscous stress tensor which is symmetric, trace-free, and independent of coordinate system and frame of reference, as follows:

$$\underset{\approx}{Q} = \rho\mu_Q\left[(\underset{\sim}{\nabla}\underset{\sim}{u}) - \frac{1}{3}\underset{\sim}{\nabla}\cdot\underset{\sim}{u}\underset{\approx}{I}\right] \tag{4.18}$$

where $(\underset{\sim}{\nabla}\underset{\sim}{u})$ is the symmetrized velocity gradient tensor, $\underset{\approx}{I}$ is the unit tensor, and μ_Q is the coefficient of artificial viscosity given by

$$\mu_Q = -c_1\ell c_s + \min(c_2\ell^2\underset{\sim}{\nabla}\cdot\underset{\sim}{u}, 0). \tag{4.19}$$

Here c_1 and c_2 are constants of order unity, c_s is the adiabatic sound speed of the gas, and 4 to 5 ℓ is the physical thickness over which the shock is spread. $\underset{\approx}{Q}$ is constrained to be trace-free so that there will be no viscous heating in a homologous contraction, a desirable property when computing the Kelvin-Helmholtz contraction of a pre-main sequence star (Tscharnuter and Winkler(1979)).

The reader will recognize Eqn. (4.18) as the standard expression for the shear stress in a fluid of kinematic viscosity μ_Q. Therefore our artificial viscosity treatment is "artificial" only as regards to our choice of μ_Q. The coefficient of artificial viscosity given in Eqn. (4.19) is a sum of two terms generally referred to as linear and quadratic multipliers in the sense that substituting Eqn. (4.19) for μ_Q in Eqn. (4.18) yields a viscous stress tensor composed of terms which are linear and quadratic in the rate of strain (see e.g. Noh 1976). Notice that μ_Q is a scaler invariant (i.e. independent of coordinate system and frame reference) as required by the adaptive mesh. The quadratic multiplier vanishes in regions of expansion in analogy with the von Neumann-Richtmyer approach. The linear multiplier is chosen to be small but nonzero in order to damp small amplitude oscillations that occur predominantly in stagnant (with respect to the mesh) parts of the flow.

The viscous momentum transfer $\underset{\sim}{u}_Q$ and heating ε_Q are given by

$$\underset{\sim}{u}_Q \equiv -\underset{\sim}{\nabla}\cdot\underset{\approx}{Q} \qquad \text{(per unit volume)}, \tag{4.20}$$

$$\varepsilon_Q \equiv -\frac{1}{\rho}\,\underset{\approx}{Q}\cdot(\underset{\sim\sim}{\nabla u}) \qquad \text{(per gram)} \tag{4.21}$$

Specializing to spherical symmetry, we have the following expressions for the quantities appearing in the momentum and energy equation (for a derivation see Tscharnuter and Winkler(1979)):

$$u_Q = -\frac{2}{3}\frac{1}{r}\frac{\Delta\left[r^3\rho\mu_Q\left(\frac{\Delta u}{\Delta r}-\frac{u}{r}\right)\right]}{\Delta \mathrm{Vol}} \tag{4.22}$$

and

$$\varepsilon_Q = -\frac{2}{3}\mu_Q\left(\frac{\Delta u}{\Delta r}-\frac{u}{r}\right)^2 \tag{4.23}$$

where

$$\mu_Q = - c_1 \ell c_s + \min[c_2 \ell^2 \frac{\Delta(r^2 u)}{\Delta Vol}, 0]. \tag{4.24}$$

Here we have discretized in space by substituting the finite difference symbol Δ for the partial derivative symbol ∂. The centering of u_Q and ε_Q on the grid, as well as the rules for space and time averaging of r, ρ, and u are given Section V.

The length ℓ appearing in Eqn. (4.24) governs the thickness of the shock front, and is chosen to either a "fixed length" (i.e. ℓ/R_{scale} = constant) or a fixed "relative length" (i.e. ℓ/r = constant) depending on whether the domain of interest extends over one or many orders of magnitude in radius, respectively. The constant is chosen to be as small as possible in order to produce sharp shock fronts, yet large enough to avoid convergence difficulties. When used in conjunction with the adaptive mesh described above, our artificial viscosity method yields well-resolved (in a numerical sense) shock transitions with a typical thickness of 10^{-2} to 10^{-3} of an average zone size. This is possible because the adaptive mesh can achieve a local mesh refinement of up to a factor of a million. This high resolution is also achieved for a spherical shock reflecting off its center, hence the usual substitution in 1-D problems of $\Delta u/\Delta r$ for $\tilde{\Delta}\cdot\tilde{u}$ in the coefficient of viscosity, in Eqn. (4.24), is not required. For completeness we also give the expansion used for P_Q in the grid distribution Eqn. (4.16)

$$P_Q = \rho \mu_Q \frac{\Delta(r^2 u)}{\Delta Vol} \tag{4.25}$$

C. Artificial Mass and Heat Diffusion

It is well known that the use of artificial viscosity in the solution of inviscid gas flow problems leads to spurious results whenever strong shocks interact with walls, contact discontinuities, and other strong shocks (e.g. Cameron (1966)). This is a simple consequence of using viscous equations to solve an inviscid problem, and no amount of manipulation of the finite-difference approximations to the viscous equations can alter this fundamental fact. What can be done is to construct special-purpose fixes which are invariably calibrated by knowledge of the desired solution.

The errors produced in regions containing shock interactions are errors in the final distribution of entropy after the waves have ceased to interact. Because entropy production in weak shocks scales as the third power of the pressure difference across the front (e.g. Section 83) in Landau and Lifshitz (1959), these entropy errors are noticeable only in interactions involving strong shocks, i.e. $\eta \equiv P_2/P_1 >> 1$. In such cases, the error typically appears as a spike of higher temperature gas in pressure equilibrium with the surrounding gas. The classical example of this effect is known as wall-heating — the spurious overheating and consequent decompression when a strong shock reflects from a rigid wall (e.g. Noh 1976). The spatial width of the feature is approximately equal to the thickness of the smeared-out shock wave.

Our use of a fixed-length artificial viscosity in combination with the adaptive grid allows us to reduce the smeared-out shock thickness to a small fraction of an average zone size, typically $\ell = 10^{-2}$ to 10^{-3} $\overline{\Delta x}$. This means that the fractional amount of mass involved in the erroneous region is extremely small. Sometimes the width of the shock can be kept small compared to the width of a slightly smeared out contact discontinuity. In this case the spike is practically speaking lost in the steep gradient of the contact discontinuity and does not show up prominently. However, if the above condition is not fulfilled, the error shows up in the solution as a conspicous and undesirable spike. Our strategy to eliminate entropy spikes is to diffuse them away. Often the numerical diffusion implicit in the advection procedure is enough to prevent the spike from appearing in the first place. In extreme cases, however, this is not sufficient (such is the case in the interacting blast-wave test problem of Sec. VII. We then resort to introducing mass-diffusion and heat-conduction terms explicitly into the continuity and energy equations, respectively. The form of these terms is

$$D_\rho \equiv \underline{\nabla}\cdot\left(\sigma_\rho\ \underline{\nabla}\rho\right)\ \mathrm{dVol} \tag{4.26}$$

in the equation of continuity (3.25), and

$$D_e \equiv \underline{\nabla}\cdot\left(\sigma_e\ \rho\underline{\nabla}e\right)\ \mathrm{dVol} \tag{4.27}$$

in the internal energy equation (3.26).

In most applications the transport coefficients σ_ρ and σ_e are set to zero. Otherwise, we use the simple prescription $\sigma_\rho = \sigma_e = \ell^2/\tau$, where ℓ is the shock thickness, and τ is the characteristic time for the shock to propagate an average zone-width $\overline{\Delta x}$. Since $\ell << \overline{\Delta x}$, this ensures that steady shock solutions are unaffected.

We have found that the use of the asymmetric time filter in the adaptive mesh virtually eliminates the effect of wall-heating in our calculations. The way this is accomplished is that during the moment of shock reflection, the adaptive mesh automatically refines the zoning adjacent to the wall to a size of order ℓ. The effect of the time filter is to maintain this fine zoning for the duration of the reflection process. Thus the entropy spike has a width of a few ℓ when it is produced. As the reflected shock recedes, however, the zoning next to wall expands back up to an average zone size of 10^2 to 10^3 ℓ. In the process the miniscule amount of mass with incorrect entropy is averaged with 10^2 to 10^3 times as much mass with the correct entropy, thus reducing the relative error accordingly and confining it to the single zone adjacent to the reflecting boundary. Thus, no explicit diffusion or heat conduction is required to eliminate numerical wall-heating. In principle, the same procedure should work for internal boundaries, e.g. contact dicontinuities represented numerically as discontinous jumps.

V. NUMERICAL RADIATION HYDRODYNAMICS

A. Symbolic Equations

We are finally prepared to write down the difference-equation representations of the adaptive-mesh equations (3.22)-(3.27), (4.1) in symbolic form.

$$\Delta m - \rho\, \Delta \mathrm{Vol} = 0. \tag{5.1}$$

$$\frac{\delta[\rho \Delta \mathrm{Vol}]}{\delta t} + \Delta[r^\mu\, u_{rel}\rho] = \Delta[r^\mu\, \sigma_\rho\, \frac{\Delta\rho}{\Delta r}] \tag{5.2}$$

$$\frac{\delta[(e + \frac{E_o}{\rho})\Delta\xi]}{\delta t} - \Delta[\frac{\delta m}{\delta t}(e + \frac{E_o}{\rho}) - r^\mu(F_o + F_c)] + (P + P_o)\Delta(r^\mu u)$$

$$+ \left[(E_0 - 3P_0) \frac{u}{r} + \frac{2a\ F_0}{c^2} \right] \Delta \mathrm{Vol} = (\varepsilon_N + \varepsilon_Q)\Delta\xi + \Delta\left[r^\mu\ \sigma_e \rho \frac{\Delta e}{\Delta r} \right] \qquad (5.3)$$

$$\frac{\delta\left[(\frac{E_0}{\rho})\Delta\xi \right]}{\delta t} - \Delta\left[\frac{\delta m}{\delta t}(\frac{E_0}{\rho}) - r^\mu F_0 \right] + P_0\ \Delta(r^\mu\ u)$$

$$+ \left[(E_0 - 3P_0) \frac{u}{r} + \frac{2a\ F_0}{c^2} \right] \Delta \mathrm{Vol} = \left[4\pi\ \kappa_P B_0 - c\kappa_E E_0 \right] \Delta\xi \qquad (5.4)$$

$$\frac{\delta [u\ \Delta\xi]}{\delta t} - \Delta\left[\frac{\delta m}{\delta t} u \right] + r^\mu \Delta P + \frac{4\pi G m}{r^\mu} \Delta\xi = \frac{\kappa_F}{c} F_0 \Delta\xi + u_Q\ \Delta \mathrm{Vol} \qquad (5.5)$$

$$\frac{\delta\left[(\frac{F_0}{\rho c^2})\ \Delta\xi \right]}{\delta t} - \Delta\left[\frac{\delta m}{\delta t} (\frac{F_0}{\rho c^2}) \right] + r^\mu (\Delta P_0 + \frac{F_0}{c^2} \Delta u)$$

$$+ \left[\frac{3P_0 - E_0}{r} + \frac{a}{c^2} (E_0 + P_0) \right] \Delta \mathrm{Vol} = - \frac{\kappa_F}{c} F_0 \Delta\xi \qquad (5.6)$$

$$-\sigma_r \frac{\delta r}{\delta t} + \Delta\left[f^r + f^s \right] = 0. \qquad (5.7)$$

The independent variables in these equations are time-index n and adaptive-mesh index j; δ denotes a time difference at fixed j, and Δ denotes a spatial difference with respect to j at fixed t. The primary dependent variables are r, m, u, ρ, e, E_0, and F_0. Secondary variables derivable from the primary set are u_{rel}, a, P, P_0, T, B_0, and the opacities. In Eqns. (5.2) - (5.6) $\Delta\xi$ is a shorthand notation for $\xi \equiv \rho\Delta\mathrm{Vol}$ where $\Delta\mathrm{Vol} \equiv (1/(\mu+1))\Delta(r^{\mu+1})$. The choice μ = 0, 1, 2 describes correctly pure hydrodynamics flows for planar, cylindrical, and spherical geometry. For μ = 2, radiation effects are properly included; for μ = 0, we have to assume in addition the transition $r \to \infty$ to recover the correct planar radiation moment equations. Gravity can be shut off by setting G = 0. The spatial centering of the primary and secondary (in parentheses) variables is sketched in Fig. 1. Clearly vectors and tensors of odd rank are centered at interfaces; scalars and tensors of even rank are zone-centered.

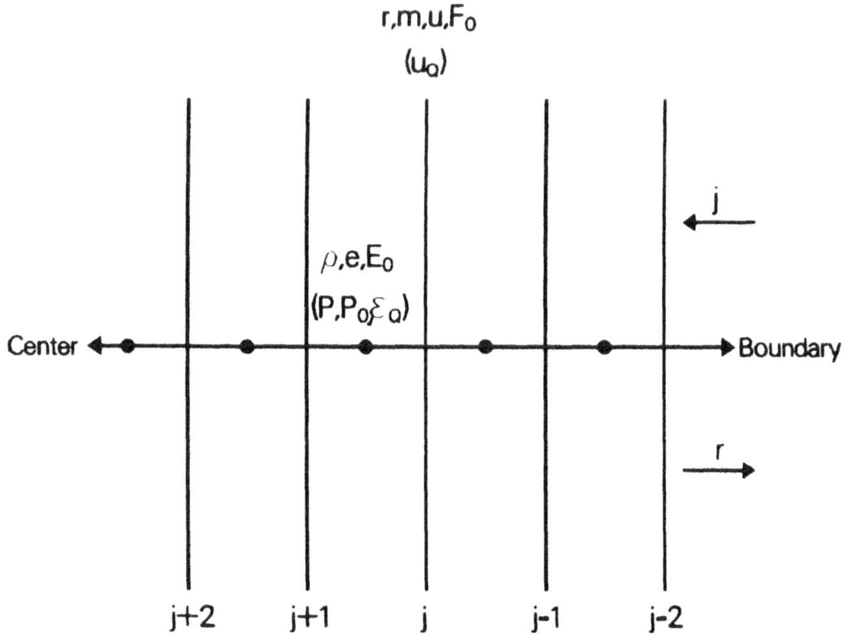

Fig. 1 Centering of physical variables on the adaptive grid.

In Eqn. (5.2), the term on the right-hand side (normally zero) is used to produce artificial mass diffusion when required. Similarly, in Eqn. (5.3), the term containing σ_e is used to produce artificial diffusion of internal energy when required. In Eqn. (5.3) and (5.5), the terms ε_Q and u_Q represent, respectively, energy dissipation and momentum deposition by artificial viscosity; they are computed according to the algorithms described in Section IV. Eqn. (5.7) is the grid equation that determines the adaptive mesh. Notice in Eq. (5.3) we evolve the total internal energy density (radiation + matter), rather than a difference approximation to the gas internal energy equation (3.26).

With the exceptions of Eqns. (5.1) and (5.7), which both apply at the advanced time level, all physical variables are first centered in time as

$$\langle x\rangle_t \equiv z\ x^{n+1} + (1 - z)\ x^n; \tag{5.8}$$

z lies on the range $0.5 < z \leq 1.0$, with a typical value of 0.55. Thus the code uses the differential operators on time-averages rather than time-averages of the differential operators. With the exception of advected quantities (see below), simple spatial averages are used where necessary for centering; nonlinear terms are represented as products of averages rather than averages of products.

The advection scheme is described in detail in Section V.C including reasons for using $(-\delta m/\delta t)$ instead of $r^\mu u_{rel}\rho$, which is formally indentical, in Eqns. (5.3) - (5.6). Suffice it to say here that in Eqns. (5.3) and (5.4), $e + (E_0/\rho)$ and (E_0/ρ) are evaluated at interfaces using a monotonized interpolation scheme, as are u and $(F_0/c^2\rho)$ in Eqns. (5.5) and (5.6). These quantities are then advected according to the sign of $(-\delta m/\delta t)$ at the interface in Eqns. (5.3) and (5.4), and according to the sign of the zone-centered spatial average of $(-\delta m/\delta t)$ in Eqns. (5.5) and (5.6). In Eqn. (5.6), the zone centered density ρ_j is used at interface j (instead of an interface-centered spatial average) in order to avoid enlarging the stencil from 5 to 7 points. In non-relativistic flows this term is too small for the miscentering to have any practical consequences. In Eqns. (5.3) and (5.4) we use the spatial averages of $\Delta\xi$, ΔVol, E_0, P_0, and κ_F. All other quantities are automatically correctly centered on the staggered mesh.

B. On the Use of $\Delta\xi$, Δm, and $(\delta m/\delta t)$

At first glance, adopting both the gas density ρ and the integrated mass m as primary dependent variables, and solving the implicit Eqns. (5.1) and (5.2) for them may seem unnecessary and redundant, since ρ can easily be derived from m given that the radius r as a function of the zone index j is known. There are practical reasons, however, for doing so which are largely specific to the computation of gravitational collapse problems describing accretion onto compact objects.

Integrated mass is carried as a variable in addition to ρ, otherwise $m(r)$ would have to be evaluated by the spatial integration $m(r) \equiv \int_0^r \rho\ x^\mu dx$, which would destroy the band structure of the matrix to be inverted for the solution of the implicit equations. On the other hand, ρ is carried as a variable in addition to m, otherwise the differential mass of a zone $\Delta\xi \equiv \rho\Delta\text{Vol}$ would have to be approximated by Δm, and would therefore be susceptible to roundoff errors. For example, it is often the case that $\Delta\xi < 10^{-14}$ m over part of an accreting envelope in free fall. Differencing m to get the mass of a zone in such a region for use in the time derivatives of Eqns. (5.2) - (5.7) would lead to unacceptably large errors on a machine such as a CRAY-1, which has only 14 significant digits in a single-precision floating point word. Thus the spatial properties of gravitational flow require both ρ and m to be carried as dependent variables.

Likewise, the temporal properties of gravitational flows necessitate using the quantity $(-\delta m/\delta t)$ in the advection terms of Eqn. (5.3) - (5.6) instead of $r^{\mu}u_{rel}\rho$, which is algebraically, but not numerically, equivalent. We find this to be essential for numerical convergence whenever one is computing with a time-step such that the flux of mass through a zone in one timestep, δm, greatly exceeds the mass in the zone, $\Delta\xi$. For example, when integrating on the evolutionary timescale of an accreting compact object such as a protostellar core, it is not unusual to encounter $\delta m \simeq 10^{12}\Delta\xi$ in certain parts of the freefalling envelope. In such a circumstance, the physical solution to which one is attempting to converge to is quasi-steady, characterized by δm_j = contant to at least 12 decimal places from zone to zone.

Now consider what would happen if, during the Newton-Raphson iteration procedure, the net mass flux in a single zone were computed to only 11 decimal places of accuracy. Then, in a single iteration the mass of that zone could change by a factor of ten, and prevent numerical convergence. Our experience is that representing $(-\delta m/\delta t)$ by the product $r^{\mu}u_{rel}\rho$ in the difference equations leads to convergence difficulties of this nature. On the other hand, we find that using the corrections δm to the integrated mass directly avoids these problems.

C. Advection: Donor Cell, van Leer, PPM

The second terms in Eqns. (5.2) - (5.6) describe the advection of mass, total internal energy, radiation energy, gas momentum, and radiation momentum, respectively. In this section we describe in detail three different methods to perform the advection numerically. All three schemes are implicit, but differ in the order of accuracy with which they describe the solution. The first method, Donor Cell, is first-order accurate and requires a 3-point difference molecule. The second one, an implicit version of van Leer's second-order accurate, monotonic, upwind scheme requires a 5-point difference molecule. The third one, an implicit version of Woodward's (this book) third-order accurate, piecewise-parabolic method (PPM) requires a 7-point difference molecule for a fixed mesh, and a 9-point difference molecule for the adaptive grid of our staggered mesh difference scheme.

We begin by noticing that Eqns. (5.2) - (5.6) have one of two basic forms

$$\delta(q\Delta Vol) + \Delta(\delta t\, r^{\mu} u_{rel}\, q_a) + \text{ other terms} = 0, \quad \text{or} \qquad (5.9a)$$

$$\delta(q\Delta\xi) + \Delta(-\delta m\, q_a) + \text{ other terms} = 0, \qquad (5.9b)$$

The numerical reasons for the form used in Eqn. (5.9b) have been given in Section Vb. Here q is the zone average of any quantity such as ρ, u, E_0/ρ etc., and q_a is the corresponding advected value to be defined below. Equations (5.9) say that, apart from other terms, the total amound of "q" in a zone is updated in time by differencing the fluxes at the zone boundaries.

The volume flux $\delta t r^{\mu} u_{rel}$ and mass flux $-\delta m$ are time-centered according to

$$[\delta t\, r^{\mu} u_{rel}]_j \equiv \delta t\, \langle r_j^{\mu}\rangle_t \left(\langle u_j\rangle_t - \frac{r_j^{n+1} - r_j^{n}}{\delta t}\right), \quad \text{and} \qquad (5.10a)$$

$$[\delta m_j] \equiv m_j^{n+1} - m_j^{n} \qquad (5.10b)$$

where $\langle x_j\rangle_t$ is defined in Eqn. (5.8). To this point there is no difference in the algorithm for the three different methods. The methods differ only in the way in which q_a is determined.

Donor Cell is the most simple (and least accurate) algorithm of the three, and proceeds according to Fig. 2. In this scheme q_a is simply set to the upstream, time-averaged, zone-averaged value.

$$q_{a_j} = (0.5 + s_j)\,(z\, q_j^{n+1} + (1-z)\, q_j^{n}) + (0.5 - s_j)\,(z\, q_{j-1}^{n+1} + (1-z)\, q_{j-1}^{n}) \qquad (5.11)$$

Here s_j is a logical switch which chooses the upwind-sided value of the q_a in question according to

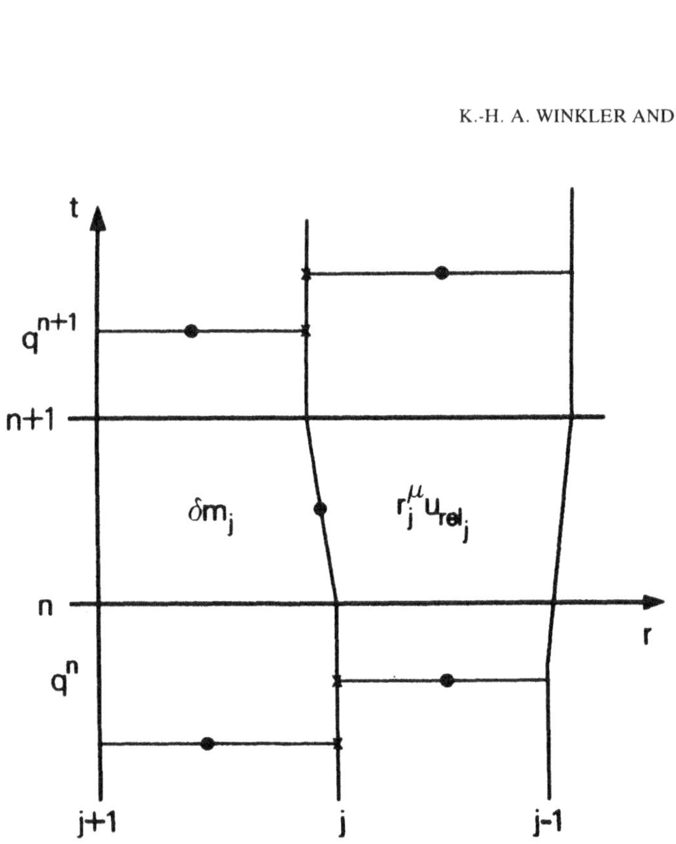

Fig. 2 Schematic Donor Cell advection.

$$s_j = \text{CVMGP}(+0.5,-0.5,u_{rel_j}) \quad \text{or} \tag{5.12a}$$

$$s_j = \text{CVMGP}(+0.5,-0.5,-\delta m_j) \tag{5.12b}$$

depending on the case.

When updating the face-centered quantities u and F_Q (cf. Fig. 1), $\Delta\xi$ and δm appearing in Eqn. (5.9b) are appropriately centered with respect to the "momentum" cell by straight arithmetic averages.

Going now from a piecewise constant subgrid representation of q to a monotonic piecewise linear description, we recover van Leer's (1979) prescription for the slopes, which is shown in Fig. 3. For this scheme we find

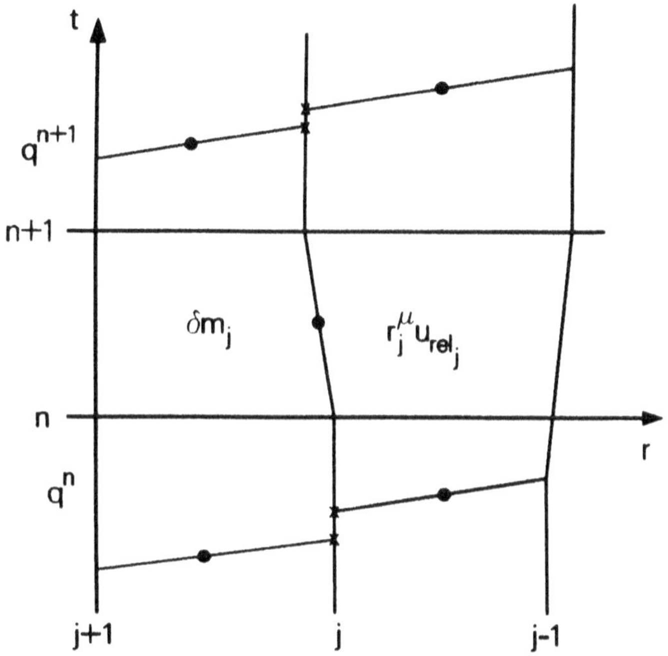

Fig. 3 Schematic of second-order, upwind Van Leer advection scheme.

$$dq_j = \begin{cases} \dfrac{c_3 \, \Delta q_j \Delta q_{j+1}}{\Delta q_j + \Delta q_{j+1}}, & \text{if } \Delta q_j \Delta q_{j+1} > 0 \\ 0 & \text{otherwise} \end{cases} \tag{5.13}$$

Here $\Delta q_j \equiv q_{j-1} - q_j$, and the constant c_3, which controls the order of the interpolation, is set to 2.

q_a is then computed by time-averaging old and new interpolated values evaluated on the upwind side of the zone interface in question, thus

$$q_{aj} = (0.5+s_j)\left[z\left(q_j^{n+1} + 0.5 dq_j^{n+1} \right) + (1-z)\left(q_j^n \quad + 0.5 dq_j^n \quad \right)\right]$$

$$+ (0.5-s_j)\left[z\left(q_{j-1}^{n+1}-0.5dq_{j-1}^{n+1}\right) + (1-z)\left(q_{j-1}^{n}-0.5dq_{j-1}^{n}\right)\right] \qquad (5.14)$$

Again, s_j is given by Eqn. (5.12a) or (5.12b). If we set the slopes to zero, that is, set $c_3 = 0$, we of course recover the Donor Cell scheme. One feature to notice is that the actual (non-equidistant) spacing of the mesh does not enter explicitly in van Leer's original prescription for the slopes.

Although van Leer's formula can be modified to account for a nonequidistant mesh, we won't go that route, but we will use instead the monotonic piecewise-parabolic description of PPM, illustrated in Fig. 4. The explicit version of the PPM advection scheme for nonequidistant zoning is described in Colella and Woodward (1983). For equidistant zoning an excellent description can be found in Woodward (this book). We have taken their formulae and adapted them to our implicit method.

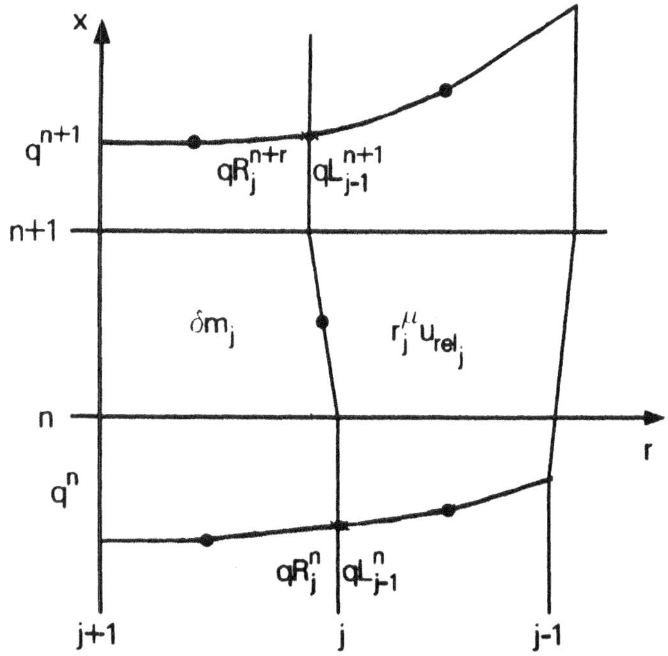

Fig. 4. Schematic of third order upwind PPM advection scheme.

We find

$$q_{a_j} = (0.5 + s_j)\left[z\, qR_j^{n+1} + (1-z)\, qR_j^n\right]$$
$$+ (0.5 - s_j)\left[z\, qL_{j-1}^{n+1} + (1-z)\, qL^n{}_{j-1}\right] \tag{5.15}$$

where qR_j and qL_j in the following way. We first evaluate Δq_j according to

$$\Delta q_j = \frac{\Delta\xi_j}{\Delta\xi_{j-1}+\Delta\xi_j+\Delta\xi_{j+1}}\left[\frac{2\Delta\xi_{j+1}+\Delta\xi_j}{\Delta\xi_{j-1}+\Delta\xi_j}(q_{j-1}-q_j) + \frac{2\Delta\xi_{j-1}+\Delta\xi_j}{\Delta\xi_{j+1}+\Delta\xi_j}(q_j-q_{j+1})\right] \tag{5.16}$$

and then monotonize it

$$\Delta_m q_j = \begin{cases} \min(|\Delta q_j|, 2|q_j-q_{j+1}|, 2|q_j-q_{j-1}|) \times \mathrm{sign}(\Delta q_j), & \text{if } (q_{j-1}-q_j)(q_j-q_{j+1}) > 0 \\ 0, & \text{otherwise} \end{cases} \tag{5.17}$$

Now we are prepared to compute provisional interface values

$$qI_j = q_j + \frac{\Delta\xi}{\Delta\xi_j+\Delta\xi_{j-1}}(q_{j-1} - q_j) + \frac{1}{\Delta\xi_{j-2}+\Delta\xi_{j-1}+\Delta\xi_j+\Delta\xi_{j+1}}$$
$$\times \left\{\frac{2\Delta\xi_{j-1}\Delta\xi_j}{\Delta\xi_j+\Delta\xi_{j+1}}\left[\frac{\Delta\xi_{j+1}+\Delta\xi_j}{2\Delta\xi_j+\Delta\xi_{j-1}} - \frac{\Delta\xi_{j-2}+\Delta\xi_{j-1}}{2\Delta\xi_{j-1}+\Delta\xi_j}\right](q_{j-1} - q_j)\right.$$
$$\left. - \Delta\xi_j\frac{\Delta\xi_{j+1}+\Delta\xi_j}{2\Delta\xi_j+\Delta\xi_{j-1}}\Delta_m q_{j-1} + \Delta\xi_{j-1}\frac{\Delta\xi_{j-1}+\Delta\xi_{j-2}}{\Delta\xi_j+2\Delta\xi_{j-1}}\Delta_m q_j\right\}, \tag{5.18}$$

at, respectively, the left and right interface of zone j:

$$qL_j = qI_{j+1}, \qquad qR_j = qI_j \tag{5.19}$$

These have to be monotonized one more time according to the prescription

$$qL_j = qR_j = q_j, \text{ if } (qR_j - q_j)(q_j - qL_j) \leqslant 0.,$$

$$qL_j = 3q_j - 2qR_j, \text{ if } +(qR_j - qL_j)\big(q_j - 0.5(qL_j + qR_j)\big) > \frac{(qR_j - qL_j)^2}{6},$$

$$qR_j = 3q_j - 2qL_j, \text{ if } -(qR_j - qL_j)\big(q_j - 0.5(qL_j + qR_j)\big) > \frac{(qR_j - qL_j)^2}{6} \qquad (5.20)$$

The values of qL_j and qR_j given by Eqns. (5.16) - (5.20) are the regular version of PPM. In case we want to get a sharper profile around a contact discontinuity, the interface values qL_j and qR_j have to be modified as follows before applying the monotonization procedure (5.20)

$$qL_j = qL_j(1-\eta_j) + \eta_j(q_{j+1} + 0.5\Delta_m q_{j+1})$$

$$qR_j = qR_j(1-\eta_j) + \eta_j(q_{j-1} - 0.5\Delta_m q_{j-1}) \qquad (5.21)$$

where

$$\eta_j = g\big[\max\{0.,\min(\eta_1\{\tilde{\eta}_j - \eta_2\},1.0)\}\big]$$

$$g(x) \equiv x^2(3 - 2x)$$

$$\eta_1 = \text{constant} = 20.0 \text{ (for example)}$$

$$\eta_2 = \text{constant} = \frac{1.}{20.0} \text{ (for example)} \qquad (5.22)$$

The function $g(x)$, not present in the explicit version of the method, is used to construct a continuously differentiable switch for the implicit scheme described here. $\tilde{\eta}_j$ itself is computed from the following relation

$$- \left(\frac{\Delta^2 q_{j-1} - \Delta^2 q_{j+1}}{\xi_{j-1} - \xi_{j+1}}\right)\left(\frac{(\xi_j - \xi_{j+1})^3 + (\xi_{j-1} - \xi_j)^3}{q_{j-1} - q_{j+1}}\right),$$

$$\tilde{\eta}_j = \begin{cases} \text{if } -\Delta^2 q_{j-1}\, \Delta^2 q_{j+1} > 0 \text{ and } \\ \qquad |q_{j-1} - q_{j+1}| - \varepsilon \min(|q_{j-1}|, |q_{j+1}|) > 0. \\ 0, \qquad \text{otherwise,} \end{cases} \tag{5.23}$$

where ε = constant = 0.01 (for example), and where $\Delta^2 q_j$ is given by

$$\Delta^2 q_j = \frac{1.}{\Delta\xi_{j+1} + \Delta\xi_j + \Delta\xi_{j-1}} \left[\frac{q_{j-1} - q_j}{\Delta\xi_{j-1} + \Delta\xi_j} - \frac{q_j - q_{j+1}}{\Delta\xi_j + \Delta\xi_{j+1}}\right] \tag{5.24}$$

We would like to point out that the need for the contact sharpener, given by Eqns. (5.21)-(5.24), is greatly reduced by our adaptive mesh. It would switch on only if a contact discontinuity was resolved by only 2 gridpoints.

The basic difference between the explicit versions of van Leer's and Woodward's advection scheme and our implicit advection scheme is the following: They track one characteristic of the flow, i.e. the streamline, in order to find that section of the flow which will be carried across the interface in one time-step. Then the subgrid distribution is averaged over that section at the old time level, and is taken as the time-average for the advected quantity to be used in the computation of the fluxes needed in the advection terms of Eqns. (5.2) to (5.6).

In contrast, we average directly the distribution of the different variables along the zone interfaces in time, i.e. along that line where they are actually needed in the computation of the fluxes.

$$\int_t^{t+\delta t} dt \int_{\partial V} q\, \underset{\sim}{u}_{rel} \cdot d\underset{\sim}{S}. \tag{5.25}$$

We keep the time-centering constant z as close as possible to 0.5 for reasons of accuracy, but typically no smaller than 0.55 for reasons of numerical stability. Of course, each flux has to be taken upwind with respect to the interface to assure numerical stability.

Examples of the various advection schemes are shown in Fig. 5. In Fig. 5a the initial density distribution is given for 100 gridpoints. We assume a constant advection velocity of u = 1.0 and solve the continuity equation for density. Figs. 5b, c, d, and e show the solution at t = 0.6 for an Eulerian grid distribution using, b) Donor Cell, c) van Leer, d) PPM, and e) PPM + contact sharpener for the advection term. Fig. 5f is computed with an adaptive mesh using only 20 gridpoints, and

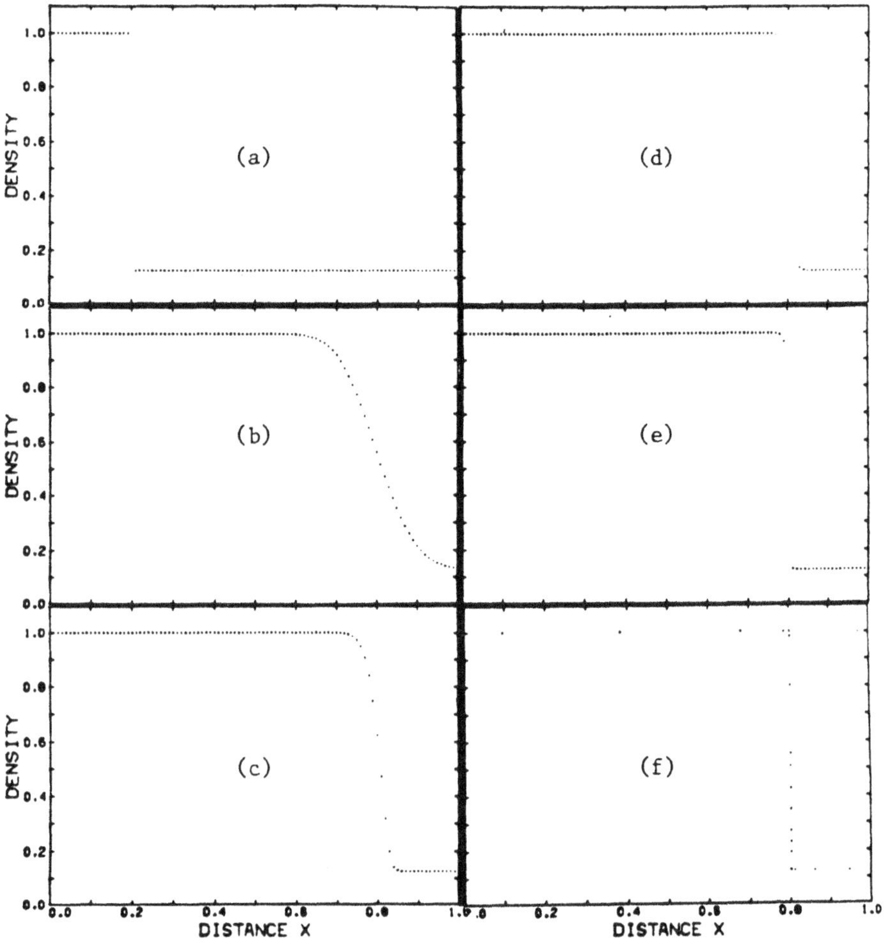

Fig. 5. Comparison of 5 advection schemes. (a) Initial distribution. (b) Donor cell. (c) van Leer. (d) PPM. (e) PPM + contact sharpener. (f) Adaptive mesh.

essentially looks the same for all the above described advection schemes. We conclude that it is advantageous to use the PPM advection scheme, especially in smooth parts of the flow, where of course the adaptive mesh is a lot coarser spaced than near discontinuities.

D. Solution Procedure and Control

The implicit finite-difference equations for the radiation hydrodynamics [Eqns. (5.1) - (5.6)] and the adaptive mesh [Eqns. (4.1) - (4.17)], together with closures, constitute a complete set of coupled, nonlinear algebraic equations for the vector of seven unknowns

$$\underset{\sim}{x}_j \equiv x_{i,j} \equiv (r_j, m_j, \rho_j, e_j, u_j, E_{0j}, F_{0j})$$

as a function of zone index j and time level n (we are not considering convection here). This nonlinear system is solved iteratively by the Newton-Raphson method. The functional dependence of the advection terms in Eqns. (5.2) - (5.6) on the neighboring zones, and the spatial structure of the adaptive mesh equations define a five-, seven-, or nine-point spatial stencil at both the old and new time levels. Thus, upon linearizing, the matrix equation for the solution corrections $\delta\underset{\sim}{x}_j \equiv \delta x_{i,j}$ will have a block 5-, 7-, or 9-diagonal structure, where each block is a (7 × 7) full submatrix. Exploiting the band struture of the complete matrix, we make an LU decomposition, and solve the linear system directly by block Gaussian elimination. We iterate over the entire system until numerical convergence, as defined below, is achieved.

The Newton-Raphson iterations are controlled in the following way. If ℓ is the iteration count and $\underset{\sim}{x}_j^\ell$ is the previous iterate, then the ℓth correction $\delta\underset{\sim}{x}_j^\ell$ is applied to

$$\underset{\sim}{x}_j^{\ell+1} = \underset{\sim}{x}_j^\ell + f_1^\ell \, \delta\underset{\sim}{x}_j^\ell , \tag{5.26}$$

where f_1^ℓ is the fraction by which the corrections are reduced if they are too large. f_1^ℓ is given by the formula

$$f_1^\ell = \min[1, \min_{i,j} (\frac{X_i^{max}}{X_{i,j}^\ell})]. \tag{5.27}$$

Here i is the variable index, allowed to range only over the unsigned variables. X_i^{max} is the desired maximum relative change for the variable x_i, and $X_{i,j}^\ell \equiv \delta x_{i,j}^\ell / x^\ell_{i,j}$ is the computed relative change. We furthermore limit f_1^ℓ by the condition, that anywhere in the mesh the spacing of a zone (however not the radial position of a grid-point) cannot change by more than 70% or 80%. In this way the Newton-Raphson iteration cannot jump into the other (non-physical) branch of solution space and the stabilizing operators S2 and S4 in Eqns. (4.4a) and (4.4b) can go highly nonlinear and avoid any tangling of the mesh.

We say that the solution has converged when

$$\max_{i,j} \left(X_{i,j}^\ell\right) < \varepsilon, \text{ and } \left(\ell_{min} \leq \ell \leq \ell_{max}\right). \tag{5.28}$$

Typical values are $\varepsilon = 10^{-5}$, $\ell_{min} = 2$, and $\ell_{max} = 20$. Under normal conditions convergence is reached in 4 to 6 iterations. We then apply a "boost" iteration, which boosts the final relative corrections into the range of 10^{-8} to 10^{-10}. As the Newton-Raphson method has second-order convergence, the equations themselves have also been solved to this level of accuracy.

As in any implicit system, the timestep is determined by accuracy, not stability, considerations. To compute the new timestep, we multiply the old timestep by some factor f_2, which is itself recomputed at each timestep. The reason for this approach is that in applications involving gravitational collapse or explosions, the characteristic timescales typically decrease as a function of time in some smooth way. It is when the timestep makes deviations from this systematic trend that changes show up in the iteration procedure. We therefore increase or decrease the timestep factor depending on whether the largest cumulative relative correction X is smaller or larger, respectively, than X_{max}, the maximum relative correction allowed. Thus, defining

$$X = \max_{i,j} \left(\sum_{\ell} \frac{|\delta x^{\ell}_{i,j}|}{x^{n}_{i,j}} \right) \tag{5.29}$$

and

$$\phi_1 = \begin{cases} \left(\frac{X_{max}}{X}\right)^{1/2} & \text{if } \left(\frac{X}{X_{max}}\right) < 1 \\ \left(\frac{X_{max}}{X}\right)^{2} & \text{if } \left(\frac{X}{X_{max}}\right) > 1, \end{cases} \tag{5.30}$$

then we set

$$\delta t^{n+1} = f_2^{n+1} \, \delta t^{n}, \tag{5.31}$$

where

$$f_2^{n+1} = \min\{ [f_2^{n} \max(0.1, \phi_2)]^{1/2}, \; 10 \}, \tag{5.32}$$

and

$$\phi_2 = \min[2, \min(\phi_1, 1.25 f_2^{n})]. \tag{5.33}$$

Typical applications are run with $0.05 < X_{max} < 0.15$. The nonlinearity in Eqn. (5.30) prevents f_2 from jumping between two extremes. The seemingly complicated procedure given above can be thought of as a nonlinear algorithm for computing time as a function of time index n.

If the solution fails to converge in the sense of Eqn. (5.28), the entire iteration procedure is repeated with a smaller timestep determined according to the following simple prescription:

$$\delta t_k = \max(0.1, \; 0.8 - 0.2k) \; \delta t_{K-1}; \quad (k = 1, \ldots, 6) \tag{5.34}$$

where k counts the number of attempts at finding the converged solution. The effect of this procedure is to cut back the timestep by a small factor for the first few attempts, but thereafter to cut back by factors of 0.1, for a maximum of six attempts. If the solution has failed to converge after six attempts, we say that it has diverged and we terminate the computation.

A characteristic of divergence is the unbounded growth of the corrections $\delta \underset{\sim}{x}^{\ell}$ during the Newton-Raphson iteration procedure. This situation usually arises when, for either physical or numerical reasons, the solution is not sufficiently steady with respect to the adaptive mesh. Examples of difficulties related to physics are: (1) The equation-of-state and opacity tables are not smooth enough; (2) the space-time resolution of the physical phenomena is not fine enough; and (3) the physical nonlinearities induce a sudden change in the character of the solution. Examples of difficulties related to numerics are: (4) the adaptive mesh is not stabilized or is not tracking the flow features well enough; and (5) roundoff errors and/or unsophisticated mathematical library subroutines prevent the solution of linear systems with enormous condition numbers.

We have found that the use of monotonic splines reduces difficulties associated with (1). Divergence caused by difficulties (2)-(4) can usually be overcome by suitable adjustment of input constants and weight factors. The use of better linear algebra routines for matrix operations (e.g. the robust implementation of LINPACK by T. L. Jordan at Los Alamos National Laboratory) eases some of the difficulties associated with (5), although roundoff errors seem to set a practical limit to what can be accomplished, especially regarding mesh refinement.

VI. PRINCIPLES OF CODING

Of course, the final step of algorithm development is the implementation of the numerical scheme in a computer program so that it can be applied successfully to problems of interest. Once such programs grow in size and complexity beyond certain limits particular care is required for their generation and maintenance. Otherwise, they quickly become unmanageable.

Implicit radiation hydrodynamic programs are certainly very complex, in particular because a large number of derivatives for evaluating the Jacobian matrix of the system of equations has to be computed reliably. Any error in these derivatives will eventually slow down or even prohibit the convergence of the Newton-Raphson iterations towards the desired solution of the numerical equations. However, it is often very difficult to find out whether the reason for poor convergence is due to an error in the Jacobian or due to a numerically unstable algorithm. Good software engineering concepts will hopefully ease the burden somewhat in dealing with these problems. Therefore, in this section we would like to discuss the software engineering principles which we used for the generation of WH80s and address in some detail how the computer program has been implemented in Fortran.

Let's start with a prioritized list of what we consider the 13 deadly mistakes of efficient Fortran programming:

1) No visible program structure
2) Unclear data structure
3) Poor flow of program control
4) Flow of program control mixed with physics or mathematics
5) Physical model equations mixed with mathematical algorithms
6) Special purpose fixes rather than general purpose coding
7) Non-standard Fortran
8) Confusing lay out
9) Bad notation
10) Non-invariant coding
11) Unautomated book keeping
12) Insufficient use of available memory
13) Not taking advantage of particular machine structure

Points 1) to 6) are concerned with the overall organization of the computer code, whereas 7) to 11) relate to the specifics of the implementation. Points 12) and 13) reflect the consideration of a particular hardware environment. In the following we will now explain these points in some detail.

In the same way as in real estate the value of a property is determined by the following three factors: LOCATION, LOCATION, LOCATION; the overall design, readability, verifiability, portability and maintainability of a Fortran code is determined by these 3 factors: STRUCTURE, STRUCTURE, STRUCTURE. By this we mean that the use of different levels of abstraction in creating a large computer code is of utmost importance. Using different levels of abstraction allows one to work on global aspects of the code without having to worry about specific details. For example, if we work on the solution procedure for partial differential equations (PDEs), we are not concerned at all with the details of a particular implementation of a specific set of non-linear PDEs. The concept of different levels of abstraction, intended to separate different hierarchal layers of the code, is complementary to the concept of modularity. By modularity we mean to group specific functions into specific subroutines, so that they are well separated and can be used (e.g. in parallel) as functional subunits. In WH80s the following 9 vertical program layers are used:

Level	Functionality	Example Program Names
1	Assemble the code	MAIN
2	Control the code	DRIVE1
3	Assemble the PDEs solution procedure	HYDRO1
4	Control the non-linear solution procedure	HIT001
5	Control the linear solution procedure	BK7D01
6	Assemble the PDEs and their Jacobian Matrix	EQTH01
7	Assemble a single equation	E06H01
8	Assemble a single term in an equation	E06A01
9	Assemble a single part of a term in an equation	IFACE4

The MAIN program's primary function is to select from a subroutine pool specific names and pass them to the main driver DRIVE1. DRIVE1 itself is actually written as an empty shell. It operates on generic subroutine names only and manages the overall execution of the entire program. It calls the generic physics modules, has a time-step and quality (i.e. accuracy) control, a dump, output, and documentation management, and furthermore knows how to recover from aborted iteration attempts of the non-linear solution procedure for the system of PDEs. However, DRIVE1 is only as powerful as the specific subroutines are whose names where passed from MAIN. In the same manner as MAIN selects the overall program flow, HYDRO1 draws names from another subroutine pool and chooses the specific linear and non-linear solution procedures, e.g. HIT001 and BK7D01. It also passes the name of a specific system of non-linear PDEs, e.g. EQTH01, to the solution procedures. Subroutine EQTH01 itself supervises the computation of the equations and their Jacobian Matrix of the system of PDEs. For this purpose it draws on another large pool of subroutines where e.g. specific difference methods for advection, diffusion, viscosity, and other operators are implemented. EQTH01 also adds to the system of physical equations the adaptive grid equation.

A clear data structure should accompany the above outlined program structure. Modern languages, such as ADA, automatically separate the program from the data structure completely. That way, execution of the program cannot lethally damage the data structure. Unfortunately, the absence of this feature in Fortran can lead to disastrous consequences. The problem typically arises when unintentionally memory is overwritten somewhere during the execution of the code, which in turn later forces a floating point error. Isolating the memory-damaging piece of program in the code can be a frustrating and time-consuming experience. However, within the Fortran programming context certain precautions can be taken to minimize this type of problem. An ounce of prevention is worth a pound of cure. We proceed by making extensive use of precompilers. All the coding for global storage is kept in one place and stuck into the actual subroutines through include statements. All do-loop index boundaries and the names of storage locations passed through calling sequences are extremely carefully checked before attempting compilation of the program. Seemingly little things like this actually make a big difference in the overall coding efficiency.

Points 4) and 5) of our prioritized list (to separate program control, mathematical solution procedure, and a particular numerical difference representation of a specific set of physical equations) are self-explanatory. Point 6) simply indicates to try to avoid either too specific or too general Fortran statements. In the first case, one is not flexible enough to respond to different but similar situations, in the second case one can get lost in time-consuming generalities. Obviously experience and intuition will have a profound impact on the personal judgement in these issues.

Porting codes between different machines is becoming increasingly more important as improved networking capabilities allow one to switch between powerful workstations and supercomputers to utilize their unique advantages. One way to ease this mode of operation is to rely as much as possible on standard Fortran77 with implicit type definitions. When using machines with different word lengths one can be forced to go from an 8 byte single precision word to an 8 byte double precision word. In this case, exclusive use of generic names for the intrinsic functions facilitates an automatic conversion. Furthermore, one should not pass real constants in the calling sequences of subroutines or functions in order to avoid conversion problems. Particular attention has to be paid to Cray vector merge functions because they do not discriminate between integer and real arguments. This becomes a problem when on the target machine the real word used differs in length from the integer word. Use of different names on the Cray through inline function definitions for the two cases does not prohibit vectorization and allows for automatic conversion when coupled with appropriate function definitions on the target machine.

In the following, we will give an example of the coding of a mathematical algorithm completely decoupled from the program control and the physical model. We show how the solution procedure for a linear system with a block-7-diagonal matrix can be derived and implemented in Fortran in a very compact way. Let's think of $Ep3_j, \ldots, Em3_j$ as the Jacobian derivatives of the system of nonlinear equations equ_j and of x_j as the solution to the linear system (or as the corrections to the previous approximation of the non-linear system).

```
        Ep3   Ep2   Ep1   E00   Em1   Em2   Em3
                          equ
                           j
    --- + --- + --- + --- + --- + --- + --- + ---
        x     x     x     x     x     x     x
       j+3   j+2   j+1    j    j-1   j-2   j-3
```

Then the sytem of linear equations is

$$\begin{aligned} &Ep3_j*x_{j+3} + Ep2_j*x_{j+2} + Ep1_j*x_{j+1} + E00_j*x_j \\ +\, &Em3_j*x_{j-3} + Em2_j*x_{j-2} + Em1_j*x_{j-1} + equ_j = 0. \end{aligned}$$

If we make the following Ansatz

$$x_j = hv_j + H1_j*x_{j+1} + H2_j*x_{j+2} + H3_j*x_{j+3}$$

and shift the grid index j by -1, -2, and -3

$$\begin{aligned} x_{j-1} &= hv_{j-1} + H1_{j-1}*x_j + H2_{j-1}*x_{j+1} + H3_{j-1}*x_{j+2} \\ x_{j-2} &= hv_{j-2} + H1_{j-2}*x_{j-1} + H2_{j-2}*x_j + H3_{j-2}*x_{j+1} \\ x_{j-3} &= hv_{j-3} + H1_{j-3}*x_{j-2} + H2_{j-3}*x_{j-1} + H3_{j-3}*x_j \end{aligned}$$

we find after substitution the following system of equations

$$S0_j*x_j = sv_j + S1_j*x_{j+1} + S2_j*x_{j+2} + S3_j*x_{j+3}, \quad \text{where}$$

$$\begin{aligned} S0_j = -\,[\, &Em3_j*[\, H1_{j-3}*[\, H1_{j-2}*H1_{j-1}+H2_{j-2}\,]+H2_{j-3}*H1_{j-1}+H3_{j-3}\,] \\ +&Em2_j*[\, H1_{j-2}*H1_{j-1}+H2_{j-2}\,] \\ +&Em1_j*H1_{j-1} \\ +&E00_j\,], \end{aligned}$$

$$\begin{aligned} sv_j = [\, &Em3_j*[\, H1_{j-3}*[\, H1_{j-2}*hv_{j-1}+hv_{j-2}\,]+H2_{j-3}*hv_{j-1}+hv_{j-3}\,] \\ +&Em2_j*[\, H1_{j-2}*hv_{j-1}+hv_{j-2}\,] \\ +&Em1_j*hv_{j-1} \\ +&equ_j\,], \end{aligned}$$

$$\begin{aligned} S1_j = [\, &Em3_j*[\, H1_{j-3}*[\, H1_{j-2}*H2_{j-1}+H3_{j-2}\,]+H2_{j-3}*H2_{j-1}\,] \\ +&Em2_j*[\, H1_{j-2}*H2_{j-1}+H3_{j-2}\,] \\ +&Em1_j*H2_{j-1} \\ +&Ep1_j\,], \end{aligned}$$

$$\begin{aligned} S2_j = [\, &Em3_j*[\, H1_{j-3}*[\, H1_{j-2}*H3_{j-1}\,]+H2_{j-3}*H3_{j-1}\,] \\ +&Em2_j*[\, H1_{j-2}*H3_{j-1}\,] \\ +&Em1_j*H3_{j-1} \\ +&Ep2_j\,], \text{ and} \end{aligned}$$

$$S3_j = Ep3_j.$$

Comparison with the Ansatz results in the determining equations for the Ansatz vector sv_j and the Ansatz matrices $H1_j$, $H2_j$, $H3_j$. We find

$$S0_j * hv_j = sv_j$$
$$S0_j * H1_j = S1_j$$
$$S0_j * H2_j = S2_j$$
$$S0_j * H3_j = S3_j$$

or in more compact notation

$$S0_j * (hv_j, H1_j, H2_j, H3_j) = (sv_j, S1_j, S2_j, S3_j).$$

By making extensive use of linear algebra library subroutines the just derived algorithm can be implemented in Fortran in the following way. First (not shown below) we set 3 scratch zones out side the normal matrix to zero so that application of the full algorithm everywhere will recover the boundary conditions correctly. Then we compute the start addresses m0 through m4 for the vector sv_j and the matrices $S0_j$ through $S3_j$ in the temporary matrix SHEN.

```
C --- START ADDR. IN SHEN : M0=LHS, M1=VCT, M2=1-MTX, M3=2-MTX, M4=3-MTX
      M0   = 1
      M1   = 1 + N
      M2   = 1 + N + 1
      M3   = 1 + N + 1 + N
      M4   = 1 + N + 1 + N + N
C --- M5   = 1 + N + 1 + N + N + N  NEXT FREE START ADDR. ---
C --- NUMBER OF POINTS IN SUB-MATRIX ---
      NQ   = N ** 2
C --- NUMBER OF RHS (RIGHT HAND SIDES) OF THE LINEAR SYSTEM ---
      NRHS = 1 + 3*N
C
C
C---- DOWN: COMPUTE LINEAR SYSTEMS DETERMINING HVJ, H1J,...,H3J --------
C
C
      DO  100  J=NS,NE
C --- STORE MULTIPLE USED TERM IN T2
      CALL MXM  ( H1  (1, 1,J-2),N, H1  (1, 1,J-1),N, T2,            N )
      CALL SAXPY( NQ,+ONE,          H2  (1, 1,J-2),1, T2,            1 )
C --- COMPUTE MATRIX OF LEFT HAND SIDE OF LINEAR SYSTEM
      CALL SCOPY( NQ,               E00 (1, 1,J  ),1, SHEN(1,M0   ),1 )
      CALL MXM  ( EM1 (1, 1,J  ),N, H1  (1, 1,J-1),N, T1,            N )
      CALL SAXPY( NQ,+ONE,          T1,            1, SHEN(1,M0   ),1 )
      CALL MXM  ( EM2 (1, 1,J  ),N, T2,            N, T1,            N )
      CALL SAXPY( NQ,+ONE,          T1,            1, SHEN(1,M0   ),1 )
      CALL SCOPY( NQ,               H3  (1, 1,J-3),1, T1,            1 )
      CALL MXM  ( H2  (1, 1,J-3),N, H1  (1, 1,J-1),N, T3,            N )
      CALL SAXPY( NQ,+ONE,          T3,            1, T1,            1 )
```

```
      CALL MXM  ( H1  (1, 1,J-3),N, T2,            N, T3,            N )
      CALL SAXPY( NQ,+ONE,          T3,            1, T1,            1 )
      CALL MXM  ( EM3 (1, 1,J  ),N, T1,            N, T3,            N )
      CALL SAXPY( NQ,+ONE,          T3,            1, SHEN(1,M0    ),1 )
      CALL SSCAL( NQ,-ONE,                            SHEN(1,M0    ),1 )
C
C --- STORE MULTIPLE USED TERM IN T2
      CALL MXM  ( H1  (1, 1,J-2),N, HV  (1   ,J-1),N, T2,            1 )
      CALL SAXPY( N ,+ONE,          HV  (1   ,J-2),1, T2,            1 )
C --- COMPUTE VECTOR OF RHS OF LINEAR SYSTEM
      CALL SCOPY( N ,               EQU (1,   J  ),1, SHEN(1,M1    ),1 )
      CALL MXM  ( EM1 (1, 1,J  ),N, HV  (1   ,J-1),N, T1,            1 )
      CALL SAXPY( N ,+ONE,          T1,            1, SHEN(1,M1    ),1 )
      CALL MXM  ( EM2 (1, 1,J  ),N, T2,            N, T1,            1 )
      CALL SAXPY( N ,+ONE,          T1,            1, SHEN(1,M1    ),1 )
      CALL SCOPY( N ,               HV  (1,   J-3),1, T1,            1 )
      CALL MXM  ( H2  (1, 1,J-3),N, HV  (1   ,J-1),N, T3,            1 )
      CALL SAXPY( N ,+ONE,          T3,            1, T1,            1 )
      CALL MXM  ( H1  (1, 1,J-3),N, T2,            N, T3,            1 )
      CALL SAXPY( N ,+ONE,          T3,            1, T1,            1 )
      CALL MXM  ( EM3 (1, 1,J  ),N, T1,            N, T3,            1 )
      CALL SAXPY( N ,+ONE,          T3,            1, SHEN(1,M1    ),1 )
C
C --- STORE MULTIPLE USED TERM IN T2
      CALL MXM  ( H1  (1, 1,J-2),N, H2  (1, 1,J-1),N, T2,            N )
      CALL SAXPY( NQ,+ONE,          H3  (1, 1,J-2),1, T2,            1 )
C --- COMPUTE 1-MTX  OF RHS OF LINEAR SYSTEM
      CALL SCOPY( NQ,               EP1 (1, 1,J  ),1, SHEN(1,M2    ),1 )
      CALL MXM  ( EM1 (1, 1,J  ),N, H2  (1, 1,J-1),N, T1,            N )
      CALL SAXPY( NQ,+ONE,          T1,            1, SHEN(1,M2    ),1 )
      CALL MXM  ( EM2 (1, 1,J  ),N, T2,            N, T1,            N )
      CALL SAXPY( NQ,+ONE,          T1,            1, SHEN(1,M2    ),1 )
      CALL MXM  ( H2  (1, 1,J-3),N, H2  (1, 1,J-1),N, T1,            N )
      CALL MXM  ( H1  (1, 1,J-3),N, T2,            N, T3,            N )
      CALL SAXPY( NQ,+ONE,          T3,            1, T1,            1 )
      CALL MXM  ( EM3 (1, 1,J  ),N, T1,            N, T3,            N )
      CALL SAXPY( NQ,+ONE,          T3,            1, SHEN(1,M2    ),1 )
C
C --- STORE MULTIPLE USED TERM IN T2
      CALL MXM  ( H1  (1, 1,J-2),N, H3  (1, 1,J-1),N, T2,            N )
C --- COMPUTE 2-MTX  OF RHS OF LINEAR SYSTEM
      CALL SCOPY( NQ,               EP2 (1, 1,J  ),1, SHEN(1,M3    ),1 )
      CALL MXM  ( EM1 (1, 1,J  ),N, H3  (1, 1,J-1),N, T1,            N )
      CALL SAXPY( NQ,+ONE,          T1,            1, SHEN(1,M3    ),1 )
      CALL MXM  ( EM2 (1, 1,J  ),N, T2,            N, T1,            N )
      CALL SAXPY( NQ,+ONE,          T1,            1, SHEN(1,M3    ),1 )
      CALL MXM  ( H2  (1, 1,J-3),N, H3  (1, 1,J-1),N, T1,            N )
      CALL MXM  ( H1  (1, 1,J-3),N, T2,            N, T3,            N )
      CALL SAXPY( NQ,+ONE,          T3,            1, T1,            1 )
      CALL MXM  ( EM3 (1, 1,J  ),N, T1,            N, T3,            N )
```

```
      CALL SAXPY( NQ,+ONE,         T3,            1, SHEN(1,M3    ),1 )
C
C --- COMPUTE 3-MTX  OF RHS OF LINEAR SYSTEM
      CALL SCOPY( NQ,              EP3 (1, 1,J  ),1, SHEN(1,M4    ),1 )
C
C --- SOLVE LINEAR SYSTEM WITH  NRHS  RIGHT-HAND-SIDES.
      CALL LINSLV ( SHEN, N, N, IPVT, NRHS )
      CALL SCOPY  ( N , SHEN(1,M1),1, HV  (1    ,J  ),1 )
      CALL SCOPY  ( NQ, SHEN(1,M2),1, H1  (1, 1,J  ),1 )
      CALL SCOPY  ( NQ, SHEN(1,M3),1, H2  (1, 1,J  ),1 )
      CALL SCOPY  ( NQ, SHEN(1,M4),1, H3  (1, 1,J  ),1 )
100   CONTINUE
```

N is the number of unknowns per gridpoint, ONE=1. is locally defined, MXM multiplies two matrices, and LINSLV solves a full linear system. The other subroutines are basic linear algebra routines and can be found e.g. in LINPACK. In the following piece of Fortran SOLU is the desired solution vector x_j.

```
C
C
C---- UP: RESOLVING THE ANSATZ -----------------------------------------
C
C
      DO  200  K=1,N
      SOLU (K,NE+2) = 0.
      SOLU (K,NE+1) = 0.
200   CONTINUE
      CALL SCOPY( N ,              HV  (1    ,NE  ),1, SOLU(1, NE),1 )
C
      DO  201  J=NE-1,NS,-1
      CALL SCOPY( N ,              HV  (1    ,J   ),1, SOLU(1, J ),1 )
      CALL MXM  ( H1  (1, 1,J  ),N, SOLU(1,   J+1 ),N, T1,          1 )
      CALL SAXPY( N ,+ONE,         T1,               1, SOLU(1, J ),1 )
      CALL MXM  ( H2  (1, 1,J  ),N, SOLU(1,   J+2 ),N, T1,          1 )
      CALL SAXPY( N ,+ONE,         T1,               1, SOLU(1, J ),1 )
      CALL MXM  ( H3  (1, 1,J  ),N, SOLU(1,   J+3 ),N, T1,          1 )
      CALL SAXPY( N ,+ONE,         T1,               1, SOLU(1, J ),1 )
201   CONTINUE
```

In the above example, the lay out of the code on the page is an important consideration. We try to organize the code in such a way that patterns can be easily recognized by the human brain eye system. In particular, we arrange that similar things in different lines appear in the same column. This allows one to rapidly scan many lines of coding in the vertical direction for verification. Breaking up mnemonic names by blanks allows for further structuring of the code in the vertical direction. It also enhances the mnemonic character of the name itself. The following code example shows a piece from the computation of the physical equations and their Jacobian derivatives. Here we choose the upstream side in an advection computation.

```
C---- VOLUME - FLOW --- CHOOSE    :  UPSTREAM SIDE FOR ADVECTION --------
C---- ACTUALLY ADVECTED QUANTITY :  AA  +  DERIVATIVES  AA D XX --------
C
      DO  400  J=NS,NE+1
      X FLOW   (J) =  RMY(J) * V (J) * DTIME
      S            =  CVMGP( +HALF, -HALF, X FLOW(J) )
C
      HP1          =  ( .5 + S )
      HM1          =  ( .5 - S )
C
      AA       (J) =  HP1 * AQ R      (J)  +  HM1 * AQ L      (J-1)
C
      AA D QM3 (J) =                          HM1 * AQ L QM2 (J-1)
      AA D QM2 (J) =  HP1 * AQ R QM2 (J)  +  HM1 * AQ L QM1 (J-1)
      AA D QM1 (J) =  HP1 * AQ R QM1 (J)  +  HM1 * AQ L Q00 (J-1)
      AA D Q00 (J) =  HP1 * AQ R Q00 (J)  +  HM1 * AQ L QP1 (J-1)
      AA D QP1 (J) =  HP1 * AQ R QP1 (J)  +  HM1 * AQ L QP2 (J-1)
      AA D QP2 (J) =  HP1 * AQ R QP2 (J)
C
      AA D VM3 (J) =                          HM1 * AQ L VM2 (J-1)
      AA D VM2 (J) =  HP1 * AQ R VM2 (J)  +  HM1 * AQ L VM1 (J-1)
      AA D VM1 (J) =  HP1 * AQ R VM1 (J)  +  HM1 * AQ L V00 (J-1)
      AA D V00 (J) =  HP1 * AQ R V00 (J)  +  HM1 * AQ L VP1 (J-1)
      AA D VP1 (J) =  HP1 * AQ R VP1 (J)  +  HM1 * AQ L VP2 (J-1)
      AA D VP2 (J) =  HP1 * AQ R VP2 (J)
  400 CONTINUE
C
C---- TRANSFORM  AA D QXX -> AA D DXX  --------------------------------
C---- TRANSFORM  AA D VXX -> AA D RXX  --------------------------------
C
      DO  500  J=NS,NE+1
      AA D DM3 (J) =  AA D QM3 (J) * AQ D00   (J-3)
      AA D DM2 (J) =  AA D QM2 (J) * AQ D00   (J-2)
      AA D DM1 (J) =  AA D QM1 (J) * AQ D00   (J-1)
      AA D D00 (J) =  AA D Q00 (J) * AQ D00   (J  )
      AA D DP1 (J) =  AA D QP1 (J) * AQ D00   (J+1)
      AA D DP2 (J) =  AA D QP2 (J) * AQ D00   (J+2)
C
      AA D RM3 (J) =  AA D VM3 (J) * DV R00 N (J-3)
      AA D RM2 (J) =  AA D VM2 (J) * DV R00 N (J-2)
     1              + AA D VM3 (J) * DV RP1 N (J-3)
      AA D RM1 (J) =  AA D VM1 (J) * DV R00 N (J-1)
     1              + AA D VM2 (J) * DV RP1 N (J-2)
```

```
44        AA D R00 (J) =   AA D V00 (J) * DV R00 N (J  )
45       1               + AA D VM1 (J) * DV RP1 N (J-1)
46        AA D RP1 (J) =   AA D VP1 (J) * DV R00 N (J+1)
47       1               + AA D V00 (J) * DV RP1 N (J  )
48        AA D RP2 (J) =   AA D VP2 (J) * DV R00 N (J+2)
49       1               + AA D VP1 (J) * DV RP1 N (J+1)
50        AA D RP3 (J) =   AA D VP2 (J) * DV RP1 N (J+2)
51 500    CONTINUE
```

Concerning the notation used, S stands for switch, AA for actually advected quantity, AQ for advected quantity, i.e the variable we want to advect, AQ R for advected quantity from right, AQ L for advected quantity from left, AA D QM3 through AA D QP2 for the derivative of AA with respect to the generically advected quantity at gridpoints j-3 through j+2. AA D VM3 through AA D VP2 stand for the Jacobian derivatives of AA with respect to the volume coordinate at gridpoints j-3 through j+2. In do-loop 500 we finally apply the chain rule one last time to calculate the Jacobian derivatives with respect to the primary dependent variables D for density and R for radius. Taking line 50 as an example we will now see that the variable names are not only chosen for mnemonic reasons but that they also carry algorithmic significance for the computation of the chain rule. In this line we compute the derivative of AA(J) with respect to R(J+3). We first compute AA D VP2(J), i.e. the derivative of AA(J) with respect to the volume coordinate V(J+2). Then we compute the derivative of the volume coordinate V(J+2) with respect to the relative position R(J+1), i.e. a total of R(J+3), and we are done. Put another way, in order to get the derivative with respect to (J+3) we add the 2 from VP2 to the J from VP2(J) and find that we have to take the derivative of volume V at (J+2) with respect to the radius RP1(j+2)=R(1+J+2)=R(J+3). Automatic application of this little trick in the computation of chain rules greatly reduces the effort of coding and therefore reduces significantly the introduction of errors into the actual Fortran coding. Not having to debug in the first place is by far the best debugging aid.

Another concept we used in the above example is the one of invariant coding. By that we mean to not exploit the specific character of an object in contrast to its generic character. In the above example we could have expressed e.g. early in the loop the derivatives with respect to density D and not with respect to the generic name AQ. In doing so we would have hard wired a specific functional dependency and would not have been able to use this particular piece of code for the computation of all advected quantities of the problem at hand. To get a better understanding of the concept of invariant coding let´s look at the following example.

```
*DECK MSAVE1
      SUBROUTINE  MSAVE1
C
C  WH80S/HWSUBS/HWSUBS1.MSAVE1  <-------------- WRITE, READ OF DUMP-FILE
C
C======================================================================
C
*CALL PID
*CALL COM
*CALL EQV
C
C======================================================================
C
      WRITE ( IWMOD )                          I2,P2, SLO,SLN
      RETURN
      ENTRY   MGET1
      READ  ( IRMOD )                          I2,P2, SLO,SLN
      RETURN
      ENTRY   MGET2 ( L END )
      READ  ( IRMOD , ERR=100 , END=101 )      I2,P2, SLO,SLN
      L END = .FALSE.
      RETURN
100   STOP
101   L END = .TRUE.
      RETURN
      END
```

This particular subroutine is free of any specifics because it only references generic names like SLO for solution old, or SLN for solution new, I2 for integer parameters, and P2 for real parameters. We are free to change the specific names and storage locations in PID, COM, and EQV as often as we want, without having to rewrite a single line of the subroutine MSAVE1, which writes and reads the dump files of the entire computation. MSAVE1 also represents another example of automatic book keeping through the extensive use of precompilers.

One final example of invariant coding is shown in the following, where we have separated the definition of the adaptive mesh stabilizing operators from the objects they operate on. We can change the specifics of an operator, without having to worry about the rest of the subroutine as long as we don't increase the functional dependence of the expression. We also would like to point out that in the inline definition of the derivative operators use of the symbolic names of the operators themselves, greatly simplifies the expressions. Of course, it also is another example of invariant coding.

```
*DECK RF0L01
      SUBROUTINE  RF0L01 ( NS , NE )
C
C  WH80S/HWSUBS/HWSUBS1.RF0L01  <------------- RADIAL FUNCTION LOGARITHM
C
C=======================================================================
C
*CALL PID
*CALL DIM001
*CALL DIM002
C
C-----------------------------------------------------------------------
C
      DIMENSION  DLR    (IGR),  DLR RJ (IGR),  DLR RP (IGR)
C
*CALL COM
*CALL COM002
*CALL EQV
*CALL EQV001
*CALL EQV002
C
C-----------------------------------------------------------------------
C
      EQUIVALENCE  (SC2(1,001),DLR    (1))
     1           ,  (SC2(1,002),DLR RJ (1))
     2           ,  (SC2(1,003),DLR RP (1))
C
C-----------------------------------------------------------------------
C
C --- X := R N
C
      F1    (       XJ,XP    ) =                 (XJ-XP)    /(XJ+XP)
      F1 XJ (       XJ,XP    ) = (+1. - F1 (      XJ,XP   ))/(XJ+XP)
      F1 XP (       XJ,XP    ) = (-1. - F1 (      XJ,XP   ))/(XJ+XP)
C
C --- X := DLR := D LN R
C
      F2    (    XM,XJ,XP    ) =                  XJ**2      /(XM*XP)
      F2 XM (    XM,XJ,XP    ) = - 1. * F2 (   XM,XJ,XP    ) / XM
      F2 XJ (    XM,XJ,XP    ) = + 2. * F2 (   XM,XJ,XP    ) / XJ
      F2 XP (    XM,XJ,XP    ) = - 1. * F2 (   XM,XJ,XP    ) / XP
C
      F3    ( XI,XM,XJ,XP,XL ) =               (XI*XJ**6*XL) /(XM*XP)**4
      F3 XI ( XI,XM,XJ,XP,XL ) = + 1. * F3 (XI,XM,XJ,XP,XL) / XI
      F3 XM ( XI,XM,XJ,XP,XL ) = - 4. * F3 (XI,XM,XJ,XP,XL) / XM
      F3 XJ ( XI,XM,XJ,XP,XL ) = + 6. * F3 (XI,XM,XJ,XP,XL) / XJ
      F3 XP ( XI,XM,XJ,XP,XL ) = - 4. * F3 (XI,XM,XJ,XP,XL) / XP
      F3 XL ( XI,XM,XJ,XP,XL ) = + 1. * F3 (XI,XM,XJ,XP,XL) / XL
C
      F4    ( XJ ) =                ( XJ       / G LR MAX )**N LR POW
```

```
      F4 XJ ( XJ ) = + N LR POW * F4 ( XJ )      / XJ
C
      F5    ( XJ ) =                    ( G LR MIN / XJ       )**N LR POW
      F5 XJ ( XJ ) = - N LR POW * F5 ( XJ )      / XJ
C
      F6    ( XM,XJ,XP ) =                       ( (XM-XP) /  XJ )
      F7    ( XM,XJ,XP ) =                       ( (XM-XP) /  XJ )**2
      F7 XM ( XM,XJ,XP ) = + 2.       * F6 (XM,XJ,XP) /  XJ
      F7 XJ ( XM,XJ,XP ) = - 2.       * F7 (XM,XJ,XP) /  XJ
      F7 XP ( XM,XJ,XP ) = - 2.       * F6 (XM,XJ,XP) /  XJ
C
      F8    ( XM,XJ,XP ) =   ( (XM-XP)/(XJ*G LDLR MAX) ) ** (N LR POW-1)
      F9    ( XM,XJ,XP ) =   ( (XM-XP)/(XJ*G LDLR MAX) ) **  N LR POW
      F9 XM ( XM,XJ,XP ) = + N LR POW * F8 (XM,XJ,XP) / (XJ*G LDLR MAX)
      F9 XJ ( XM,XJ,XP ) = - N LR POW * F9 (XM,XJ,XP) /  XJ
      F9 XP ( XM,XJ,XP ) = - N LR POW * F8 (XM,XJ,XP) / (XJ*G LDLR MAX)
C
C
C======================================================================
C==== R F : RADIAL FUNCTION <=================== GRID STABILIZING TERMS
C
      DO  100  J=NS-3,NE+3
      RF        (J) = 0.
      RF D RM2 (J) = 0.
      RF D RM1 (J) = 0.
      RF D R00 (J) = 0.
      RF D RP1 (J) = 0.
      RF D RP2 (J) = 0.
      RF D RP3 (J) = 0.
  100 CONTINUE
C
C
C============================================== STABILIZING OPERATOR S0
C==== DLR(J) = D LN R(J) ==============================================
C
      DO  200  J=NS-2,NE+1
      DLR    (J) = F1    (                   R N(J), R N(J+1)           )
      DLR RJ (J) = F1 XJ (                   R N(J), R N(J+1)           )
      DLR RP (J) = F1 XP (                   R N(J), R N(J+1)           )
C
      RF        (J) = RF        (J) + W LR 0 * DLR    (J)
      RF D R00 (J) = RF D R00 (J) + W LR 0 * DLR RJ (J)
      RF D RP1 (J) = RF D RP1 (J) + W LR 0 * DLR RP (J)
  200 CONTINUE
C
C============================================== STABILIZING OPERATOR S2
C==== + DLR(J)**2 / ( DLR(J-1) * DLR(J+1) ) ===========================
C
      DO  300  J=NS,NE-1
      RZ       = F2    (          DLR(J-1), DLR(J), DLR(J+1)            )
```

```
      RZ LR M = F2 XM (           DLR(J-1), DLR(J), DLR(J+1)           )
      RZ LR J = F2 XJ (           DLR(J-1), DLR(J), DLR(J+1)           )
      RZ LR P = F2 XP (           DLR(J-1), DLR(J), DLR(J+1)           )
C
      RF       (J) = RF       (J) + W LR 2 * RZ
      RF D RM1 (J) = RF D RM1 (J) + W LR 2 * RZ LR M * DLR RJ (J-1)
      RF D R00 (J) = RF D R00 (J) + W LR 2 * RZ LR M * DLR RP (J-1)
     1                            + W LR 2 * RZ LR J * DLR RJ (J  )
      RF D RP1 (J) = RF D RP1 (J) + W LR 2 * RZ LR J * DLR RP (J  )
     1                            + W LR 2 * RZ LR P * DLR RJ (J+1)
      RF D RP2 (J) = RF D RP2 (J) + W LR 2 * RZ LR P * DLR RP (J+1)
  300 CONTINUE
C
C================================================= STABILIZING OPERATOR S4
C==== + ( DLR(J-2)*DLR(J)**6*DLR(J+2) ) / ( DLR(J-1)*DLR(J+1) )**4 =====
C
      DO  400  J=NS,NE-1
      RZ       = F3    ( DLR(J-2), DLR(J-1), DLR(J), DLR(J+1), DLR(J+2) )
      RZ LR I = F3 XI ( DLR(J-2), DLR(J-1), DLR(J), DLR(J+1), DLR(J+2) )
      RZ LR M = F3 XM ( DLR(J-2), DLR(J-1), DLR(J), DLR(J+1), DLR(J+2) )
      RZ LR J = F3 XJ ( DLR(J-2), DLR(J-1), DLR(J), DLR(J+1), DLR(J+2) )
      RZ LR P = F3 XP ( DLR(J-2), DLR(J-1), DLR(J), DLR(J+1), DLR(J+2) )
      RZ LR L = F3 XL ( DLR(J-2), DLR(J-1), DLR(J), DLR(J+1), DLR(J+2) )
C
      RF       (J) = RF       (J) + W LR 4 * RZ
      RF D RM2 (J) = RF D RM2 (J) + W LR 4 * RZ LR I * DLR RJ (J-2)
      RF D RM1 (J) = RF D RM1 (J) + W LR 4 * RZ LR I * DLR RP (J-2)
     1                            + W LR 4 * RZ LR M * DLR RJ (J-1)
      RF D R00 (J) = RF D R00 (J) + W LR 4 * RZ LR M * DLR RP (J-1)
     1                            + W LR 4 * RZ LR J * DLR RJ (J  )
      RF D RP1 (J) = RF D RP1 (J) + W LR 4 * RZ LR J * DLR RP (J  )
     1                            + W LR 4 * RZ LR P * DLR RJ (J+1)
      RF D RP2 (J) = RF D RP2 (J) + W LR 4 * RZ LR P * DLR RP (J+1)
     1                            + W LR 4 * RZ LR L * DLR RJ (J+2)
      RF D RP3 (J) = RF D RP3 (J) + W LR 4 * RZ LR L * DLR RP (J+2)
  400 CONTINUE
C
C======================================================= MAXIMUM OPERATOR
C==== + ( DLR(J) / G LR MAX )**N LR POW ==============================
C
      DO  500  J=NS,NE-1
      RZ       = F4    ( DLR(J) )
      RZ RJ    = F4 XJ ( DLR(J) ) * DLR RJ (J)
      RZ RP    = F4 XJ ( DLR(J) ) * DLR RP (J)
C
      RF       (J) = RF       (J) + W LR MAX * RZ
      RF D R00 (J) = RF D R00 (J) + W LR MAX * RZ RJ
      RF D RP1 (J) = RF D RP1 (J) + W LR MAX * RZ RP
  500 CONTINUE
C
```

```
C=================================================== MINIMUM OPERATOR
C==== - ( G LR MIN / DLR(J) )**N LR POW =============================
C
      DO  600  J=NS,NE-1
      RZ      = F5    ( DLR(J) )
      RZ RJ   = F5 XJ ( DLR(J) ) * DLR RJ (J)
      RZ RP   = F5 XJ ( DLR(J) ) * DLR RP (J)
C
      RF       (J) = RF       (J) - W LR MIN * RZ
      RF D R00 (J) = RF D R00 (J) - W LR MIN * RZ RJ
      RF D RP1 (J) = RF D RP1 (J) - W LR MIN * RZ RP
600   CONTINUE
C
C===================================================== RATIO OPERATOR
C==== + ( (DLR(J-1)-DLR(J+1)) /  DLR(J) )**2 ==========================
C
      DO  700  J=NS,NE-1
      RZ       = F7    (           DLR(J-1), DLR(J), DLR(J+1)           )
      RZ LR M = F7 XM (           DLR(J-1), DLR(J), DLR(J+1)           )
      RZ LR J = F7 XJ (           DLR(J-1), DLR(J), DLR(J+1)           )
      RZ LR P = F7 XP (           DLR(J-1), DLR(J), DLR(J+1)           )
C
      RF       (J) = RF       (J) + W LDLR     * RZ
      RF D RM1 (J) = RF D RM1 (J) + W LDLR     * RZ LR M * DLR RJ (J-1)
      RF D R00 (J) = RF D R00 (J) + W LDLR     * RZ LR M * DLR RP (J-1)
     1                            + W LDLR     * RZ LR J * DLR RJ (J  )
      RF D RP1 (J) = RF D RP1 (J) + W LDLR     * RZ LR J * DLR RP (J  )
     1                            + W LDLR     * RZ LR P * DLR RJ (J+1)
      RF D RP2 (J) = RF D RP2 (J) + W LDLR     * RZ LR P * DLR RP (J+1)
700   CONTINUE
C
C============================================ MAXIMUM RATIO OPERATOR
C==== + ( (DLR(J-1)-DLR(J+1)) / (DLR(J)*G LDLR MAX) )**N LR POW ========
C
      DO  800  J=NS,NE-1
      RZ       = F9    (           DLR(J-1), DLR(J), DLR(J+1)           )
      RZ LR M = F9 XM (           DLR(J-1), DLR(J), DLR(J+1)           )
      RZ LR J = F9 XJ (           DLR(J-1), DLR(J), DLR(J+1)           )
      RZ LR P = F9 XP (           DLR(J-1), DLR(J), DLR(J+1)           )
C
      RF       (J) = RF       (J) + W LDLR MAX * RZ
      RF D RM1 (J) = RF D RM1 (J) + W LDLR MAX * RZ LR M * DLR RJ (J-1)
      RF D R00 (J) = RF D R00 (J) + W LDLR MAX * RZ LR M * DLR RP (J-1)
     1                            + W LDLR MAX * RZ LR J * DLR RJ (J  )
      RF D RP1 (J) = RF D RP1 (J) + W LDLR MAX * RZ LR J * DLR RP (J  )
     1                            + W LDLR MAX * RZ LR P * DLR RJ (J+1)
      RF D RP2 (J) = RF D RP2 (J) + W LDLR MAX * RZ LR P * DLR RP (J+1)
800   CONTINUE
      RETURN
      END
```

Concluding this section, we would like to remark briefly on memory usage and vectorization. They are the only places in our considerations were economic issues of a particular hardware configuration enter. Excessive memory usage can greatly speed up a code (one computes only once often-used results and stores them for future reference) and is certainly justified by the ready availability of massive amounts of cheap memory. This certainly is a dramatic deviation from the past when the opposite was true. Following the above given examples of structured coding naturally leads to the exploitation of machine specific advantages, in particular vectorization and parallel processing. In fact, the beginnings of this coding style date back to a time when we first wanted to take advantage of the vectorization on a Cray.

VII. EXAMPLES

As an illustration of the use of implicit adaptive-mesh techniques on a problem of some familiarity, we consider the interaction of two blast waves as studied by Woodward and Colella (1983) and discussed by Woodward in this book. For this computation, radiation and self-gravity are ignored, so that the flow is described by the equation of ideal gas dynamics. Planar geometry and a gamma-law equation of state $P = (\gamma-1)\rho e$ is assumed, with $\gamma = 7/5$.

The initial conditions consist of a uniform, stationary gas of unit density on the domain $0 \leq x \leq 1$. The pressure distribution consists of three constant states: $P = 10^3$ for $x < 0.1$; $P = 10^2$ for $x > 0.9$; and $P = 10^{-2}$ for $0.1 \leq x \leq 0.9$. Reflecting boundary conditions are assumed at either end of the domain. The high pressure regions near the boundaries drive two strong shock waves into the interior of the domain, which eventually collide and produce a new contact discontinuity, as displayed in Fig. 10 of Woodward in these proceedings. This wave interaction is difficult to compute correctly on a fixed mesh, but poses no difficulties on an adaptive mesh. Four snapshots of this evolution are shown in Fig. 6.

Although only 400 zones were used in this computation, the adaptivity of the mesh places a large number of extremely fine zones in the fluid discontinuities, making them extremely sharp in physical space (cf. Figs. 6a-c), nevertheless well-resolved in numerical space. Fig. 6d shows the density distribution at $t = 0.038$, after the shocks have interacted and separated. This plot can be compared directly to Fig. 11 in Woodward's contribution to this volume, and is seen to agree quite well to his PPM result using 1200 Eulerian zones.

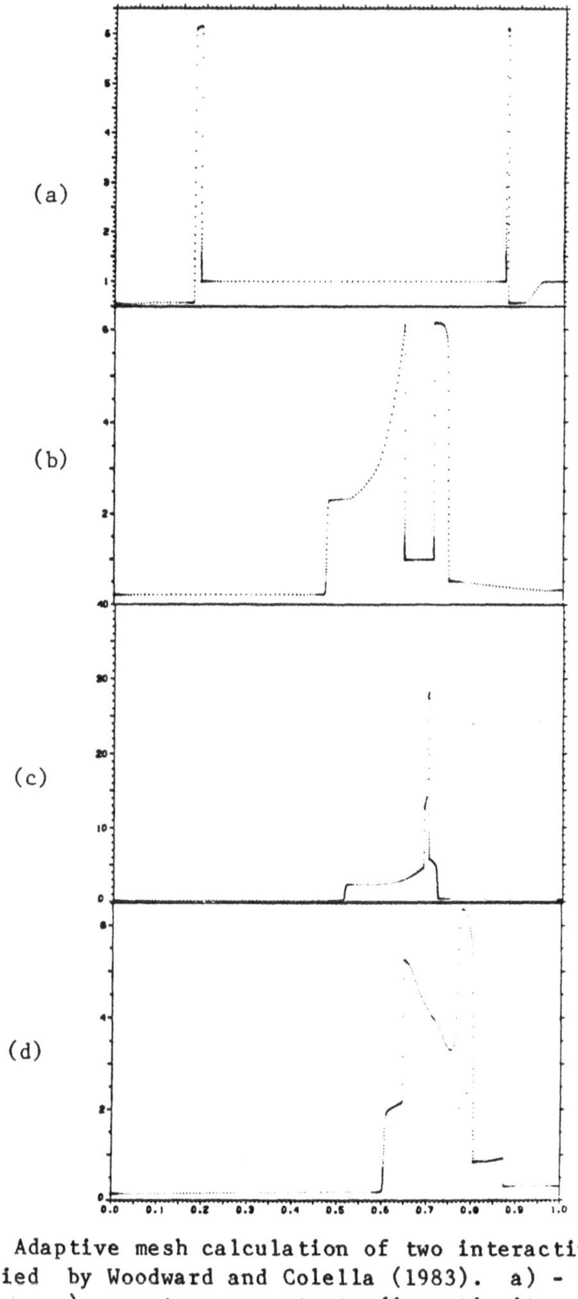

Fig. 6. Adaptive mesh calculation of two interacting blastwaves, as studied by Woodward and Colella (1983). a) - b) approaching blastwaves; c) moment on contact discontinuity formation; d) solution at t = 0.038 after shocks have interacted and separated. Density is plotted.

Although excellent results were obtained with WH80s on this challenging test problem using fewer zones than the sophisticated PPM algorithm, the computational cost is much higher (10-100) because of time-consuming matrix inversions dictated by an implicit formulation. In modeling ideal gas flows, fluid discontinuities need not be resolved - they have no internal structure. Explicit PPM is far more cost-effective than WH80s in such cases. Radiation, self-gravity and realistic material properties introduce new length- and time scales, however, which render explicit, fixed-grid methods entirely inadequate in many cases, as the following example illustrates. The problem involves the formation of a supercritical radiating shock front in an accreting protostellar envelope.

Consider a homogeneous isothermal interstellar cloud of $100M_\odot$, embedded in a sphere of radius 10^{18} cm, in radiative equilibrium with an imposed external radiation field having a radiation temperature of 60 K. In this initially transparent configuration gravitational forces overpower pressure gradients. There ensues a gravitational collapse, which is finally stopped when the density in the central part of the cloud becomes high enough to form an opaque core and trap radiation. From that time onward, the energy gained from the fall into the gravitational potential well is no longer directly radiated away but goes instead into heating the gas. Pressure then builds up until a quasi-hydrostatic core forms, surrounded by an accretion shock.

The total luminosity of such an object has two contributions. Part of it is generated in a slow contraction of the quasi-hydrostatic core. The remainder is emitted directly from the accretion shock within which the kinetic energy of the infalling material is transformed primarily into outgoing radiation.

In order to get a clean test problem for our radiation hydrodynamics method, we avoided complicated equation-of-state and opacity tables and used instead the ideal equation of state, with $\gamma = 5/3$ and the following opacity law:

$$\chi = 10^{\{2\tanh[(T-5\times10^4)/(2.5\times10^4)]-2\}} \tag{7.1}$$

Although this highly idealized problem does not provide a realistic model of the star-forming process, it nevertheless gives us the flavor of the correct physical phenomena with their multiple length and time scales. In this problem the range of relevant length scales varies by eight orders of magnitude (= width of radiative shock front / cloud radius) and the range of relevant time scales varies by nine orders of magnitude

(= core dynamical timescale / Kelvin-Helmholtz contraction timescale).

The flow structure in the vicinity of the accretion shock at 1.236 initial freefall times is displayed in Fig. 7. 300 zones were used in the computational domain spanning $10^8 \leqslant r(\mathrm{cm}) \leqslant 10^{18}$. The shock transition in velocity shown in Fig. 7a is at the same position where the gas temperature overshoots by a factor of 2.56, as shown in Fig. 7b. This optically thin postshock temperature spike is characteristic of a supercritical radiating shock front, first recognized by Heaslet and Balwin (1963).

An immense luminosity is generated within the accretion shock. As shown in Fig. 7c, it rises from about $600 L_\odot$, emitted by the contracting core, to about $1400 L_\odot$ in the transparent envelope, where it remains constant. In these early phases the evolution, the optical depth, shown in Fig. 7d, increases primarily as a result of increases in density, shown in Fig. 7h, and not as a result of changes in opacity.

Figures 7e,f,g show that except in the shock itself, gas pressure is dominant over radiation and viscous pressure. The radiation never plays a significant role in these early phases of evolution, and viscous pressure (from the artificial viscosity) is important only within the shock front where it sets the spatial scale of the shock's structure, which is fully resolved with approximately 30 zones of size $\Delta r = 5 \times 10^8$ cm.

The solution shown in Fig. 7 was computed with the following adaptive-mesh weight factors: $W_{r10} = 1$, $W_{r12} = 60$, $W_{r1max} = 1$, $n = 4$, $\Delta r1_{max} = 0.006$, $W_{r1min} = 1$, $\Delta r1_{min} = 10^{-8}$, $W_m = 30$, $W_{m1} = 3$, $W_{q1} = 15$, $W_{e1} = 12$, $W_{\rho 1} = 6$, $W_{P1} = 6$, $W_{E0 1} = 9$, $W_{\chi\rho 1} = 30$, and $q_{floor} = 1$. All other parameters were set to zero. The parameters used for artificial viscosity were a fixed relative length $l = 10^{-3} r$ and constants $c_1 = 0.3$, $c_2 = 3$.

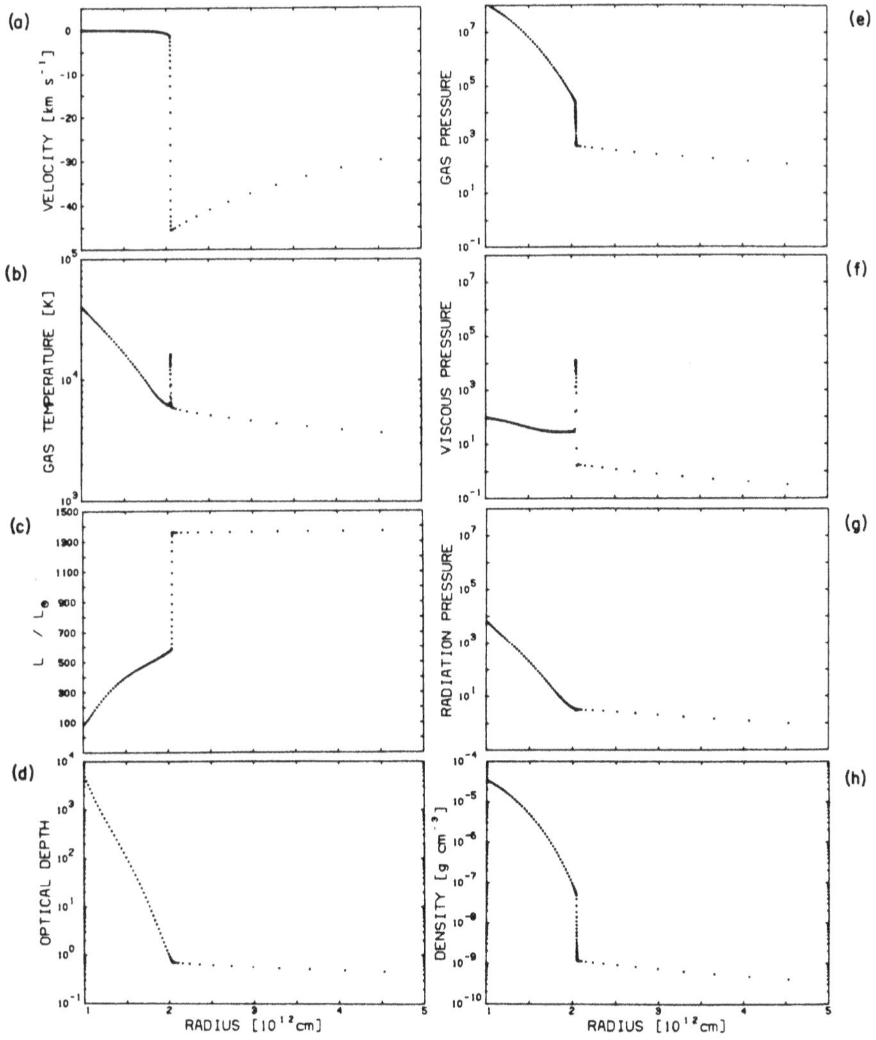

Fig. 7. Medium scale flow structure in vicinity of supercritical shock. (a) Velocity. (b) Gas temperature showing optically thin postshock temperature spike. (c) Luminosity variation showing "discontinuity" from emission by radiating shock. (d) Optical depth. (e) Gas pressure showing discontinuity across shock. (f) Viscous pressure showing spike within shock front. (g) Radiation pressure. (h) Density showing jump of factor of 60 across shock.

REFERENCES

1.) Böhm-Vitense, E., 1958, Zs. f. Ap., 46, p. 108.

2.) Cameron, I. G., 1966, J. Comput. Phys., 1, p. 1.

3.) Castor, J. I., 1972, Ap. J., 178, p. 779.

4.) Castor, J. I., Davis, C. G., and Davison, D. K., 1977, "Dynamic Zoning Within a Lagrangean Mesh by Use of DYN, a Stellar Pulsation Code", Los Alamos Sci. Lab. Report LA-6644, Univ. of Calif., Los Alamos. 5.) Colella, P., and Woodward, P., 1983, J. Comput. Phys., 54, p. 174.

6.) Cox, J. P., and Giuli, R. T., 1968, "Principles of Stellar Structure," New York: Gordon and Breach.

7.) Freeman, B. E., Hauser, L. E., Palmer, J. T., Pickard, S. O., Simmons, G. M., Williston, D. G., and Zerkle, J. E., 1968, "The VERA Code: A One-Dimensional Radiative Hydrodynamics Program", Vol. I, Defense Atomic Support Agency Rep. No. 2135, Systems, Science, and Software, Inc., La Jolla, Calif.

8.) Heaslet, M. A., and Baldwin, B. S., 1963, Phys. Fluids, 6, p. 781. 9.) Huebner, W. F., Merts, A. L., Magee Jr., N. H., and Argo, M. F., 1977, "Astrophysical Opacity Library," Los Alamos Scientific Lab. Rep. LA-7934-MS.

10.) Landau, L. D., and Lifshitz, E. M., 1959, "Fluid Mechanics," Pergamon Press, Oxford and New York.

11.) Mihalas, D., and Mihalas, B. W ., 1984, "Foundations of Radiation Hydrodynamics," Oxford Univ. Press, London and New York.

12.) Noh, W. F., 1976, "Numerical Methods in Hydrodynamics Calculations," Lawrence Livermore Lab. Rep. UCRL-52112, Univ. of Calif., Livermore.

13.) Tscharnuter, W., and Winkler, K.-H. A., 1979, Comput. Phys. Commun, 18, p. 171.

14.) von Neumann, J., and Richtmyer, R. D., 1950, J. Appl. Phys., 21, p. 232.

15.) Winkler, K.-H. A., 1976, A numerical procedure for the calculation of nonsteady spherical shock fronts with radiation. Ph.D. Thesis, University of Gottingen. Issued as technical report MPI-Astro 90 by the Max Planck Institut fur Physik und Astrophysik, Munich; English translation: Lawrence Livermore Lab. Rep. UCLR-Trans 11206, Univ. of Calif., Livermore, 1977.

16.) Winkler, K.-H. A., and Newman, M. J., 1980a, Ap. J., 236, p. 201.

17.) Winkler, K.-H. A., and Newman, M. J., 1980b, Ap. J., 238, p. 311.

18.) Winkler, K.-H. A., Mihalas, D., and Norman, M. L., 1985, Comput. Phys. Commun., 36, p. 121.

19.) Winkler, K.-H. A., Norman, M. L., and Mihalas, D., 1984, J. Quant. Spectrosc. Radiat. Transfer, 31, No. 6, p. 473.

20.) Winkler, K.-H. A., Norman, M. L., and Newman, M. J., 1984, Physica, 12D, p. 408.

21.) Woodward, P., and Colella, P., 1983, J. Comput. Phys., 54, p. 115.

NUMERICAL SOLUTION OF THE EQUATION OF RADIATION TRANSFER

H.W. Yorke

Universitätssternwarte
Geismarlandstraße 11
D-3400 Göttingen

ABSTRACT: We discuss the numerical solution of the equation of radiation transfer in an astrophysical context. Rather than try to review this complex subject, we will be principally concerned with including radiation transfer effects in numerical "model" calculations. We will not be trying to present an exact mathematical theory. Instead, emphasis will be on practical considerations with regard to storage and time requirements, stability and convergence properties and numerical accuracy.

We demonstrate how to obtain the "exact" numerical solution of the continuum and line transfer equations in spherical symmetry. We also show how to calculate the expected appearance of models for which the source function and extinction coefficient are known. Examples of such calculations are presented and discussed.

1. INTRODUCTION

With the exception of neutrinos, cosmic rays, meteorites and the non-detection of gravitational radiation, all observable information on the universe outside our solar system comes to us in the form of electromagnetic radiation. These photons are the witnesses of astrophysical events. Furthermore, the absorption and emission of photons often influence these events. Therefore our theoretical understanding of astrophysical processes is intimately related to our understanding radiation transfer.

The following is not meant to be a comprehensive review of the subject of numerical radiation transfer, but rather a guide for including radiation transfer effects in numerical astrophysi-

K.-H. A. Winkler and M. L. Norman (eds.), Astrophysical Radiation Hydrodynamics, 141–179.

cal calculations. In section 2 we shall deal with the transfer of radiation from one point to the next when the absorption and emission are dominated by atomic or molecular lines and/or continuum processes. In section 3 we shall discuss a numerical method for obtaining an "exact" solution in the case of continuum radiation in spherically symmetric geometry. Here we show how to couple this exact solution with the hydrodynamic equations of motion. In section 4 the line transfer problem for molecular hydrogen in a spherically symmetric non-stationary envelope is solved numerically. Finally, in section 5 we speculate on the feasibility of solving the equation of continuum radiation transfer in axially symmetric geometry.

1.1 Radiation intensity

The radiation intensity I is defined as the electromagnetic energy δE which passes through an area dA (with normal $\hat{n}$) at the point $\underline{x}$ from a direction $\hat{s}$ within the solid angle $d\Omega$, in the frequency interval $\nu, \nu + d\nu$ during the time dt

$$\delta E = I(\underline{x}, \hat{s}, \nu, t)\hat{n}\cdot\hat{s}\ dA\ d\Omega\ d\nu\ dt \tag{1.1}$$

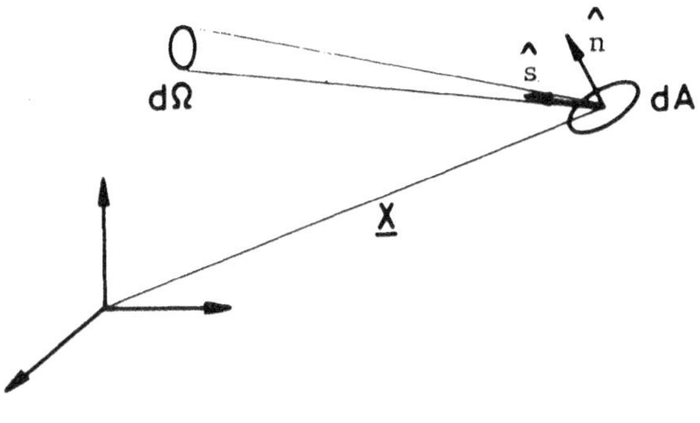

Figure 1. Definition of the radiation intensity

In general I is a function of seven variables, three spatial, two directional, frequency and time. It is related to the density of photons in phase space. Numerically, one can see that in order to store the intensity in a computer with reasonable numerical resolution (about 100 grid points for each variable), one needs to store 10^{12} numbers for each time step. Obviously this number is too large to handle with present day computers. Simplifications have to be made. Usually not all previous time steps need to be stored; only the most recent time step is necessary even when doing time-dependent problems. For the time-independent case in 1-D (slab or spherical symmetry), 2-D (axial symmetry) or 3-D (no symmetry assumed) the radiation intensity is a function of three, five or six variables, respectively.

	1-D	2-D	3-D
$I =$	$I_\nu(r,\theta)$	$I_\nu(r,\theta,\theta',\phi')$	$I_\nu(r,\theta,\phi,\theta',\phi')$

1.2 Equation of radiation transfer

The equation of the transfer of radiation (time dependent but non-relativistic, see i.e. (1,2)) can be written:

$$1/c\ \partial I_\nu/\partial t + \partial I_\nu/\partial s = -\kappa_\nu^e I_\nu + \epsilon_\nu \qquad (1.2)$$

where s is the length along the line of sight in the direction considered, κ_ν^e is the extinction coefficient and the source term ϵ_ν is the sum of the intrinsic emissivity of the medium and the radiation scattered into the line of sight out of all other directions. With the definition of the source function $S_\nu = \epsilon_\nu/\kappa_\nu^e$ the time-independent form of (1.2) may be written:

$$dI_\nu/ds = -\kappa_\nu^e (I_\nu - S_\nu) \qquad (1.3)$$

which has the formal solution (integrating from point (a) to (b))

$$I_\nu^b = I_\nu^a \exp\{-\int_a^b \kappa_\nu^e ds + \int_a^b S_\nu \exp\{-\int_s^b \kappa_\nu^e ds'\}\kappa_\nu^e ds \qquad (1.4)$$

This formal solution must be applied to all directions and each point in space. Difficulties in using the formal solution (1.4) for physical problems arise when either S_ν or κ_ν^e depends on the radiation intensity I_ν. In this case equation (1.4) must be solved iteratively together with the equations for S_ν and κ_ν^e.

1.3 Moment equations

In an orthogonal coordinate system the operator d/ds can be written:

$$d/ds = \hat{\alpha}\cdot\underline{\nabla} = \alpha_i \partial_i \qquad (1.5)$$

(summation of repeated indices i = 1,2,3 is implied), where $\hat{\alpha}$ is a unit vector in the direction considered (α_i are the direction cosines.) and $\partial_i = \partial/\partial x_i$ is the partial differential operator. It is useful to introduce definitions of the moments of the radiation field:

$$J = 1/4\pi \int_{4\pi} {}_i\ I\ d\Omega \qquad \text{0-th moment} \qquad (1.6a)$$

$$H_i = 1/4\pi \int_{4\pi} \alpha_i\ I\ d\Omega \qquad \text{1-st moment} \qquad (1.6b)$$

$$K_{ij} = 1/4\pi\int_{4\pi} \alpha_i\alpha_j \; I \; d\Omega \quad \text{2-nd moment} \tag{1.6c}$$

where $\int_{4\pi}()d\Omega$ means an integration of () over all directions. Note that not all nine components of the 2-nd moment tensor are linearly independent of the other moments. From inspection of equation (1.6c), $K_{ij} = K_{ji}$ and the trace of K_{ij} is $K_{ii} = J$ due to the identity $\alpha_i\alpha_i = 1$. Thus in general we have one 0-th moment, three 1-st moment and five 2-nd moment components. Frequency subscripts have been dropped for clarity.

If we consider an orthogonal coordinate system with axes fixed in space (i.e. a cartesian coordinate system) and integrate equation (1.2) over all directions, we obtain:

$$1/c \; \partial J/\partial t + \partial_i H_i + \frac{1}{4\pi}\int_{4\pi} \kappa^e \; (I - S) \; d\Omega = 0 \tag{1.7}$$

Multiplying equation (1.2) by α_j and integrating over all directions leads to

$$1/c \; \partial H_j/\partial t + \partial_i K_{ij} + \frac{1}{4\pi}\int_{4\pi} \alpha_j \kappa^e \; (I - S) \; d\Omega = 0 \tag{1.8}$$

Restricting ourselves for the moment to isotropic scattering, absorption and emission of thermal continuum radiation we can write

$$\kappa^e S = \kappa^a \; B + \kappa^s \; J \tag{1.9}$$

where κ^a is the absorption coefficient and κ^s the scattering coefficient, and $\kappa^a + \kappa^s = \kappa^e$. B(T) is Planck's radiation function. (1.9) allows us to simplify the moment equations (1.7) and (1.8) to:

$$1/c \; \partial J/\partial t + \partial_i H_i + \kappa^a \; (J - B) = 0 \tag{1.10a}$$

$$1/c \; \partial H_j/\partial t + \partial_i K_{ij} + \kappa^e \; H_j = 0 \tag{1.10b}$$

where, as before, repeated indices imply summation.

It is interesting to consider other coordinate systems, the axes of which are not fixed in space. In spherical polar coordinates, for example, the axes $\hat{e}_r$, $\hat{e}_\theta$ and $\hat{e}_\phi$ change as a function of the position (r,θ,ϕ) considered. Expressing the line of sight direction $\hat{s}$ in terms of this coordinate system, one could specify θ' as the angle between $\hat{e}_r$ and $\hat{s}$ and ϕ' as the angle between $\hat{e}_\theta$ and the projection of $\hat{s}$ onto the plane normal to $\hat{e}_r$. In this case the moments are defined by:

$$J\ (r,\theta,\phi) = 1/4\pi \int_0^{2\pi}\int_0^{\pi} I(r,\theta,\phi,\theta',\phi')\ \sin\theta' d\theta' d\phi' \qquad (1.11a)$$

$$H_i\ (r,\theta,\phi) = 1/4\pi \iint \alpha_i\ I(r,\theta,\phi,\theta',\phi')\ \sin\theta' d\theta' d\phi' \qquad (1.11b)$$

$$K_{ij}(r,\theta,\phi) = 1/4\pi \iint \alpha_i \alpha_j I(r,\theta,\phi,\theta',\phi')\ \sin\theta' d\theta' d\phi' \qquad (1.11c)$$

where α_i are given by

$$\alpha_\theta = \sin\theta'\ \cos\phi' \qquad (1.12a)$$

$$\alpha_\phi = \sin\theta'\ \sin\phi' \qquad (1.12b)$$

$$\alpha_r = \cos\theta' \qquad (1.12c)$$

The corresponding moment equations are:

$$\frac{1}{c}\frac{\partial J}{\partial t} + \frac{\partial H_r}{\partial r} + \frac{2H_r}{r} + \frac{1}{r\sin\theta}\frac{\partial}{\partial\theta}(\sin\theta\ H_\theta) + \frac{1}{r\sin\theta}\frac{\partial H_\phi}{\partial\phi} =$$

$$= -\kappa^a(J-B) \qquad (1.13)$$

$$\frac{1}{c}\frac{\partial H_r}{\partial t} + \frac{\partial K_{rr}}{\partial r} + \frac{3K_{rr} - J}{r} + \frac{1}{r\sin\theta}\frac{\partial}{\partial\theta}(\sin\theta\ K_{r\theta}) +$$

$$+ \frac{1}{r\sin\theta}\frac{\partial K_{r\phi}}{\partial\phi} = -\kappa^e H_r \qquad (1.14a)$$

$$\frac{1}{c}\frac{\partial H_\theta}{\partial t} + \frac{\partial K_{r\theta}}{\partial r} + \frac{3K_{r\theta}}{r} + \frac{1}{r\sin\theta}\frac{\partial}{\partial\theta}(\sin\theta\ K_{\theta\theta}) +$$

$$+ \frac{\cos\theta}{r\sin\theta}\left[K_{rr} + K_{\theta\theta} - J\right] + \frac{1}{r\sin\theta}\frac{\partial K_{\theta\phi}}{\partial\phi} = -\kappa^e H_\theta \qquad (1.14b)$$

$$\frac{1}{c}\frac{\partial H_\phi}{\partial t} + \frac{\partial K_{r\phi}}{\partial r} + \frac{3K_{r\phi}}{r} + \frac{1}{r\sin\theta}\frac{\partial}{\partial\theta}(\sin\theta\ K_{\theta\phi}) +$$

$$+ \frac{\cos\theta}{r\sin\theta}K_{\theta\phi} + \frac{1}{r\sin\theta}\frac{\partial}{\partial\phi}\left[K_{rr} + K_{\theta\theta} - J\right] = -\kappa^e H_\phi \qquad (1.14c)$$

For completeness we also give the corresponding equations for cylindrical coordinates with axes $(\hat{e}_\rho, \hat{e}_\phi, \hat{e}_z)$. Defining θ' as the angle between $\hat{e}_z$ and $\hat{s}$ and ϕ' as the angle between $\hat{e}_\rho$ and the projection of $\hat{s}$ onto the plane normal to $\hat{e}_z$, the direction cosines are given by:

$$\alpha_\theta = \sin\theta' \cos\phi' \tag{1.15a}$$

$$\alpha_\phi = \sin\theta' \sin\phi' \tag{1.15b}$$

$$\alpha_z = \cos\theta' \tag{1.15c}$$

The corresponding moment equations are:

$$\frac{1}{c}\frac{\partial J}{\partial t} + \frac{\partial H_\rho}{\partial \rho} + \frac{H_\rho}{\rho} + \frac{1}{\rho}\frac{\partial H_\phi}{\partial \phi} + \frac{\partial H_z}{\partial z} = -\kappa^a (J - B) \tag{1.16}$$

$$\frac{1}{c}\frac{\partial H_\rho}{\partial t} + \frac{\partial K_{\rho\rho}}{\partial \rho} + \frac{2K_{\rho\rho} + K_{zz} - J}{\rho} + \frac{1}{\rho}\frac{\partial K_{\rho\phi}}{\partial \phi} + \frac{\partial K_{\rho z}}{\partial z} = -\kappa^e H_\rho \tag{1.17a}$$

$$\frac{1}{c}\frac{\partial H_\phi}{\partial t} + \frac{\partial K_{\rho\phi}}{\partial \rho} + \frac{2K_{\rho\phi}}{\rho} + \frac{1}{\rho}\frac{\partial}{\partial \phi}\left[J - K_{\rho\rho} - K_{zz}\right] + \frac{\partial K_{\phi z}}{\partial z} = -\kappa^e H_\phi \tag{1.17b}$$

$$\frac{1}{c}\frac{\partial H_z}{\partial t} + \frac{\partial K_{\rho z}}{\partial \rho} + \frac{K_{\rho z}}{\rho} + \frac{1}{\rho}\frac{\partial K_{\phi z}}{\partial \phi} + \frac{\partial K_{zz}}{\partial z} = -\kappa^e H_z \tag{1.17c}$$

2. NUMERICAL INTEGRATION OF THE RAY EQUATION

In the following we shall designate equations of the type (1.2) and (1.3) as "ray" equations to distinguish them from the moment equations. The numerical integration of the ray equation (1.3) for a known source function S is straightforward although perhaps somewhat tedious due to the large number of directions involved.

2.1 Continuum radiation transfer

Considering the formal solution (1.4), we define an optical depth $\tau(s)$ such that $d\tau = \kappa^e ds$:

$$\tau = \int_a^s \kappa^e ds' \tag{2.1}$$

where the integration is from the point (a) to a distance s along the line of sight $\hat{s}$. Assuming that the source function varies linearly as a function of τ from (a) to (b) we may write

$$S(\tau) = S_a + (S_b - S_a)\,\tau/\tau_b \tag{2.2}$$

where $\tau_b = \tau(b)$. With this ansatz we find the formal solution:

$$I_b = e^{-\tau_b} I_a + \left[(1 - e^{-\tau_b})/\tau_b - e^{-\tau_b}\right] S_a$$
$$+ \left[1 - (1 - e^{-\tau_b})/\tau_b\right] S_b \qquad (2.3a)$$

Often in numerical calculations it is useful to take special care in the limits $\tau_b >> 1$ and $\tau_b << 1$ in order to avoid calculating $e^{-\tau_b}$ or to avoid subtracting two nearly equal numbers:

$$I_b = (1 - \tau_b + \tau_b/2)\ I_a + (\tau_b/2 - \tau_b/3 + \tau_b/8) S_a$$
$$+ (\tau_b/2 - \tau_b/6 + \tau_b/24) S_b \qquad (2.3b)$$

for $\tau_b << 1$ and

$$I_b = (1/\tau_b)\ S_a + (1 - 1/\tau_b) S_b \qquad (2.3c)$$

for $\tau_b >> 1$. The corresponding formulae assuming a higher order ansatz for $S(\tau)$ can be calculated in a straightforward manner.

2.2 Transfer of line radiation

The formulae (2.3abc) may not be appropriate for the numerical solution of the ray equation when the absorption and emission processes are dominated by a few atomic or molecular transitions. Consider, for example the case of a single transition at frequency ν_o, broadened by $\Delta\nu_{th}$ due to the thermal motions of the gas. The absorption and emission properties at frequency ν in the rest frame of the observer therefore depend on the component of velocity $\hat{s} \cdot \underline{v}$ along the line of sight. Suppose now that at point (a) the doppler shifted frequency $\nu' = (1 - \hat{s} \cdot \underline{v}/c)\nu$ in the rest frame of the atom is on the far red side of line center and at point (b) on the far blue side so that the contribution of this transition is negligible at both (a) and (b). In this case the formulae (2.3abc) would not include the effect of the transition at an intermediate position between (a) and (b). For this extreme case of large velocity changes from grid point to grid point the approximation of Sobolev (3,4) can be applied to calculate the contribution from the line. For many astrophysical applications, however, the velocity gradients are neither too large nor too small. Even in the case of large velocity gradients, there are always directions for which the projected velocity only changes slowly. Thus, we need to solve the ray equation for line transfer for arbitrary velocities.

Because of the doppler shift, both the absorption coefficient and the source term of equation (1.2) are highly direction dependent. We shall consider the contributions of the line and (isotropic) continuum separately. Thus, we may write for the ray equation:

$$dI/ds = \kappa^{C}(S^{C} - I) + \kappa^{L}(S^{L} - I) \tag{2.4}$$

where we use C superscripts to denote continuum quantities and L superscripts for line quantities. We shall denote the lower and upper levels of the transition considered by subscripts "1" and "2", respectively. Thus, we may express the line absorption coefficient κ^L and source term $\kappa^L S^L$ by (5):

$$\kappa^{L} = h\nu_{o}/c\; n_{1}\; B_{12}\left[1 - (n_{2}g_{1})/(n_{1}g_{2})\right]\phi(\nu') \tag{2.5}$$

$$\kappa^{L}S^{L} = n_{2}A_{21}\; h\nu_{o}\; \phi(\nu') \tag{2.6}$$

where $\phi(\nu')$ is the profile function, normalized so that $\int\phi(x)\,dx = 1$. Note that in the expressions (2.5) and (2.6) complete redistribution is implicitely assumed. Defining $\hat{I}$ and $\hat{S}$ by

$$\hat{I} = e^{\tau^{C}}\, I \tag{2.7a}$$

$$\hat{S} = e^{\tau^{C}}\, S^{L} \tag{2.7b}$$

where $\tau^{C} = \int_{a}^{s} \kappa^{C}\, ds'$, we may rewrite equation (2.4) in the form:

$$d\hat{I}/ds = -\kappa^{L}\,\hat{I} + \kappa^{L}\,\hat{S} + \kappa^{C}\,S^{C}\,e^{\tau^{C}} \tag{2.8}$$

Brute force integration of (2.8), taking care not to lose part of the line and making certain assumptions about the distribution of the gas between (a) and (b), yields:

$$I_{b} = \{[e^{-\tau_{b}^{C}} - p]\; I_{a} + p\; S_{a}^{L} + q\; S_{b}^{L} + S^{k}\}/(1 + q) \tag{2.9}$$

where the quantities p, q, S^k and τ_b^L are given by

$$q = \tau_{b}^{L}/(1 + \exp\{-\tau_{b}^{L}\}) \tag{2.10}$$

$$p = q\; \exp\{-\tau_{b}^{L}-\tau_{b}^{C}\} \tag{2.11}$$

$$S^{k} = \exp\{-\tau_{b}^{C}\} \int_{a}^{b} \kappa^{C}\, S^{C}\, \exp\{ \int_{a}^{s} \kappa^{C} ds'\}\, ds \tag{2.12}$$

$$\tau_{b}^{L} = \int_{a}^{b} \kappa^{L}\, ds \tag{2.13}$$

Note that equation (2.13) involves integration of a function which depends on the population densities n_1 and n_2 of the two levels and on the doppler shifted frequency ν', all of which vary between (a) and (b). The statistical weights g_1 and g_2 and Einstein coefficients A_{21}, B_{12} do not vary from grid point to grid point. In order to numerically integrate (2.13) we have chosen to use mean values for n_1 and n_2 so that only the profile function remains under the integral:

$$\tau_b^L = (\bar{n}_1 - \bar{n}_2 g_1/g_2)\, B_{12}\, h\nu_o/c \int_a^b \phi(\nu')\, ds \qquad (2.14)$$

The integral of the profile function may be written:

$$\int_a^b \phi(\nu')ds = (s_b - s_a)/(v_b\mu_b - v_a\mu_a)\; c/\nu\; (P_a - P_b) \qquad (2.15)$$

where $v_a\mu_a$ and $v_b\mu_b$ are the components of velocity along the line of sight (μ is the cosine of the angle between $\hat{s}$ and $\underline{v}$). The values of $P_a = P(y_a)$ and $P_b = P(y_b)$ may be found as follows. We define the dimensionless frequency y by:

$$y = (\nu' - \nu_o)/\Delta\nu_{th} = \nu/\Delta\nu_{th}\,(1 - v\mu/c) - \nu_o/\Delta\nu_{th} \qquad (2.16)$$

and the dimensionless profile integral P(y) by:

$$P(y) = \int_o^y \phi(\nu_o + y\Delta\nu_{th})\Delta\nu_{th}dy \qquad (2.17)$$

Note that in the case of a gaussian profile

$$\phi(\nu') = \exp\{-(\nu' - \nu_o)^2/\Delta\nu_{th}^2\}/(\Delta\nu_{th}\sqrt{\pi}) \qquad (2.18)$$

we may integrate (2.17) directly to obtain

$$P(y) = \mathrm{erf}(y)/2 \qquad (2.19)$$

Combining (2.14) and (2.15) we find:

$$\tau_b^L = (\bar{n}_1 - \bar{n}_2 g_1/g_2)\, B_{12} h\, (s_b - s_a)/(v_b\mu_b - v_a\mu_a)(P_a - P_b) \qquad (2.20)$$

where we have assumed $\nu_o/\nu \approx 1$ valid for $v \ll c$. This assumption simplifies the discussion, but it is not mandatory.

For the mean values $\bar{n}_1$, $\bar{n}_2$ one should weight the contributions from (a) and (b) by an algorithm which takes into account the relative doppler shifted frequency intervals from the line center at the grid points. Such an algorithm should also depend on the specific form of $\phi(\nu')$. We suggest using the profile integral P(y). Defining $P_m = P([y_a + y_b]/2)$ we may write for $\bar{n}_1$ or $\bar{n}_2$

$$\bar{n} = \{(P_m - P_a)n_a + (P_b - P_m)\, n_b\}/(P_b - P_a) \qquad (2.21)$$

For the integration of (2.12) for S^k we suggest the ansatz (2.2), substituting S^C and τ^C for S and τ. Thus, similar coefficients for S_a^C and S_b^C as appear in (2.3abc) will be obtained. This has the advantage that the formulism presented here for line transfer reduces exactly to that for the continuum case in the limit $\tau_b^C \rightarrow 0$.

2.3 Numerical examples with known source function

There are many astrophysical applications of radiation transfer for which the source terms and extinction (absorption) coefficients are known sufficiently accurately as to preclude a detailed numerical calculation, solving for the radiation intensity, the source terms and extinction coefficients simultaneously. Such exact solutions are in general feasible for the spherically symmetric case only. However, due to the use of faster and larger computers, 2-D and 3-D hydrodynamical calculations are not only feasible, but they are becoming an increasingly important investigative tool. Thus, it is useful to calculate the expected appearance of numerical hydrodynamical models in order to compare these with observations of astrophysical objects. In the following we present a few numerical examples, showing how this can be done.

The appearance of rotating protostars: In discussing the evolution of a rotating, collapsing protostellar cloud, Tscharnuter (6,7) has included the effects of angular momentum transfer by turbulence by defining a simple viscosity parameter. Radiation transport was calculated using the "Eddington approximation", i.e. $K_{ij} = J/3$ for i=j and $K_{ij} = 0$ for $i \neq j$, together with the time dependent grey moment equation. The grey opacity for temperatures $T < 1800$ K was given by the Rosseland mean of the frequency dependent absorption of a mixture of graphites, silicates and ice-coated grains - as discussed in (8,9).

Using Yorke's (8,9) grain model and Tscharnuter's values for the density and temperature at an evolutionary age 32500 yr. after formation of the disk (see Fig. 2), it is possible to calculate a frequency dependent source function and extinction coefficient at each point in space. Thus, we are able to solve the frequency dependent continuum ray equation for any given line of sight through Tscharnuter's model. This allows us to construct theoretical IR maps at any frequency and theoretical IR spectra both for any given viewing angle θ (see Figure 3).

These calculations show us that during the early phases of protostellar evolution a disk would be difficult to detect. The spectrum depends only slightly on the angle of view (less than the size of the circles in Figure 4). Theoretical IR maps, which show detail at a finer scale at $\lambda = 14\mu$ and 100μ (Figure 5) than is possible to observe, depend strongly on viewing angle. However, such objects would be difficult to distinguish from multi-

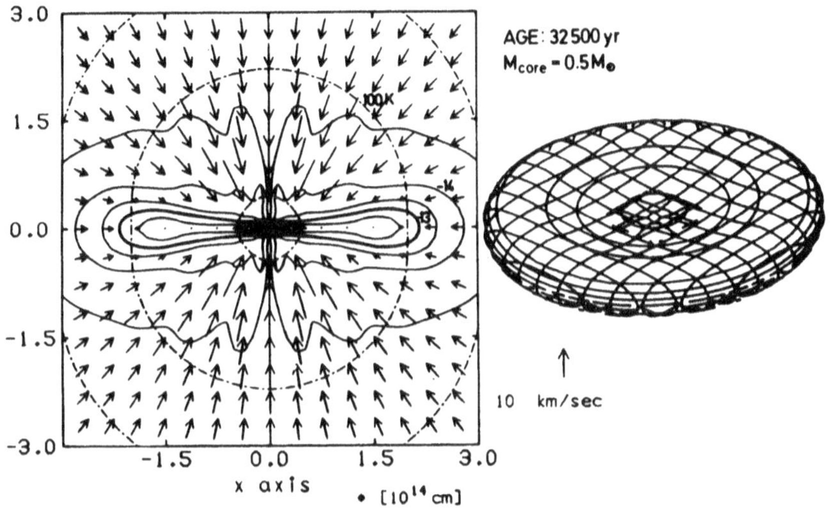

Figure 2. Meridional cross-section of Tscharnuter's $3M_{\odot}$ rotating cloud after 0.5 $M_{\odot}$ have accreted onto the core. The iso-density contours are given by solid lines, the isothermal contours by dashed-dotted lines. The logarithmic values for density (in g cm^{-3}) are indicated. The axis of symmetry (and rotation) is at X = 0, the equator at Z = 0. Arrows show the magnitude and direction of the velocity at the tips; the velocity scale is at the lower right. A 3-D representation of the isodensity surface log (density) = -13 is given to the right.

ple sources. At least for the early phases of protostellar evolution, spherically symmetric calculations are adequate for understanding observed spectra and IR maps, and they may be used in conjunction with observations to derive improved grain opacities in the far IR.

Theoretical radio maps of the "champagne phase": Two dimensional (2-D), axially symmetric hydrodynamical calculations have been made (10) describing the evolution of an HII region formed near the edge but inside of a molecular cloud. These calculations have shown the existence of a new evolutionary phase of expanding HII regions. Expansion is preferential in directions of decreasing gas density, a fact which leads to rather high velocity flow ($\approx$ 30 km/s) of ionized material away from the molecular cloud, when the ionization front (I-front) crosses the cloud boundary. Radiation transfer of the hydrogen-ionizing Lyc photons was done

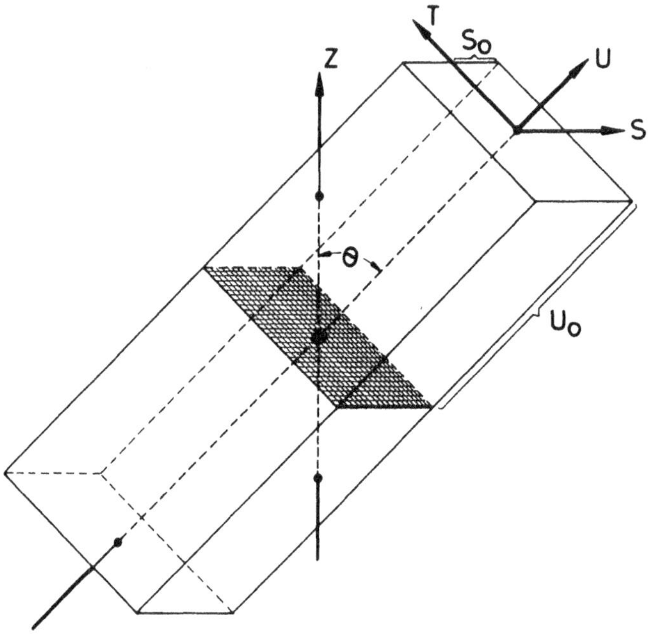

Figure 3. The (S,T,U) coordinate system, tilted by an angle θ with respect to the symmetry axis (denoted by "Z"), was used for the solution of the ray equations. Starting at a value $U = -U_o$ (U_o chosen so that the starting point is outside the region of significant emission), the ray equation is solved along lines of constant (S,T) where U varies from $-U_o$ to $+U_o$ (also far outside the region of significant emission). These calculations are repeated for different values of (S,T), until the intensity I_ν is known for a grid of (S,T) values at $U = +U_o$. Summing $I_\nu(S,T,U_o,\theta)$ over (S,T) yields one frequency point of the spectrum. Using $I_\nu(S,T,U_o,\theta)$ to construct isointensity plots in the (S,T) plane produces maps at the given frequency.

in an approximate manner; scattering of Lyc photons and recombination photons of helium and (for n=1) hydrogen were neglected. For this reason sharp "Lyc shadows" were cast by neutral material. These shadows would not have been present if the full 2-D Lyc radiation transfer problem had been solved (see i.e. 11). In spite of this shortcoming one can use the density and temperature distribution obtained in the hydrodynamic calculations to calcu-

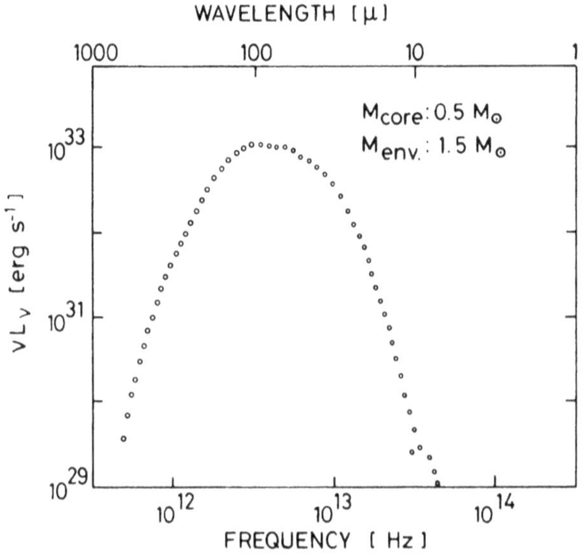

Figure 4. Calculated IR spectrum of Tscharnuter's model of a rotating protostellar cloud shown in Figure 2.

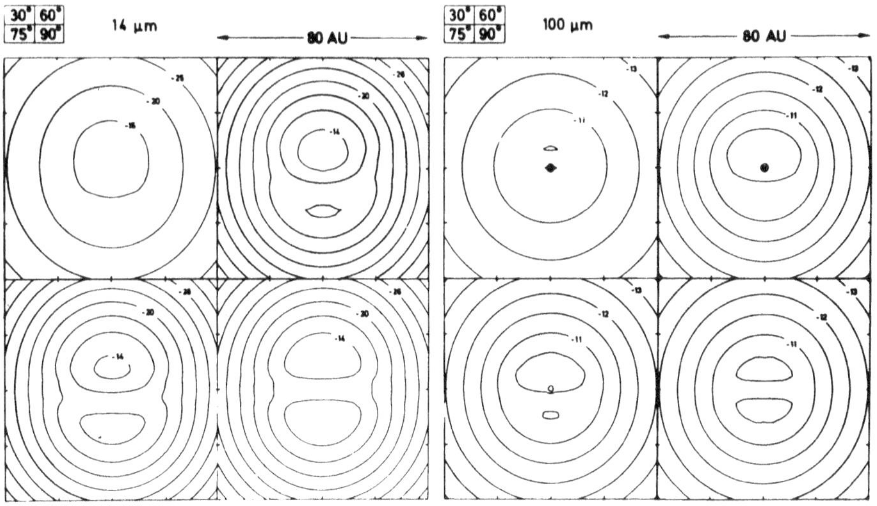

Figure 5. Calculated IR maps of Tscharnuter's model at $\lambda = 14\mu$ and $\lambda = 100\mu$ (see also Figures 2 and 4) at viewing angles $\theta = 30^o$, 60°, 75° and 90° as indicated.

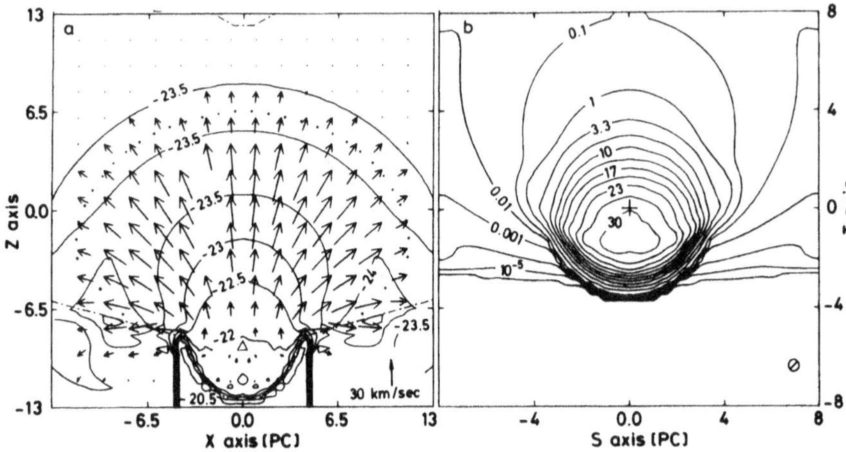

Figure 6a. Meridional view of the numerical hydrodynamical results of an HII region originally embedded inside a molecular cloud, which itself is embedded in an ambient low density medium. When the ionization front crosses the molecular cloud/intercloud boundary a flow pattern results as shown for this particular choice of parameters and at an evolutionary age $6.6 \cdot 10^5$ yr. The triangle marks the position of the ionizing source on the assumed symmetry axis (X = 0). The dashed-dotted line indicates the location of the I-front. The small crosses mark the interface between the material originally from the molecular cloud and from the ambient medium. All other symbols and lines have the meaning as in Figure 2.

Figure 6b. Theoretical radio maps at $\lambda = 11$ cm for the hydrodynamical model shown in 6a as "seen" by an observer at $\theta = 90^o$. Some smoothing of the intensity distribution has been done, simulating a radio antenna of beam size as indicated in the lower right corner. Numbers give the surface brightness in units of 10^{-18} erg s^{-1} cm^{-2} Hz^{-1}. The projected position of the ionizing star is indicated by a cross.

late the expected radio appearance, keeping in mind that in the peripheral zones any shadows which appear may be "softened" by the above mentioned effects.

Using well known formulae (12) for the free-free extinction

coefficient and source function of a warm ($T = 10^4$K) ionized gas we can calculate radio maps in a manner described in the previous subsection. Note that the resultant radio emission has very little effect on the energetics of the HII region, so that the source function and extinction coefficient do not have to be corrected after the solution of the ray equation is known. The solution obtained for two hydrodynamic cases considered by (10) are displayed in Figures 6 and 7. A detailed discussion of these results is forthcoming (13).

Theoretical X-ray maps of supernova remnants: The procedure described above for calculating IR maps of rotating protostars and radio maps of non-spherical HII regions may also be applied to the problem of the expected X-ray appearance of supernova remnants. The hydrodynamical evolution calculations do not require an exact solution of the radiation transfer problem, because the non-adiabatic cooling of the hot "Sedov" gas is dominated by optically thin free-free emission. This emission will be observable as thermal bremsstrahlung radiation in the soft X-ray or hard UV spectral range. As before we may determine the source function and extinction coefficient in the X-ray regime from the density and temperature distributions of the hydrodynamical models. Examples of the resultant X-ray maps obtained at various evolutionary phases of a non-spherical remnant is discussed elsewhere (14).

It would be possible to also calculate the thermal radio emission from the bremsstrahlung process, but this would be a purely academic exercise. Observations of supernova remnants show that the emission in the radio regime is dominated by non-thermal processes, contrary to the case in the soft X-ray range. Calculating the synchrotron emission in a supernova remnant would require detailed knowledge of the magnitude and configuration of the magnetic field as well as the distribution of relativistic electrons. These electrons are accelerated to relativistic energies at the supernova shock front by a process which is not yet well understood. Thus, it is difficult at present to make accurate predictions of the expected radio appearance of supernova remnants.

Line transfer in HII regions during the champagne phase: The numerical hydrodynamical calculations used to produce the theoretical radio maps shown in Figures 6 and 7 also supply the velocity distribution of the ionized gas. It would be interesting to consider line profiles of forbidden transitions of oxygen and recombination transitions of hydrogen for these hydrodynamic models, as an example of line transfer calculations with a known source function. For these two types of transitions the source terms are under certain circumstances proportional to n^2 inside the HII region and zero outside. Furthermore, if the density n in the ionized gas is sufficiently low, then self-absorption of the line radiation can be neglected.

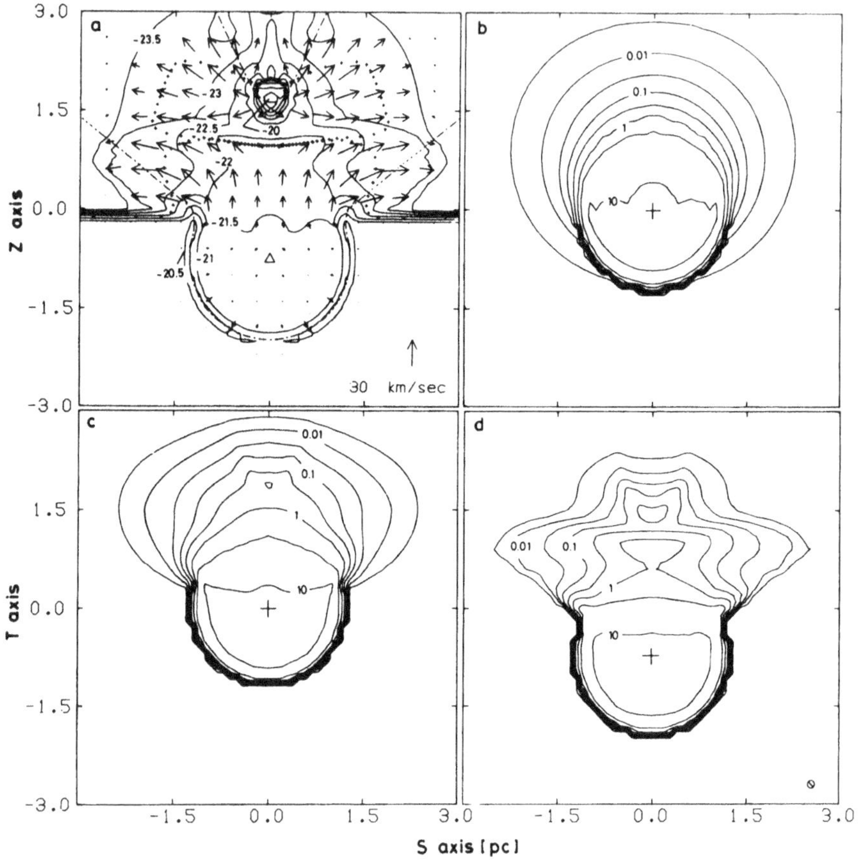

Figure 7a. Meridional view of the hydrodynamical results describing an HII region in the champagne phase interacting with a neutral globule, $8.6 \cdot 10^4$ yr. after the ionizing source was turned on. Symbols and lines have the same meaning as in Figure 6. Lagrangian tracers indicating the interfaces between material originally from the globule, the molecular cloud and the ambient intercloud medium are shown as crosses.

7b. Radio continuum isophotes at $\lambda = 11$ cm for an observer at $\theta = 30^o$ in units of 10^{-17} erg s^{-1} cm^{-2} Hz^{-1}. The map has been shifted to center the projected position of the star.

7c. Same as 7b but for $\theta = 60^o$.

7d. Same as 7b but for $\theta = 90^o$.

Consider first the forbidden transitions of singly ionized oxygen at $\lambda\lambda$ 3726,3728. Because the ionization potential of oxygen is almost exactly the same as for hydrogen, the two elements should be ionized or neutral in the same regions of space. Restricting oneself to somewhat cooler exciting stars, the ratio O^+/H^+ may be considered constant throughout the HII region. O^+ in the $^4S^o$ ground state can easily be excited to the metastabile $^2D^o$ and $^2P^o$ states by inelastic collisions. For hydrogen densities lower than about 10^8 cm^{-3} the timescale for collisional deexcitation is much longer than that for spontaneous emission, and emission of a photon usually results. The emissivity is therefore given by the collisional excitation rate $C_{12}\ n_o{+}n_e$ (C_{12} is the collisional excitation coefficient). For HII regions $C_{12} \approx$ const because $T \approx$ const. In our numerical example (Figure 8) we have considered a single isolated oxygen line only; a realistic calculation would have to convolve the two $\lambda\lambda$ 3726,3728 lines with the correct branching ratio.

The recombination transitions of hydrogen are proportional to the total recombination rate $\alpha n_e n_{H^+}$, where α is the recombination coefficient for recombinations into level 2 or higher (on the spot approximation is assumed). Here again, $\alpha \approx$ const throughout the HII region, so that the production rate of, say H_α photons is proportional to n_e^2. Extinction due to dust was included in the numerical example shown in Figure 8.

Because no strong velocity gradients were present, we used the formalism in (2.3abc) for the solution of the ray equation. The line profile $\phi(\nu)$ was assumed to be gaussian, (2.18), where the thermal line broadening was given by

$$\Delta\nu_{th}/\nu_o = v_{th}/c = (2kT/m)^{1/2}/c \qquad (2.22)$$

For hydrogen at $T = 10^4$ K v_{th} = 12.9 km/s and for oxygen v_{th} = 3.21 km/s. For this reason the oxygen line shows more velocity detail in its line profiles. In particular line splitting occurs in the oxygen lines, whereas the hydrogen lines are merely broadened. Although rather complex profile structures are obtained the peripheral zones of this particular model, the integrated profiles (not shown) look very much like the profiles centered on the star (indicated by an asterix in Figure 8).

3. AN EXACT SOLUTION OF CONTINUUM TRANSFER IN SPHERICAL SYMMETRY

An efficient numerical method for solving the detailed radiation transfer problem in extended spherical envelopes, coupled with the hydrodynamical equations of motion is described by (15). Here, we wish to illustrate this method by considering the time independent transfer equation only. There is no principal dif-

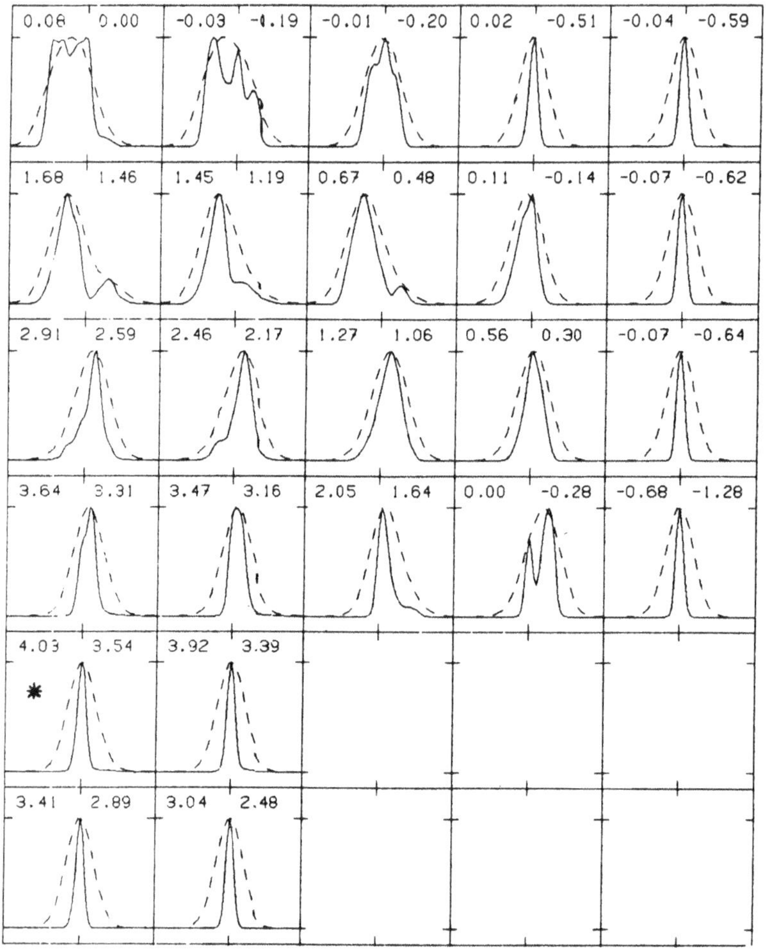

Figure 8. Theoretical line profiles of oxygen (solid lines) and hydrogen (dotted lines) at selected positions in the (S,T) plane spaced 0.8 pc apart for an observer at $\theta = 60^{\circ}$. The velocity and density distribution for the ionized gas assumed for these line transition calculations is shown in Figure 7a (neutral globule interacting with the champagne flow of an HII region). Each box shows both line profiles averaged over an assumed beam of radius 0.6 pc in the (S,T) plane. The ordinate of the box is the normalized flux in the line; the abcissa is the red-shifted velocity ranging from -45 km/s to +45 km/s. Zero red-shift (line center) is indicated. The logarithm of the normalization factors is given in the upper left (right) of each box for oxygen (hydrogen). The line of sight position centered on the star is indicated by an asterix; the boxes above and below this position lie along the projection of the symmetry axis. Empty boxes correspond to positions for which little line radiation was found.

ficulty in solving the time dependent equation as shown by (15). However, such a procedure involves somewhat more complicated equations and requires much more storage.

3.1 Numerical strategy of solution

By considering the variables (r,p) where $p = r(1-\mu^2)^{1/2}$ (μ is the cosine of the angle between $\hat{s}$ and $\hat{e}_r$) equation (1.3) can be written (16):

$$\mu \partial I_\nu^\pm / \partial r = \mp \kappa_\nu^e (I_\nu^\pm - S_\nu) \tag{3.1}$$

where the intensity I_ν has to be split into two components I_ν^+ and I_ν^-, because the variable p does not distinguish between $\mu > o$ and $\mu < o$. The corresponding moment equations, integrated over all frequencies can be written:

$$0 = \partial(fJ)/\partial r + (3f-1)J/r + \kappa_H H \tag{3.2}$$

$$0 = \partial H/\partial r + 2H/r + \kappa_J J - \kappa_B B \tag{3.3}$$

where the 0-th and 1-st moments J, H and the Eddington factor f are given by

$$J = \int_0^\infty J_\nu d\nu \tag{3.4}$$

$$H = \int_0^\infty H_\nu d\nu \tag{3.5}$$

$$f = \int_0^\infty K_\nu d\nu / J \tag{3.6}$$

and where

$$\kappa_H = \int_0^\infty \kappa_\nu^e H_\nu d\nu / H \tag{3.7}$$

$$\kappa_J = \int_0^\infty \kappa_\nu^a J_\nu d\nu / J \tag{3.8}$$

$$\kappa_B = \int_0^\infty \kappa_\nu^a B_\nu (T) d\nu / B \tag{3.9}$$

$$B = \int_0^\infty B_\nu (T) d\nu \tag{3.10}$$

Note that the equations (3.2) and (3.3) are the exact equations for time independent, non-relativistic, but frequency dependent radiation transfer, provided that a scheme can be devised for supplying the correct values for f, κ_H and κ_J. However, in order to do this, the solution of the exact transfer equation must be known. Our strategy of solution will be to solve the ray equations (3.1) with an approximate source function and extinction and absorption coefficients and then use the intensity thus obtained to calculate κ_H, κ_J and f. Then the moment equations can be solved simultaneously with the appropriate energy equations and (if need be) with the hydrodynamical equations of motion and continuity. In this manner improved estimates for the source function and extinction and absorption coefficients are obtained. The ray equations are solved again with improved S_ν, κ^e_ν and κ^a_ν and the intensity is used to recalculate f, κ_H and κ_J. Iterating between the moment and energy equations (and perhaps hydrodynamical equations) on the one hand and the ray equations on the other hand usually converges rather well. For an atmosphere with little scattering (i.e. less than 50 % scattering) using even rather poor starting values for S_ν, κ^e_ν and κ^a_ν, convergence is obtained in less than 10 iterations ; often two or three iterations are sufficient. In time-dependent calculations it usually suffices to solve the ray equations only once every hydrodynamical time step, because f , κ_J and κ_H usually change slowly with time.

3.2 Solution of system I equations

We shall denote the two ray equations for I^+_ν and I^-_ν (3.1) and the definitions (3.4) through (3.8) and (1.6abc) as system I equations. The moment equations (3.2) and (3.3) together with the relevant energy equations and hydrodynamic equations will be denoted as system II equations. The iteration procedure described in section 3.1 entailed solving system I and system II equations alternatively until convergence is achieved. In the following we shall describe the numerical solution of system I equations only. For the solution of system II equations the usual difference schemes can be used. Notable examples of the numerical solution of system II equations coupled with the author's computer code for solving the system I equations are given in (8,9,17) and described in more detail in (18).

The (r,p) grid is first discretized into a radial r_j grid and a p_k grid (see Figure 9), where j is the index corresponding to allowed radial values r_j and k is the index of allowed p-values: $0 \leqslant p_k \leqslant r_j$. The ray equations (3.1) are solved for each value of p between p = 0 and p = r_J using the solution (2.3abc) to integrate from $\mu_{jk} \cdot r_j$ to

$$\mu_{j+1,k} \cdot r_{j+1} \quad \text{where} \quad \mu_{jk} = (1-p_k^2/r_j^2)^{1/2}$$

The resulting matrices I^+_{jkl} and I^-_{jkl} (l is the index for the frequency grid ν_l) are then used to calculate the moments. The integration over all angles reduces to a summation over allowed p-values:

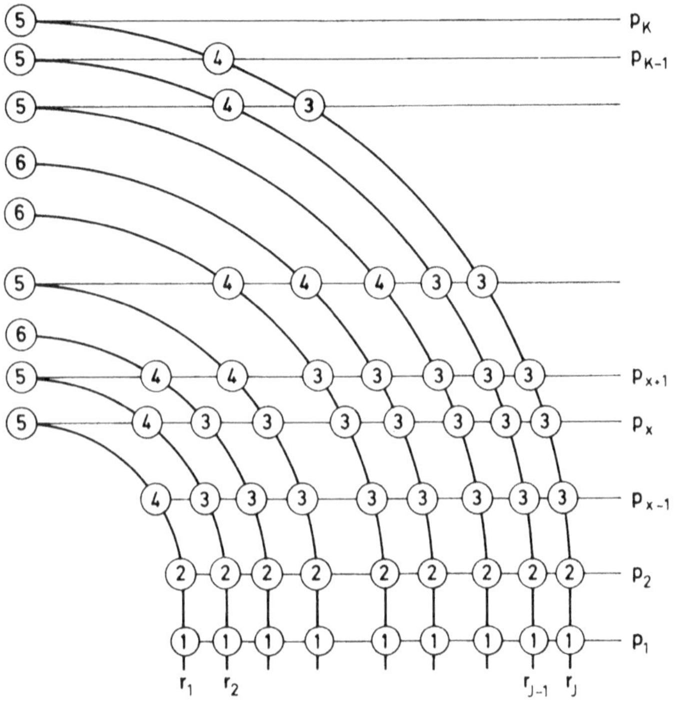

Figure 9. The discretization of the (r,p) grid used to solve the ray equations for spherical symmetry. In the example shown an inner sphere has been cut out to allow for a central source of radius $p_{x-2} = p_{x-1}$. The numbers given at the points of intersection of the lines of constant r and p correspond to different formulae used to write a_{kj}, b_{kj} and c_{kj} as a function of r_{j-1}, r_j, r_{j+1}, p_{k-1}, p_k and p_{k+1}. The ray equations (3.1) are solved along lines of constant p.

$$J_{jl} = \frac{1}{2} \sum_k (I^+_{jkl} + I^-_{jkl})\, a_{kj} \tag{3.11}$$

$$H_{jl} = \frac{1}{2} \sum_k (I^+_{jkl} - I^-_{jkl})\, b_{kj} \tag{3.12}$$

$$K_{jl} = \frac{1}{2} \sum_k (I^+_{jkl} + I^-_{jkl})\, c_{kj} \tag{3.13}$$

The integration weights a, b and c depend on the distribution of the r-grid points only and not on frequency ν_1. They are calculated by making a linear or higher order ansatz for $I^+(p)$ and $I^-(p)$ and integrating from p_k to p_{k+1} using this same ansatz for all three weights a, b and c. This must be done in a careful manner to insure that the following relationships will hold.

$$\sum_k a_{kj} = 1 \tag{3.14a}$$

$$\sum_k b_{kj}\mu_{jk} = 1/2 \tag{3.14b}$$

$$\sum_k c_{kj} = 1/3 \tag{3.14c}$$

The weights a, b and c given by (15) were constructed assuming a linear relationship for $I_\nu^\pm(p)$ between (j,k) points of type 2, 3 and 4 (see Figure 9), a quadratic relationship for $I_\nu^\pm(p)$ between (j,k) points of type 1 and 2, and a quadratic relationship for $I_\nu^\pm(\mu)$ between 4 and 5 or 4 and 6. (j,k) points of type 6 do not occur when for each radial grid point r there is one impact parameter p.

The numerical integration formulae for J, H, f, κ_J and κ_H can also be written as sums:

$$J_j = \sum_l g_l J_l \tag{3.15a}$$

$$H_j = \sum_l g_l H_{jl} \tag{3.15b}$$

$$f_j = \sum_l g_l K_{jl}/J_j \tag{3.15c}$$

$$\kappa_{J,j} = \sum_l g_l \kappa_{jl}^a J_{jl}/J_j \tag{3.15d}$$

$$\kappa_{H,j} = \sum_l g_l \kappa_{jl}^e H_{jl}/H_j \tag{\&3.15e}$$

where the frequency weights g_l depend on the distributions of frequency grid points ν_l. For the numerical examples presented here we have used a Simpson-type formula to determine the weights g_l. In the example by (8,9) the frequency dependent transfer is solved coupled to hydrodynamics. In the examples by (15,19) hydrodynamics were not included. The solutions of (17, 18) solved the grey equations of radiation transfer coupled with hydrodynamics, so that the frequency quadrature formulae (3.15a) to (3.15e) were not used.

3.3 A numerical example

In Figure 10 we show some results of the frequency dependent calculation by (19) for a dusty envelope with a spherically symmetric density distribution $\rho(r) \propto r^{-3/2}$, surrounding a black-body source of temperature T = 16 600 K and luminosity $L = 10^3\ L_\odot$. The optical depth τ of the envelope at λ = 1000 Å was allowed to vary between τ = 1 and τ = 1000. These calculations simultate the situation of a young main sequence B0 star accreting material in a freely falling envelope at varying rates.

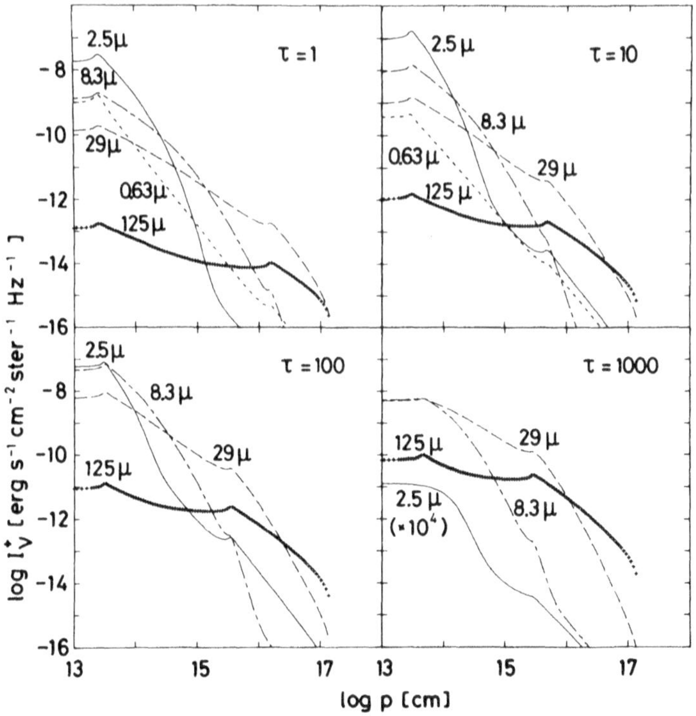

Figure 10. The distribution of the emerging flux I_ν^+ across the dusty envelope surrounding a luminous object as a function of the impact parameter p is displayed for different values of τ and for selected wavelengths as indicated. The abscissa positions of the crosses for λ = 125μ indicate some of the impact parameters for $p > 10^{13}$ cm used in the numerical calculations.

For a particular case, τ = 100, the overall spectrum and the corresponding temperature distribution are shown in Figures 15 and 16, to be discussed in section 5. Here, we wish to concentrate on the angular distribution of the resulting intensity and to discuss some practical considerations relating to these numerical calculations.

3.4 Practical considerations

We discuss in the following some practical considerations with regard to total storage and time requirements, stability and convergence properties, and numerical accuracy. The conclusions made are a result of physical reasoning combined with exploratory test calculations. Two computer programs for solving the ray equations (3.1) in spherical symmetry have been written and tested by the author. The first, described in detail in (15) and in this section, has been used with a number of computers (UNIVAC, IBM, Amdahl, PDP 11, CRAY I). The second has been designed especially for the vector pipelining capabilities of the CRAY I. Both are written in standard FORTRAN IV, except for the use of standard CRAY I vector routines in the CRAY I program. A third (stripped down) version has been written in BASIC for use in the Apple II microcomputer. The first and third were written in a manner to minimize storage, the second in a manner to minimize the time requirements. A special feature of the minimal time program is the use of large matrices to store the factors a, b, c needed in equations (3.11) to (3.13), so that for frequency dependent radiation transfer or radiation transfer with a stationary radial grid these weights do not have to be recalculated. The minimal storage code uses vectors of length J, K or L (but not their product) where J, K and L are the number of radial grid points r_j, impact parameters p_k and frequency grid points ν_l, respectively. For the examples discussed in section 3.3 about 2300 words were necessary. The minimal time code uses large matrices with a total storage of about 3xJxK for time independent problems.

The total computational time necessary for the solution of system I equations increases linearly with the product JxKxL for all versions of the code. Typical CPU execution times for different computers are given in Table 1.

The procedure of first solving system I then system II equations and iterating between the two has been found to converge rather quickly. In general, at most three to ten iterations are necessary for an accuracy in temperature of about 0.1 % - depending on the accuracy of the starting model. The results for example in the previous section are displayed after the fifth iteration, but no significant changes occurred after the third iteration. The reason for the rapid convergence is the fact that the Eddington factor f(r) depends strongly on the geometry of the spherically symmetric cloud and not on the radiation field. For a purely scattering cloud the convergence was much poorer for the cases calculated, apparently because the Eddington factor f was more strongly coupled to the radiation field itself. By a more efficient method of solution of the system II equations the convergence could be improved. In (20) the frequency dependent moment equations were used, solved by a Newton-Raphson procedure. This increased both the storage and CPU time requirements substantially, however.

In general, the stability of this computational procedure depends only on the stability of the system II equations. The convergence properties are basically more strongly dependent on the details of the method of solution of the system II equations than on the iteration procedure described here.

It should be warned that for very large optical depths $\Delta\tau$ the method presented here for solution of system I equations becomes inaccurate due to rounding errors. As $\Delta\tau$ approaches infinity I^+ and I^- become very nearly equal, so that the accuracy of the difference $(I^+ - I^-)$ is much less than that of I^+ or I^-. If $(I^+ - I^-)/(I^+ + I^-) \ll 1$ for the frequencies at which most of the radiative energy is transported, this will affect the accuracy of the final results. The critical test for this is the ratio of the first moment to the zeroth moment: H/J. In the Sun for example $H/J \simeq 10^{-12}$. Using a computer with say 17 decimal place accuracy the numerical method presented here calculates the radiative flux accurate to only five decimal places.

Table 1: Time requirements for solution of the ray equations in spherical symmetry assuming J = 100, K = 100 and L = 1

	"minimal storage" code	"minimal time" code with weights	"minimal time" code without weights
IBM 360/91	0.56s		
Amdahl 470	1.0s		
UNIVAC 1108	7.8s		
UNIVAC 1100/83	2.0s		
PDP 11/44			
(single precision	9.7s		
(double precision)	13.1s		
CRAY 1	0.1s	0.018	0.01

Finally, we would like to comment on an alternate method for solving the ray equations (3.1). In (16) the variables u and v are defined

$$u = \frac{1}{2}(I^{+} + I^{-}) \tag{3.16a}$$

$$v = \frac{1}{2}(I^{+} - I^{-}) \tag{3.16b}$$

The corresponding ray equations for these variables are obtained by adding or subtracting the ray equations for I^+ and I^-. These two new equations can be solved for u and v as they stand or, as suggested in (16), one can combine the two equations of first order into a single equation of second order:

$$\partial^2 u/\partial\tau^2 = u - S \tag{3.17}$$

where $\partial/\partial\tau = \mu/\kappa^e \, \partial/\partial r$. Using equation (3.17) for numerical calculations has the advantage that the system II moment equations can be formulated in a manner that makes them numerically compatible to (3.17). When convergence has been attained between systems I and II, one is assured that the moments obtained from the solution of the moment equations closely resemble the moments obtained from solution of (3.17). When the source function or extinction coefficient change rapidly from grid point to grid point, however, the formula (3.17) is not well-suited for numerical calculations. Experience has shown that equations (3.1) should be used in this case.

Using the ray equations (3.1) instead of (3.17) for all problems of radiation transfer has its advantages also. By comparing the moments calculated by system I using (3.1) with the moments obtained from the solution of the system II equations one has a direct numerical check for the "correctness" of the final solution. In the numerical examples discussed in (8,9,17, 18,19) J and H from system I and system II were always compatible within a factor of at least $\Delta r/r$, the largest errors occurring in regions where the extinction coefficient changed most rapidly from grid point to grid point.

4. LINE TRANSFER IN SPHERICAL SYMMETRY

Three numerical methods for solving the line transfer problem in static and moving atmospheres are discussed and compared by (21). Rather than enter into a detailed discussion of the advantages and disadvantages of each method we shall refer the reader to the above investigation. In the following we shall describe a straightforward method for the exact solution of the line transfer problem in an extended, spherical atmosphere. Applying this method to the problem of the formation of molecular lines in collapsing protostellar envelopes, we present numerical examples.

An example of the failure of this method to converge for the formation of maser lines is also discussed.

4.1 Numerical strategy of solution

When the population densities n_i of the energy levels of an atom or molecule are known one can use a ray tracing method similar to that described in section 3 for continuum radiation in order to solve for the intensity I_ν (r,p) in the rest frame of the observer at each frequency ν of interest. For the integration from grid point to grid point, it is recommended that the formulae discussed in section 2.2 be used whenever the doppler-shifted frequency ν'falls within say ten thermal doppler widths $\Delta\nu_{th}$ of one of the transitions considered. When the frequency ν' is far from any transition, the formulae of section 2.1 for continuum radiation can be used, because in the limit $\tau_b^L \rightarrow 0$ the two are completely equivalent. Our strategy of solution will entail solving for I_ν(r,p) using approximate values for n_i. We then correct the population densities by solving the time independent rate equations including effects of absorption, induced emission and spontaneous emission as well as collisional excitation and deexcitation. The intensity I_ν (r,p) enters into the rate equations via absorption and induced emission of line radiation. Thus, we obtain improved values for n_i and the ray equations can again be solved for improved values for the intensity I_ν(r,p). By alternatively solving the rate equations and the ray equations the exact solution can be obtained - provided that this iteration procedure converges. In general such a Λ-iteration converges rather slowly. For many practical applications, however, this procedure is adequate.

It is important to have good starting values for the population densities so that the above mentioned iteration procedure converges in a reasonable amount of computer time. We have therefore elected to use an escape probability formalism in the framework of the Sobolev approximation to generate the starting model. In many cases the CPU-time consuming ray equations have to be solved only once; i.e. often the source function generated by the approximate method is quite accurate even if the net spectrum produced by solving the ray equations using the Sobolev source function are for many applications very close to the actual solution. In section 4.4 we show an example of this for a hypothetical two-level atom.

4.2 The rate equations

We consider bound-bound transitions within a multi-level atom or molecule. We number the energy levels in order of increasing energy; the ground state is level 1. The rate of transitions into a given level: equals the rate of transitions from i into all other $j \neq i$ after statistical equilibrium has been reached:

$$\sum_{j\neq i} n_j(A_{ji} + B_{ji}\bar{J}_{ji} + C_{ji}) = n_i\sum_{j\neq i}(A_{ij} + B_{ij}\bar{J}_{ij} + C_{ij}) \quad (4.1)$$

where n_i is the population density of level i, A_{ij} is the Einstein coefficient for spontaneous emission from level i into j, B_{ij} the Einstein coefficient for absorption (induced emission) of a transition i to j when the level i is at a lower (higher) energy than j. C_{ij} is the coefficient for collisional excitation (de-excitation), of a transition i to j when $i < j$ $(i > j)$. As equation (4.1) stands we have made the implicit assumption that $A_{ij} = 0$ for $i < j$. The Einstein coefficients are related by the usual Einstein relations; the collision coefficients by a Boltzmann factor (c.f. 5)

$$g_jC_{ji} = g_iC_{ij}\exp\{-h\nu_{ij}/kT\} \quad (4.2)$$

valid for $i < j$ where ν_{ij} is the frequency of transition from j into i and g_i is the statistical weight of level i. Here h, k and T have their usual meanings in this context. The average radiative energy density within the line $\bar{J}_{ij}$ is obtained from the intensity by the relation:

$$\bar{J}_{ij} = \frac{1}{4\pi}\int_{4\pi}\int_0^\infty I_\nu\phi_{ij}(\nu')d\nu d\Omega \quad (4.3)$$

where as before we use ν to denote the frequency in the rest frame of the observer and ν' is the frequency in the rest frame of the atom (or molecule). ν' is a function of both the direction Ω and ν. In the framework of the Sobolev approximation $\bar{J}_{ij}$ is calculated assuming infinitely thin lines: $\phi_{ij}(\nu') = \delta(\nu'-\nu_{ij})$ so that the integral (4.3) becomes an integral over the surface of zero radial velocity as "seen" by the atom.

For the numerical integration of (4.3) we must again be careful not to "lose" part of the integrand when $\phi_{ij}(\nu')$ changes rapidly from one direction grid point to the next. This problem can usually be alleviated by integrating over a small frequency interval $\Delta\nu$ first, under the assumption that the ray equations have been solved not for the single frequency ν, but for the interval from $\nu - \Delta\nu/2$ to $\nu + \Delta\nu/2$. The angle integration can then be completed in the usual way. For spherical symmetry, a formula similar to (3.11) is used for the angle integration. The final frequency integration over the line is obtained by a simple sum of the contributions of each frequency interval.

4.3 Numerical examples using the two-level atom

In (22) an atlas of theoretical P Cygni line profiles produced using the escape probability formalism of (23) with the Sobolev approximation in a spherically symmetric expanding shell. We wish to discuss a single example of how to improve such approximate spectra. A more detailed account is given in (24). Here the following simplified model is assumed. For the expanding shell the velocity and particle density are given by

$$v(r) = 3 \cdot 10^6 \text{ cm/s } (r/r_c)^{1/2}$$

$$n(r) = 4.5 \cdot 10^{12} \text{ cm}^{-3} \; (r_c/r)^{5/2}$$

where $r_c = 10^{12}$ cm is the radius of the central source. The outer radius of the shell was assumed to be 10 r_c. A line at 25μ was assumed to have an Einstein coefficient for spontaneous emission $A_{21} = 1.07 \cdot 10^{-9}$ s^{-1} and statistical weights $g_1 = g_2 = 1$. The ratio ε, defined as $\varepsilon = (C_{21} - C_{12})/(A_{21} + C_{21} - C_{12})$ is assumed to be everywhere constant. With this simplified model the spectrum generated by the approximate method is compared to both 1) the spectrum calculated using the ray equations with the approximate source function and 2) the spectrum from the exact solution. The spectra from 1) and 2) deviated only slightly from one another, so that in Figure 11 we have only plotted the two approximate solutions for the case $\varepsilon = 10^{-4}$. In Figure 12 we show the source functions from the exact solution and from the approximate method.

Many systematic deviations of the Sobolev approximate solution from the exact solution are evident in Figures 11 and 12. The absorption minimum and the emission maximum in the Sobolev spectrum are too extreme (the maximum is too high, the minimum too low) and they occur at the wrong frequencies (the red shifted maximum is too blue, the blue minimum too red). Spurious and unrealistic fine structure is evident in the emission component; its strong assymmetry is also unrealistic. For a systematic study of the validity of the Sobolev assumption for P Cygni profiles the interested reader is referred to (25).

4.4 Formation of molecular lines in protostellar envelopes

As an example of astrophysical interest we discuss the exact numerical solution of the line transfer problem in the extended infalling envelopes surrounding protostars as considered by (24). The hydrodynamic results and dust properties of (8,9) for a 3 $M_\odot$ protostellar cloud at various stages of evolution are used for the density and velocity distribution of the infalling gas. The molecules considered are H_2 and OH. For H_2 only the seven lowest rotational levels were included; the energy of the highest (J=6) level corresponds to T = 3500 K. Wavelengths and Einstein coef-

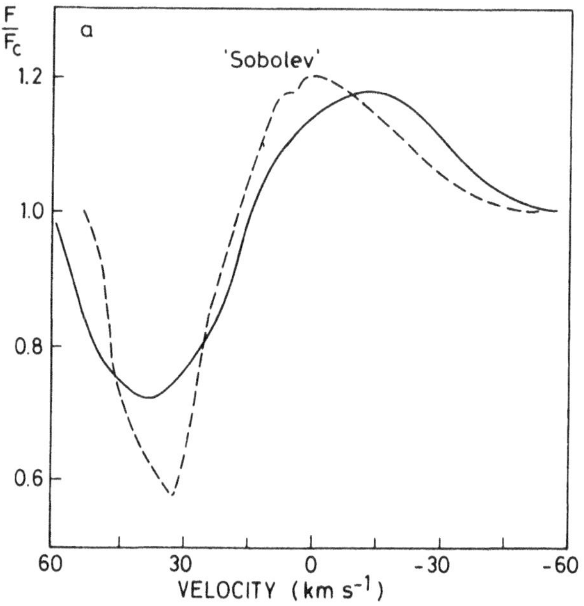

Figure 11. The two spectra obtained from the approximate source function as discussed in text. The source function is calculated using the Sobolev approximation together with an escape probability formalism. The dotted curve (labeled "Sobolev") shows the spectrum obtained assuming an infinitely thin profile function and the solid curve is the spectrum obtained from the ray equations. The exact solution (not shown) has a spectrum very similar to the solid curve.

ficients for the allowed transitions were taken from (26). The collisional cross-section for all transitions was chosen to be $4 \cdot 10^{-18}$ cm^2, consistent with numerical calculations by (27) for temperatures in the range 60K < T < 150K. Higher temperature cross-sections are unfortunately not available. Above 2000 K the H_2 density was assumed to be zero. Some of the results of the exact calculations are given in Figures 13 and 14.

For the above calculations 37 frequency intervals were used for each line when calculating the final profiles. During the iteration process, however, 21 intervals were used.
The resolution of the r- and p-grid points was 100. Approximately 5 min of UNIVAC 1100/83 CPU time were necessary for each iteration (see Table 1); the final line profiles with high frequency resolution required 10 min. Only 2 to 3 iterations were necessary for each of the evolutionary models.

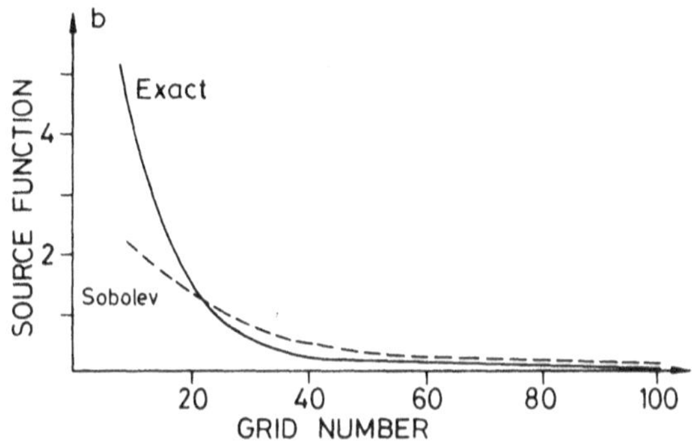

Figure 12. The source function from the exact solution (solid line) is plotted as a function of radial grid number and compared to the approximate source function (dotted line). Grid number 1 corresponds to the surface of the central source of radius r_c and grid number 100 is the outer surface at radius 10 r_c. The grid points are distributed linearly between r_c and 10 r_c.

An attempt to calculate the population densities of the first six levels of the OH molecule (without hyperfine structure) in collapsing protostellar envelopes failed due to poor convergence (c.f. 24). The first reason for this was the population inversion which occurred in the first excited level with respect to the ground level. This inversion is responsible for the 18 cm maser emission of OH. A slight change in the relative population densities of the first two levels anywhere in the region from $r = 5 \cdot 10^{13}$ cm to $r = 2.4 \cdot 10^{16}$ cm (where the inversion occurred) caused a large change in the amplified 18 cm intensity. This affected the inversion in other parts of the cloud. In spite of this the procedure did appear to be converging after 35 iterations although very slowly. Probably at least 100 iterations (corresponding to the number of radial grid points) would be necessary, provided that no efficient "convergence accelerator" can be found to improve convergence. Maser emission should be considered an extreme test of any method of calculating line profiles in extended moving atmospheres.

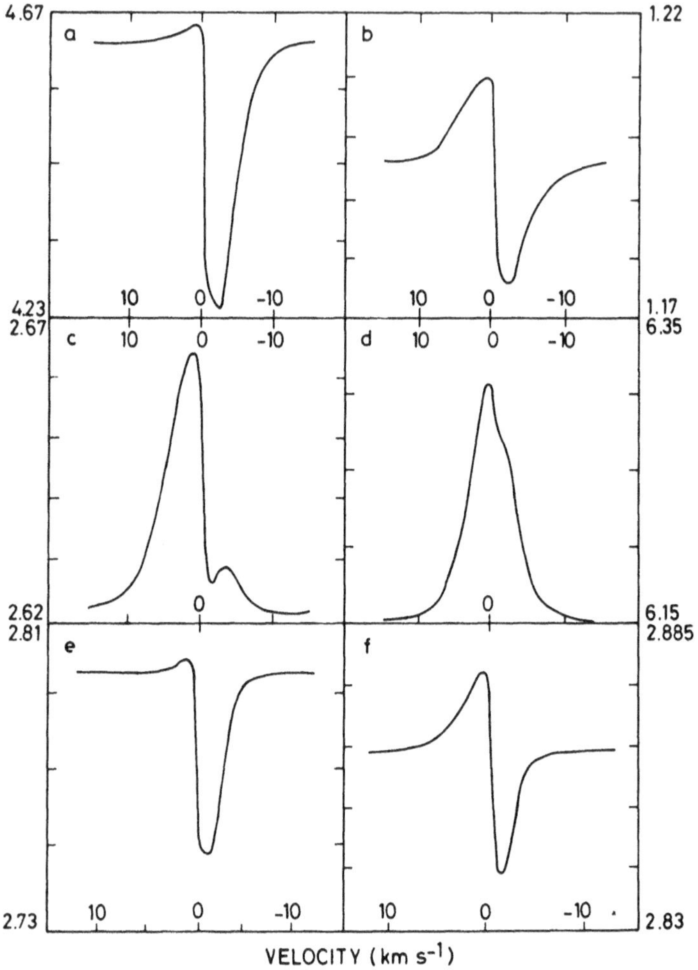

Figure 13. Normalized line profiles of para- and ortho-H_2 at various stages of evolution of a protostellar envelope after the formation of a central (accreting) protostar. The total mass of the envelope and the central core was 3 $M_\odot$. 13a. Ortho-H_2 at λ =17μ (J=3→J=1) for evolutionary model A (c.f. (8)). At this time (t=1.8·10^4yr) the central core has accreted M_*= 0.5 $M_\odot$. The range of the vertical axis is given to the left (arbitrary units). 13b. Same as 13a, except that model B was used (t=6.7·10^4yr, M_*= 1.0 $M_\odot$). The range of the vertical axis is given to the right. 13c. Same as 13a, except that model C was used (t=1.6·10^5yr, M_*= 1.5 $M_\odot$). 13d. Same as 13b, except that model D was used (t=3.3·10^5yr, M_*= 2.1 $M_\odot$). 13e. Same as 13a, except that the para-H_2 line at λ = 28.2μ (J=2→J=0) was used. 13f. Same as 13b, except that the para-H_2 line at λ = 28.2μ (J=2→J=0) was used.

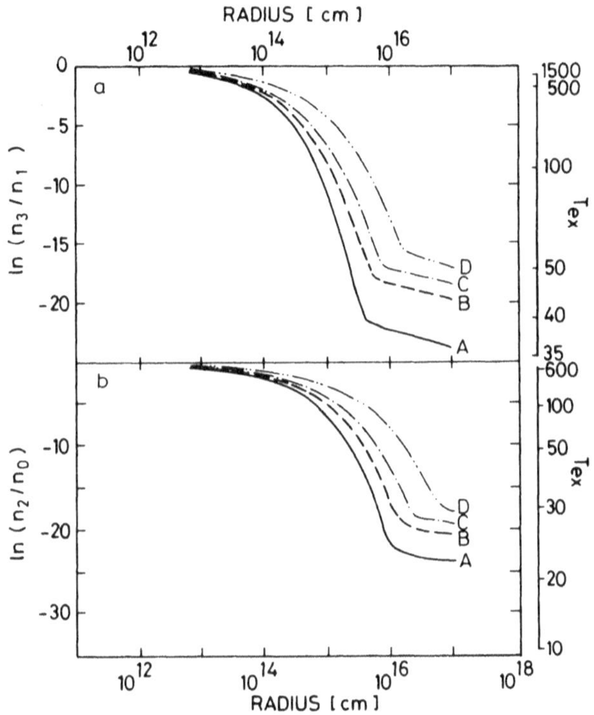

Figure 14. The ratio of population densities, a) n_3/n_1 and b) n_2/n_0 as a function of radius in the protostellar envelope is given for the four evolutionary phases A, B, C and D as indicated. The excitation temperature scale to the right shows the temperature necessary to maintain this ratio in thermodynamic equilibrium. The sharp kinks in the curves show where deviations from LTE begin to occur. In the outer regions the excitation temperatures lie above the kinetic gas temperature.

5. A FEW CONCLUDING REMARKS

From the numerical examples discussed in sections 3 and 4 we conclude that the continuum radiation transfer problem can be solved exactly in spherical symmetry and in a straightforward manner simultaneously with the hydrodynamical equations. For the transfer of radiation in a line the situation is slightly more complex. For individual non-blending lines an exact solution can be found in a reasonable amount of CPU time, provided that not too many transitions have to be considered. The rate equations can even be solved simultaneously with hydrodynamics. Complications such as partial redistribution and line blending can be included without any principal difficulties. Unfortunately, the computer code has to be strongly tailored to the particular problem at hand and it is difficult to make general conclusions, valid for all cases.

When a large number of transitions have to be considered the CPU time will increase of course. The CPU time required for the ray tracing part of the code increases linearly with the number of frequency intervals; for the solution of the rate equations the CPU time increases approximately with the cube of the number of energy levels. If we assume that the number of possible transitions increases as the square of the number of energy levels N, denote by L the number of frequency intervals per transition, by J and K the number of radial and p-grid points, respectively, we may write for the ray tracing CPU time t_{RT}

$$t_{RT} \propto N^2 \times L \times J \times K \tag{5.1}$$

and for the rate equations for statistical equilibrium

$$t_{SE} \propto N^3 \times J \tag{5.2}$$

Approximate methods for the transfer of continuum and line radiation in 1-D (spherical or slab geometry) and 2-D (axial symmetry) have been applied to astrophysical problems with varying degrees of success. Each approximate method must be tailored to fit the particular problem at hand - it is always possible to find a situation for which any given approximate method fails. For this reason it is important to check the validity of the method before overinterpreting the results. Using the approximate source function from the "Sobolev" solution together with the ray equations (see section 4.3) often yields fairly accurate spectra in cases for which the spectra from the Sobolev solution alone are poor. Although detailed calculations and tests have been performed in 1-D only, they may be extended to similar axially symmetric and 3-D cases for which no general exact solutions exist.

The question of whether it is possible to make "exact" 2-D continuum radiation transfer calculations should be answered with "in principle yes, but...". Assume that a strategy of solution for the 2-D case will be similar to that discussed in section 3.1 for the spherically symmetric case. By alternately solving the ray equations and the moment equations (using the three Eddington factors, f_{rr}, $f_{r\Theta}$ and $f_{\Theta\Theta}$ derived from the ray tracing solution) convergence should be fairly rapid, when scattering processes do not dominate. The numerical examples presented in section 2.3 can be considered the first half-step in such an iterative procedure. Here, ray tracing was done through the entire region for selected angles and frequencies. The next half-step would involve integrating the intensity multiplied by the relevant direction cosines over all angles. From this one obtains the frequency dependent moments which are then used to calculate the Eddington factors and the mean extinction coefficients: κ_J, κ_ρ and κ_z, where κ_ρ and κ_z are calculated in a manner similar to (3.7) using $H_\rho(\nu)$ and $H_z(\nu)$ instead of H_ν.

The integration over angle is not as straightforward as in the spherically symmetric case. Consider for instance a cylidrical coordinate system (ρ,z). For simplicity we drop the primes for the angles (Θ,ϕ) describing the direction. For a given angle Θ the ϕ-integration of the integrants appearing in (1.6abc) at a given point (ρ,z) in space must be transformed into a weighted sum over the corresponding subset of (S,T,U) points (see Figure 3). In general, cumbersome interpolation is unavoidable. This procedure must be repeated for a number of different angles Θ and the corresponding weighted sums for the Θ-integration must be calculated. Then the necessary frequency integrations can be completed. The latter two summations are straightforward; they are simple quadratures (without interpolation) at each (ρ,z) spatial point.

Clearly the ϕ-integration is the most difficult substep of this procedure. It appears that for each angle Θ_k, six 5-index matrices must be known of the type $a^{(k)}_{ijstu}$ so that the partial moments (partial means for the Θ-integration) $A^{(kl)}(\rho_i,z_j) = A^{(kl)}_{ij}$ can be calculated (A means J, H_ρ, H_z, $K_{\rho\rho}$, $K_{\rho z}$ or K_{zz}):

$$A^{(kl)}_{ij} = \sum_s \sum_t \sum_u a^{(k)}_{ijstu} I^{(kl)}_{stu} \tag{5.3}$$

where the matrix $I^{(kl)}_{stu}$ is the intensity for a given angle Θ_k and given frequency ν_l in the (S, T, U) coordinate system and s, t, u are the corresponding indices . Actually, most of the components of the matrices of type $A^{(k)}_{ijstu}$ are zero. For each i, j, k combination the non-zero components lie close to a tilted ring in the (S,T,U) coordinate system. Parameterizing the ϕ-variable by the index n this tilted ring can be described by arrays of the type $s = s^{(ijk)}_n$, $t = t^{(ijk)}_n$, $u = u^{(ijk)}_n$ so that the triple summation of (5.3) reduces to a single sum over n. Alternate schemes for the ϕ-integration are also possible, for example, adding the partial contributions to the partial moments $A^{(kl)}_{ij}$ for all i,j after each intensity value $I^{(kl)}_{stu}$ has been calculated. One always runs into the difficulty that a large number of computations are necessary, however.

From the previous discussion it should be clear that 2-D radiation transfer solved simultaneously with the equations of energy balance (and perhaps of hydrodynamics) is possible but very CPU-time consuming. For this reason approximate solutions

(Eddington approximation, flux limited diffusion and others) are still in general use and will be for some time to come. Comparison calculations using different methods and test calculations of these methods in cases for which the exact solution is known (i.e. in 1-D), are absolutely indispensable. As a single example of such a test calculation we shall compare the exact 1-D radiation transfer solution of (19), model (1, 100), with a simple frequency dependent flux limiter. The model (1, 100) has been described briefly in section 3.3.

We have chosen to make a very simple flux limiter. At each frequency the moment equations are solved asuming f = 1/3. At each radial grid point the first moment H_ν is truncated to the maximum allowed value J_ν. Then, these frequency dependent moments are used to calculate the mean extinction coefficients κ_J and κ_H according to (3.7) solved with κ_J, κ_H and f = 1/3 simultaneously with the energy equations for the temperatures T_r and T_v of the two types of dust grains considered. The resulting tempera- tures of volatile grains T_v (grains which are destroyed at temperatures around 150 K) and of refractory grains T_r (which are destroyed at temperatures above 1700 K) are shown in Figure 15 as a function of radius r from the central protostellar source. The resulting approximate and exact spectra are shown and compared

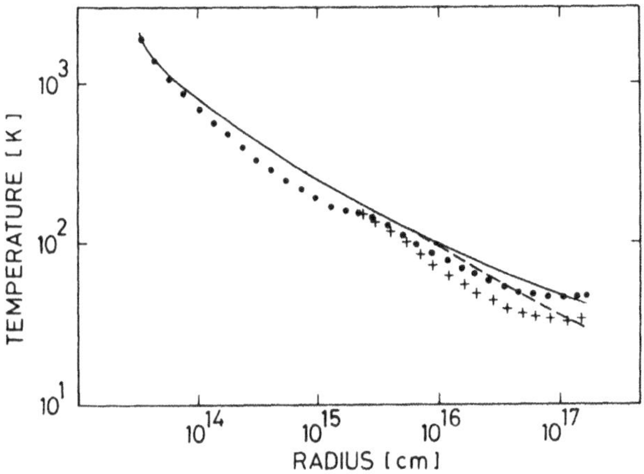

Figure 15. Temperature distributions for volatile and refractory dust grains in an infalling protostellar envelope illuminated by a central source. The total optical depth of the envelope from r = $1.5 \cdot 10^{17}$ cm to $3.3 \cdot 10^{13}$ cm (where the refractory grains are destroyed) at a wavelength λ = 1000 Å was assumed to be 100. The solid line (dashed line) shows the exact solution for refractory (volatile) grains. The dots (crosses) depict the approximate temperature of the refractory (volatile) grains.

in Figure 16. The angular distribution of radiation for this particular model has been discussed previously (see section 3.3).

Except in the outermost regions the temperatures derived by the approximate method are consistently too low, typically by a factor of 10% to 15%. The approximate spectra display the typical double-peaked spectral structure characteristic for protostellar envelopes. The ratio of the near IR and far IR peaks is very inaccurate, however. In spite of the lower calculated temperatures, the approximate solution predicts near IR fluxes that are too high, far IR fluxes that are too low. The total frequency-integrated radiation flux is conserved by the approximate solution, and the qualitative distribution of the far IR flux is also well represented. Under certain circumstances such an approximate solution could be considered sufficiently accurate. Using the approximate temperature distributions together with the ray equations one could produce more accurate spectra. Although this procedure is not necessary in the 1-D case, since the exact solution is obtained with little extra effort, such an approximate proce-

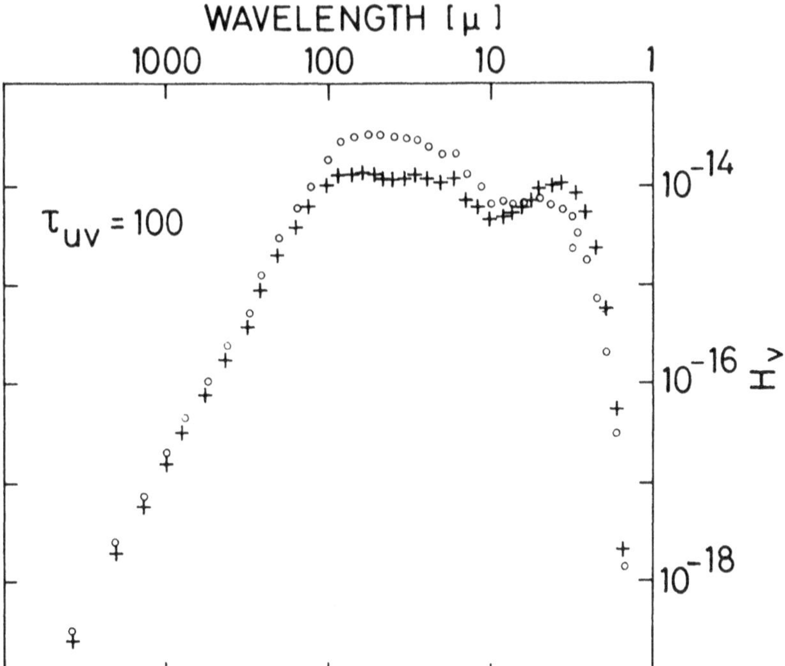

Figure 16. The approximate spectrum (crosses) is compared to the exact solution (open circles) for the case discussed in text and Figure 15. The exact solution has been taken from (19).

dure could be adopted in the 2-D case. Line transfer could also be solved approximately in 2-D, if one is willing to use the Sobolev solution for the source function in the detailed ray-tracing calculation.

Acknowledgements: I am indebted to the secretaries who typed this manuscript for their patience and to the editors of this book for their patience and encouragement.

References

1. Mihalas, D.: 1986, this volume
2. Castor, J.I.: 1986, this volume
3. Sobolev, V.V.: 1957, Soviet Astron.-AJ 1, pp. 678-689
4. Sobolev, V.V.: 1960, "Moving Envelopes of Stars", Cambridge: Harvard University Press
5. Mihalas, D.: 1978, "Stellar Atmospheres", 2nd ed., San Francisco: W.H. Freeman and Company
6. Tscharnuter, W.: 1981 in "Fundamental Problems in the Theory of Stellar Evolution", IAU Symp. No. 93, pp. 105-106
7. Tscharnuter, W.: 1986, this volume
8. Yorke, H.W.: 1979, Astron. Astrophys. 80, pp. 308-316
9. Yorke, H.W.: 1980, Astron. Astrophys. 85, pp. 215-220
10. Bodenheimer, P., Tenorio-Tagle, G., Yorke, H.W.: 1979, Astrophys. J. 233, pp. 85-96
11. Sanford II, M.T., Whitaker, R.W., Klein, R.I.: 1982, Astrophys. J. (submitted)
12. Spitzer, L. Jr.: 1978, "Physical Processes in the Interstellar Medium", New York: John Wiley & Sons
13. Yorke, H.W., Tenorio-Tagle, G., Bodenheimer, P.: 1983 (in prep.)
14. Yorke, H.W., Tenorio-Tagle, G., Bodenheimer, P.: 1983 in "Supernova Remnants and their X-ray Emission", IAU Symp. No. 101
15. Yorke, H.W.: 1980, Astron. Astrophys. 86, pp. 286-294
16. Hummer, D.G., Rybicki, G.B.: 1971, MNRAS 152, pp. 1-19
17. Winkler, K.-H., Newman, M.J.: 1980, Astrophys. J. 238, pp. 311-325
18. Winkler, K.-H.: 1986, this volume
19. Yorke, H.W., Shustov, B.M.: 1981, Astron. Astrophys. 98, pp. 125-132
20. Yorke, H.W.: 1977, Astron. Astrophys. 58, pp. 423-432
21. Bastian, U., Bertout, C., Stenholm, L., Wehrse, R.: 1980, Astron. Astrophys. 86, pp. 105-112
22. Castor, J.I., Lamers, H.J.G.L.M.: 1979, Astrophys. J. Suppl. 39, pp. 481-511
23. Castor, J.I.: 1970, MNRAS 149, pp. 111-127
24. Langbein, T.: 1982, Diplom thesis, U. Goettingen (in German)

25. Haman, W.-R.: 1981, Astron. Astrophys. 93, pp. 353-361
26. Dalgarno, A., Wright, E.L.: 1972, Astrophys. J. 174, pp. L49-51
27. Monchik, L., Schaefer, J.: 1980, J. Chem. Phys. 73, pp. 6153-61 .

IMPLICIT 2D-RADIATION HYDRODYNAMICS

Werner M. Tscharnuter
Institut für Astronomie, Universität Wien
Türkenschanzstraße 17, A-1180 Wien

I. INTRODUCTION AND GENERAL REMARKS

It is a well known fact that explicit numerical methods do not apply to evolutionary models of protostars. As was already pointed out by McNally (1964) and by Bodenheimer and Sweigart (1968), due to the conspicuous non-homology of the gravitational collapse of a (non-rotating, non-magnetic) interstellar cloud on the verge of being Jeans-unstable, both time and length scales vary by several orders of magnitude during the evolution of the cloud. To arrive at a reasonable spatial resolution and accuracy the famous Courant-Friedrichs-Lewy (CFL-) condition would demand a billion (!) time steps or so. This is why purely explicit methods are completely ruled out for tackling stellar formation models. The main difficulty to overcome is to physically and consistently describe the flow pattern which basically consists of a highly supersonic part (the freely falling envelope) and a quasi-hydrostatic part (the stellar embryo) separated by a strong shock front (the accretion shock). The accretion shock is of fundamental importance for the structure of a protostar, since there, within a very narrow zone, the kinetic energy of the matter in the envelope is almost entirely transformed into outgoing radiation, i.e. luminosity; it also determines the entropy distribution in the postshock layers, i.e. the structure of the outer layers of the stellar core. Needless to say, the problem becomes even more difficult, if the stringent condition of spherical symmetry is abandoned.

As a matter of fact, the first calculation which completely covered the protostellar evolution from the early collapse of a dispersed, Jeans-unstable cloud up to a much more compact, slowly

K.-H. A. Winkler and M. L. Norman (eds.), Astrophysical Radiation Hydrodynamics, 181–185.

contracting pre-main-sequence star (Larson, 1969), suffered from merely ad-hoc-assumptions concerning the structure of the accretion shock. However, subsequent investigations (Appenzeller and Tscharnuter, 1975; Winkler and Newman, 1980a,b) based on more and more sophisticated numerical strategies essentially confirmed Larson's early results. In a series of papers Stahler et al. (1980a,b, 1981) rediscussed these results by using a rather attractive semi-analytical approach. Their models are based on the assumption that asymptotic self-similar solutions of the dynamical equations (Shu, 1977) hold, which, indeed, is indicated by the honest numerical calculations. Particular attention has been focused on the physically correct formulation of the jump conditions for the accretion shock including the radiation field. It would be extremely helpful, if a similar method for treating non-spherical flows were able to yield convincing results, before vastly time consuming calculations have been carried out.

II. DISCUSSION OF THE BASIC EQUATIONS

The equations to be solved are the usual hydrodynamical ones, that is the equations of continuity and motion, respectively; furthermore, two energy balance equations, since the gas temperature and the radiation (equivalent-) temperature need not coincide, and Poisson's equation governing the gravitational potential. In the axially symmetric case the radiative energy transport is taken into account by using the (grey) Eddington approximation. It relates the total (frequency integrated) radiative flux to the gradient of the energy density of the radiation field by using just one fixed anisotropy factor equal 1/3. Apart from these approximations the coupling of the radiation field with the gas flow is consistent up to the order of the gas velocity divided by the speed of light (Castor, 1972). The set of equations is closed by adding constitutive relations, i.e. the equation of state, the internal energy and the opacity as functions of, say, density and temperature, as well as certain boundary conditions, e.g. the equivalent temperature of the external radiation field, the external pressure, the external mass distribution, regularity conditions at the center and, to some extent, at the symmetry axis.

The common procedure to solve this set of nonlinear equations is to replace the differential expressions by finite differences. In doing so, one ends up with a non-linear set of algebraic equations which can now be iteratively solved by a Newton-Raphson techinque. This is probably the most effective method to arrive at a sufficiently accurate solution within 4-6 iteration steps. Numerically, the matrix operations are particularly simple for 1D-problems, where the fast Henyey-algorithm applies, but for 2D- or even 3D-problems no equivalent procedure is known.

III. THE "HYBRID"-SCHEME

Protostellar collapse including rotation has to be modelled at least with axial symmetry. This constraint, together with the need for a self-consistent determination of the cloud's gravitational potential, led to the idea of matching an implicit finite-difference scheme for the r-direction with an expansion into Legendre-polynomials for the Θ-direction of a spherical polar coordinate system (r,Θ,ϕ). As a result, the block structure of the coefficient matrix representing the linearized algebraic equations for the 1D-case is retained at the expense of N times the number of equations, where N is the number of Legendre - polynomials taken into account. Thus, without further modifications, the Henyey-technique is fully applicable.

The actual calculations based on such a "hybrid"-scheme split into four distinct parts:

i) Let N_g be the number of gridpoints in the Θ-direction chosen to be suitable Gaussian integration points, let M be the number of radial gridpoints. Compute all basic physical quantities on the meshpoints of this $(M \times N_g)$-grid. It is convenient to deal with logarithms of any strictly positive quantity, like density, internal energy etc., that vary over several orders of magnitude.

ii) Constitute the equations according to a particularly chosen difference-scheme in the r-direction (plus the boundary conditions) at each point of the $(M \times N_g)$-grid. Set up their partial derivatives with respect to all Legendre-coefficients for the Newton-Raphson method.

iii) Calculate the respective N Legendre components of the equations simultaneously with their derivatives by multiplying the original physical equations by the corresponding Legendre-polynomial and integreting over the Θ-coordinate by virtue of the Gaussian quadrature formula with N_g integration points. This yields N independent equations for N unknown Legendre-coefficients, multiplied by the total number of basic physical variables taken into consideration (e.g. the density, the components of the velocity in the r-,Θ-,ϕ-direction, the gravitational potential, internal energy, radiation energy density).

iv) Apply the Henyey-method and find updated Legendre-coefficients. Depending on the accuracy check, either repeat the iteration or advance to the next time step.

A few further comments should be made: With axial symmetry matter is not allowed to cross the symmetry axis. This particular "boundary" condition can easily be fulfilled by simply choosing an appropriate Legendre-expansion of the Θ-velocity, namely, by using $P_k^1(\cos\Theta)$ instead of the usual $P_k(\cos\Theta)$, $k=0,1,\ldots N-1$. At the poles $(\Theta = \pm\pi)$ P_k^1 becomes zero, as desired. In case there exists an equatorial plane introducing an additional symmetry, k must be even, i.e. $k = 0,2,\ldots$ Since $P_0^1 = 0$, which reflects the fact that the Θ-velocity should be identically zero with spherical symmetry (N=1), the equation of motion for the Θ-direction splits into N-1 rather than N equations. Thus, one single equation is always free to our disposal and we may take the advantage of being able to naturally add any useful relation for determining the r-grid to the physical structure equations. This happy circumstance is important, since neither Eulerian nor Lagrangean systems are adequate to represent protostellar flow-patterns sufficiently well. For an extensive discussion of this point we refer to Winkler and Norman (these proceedings).

More detailed formulae pertinent to the hybrid-method are given in Tscharnuter and Winkler (1979), where also a tensor formulation of the artificial viscosity can be found. Artificial viscosity is necessary to broaden shock fronts, in particular the accretion shock, in a numerically tractable way. Further examples may be found in Tscharnuter (1978) and Bodenheimer and Tscharnuter (1979). In most cases the parameters M, N, N_g have been chosen to be 80, 8, 36, respectively.

IV. CONCLUSIONS

In spite of rather encouraging results obtained by the hybrid-method there are still important problems inherent to the method to be solved. For example, with respect to accuracy, it is not at all clear, how the Legendre-expansion, the number of polynomials and the number of Gaussian integrations points relate to the equivalent number of gridpoints of a usual finite difference scheme with respect to the Θ-direction. The hybrid-scheme is certainly useful to describe moderately flattened objects. However, its applicability to very flat disk configurations has not yet been convincingly shown. A major improvement to be made in the near future will be the (numerically stable) formulation of the structure equations in a complete conservation form. This should be done by a rather straightforward extension of Winkler and Norman's (these proceedings) ideas to the 2D-case.

Since the hybrid-scheme has been developed for the very special case of protostellar collapse, it need not at all be suited to any other 2D-flow equally well. Results of moderate or even poor accuracy for a single test example does not render a code worthless at all. The qualification of a code depends on the selected test. For example, every explicit scheme would completely fail for collapse problems, but some are very powerful for other type of flows (Woodward, these proceedings). In general, codes which, in some way or another, are based on a series expansion are not expected to yield the most accurate results.

This is particularly true for our hybrid-scheme, but as long as there are no better codes available, it may well serve as a useful tool for studying the physics of rotating (proto-) stars quite in detail.

V. REFERENCES

Appenzeller, I. and Tscharnuter, W.: 1975, Astr. Ap. 40, 397
Bodenheimer, P., and Sweigart, A.: 1968, Ap. J. 152, 515
Bodenheimer, P., and Tscharnuter, W.M.: 1979, Astr. Ap. 74, 288
Castor, J.I.: 1972, Ap. J. 178, 779
Kippenhahn, R., Weigert, A., Hofmeister, E.: 1967, Methods Comp. Phys. 7, 129
Larson, R.B.: 1969, M.N.R.A.S. 145, 271
McNally, D.: 1964, Ap. J. 140, 1088
Shu, F.M.: 1977, Ap. J. 214, 488
Stahler, S.W., Shu, F.M., Taam, R.E.: 1980a, Ap. J. 241, 637
Stahler, S.W., Shu, F.M., Taam, R.E.: 1980b, Ap. J. 242, 226
Stahler, S.W., Shu, F.M., Taam, R.E.: 1981, Ap. J. 248, 727
Tscharnuter, W.M.: 1978, Moon and Planets 19, 237
Tscharnuter, W.M., and Winkler, K.-H. A.: 1979, Computer Phys. Comm. 18, 171
Winkler, K.-H. A. and Norman, M.L.: 1986, these proceedings
Winkler, K.-H. A., and Newmann, M.J.: 1980a, Ap. J. 236, 201
Winkler, K.-H. A., and Newmann, M.J.: 1980b, Ap. J. 238, 311
Woodward, P.: 1986, these proceedings

2-D EULERIAN HYDRODYNAMICS WITH FLUID INTERFACES, SELF-GRAVITY AND ROTATION

Michael L. Norman and Karl-Heinz A. Winkler

Los Alamos National Laboratory and
Max-Planck-Institut für Physik und Astrophysik

1. INTRODUCTION

The purpose of this paper is to describe in detail the numerical approach we have developed over the past five years for solving 2-dimensional gas-dynamical problems in astrophysics involving inviscid compressible flow, self-gravitation, rotation, and fluid instabilities of the Rayleigh-Taylor and Kelvin-Helmholtz types. The computer code to be described has been applied most recently to modeling jets in radio galaxies (Norman et al. 1981, 1982) and is an outgrowth of a code developed for studying rotating protostellar collapse (Norman, Wilson and Barton 1980; Norman 1980). The basic methodology draws heavily on the techniques and experience of James R. Wilson and James M. LeBlanc of the Lawrence Livermore National Laboratory, and thus the code is designed to be a general purpose 2-D Eulerian hydrocode, and is characterized by a high degree of simplicity, robustness, modularity and speed. Particular emphases of this article are: 1) the recent improvements to the code's accuracy through the use of vanLeer's (1977) monotonic advection algorithm, 2) a discussion of the importance of what we term consistent advection, and 3) a description of a numerical technique for modeling dynamic fluid interfaces in multidimensional Eulerian calculations developed by LeBlanc.

The outline of this paper is as follows. In Sec. 2 we present the physical equations and our two-step methodology for solving them. Finite-difference equations for these two steps -the source step and transport step- are given in Secs. 3 and 4, respectively. The fluid interface technique we use is then described and discussed in Sec. 5. In Sec. 6 we summarize our iterative alternating-direction-implicit (ADI) procedure for solving the Poisson equation. Our timestep control procedure is given in Sec. 7. Finally, Sec. 8 contains several applications of this code to astrophysical problems of current interest involving fluid interfaces, self-gravity and rotation.

K.-H. A. Winkler and M. L. Norman (eds.), Astrophysical Radiation Hydrodynamics, 187–221.

2. BASIC EQUATIONS AND METHODOLOGY

2.1 Fluid Equations in moving coordinates

In applications involving gravitational collapse or explosions, a moving coordinate mesh is used to maintain adequate problem coverage and zoning resolution. We therefore begin by writing the basic equations of self-gravitating ideal gas dynamics in such a coordinate system:

continuity equation

$$d/dt \int \rho d\tau + \int \rho(\mathbf{v}-\mathbf{v}_g)\cdot \mathbf{d\Sigma} = 0, \tag{1}$$

momentum equation

$$d/dt \int \rho \mathbf{v} d\tau + \int \rho \mathbf{v}(\mathbf{v}-\mathbf{v}_g)\cdot \mathbf{d\Sigma} + \int (\nabla P + \rho \nabla \Phi) d\tau = 0, \tag{2}$$

internal energy equation

$$d/dt \int \rho \varepsilon d\tau + \int \rho \varepsilon (\mathbf{v}-\mathbf{v}_g)\cdot \mathbf{d\Sigma} + \int P \nabla \cdot \mathbf{v} d\tau = 0. \tag{3}$$

Here, the time derivatives and spatial integrations operate on the moving grid zone of volume $d\tau$ and surface area $\mathbf{d\Sigma}$ moving with velocity $\mathbf{v}_g$ with respect to a fixed (Eulerian) observer; ρ, ε and $\mathbf{v}$ are the fluid's mass density, specific internal energy and Eulerian velocity, respectively. The pressure P will usually be computed from the ideal gas law $P=(\gamma-1)\rho\varepsilon$, where γ is the ratio of specific heats, although introducing a general equation of state $P=P(\rho,\varepsilon)$ offers no principle difficulties. The gravitational potential Φ is computed from the Poisson equation

$$\nabla^2 \Phi = 4\pi G \rho. \tag{4}$$

Equations (1)-(4) form a complete set once $\mathbf{v}_g$ is specified, and are sufficient to determine the problem for given initial and boundary conditions. Note that if one sets $\mathbf{v}_g=0$ in eqs. (1)-(3), then d/dt becomes the Eulerian time derivative $\partial/\partial t$, which commutes with the volume integral. Applying the divergence theorem, one easily recovers the Eulerian differential equations of hydrodynamics.

2.2 Two-step solution procedure

Explicit multi-step solution procedures for solving the equations of hydrodynamics are generally more accurate than a single step that simply extrapolates forward in time on the basis of old data. Before we specialize eqs. (1)-(4) to a particular geometry and begin writing down difference equations, we would like to describe our two-step approach to solving the fluid equations which is independent of geometry or dimensionality.

Eqs. (1)-(3) are solved in two logically independent steps which we call the *source* and *transport* steps. In the source step, we accelerate fluid velocities and perform pressure work on the gas internal energy by solving finite-difference approximations to the following differential equations:

$$\rho d\mathbf{v}/dt = -(\nabla P + \rho\nabla\Phi) - \nabla\cdot\mathbf{Q} \tag{5}$$

$$\rho d\varepsilon/dt = -P\nabla\cdot\mathbf{v} - \mathbf{Q}:\nabla\mathbf{v} \,. \tag{6}$$

We have introduced additional terms involving $\mathbf{Q}$ in eqs. (5) and (6), which represent acceleration and heating due to artificial viscous stresses used to mediate the numerical shock transitions. In the transport step, fluid is transported through the computational mesh by solving finite-difference approximations to the following integral equations:

$$d/dt \int \rho \, d\tau = -\int \rho \, (\mathbf{v}-\mathbf{v}_g)\cdot\mathbf{d\Sigma}, \tag{7}$$

$$d/dt \int \rho\mathbf{v} d\tau = -\int \rho\mathbf{v}(\mathbf{v}-\mathbf{v}_g)\cdot\mathbf{d\Sigma}, \tag{8}$$

$$d/dt \int \rho\varepsilon d\tau = -\int \rho\varepsilon(\mathbf{v}-\mathbf{v}_g)\cdot\mathbf{d\Sigma}. \tag{9}$$

The updated values of $\mathbf{v}$ and ε from the source steps are used to evaluate the right-hand-sides of eqs. (7)-(9), and enter as the old values in the time-discretized left-hand-sides of eqs. (8) and (9).

To understand the origin of these equations, consider the momentum equation in differential form, which can be derived from eq. (2) using the identity

$$\nabla\cdot\mathbf{v}_g \equiv d/dt\,(\ln d\tau). \tag{10}$$

Letting $\mathbf{S} = \rho\mathbf{v}$, we have

$$d\mathbf{S}/dt + \mathbf{S}\nabla\cdot\mathbf{v}_g + \nabla\cdot[(\mathbf{v}-\mathbf{v}_g)\mathbf{S}] + \nabla P + \nabla\cdot\mathbf{Q} + \rho\nabla\Phi = 0 \tag{11}$$

which we solve incrementally as

$$d\mathbf{S}/dt = d\mathbf{S}/dt)_{\text{source}} + d\mathbf{S}/dt)_{\text{transport}} \tag{12}$$

where

$$d\mathbf{S}/dt)_{\text{source}} = -(\nabla P + \rho\nabla\Phi) - \nabla\cdot\mathbf{Q}, \tag{13}$$

and

$$d\mathbf{S}/dt)_{\text{transport}} = -\nabla\cdot[(\mathbf{v}-\mathbf{v}_g)\mathbf{S}] - \mathbf{S}\nabla\cdot\mathbf{v}_g. \tag{14}$$

Noting $d\rho/dt)_{\text{source}} = 0$ in eq. (13), we recover eq. (5); integrating eq. (14) over a moving volume we recover eq. (8) using eq. (10). The advantage of the integral formulation for the transport step is that it is in conservative form in a moving coordinate system, whereas eq. (14) is not due to the grid compression term. Moreover, an integral formulation is manditory for advecting fluid interfaces, which, since we treat them numerically as true discontinuities, cannot be described by differential equations.

3. SOURCE STEP

3.1 Grid and variables

Let U, V and W be the velocity components of a fluid element in the Z, R and θ directions, respectively, of a cylindrical coordinate system, and let $S=\rho U$, $T=\rho V$, and $A=\rho WR\equiv\rho\Omega R^2$ be the element's associated linear momentum density components and angular momentum density, respectively. Then, using the fact that $d\rho/dt)_{source} = 0$, we write eq. (5) in the explicit component form in which it is differenced as

$$dS/dt)_{source} = -\,\partial P/\partial Z - \rho\partial\Phi/\partial Z - \partial Q^{ZZ}/\partial Z, \tag{15}$$

$$dT/dt)_{source} = -\,\partial P/\partial R - \rho\partial\Phi/\partial R - \partial Q^{RR}/\partial R + \rho\Omega^2 R, \tag{16}$$

$$dA/dt)_{source} = 0, \tag{17}$$

and write eq. (6) as

$$dE/dt)_{source} = -\,P(\partial U/\partial Z + R^{-1}\partial RV/\partial R) - Q^{ZZ}\partial U/\partial Z - Q^{RR}\partial V/\partial R, \tag{18}$$

where $E=\rho\varepsilon$ is the internal energy density. Notice that only the diagonal elements of the artificial viscosity tensor have been retained in eqs. (15, 16 and 18), and that geometric terms such as Q^{RR}/R and $Q^{RR}V/R$ have not been included. The reasons for this are, first, we want artificial viscosity to be sensitive only to compressions to pick out shock fronts, hence we discard the off-diagonal elements, and second, we want the numerical shock width to be the same regardless of its distance from the symmetry axis, as it would be in nature on a macroscopic scale. Note, however, that a proper tensor formulation may be required for the artificial viscosity if special properties are sought (see Winkler and Norman, this volume).

The centering of the variables on the finite difference mesh and zone measurements are shown in Fig. 1. The Z and R grid lines have indices k and j, respectively. Linear momentum and velocity components are located at the zone faces; scalar densities, the gravitational potential and the angular momentum density are located at the zone center. The velocity components U and V are derived by dividing their respective momentum components by an arithmetic average of the two adjacent mass densities.

3.2 Difference equations

The location of the quantities on the mesh (Fig. 1) allow for simple centered differences and averages of the terms appearing on the right-hand-sides of eqs. (15,16 and 18). These equations are solved in steps as follows: 1) accelerate S and T due to pressure, gravitational and centrifugal forces; 2) using the updated velocities, compute the artificial viscous heating and acceleration; 3) using the updated E and velocity components from step 2), perform compressional heating on the gas. Thus, letting the superscripts a, q and p represent the updated values from the three steps and the unsuperscripted quantities represent values at the old

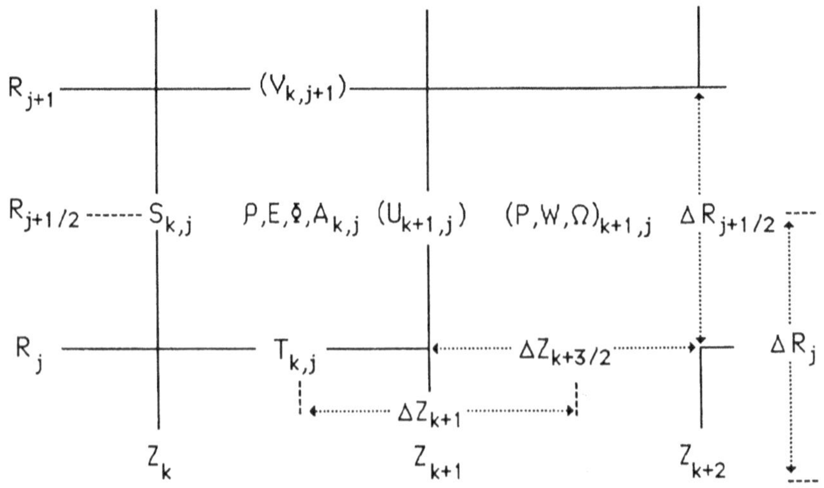

Fig. 1 Centering of the primary and secondary (derived) quantities on the mesh (Z_k,R_j).

time level, we solve the following explicit difference equations:

step 1

$$(S^a-S)_{k,j} / \delta t = -[(P_{k,j}-P_{k-1,j}) + \langle\rho\rangle_Z(\Phi_{k,j}-\Phi_{k-1,j})] / \Delta Z_k, \tag{19}$$

$$(T^a-T)_{k,j} / \delta t = -[(P_{k,j}-P_{k,j-1}) + \langle\rho\rangle_R(\Phi_{k,j}-\Phi_{k,j-1})] / \Delta R_j + \langle\rho\rangle_R \langle\Omega\rangle_R{}^2 R_j, \tag{20}$$

where the spatial averages are defined $\langle X\rangle_Z = (X_{k,j}+X_{k-1,j})/2$ and $\langle X\rangle_R = (X_{k,j}+X_{k,j-1})/2$. The form of the centrifugal force term may vary according to application; averaging on Ω as opposed to W, say, is superior when $\Omega(R)\approx$constant.

step 2

$$(S^q-S^a)_{k,j} / \delta t = -(Q^{ZZ}{}_{k,j} - Q^{ZZ}{}_{k-1,j})/\Delta Z_k, \tag{21}$$

$$(T^q-T^a)_{k,j} / \delta t = -(Q^{RR}{}_{k,j} - Q^{RR}{}_{k,j-1})/\Delta R_j, \tag{22}$$

$$(E^q - E)_{k,j} / \delta t = - Q^{ZZ}_{k,j} (U^a_{k+1,j} - U^a_{k,j})/\Delta Z_{k+1/2}$$
$$- Q^{RR}_{k,j} (V^a_{k,j+1} - V^a_{k,j})/\Delta R_{j+1/2}, \qquad (23)$$

where

$$Q^{ZZ}_{k,j} = \rho_{k,j}(U^a_{k+1,j} - U^a_{k,j})[-C_1 a + C_2 \min(U^a_{k+1,j} - U^a_{k,j}, 0)], \qquad (24)$$

$$Q^{RR}_{k,j} = \rho_{k,j}(V^a_{k,j+1} - V^a_{k,j})[-C_1 a + C_2 \min(V^a_{k,j+1} - V^a_{k,j}, 0)]. \qquad (25)$$

Here C_1 and C_2 are constants of order unity which govern the linear and quadratic artificial viscosities and a is the adiabatic speed of sound. The linear viscosity is rarely used, and then only sparingly to damp oscillations in stagnant regions of the flow.

step 3

Here, to improve energy conservation, we write an implicit difference equation involving the time-centered pressure $P^{n+1/2} = [P^n + (\gamma - 1)E^p]/2$ in $P\nabla \cdot \mathbf{v}$ which can be rearranged and solved explicitly:

$$(E^p - E^q)_{k,j} / \delta t = - [P^n + (\gamma - 1)E^p]/2 \, (\nabla \cdot \mathbf{v})_{k,j}, \qquad (26)$$

or

$$E^p_{k,j} = [E^q - P^n \delta t(\nabla \cdot \mathbf{v})/2]_{k,j} \, / \, [1 + (\gamma - 1)\delta t(\nabla \cdot \mathbf{v})/2]_{k,j}, \qquad (27)$$

where

$$(\nabla \cdot \mathbf{v})_{k,j} = (U^q_{k+1,j} - U^q_{k,j})/\Delta Z_{k+1/2}$$
$$+ (R_{j+1} V^q_{k,j+1} - R_j V^q_{k,j})/(R_{j+1/2} \Delta R_{j+1/2}). \qquad (28)$$

This procedure explicitly assumes a gamma-law gas; for a general equation of state $P = P(\rho, \varepsilon)$, we use a predictor-corrector approach to find the time-centered pressure, thus

predictor step

$$(E' - E^q)_{k,j} / \delta t = - P^n (\nabla \cdot \mathbf{v})_{k,j}, \qquad (29)$$

corrector step

$$(E^p - E^q)_{k,j} / \delta t = - [P^n + P']/2 (\nabla \cdot \mathbf{v})_{k,j}, \qquad (30)$$

where

$$P' = P(\rho, E'/\rho) \, . \qquad (31)$$

Experience has shown that energy conservation is improved by using the same pressure in eqs. (19-21) as is used for acceleration (i.e., P^n rather than, say $(\gamma-1)E^q$.)

4. TRANSPORT STEP

We now describe our numerical procedure for solving eqs. (7-9), which are all of the form

$$d/dt \int q \, d\tau = -\int q \, (\mathbf{v}-\mathbf{v}_g)\cdot \mathbf{d\Sigma}. \tag{32}$$

Equation (32) is manifestly in conservative form, and describes the advection of a quantity q on the moving mesh allowing for volumetric changes due to fluid ($\mathbf{v}$) and zonal ($\mathbf{v}_g$) convergence. This compound process we term *transport*. The obvious second-order accurate finite-difference approximation to eq. (32) is

$$(q^{n+1}\tau^{n+1}-q^n\tau^n)_{k,j}/\delta t = -(F^q_{k+1,j} - F^q_{k,j} + G^q_{k,j+1} - G^q_{k,j})^{n+1/2} \tag{33}$$

where $\tau_{k,j}$ is the zone volume and F^q and G^q are the time-centered fluxes at the faces of the zone at k,j in the axial and radial directions, respectively. Note that since q is assumed to be located at zone centers and at time-level n, interpolation and extrapolation procedures are in general required to compute the value of q at zone faces and at time-level n+1/2. A variety of such procedures have been developed over the years; indeed, the history of numerical Eulerian hydrodynamics is largely concerned with devising accurate estimates for the fluxes while insuring numerical stability. We employ Van Leer's (1977) second-order accurate monotonic interpolation scheme for the spatial centering, and extrapolate q along the relative streamline given by $d\mathbf{x}/dt = \mathbf{v}-\mathbf{v}_g$ for the temporal centering. This is illustrated below for a model one-dimensional problem, and then applied to our two-dimensional problem.

4.1 Van Leer monotonic interpolation scheme

Consider a one-dimensional strip of zones with index i, and a set of zone averages $\{q_i\}$ as in Fig. 2. Second-order accurate interpolation functions q(x) result from assuming a piecewise linear distribution of q within zones $q_i(\chi) = q_i + dq_i\chi$, $-1/2\leq\chi\leq 1/2$, where χ is the normalized distance from the zone's center. From this definition, it is clear that q_i is a zone average, since $\int q_i(\chi)d\chi=q_i$. Van Leer's (1977) monotonic interpolation scheme chooses the largest (in absolute magnitude) dq_i such that $q_i(-1/2)$ and $q_i(1/2)$ do not exceed the neighboring zone averages $q_{i\pm1}$. In the event that q_i is a local extremum, $dq_i=0$. Mathematically, letting $\Delta q_i \equiv q_i - q_{i-1}$, then

$$dq_i = \begin{cases} 2\Delta q_i \Delta q_{i+1}/\,(\Delta q_i+\Delta q_{i+1}), & \Delta q_i\Delta q_{i+1}>0, \\ 0, & \Delta q_i \Delta q_{i+1}\leq 0. \end{cases} \tag{34}$$

The flux of q at interface i is then taken to be

$$F^q_i = q^*_i (v-v_g)_i^{n+1/2} \Sigma_i^{n+1/2}, \tag{35}$$

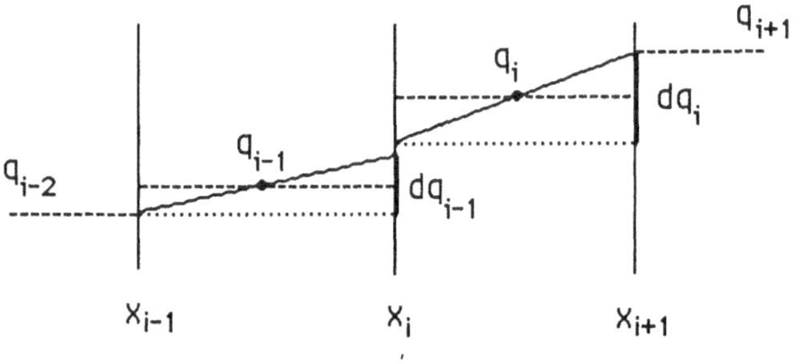

Fig. 2 Van Leer monotonic interpolation scheme. Zone interfaces are at x_i, and zone averages q_i are at zone centers. Piecewise-linear interpolation function (solid lines) is constructed such that the interface values do not exceed the neighboring zone averages (dashed lines). Zone differences dq_i are given by eq. (34).

where $\Sigma_i^{n+1/2}$ is the time-centered area of the zone face, and q^*_i is the upstream interpolated value of q given by (Fig. 2)

$$q^*_i = \begin{cases} q_{i-1} + (\Delta x_{i-1} - \delta_i) dq_{i-1}/(2\Delta x_{i-1}), & \delta_i \geq 0, \\ q_i \quad - (\Delta x_i + \delta_i) dq_i /(2\Delta x_i), & \delta_i < 0, \end{cases} \qquad (36)$$

where $\delta_i = (v-v_g)_i^{n+1/2}(\delta t/2)$, and $\Delta x_i \equiv x_{i+1}-x_i$.

The physical picture behind this prescription is illustrated in Fig. 3. To first order, the value of q on the interface at the half time-level is that obtained by passive advection of q^n for half a timestep. The relative streamline has slope $Dx/Dt-dx/dt \equiv v-v_g$, hence eq. (36). Since q^*_i appears in eq. (33) through a centered difference, the method is formally second-order accurate.

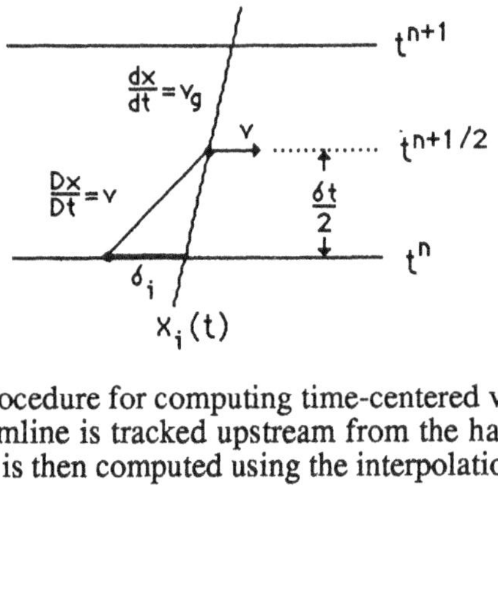

Fig. 3 Upwind procedure for computing time-centered value q^*_i in the flux F^q_i. The relative streamline is tracked upstream from the half time-level a distance $\delta_i=(v-v_g)\delta t/2$. q^*_i is then computed using the interpolation function of Fig. 2 via eq. (34).

4.2 Continuity equation

Letting $q=\rho$ in eq. (33), we have the finite difference form to eq. (7), where the fluxes are given by

$$F^{\rho}_{k,j} = \rho^{*}_{k}(U^{q}_{k,j}-U_{g,k})(R_{j+1/2}\Delta R_{j+1/2})^{n+1/2}, \tag{37}$$

$$G^{\rho}_{k,j} = \rho^{*}_{j}(V^{q}_{k,j}-V_{g,k})\, R^{*}_{k,j}\Delta Z^{\,n+1/2}_{k+1/2}. \tag{38}$$

Here, ρ^{*}_{k} and ρ^{*}_{j} are the interpolated values of density, with the index denoting the direction (i.e., Z or R) of interpolation.The time-centered coordinates are given by

$$R^{n+1/2}_{j+1/2} = R^{n}_{j+1/2} + (\delta t/4)(V_{g,j+1}+V_{g,j}), \tag{39}$$

$$\Delta R^{n+1/2}_{j+1/2} = \Delta R^{n}_{j+1/2} + (\delta t/2)(V_{g,j+1}-V_{g,j}), \tag{40}$$

$$\Delta Z^{n+1/2}_{k+1/2} = \Delta Z^{n}_{k+1/2} + (\delta t/2)(U_{g,k+1}-U_{g,k}), \tag{41}$$

and

$$R^{*}_{k,j} = R^{n}_{j} - (\delta t/2)(V^{q}_{k,j}-V_{g,j}), \tag{42}$$

the mean radius of the advected fluid element.

The new density is then simply

$$\rho^{n+1}_{k,j} = \rho^{n}_{k,j}\tau^{n}_{k,j}/\tau^{n+1}_{k,j}, \tag{43}$$

where

$$\tau^{n(+1)}_{k,j} = (R_{j+1/2}\Delta R_{j+1/2}\Delta Z_{k+1/2})^{n(+1)}. \tag{44}$$

4.3 Consistent advection and the local conservation of angular momentum

In principle, the procedure just described for transporting the mass density could be applied to all the other densities in the problem - E, S, T and A - with out any further thought, remembering only that we must define appropriate control volumes for the linear momentum components S and T. In the case of angular momentum transport, however, such an approach is far from optimal, and in some circumstances, has disasterous results on the local conservation of angular momentum (see Fig. 4). The concern about local conservation of angular momentum in rotating protostar collapse calculations led to the notion of *consistent advection* (Norman, Wilson and Barton 1980), in which the angular momentum flux is calculated by multiplying the mass flux with a best quess for the specific angular momentum of the advected fluid element. Thus, angular momentum is transported consistently with the mass. The physical rationale for this is that $K\equiv A/\rho$ is conserved along a fluid streamline in axial symmetry in the absence of viscous torques, and therefore the spatial interpolation should be

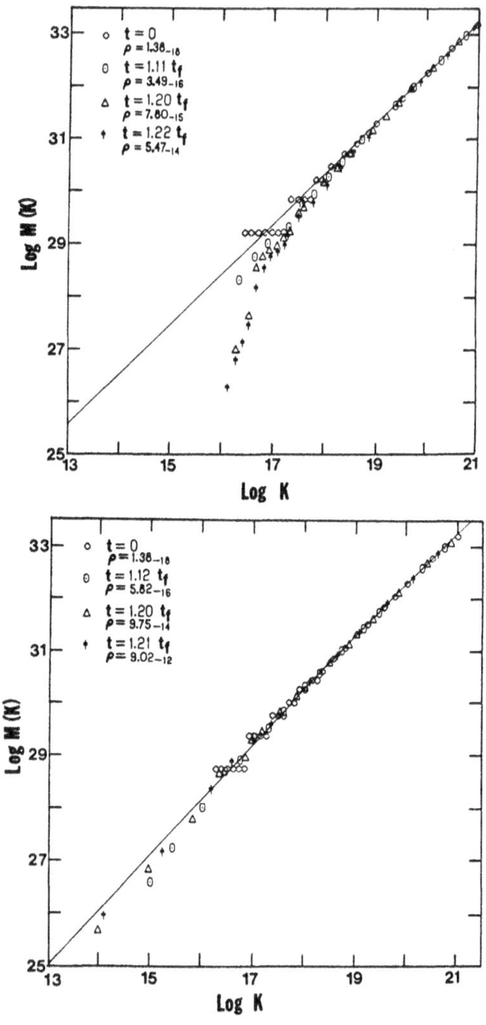

Fig. 4 Numerical diagnostic of the local conservation of angular momentum in a gravitationally collapsing rotating protostellar cloud showing the importance of consistent advection. Local conservation of angular momentum is monitored through changes in the specific angular momentum spectrum, defined as $M(K)=\int^K dM(k)$, where $K=\Omega R^2$ and $dM(k)$ is the mass at specific angular momentum k. M(K) is a constant of the motion for inviscid axisymmetric flow, therefore any changes in the spectrum show numerical redistribution of angular momentum. **a**) Significant evolution of the specific angular momentum spectrum results from using the highly innacurate donor-cell procedure. Various symbols correspond to the indicated times, measured in initial freefall times. **b**) Same collapse problem computed using second-order accurate consistent advection procedure described in Sec. 4.3 showing improved local conservation of angular momentum. From Norman, Wilson and Barton (1980).

performed on K rather than A, since K is the physically more relevant quantity.

Letting q=A in eq. (33), we have the finite-difference form of

$$d/dt \int A\, d\tau = -\int A\, (\mathbf{v}-\mathbf{v}_g)\cdot \mathbf{d\Sigma}, \tag{45}$$

describing conservation of angular momentum, where the fluxes are given by

$$F^{A}_{k,j} = K^{*}_{k}\, F^{\rho}_{k,j}, \tag{46}$$

$$G^{A}_{k,j} = K^{**}_{j}\, G^{\rho}_{k,j}, \tag{47}$$

with K^{*}_{k} being the interpolated value of specific angular momentum in the axial direction, and K^{**}_{j} is computed by interpolating on the flattest of three angular quantities. Defining

$$K_{k,j} = A_{k,j} / \rho_{k,j}, \tag{48}$$

$$W_{k,j} = K_{k,j}\, [3R_{j+1/2} / (R^2_{j+1} + R_{j+1}R_j + R^2_j)], \tag{49}$$

$$\Omega_{k,j} = K_{k,j}\, [2 / (R^2_{j+1} + R^2_j)], \tag{50}$$

that is, the values of K, W and Ω assuming they are uniform in a zone, then we take

$$K^{**}_{j} = \begin{cases} \Omega^{*}_{j}\, (R^{*}_{k,j})^2 & \text{if } | d\Omega_j / \Omega_{k,j} | \text{ smallest,} \\ W^{*}_{j}\, (R^{*}_{k,j}) & \text{if } | dW_j / W_{k,j} | \text{ smallest,} \\ K^{*}_{j} & \text{if } | dK_j / K_{k,j} | \text{ smallest.} \end{cases} \tag{51}$$

Here, the single asterisk means values determined by monotonic interpolation as described in Sec. 4.1.

An equation analogous to eq. (43) is then solved to find A^{n+1}.

Likewise are temperature and velocity intrinsic properties of fluid elements, and therefore it makes physical sense to construct fluxes of energy and momentum density by multiplying the mass flux by the appropriate interpolated values of ε, U and V, even though these quantities are not conserved. We follow this procedure here. A numerical justification is that a product of monotonic functions is monotonic; e.g., $E^*=\rho^*\varepsilon^*$, while the same is not true of the quotient of monotonic functions; e.g., $\varepsilon^*=E^*/\rho^*$, which could lead to difficulties if, for example, the physical model contained a source term with a strong nonlinear dependence on ε. As we shall see in the next section, consistent advection of momentum is manditory in the vicinity of fluid interfaces, where the momentum density may jump by orders of magnitude but the normal velocity component is continuous.

Letting q=E in eq. (33), we have the finite-difference form of eq. (9), where the fluxes are given by

$$F^E_{k,j} = \varepsilon^*_k \, F^\rho_{k,j}, \tag{52}$$

$$G^E_{k,j} = \varepsilon^*_j \, G^\rho_{k,j}, \tag{53}$$

where ε^*_k and ε^*_j are computed in the same fashion as the interpolated densities. An equation analogous to eq. (43) is then solved to find E^{n+1}.

4.4 Momentum transport

Letting q=S and T in eq. (33), we have the finite-difference approximation to eq. (8), where now S and T are interpreted as zone averages over their respective control volumes. Since S and T are face-centered quantities, their control volumes are offset by a half zone-width in the Z and R directions, respectively, from the control volume centered on ρ. The situation is illustrated in Fig. 5. The momentum fluxes are computed by multiplying an appropriate average of the mass flux by the appropriate velocity component interpolated to the zone face. Thus, to transport S, we have (cf. Fig. 5a)

$$F^S_{k+1,j} = (F^\rho_{k,j} + F^\rho_{k+1,j}) \, U^*_{k+1} / 2, \tag{54}$$

$$G^S_{k,j+1} = (G^\rho_{k-1,j+1} + G^\rho_{k,j+1}) \, U^*_{j+1} / 2. \tag{55}$$

An additional step is involved in the radial momentum transport calculation. Specifically, because the control volume for T is offset in the radial direction from the mass control volume (cf. Fig 5b), the radial area factors are removed from the mass fluxes prior to averaging, and then the offset radial area factors are multiplied back in. Thus, we have

$$F^T_{k+1,j} = [(F^\rho/\Sigma)_{k+1,j} + (F^\rho/\Sigma)_{k+1,j-1}] \, \Sigma^T_j V^*_{k+1} / 2, \tag{56}$$

$$G^T_{k,j+1} = [(G^\rho/R^*)_{k,j+1} + (G^\rho/R^*)_{k,j}] \, R^*_{k,j+1/2} V^*_{j+1} / 2, \tag{57}$$

where

$$\Sigma^T_j = (R_j \, \Delta R_j)^{n+1/2}, \tag{58}$$

$$R^*_{k,j+1/2} = R^n_{j+1/2} - (\delta t/4)(V^q_{k,j} + V^q_{k,j+1} - V_{g,j} - V_{g,j+1}), \tag{59}$$

with analogous expressions to eqs. (39,40) for the time-centered quantities appearing in eq. (58).

The new momentum densities S^{n+1} and T^{n+1} are computed in analogy to eq. (43) using the appropriate momentum control volumes.

5. FLUID INTERFACES

A fluid interface is a numerical representation of a boundary between fluids

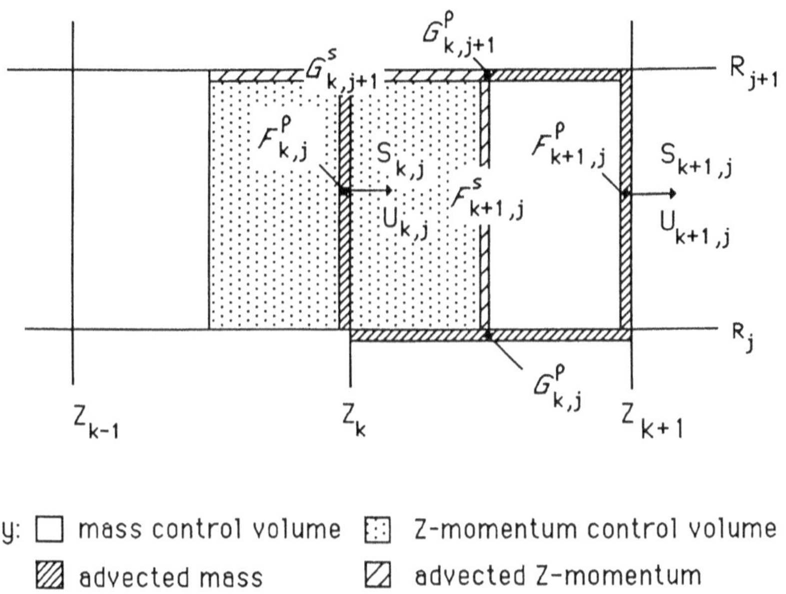

Fig. 5a Mass and momentum control volumes and fluxes for the transport of S, the Z-momentum density.

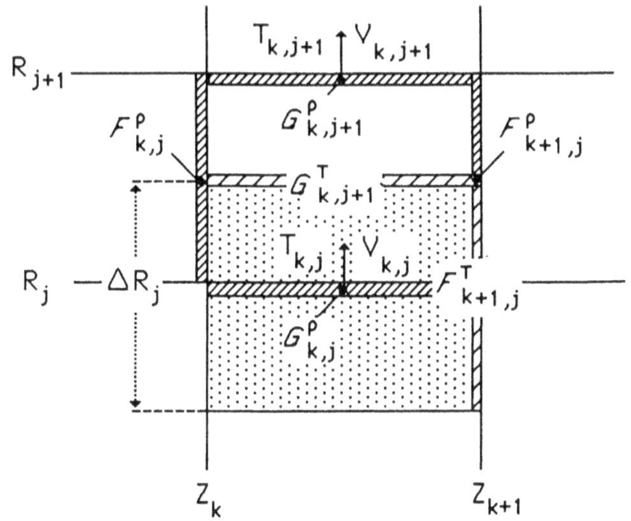

Fig. 5b Mass and momentum control volumes and fluxes for the transport of T, the R-momentum density. Otherwise, key as above.

of different material properties in Eulerian hydrodynamics computations. Some examples of material properties that one might like to distinquish using interfaces are the constitutive properties of the fluid (e.g., equation of state and opacity), the underlying physical model, or simply density or temperature. Since we are modeling ideal (i.e., inviscid) gas flow, such boundaries are idealized as contact discontinuites, and the function of the interface is to prevent the numerical diffusion of the adjacent gas elements into one another. Examples of this technique's use are given in Sec. 8 on several problems in astrophysics where we would like to preserve and track the interface between a hot diffuse medium and a cold dense medium.

Operationally, each material in the calculation is labeled. The label is used as an indicator of material properties. A *mixed* zone is a zone containing more than one material. Zones containing a single material are called *clean* zones, which are advanced in time as described in Secs. 3 and 4. In this section we describe the algorithms we use to advance mixed zones, which were developed by J.M. LeBlanc of the Lawrence Livermore National Laboratory. But first, we give some background on interface methods in general.

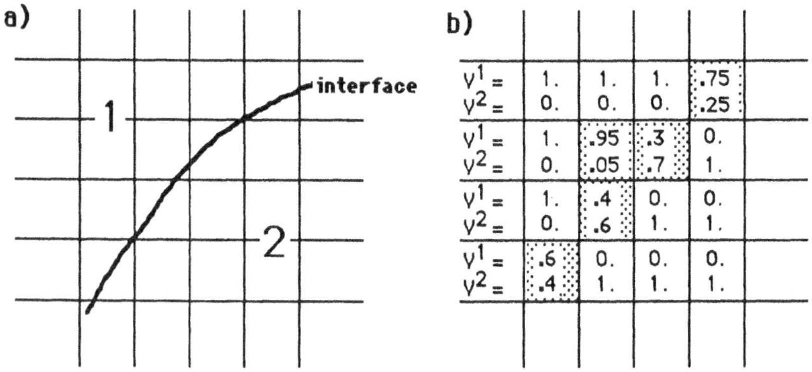

Fig. 6 Numerical representation of a fluid interface on the mesh **a**) by an array of fractional volumes **b**). Mixed zones are shaded, clean zones are not.

5.1 Method of fractional volumes

The most obvious approach to modeling fluid discontinuites in multidimensional Eulerian calculations is to discretize the surface of reduced dimensionality, such as a set of points approximating a line in a 2-D computation, and then to evolve this surface by solving additional numerical equations approximating the Rankine-Hugoniot jump conditions in its immediate vicinity (see, for example, Richtmyer and Morton 1967, p378). While potentially powerful, this approach has been difficult to implement in an efficient and robust fashion in general-purpose hydro codes, and thus is not commonly used. Recent progress using advanced programming languages (Glimm 1985) may change this state of affairs. however.

A second and more approximate approach, first developed by deBar (1974) and in extensive use today for handling contact discontinuities, is to represent the global structure of the interface by a function that is defined locally. This function is the fractional volume occupied by each material in a zone, and is denoted V^i, where i is the material index. V^i is a vector of unknowns defined at every zone k,j satisfying the constraint $\Sigma V^i = 1$. In a clean zone containing material with index j, $V^i = \delta^{ij}$, where δ^{ij} is the Kronecker delta. In a mixed zone, more than one fractional volume is nonzero. Fig. 6 illustrates how an interface between two different fluids would be represented on the computional mesh using fractional volumes.

In addition to specifying the fractional volumes of a mixed zone, one also specifies the composition of the mixed zone through its fractional densities of mass, energy and any other fluid property (e.g. specific angular momentum) that may be discontinuous at the interface. The basic tasks of this approach are 1) to reconstruct the global structure of the interface given $V^i_{k,j}$, and 2) to find equations of motion for the fractional volumes and densities that are simple and easy to program, and which give a reasonably accurate discription of the evolution of the interface in a variety of circumstances.

Two basic paths have been followed over the past decade addressing task 1. The first follows the work of deBar (1974) as implemented in the KRAKEN code, in which the position and orientation of the interface within a mixed zone is reconstructed using the distribution of fractional volumes in *all* the adjacent zones (e.g., in a 3 x 3 block of zones in 2-D). The method of LeBlanc is an example of the second approach, whereby the multidimensional problem is reduced to a series of 1-dimensional problems, and only the adjacent zones in 1-D are used to determine the interface position and orientation. A consequence of this reduction is that the interface geometry is no longer unique; that is, its representation within a zone is different in the X and Y passes. The disadvantage of the directional-splitting approach, of which the SLIC method of Noh and Woodward is another example (see Woodward, these proceedings), is a potential loss of accuracy. The advantage is one of considerably simplifying the algorithm and hence the programming task. Surprisingly, the results obtained with splitting compare quite favorably to the KRAKEN approach (Noh and Woodward 1976).

5.2 Method of LeBlanc

We now describe the interface method of LeBlanc as it is implemented in our code. The following quantities are stored for each material i present in a mixed zone:

$$V^i = \tau^i / \tau, \qquad \text{fractional volume,} \tag{60}$$

$$D^i = \rho^i V^i, \qquad \text{fractional mass density,}$$

$$E^i = D^i \varepsilon^i, \qquad \text{fractional internal energy density,}$$

$$A^i = D^i K^i, \qquad \text{fractional angular momentum density,}$$

where τ is the volume of the zone, and τ^i, ρ^i, ε^i and K^i are the volume, density, specific internal energy, and specific angular momentum of material i(in this and in subsequent equations, we will suppress the dependence on zone indices k and j).The quantities D^i, E^i and A^i are therefore the densities material i would have if it occupied the entire volume of the zone. It follows from these definitions that

$$1 = \sum_i V^i, \tag{61}$$

$$\rho = \sum_i D^i,$$

$$E = \sum_i E^i,$$

$$A = \sum_i A^i,$$

where the summation is over the material index i. Only a single set of velocity and momentum density components are carried for a mixed zone, as they are vector quantities.

5.2.1 source step

The pressure in a mixed zone is found by adding the partial pressures:

$$P = \sum_i (\gamma^i - 1) E^i. \tag{62}$$

The angular velocity of a mixed zone is computed as a mass-weighted average of the fractional angular velocities. Thereafter, mixed zones are accelerated like clean zones [cf. eqs. (19,20).]

Heating from artificial viscosity and compressional work is equally partitioned to each material i in a mixed zone:

$$E^{i,q} = (E^q/E)\, E^i,$$

$$E^{i,p} = (E^p/E^q)\, E^{i,q}\,, \tag{63}$$

where the superscripts refer to steps 2 and 3 of Sec. 3.

5.2.2 transport step

The motion of the interface comes about by transporting the fractional volumes along with the other fractional densities. As in the clean calculation (Sec. 4), transport is done in 1-D sweeps, however a first-order method (donor dell) is used in the vicinity of the interface as an aid to numerical stability.

Consider the triad of zones containing at least one mixed zone as shown in Fig. 7a, and suppose we wish to update the middle zone. Define $\mathfrak{I}^i_{L,R}$ as the fractional fluxes of material i on the left and right, respectively, of this zone, and

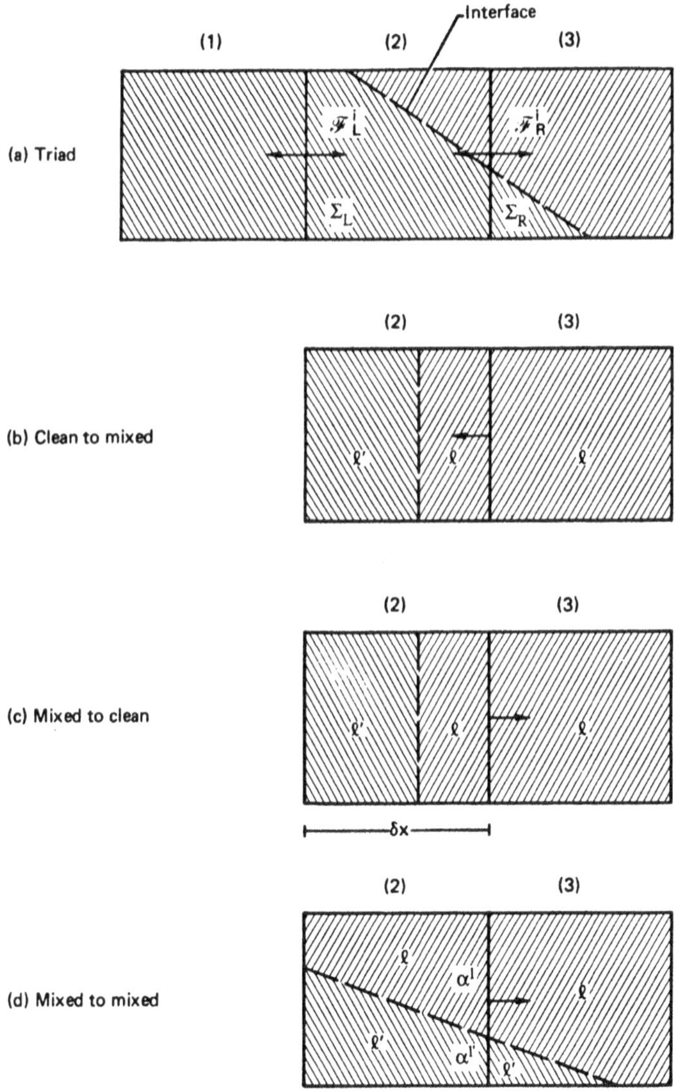

Fig. 7 Advection procedure in a triad of zones containing at least one mixed zone. **a**) Interface geometry is inferred from the distribution of fractional volumes (cf. Fig. 6). The fractional fluxes $\mathfrak{I}^i$ are derived according to the following three situations. **b**) Advection from a clean zone to a mixed zone. **c**) Advection from a mixed zone to a clean zone. **d**) Advection from a mixed zone to a mixed zone. α^l and $\alpha^{l'}$ are the aperture as defined in eq. (70).

$\Sigma_{L,R}$ as the areas of the left and right zone faces, respectively.The difference equations for the transport of the fractional densities and fractional volumes are then

$$V^{i,n+1} - V^{i,n} = (\Delta^i_R \Sigma_R - \Delta^i_L \Sigma_L) / \tau \,, \tag{64}$$

$$D^{i,n+1} - D^{i,n} = (\mathfrak{J}^i_R \Sigma_R - \mathfrak{J}^i_L \Sigma_L) / \tau \,, \tag{65}$$

and similarly for E^i and A^i. Here, the mass flux is defined

$$\mathfrak{J}^i_{L,R} = \Delta^i_{L,R} \, (D^i/V^i)_d, \tag{66}$$

where the subscript d stands for the donor cell values and the $\Delta^i_{L,R}$ are computed according to the following three cases.

case 1: clean to mixed

Referring to Fig. 7b, in the case of flow from a clean zone containing material l to a mixed zone, we have simply

$$\Delta^i = (U\text{-}U_g)\delta t \, \delta^{il} \,. \tag{67}$$

The donor cell remains clean and the acceptor cell remains mixed.

case 2: mixed to clean

Referring to Fig. 7c, we see that material l may become negative in the donor cell if $|U\text{-}U_g|\delta t > (V^l \Delta x)_d$. Therefore

$$\Delta^i = \text{sign}\,(U\text{-}U_g)\,\min[\,|U\text{-}U_g|\delta t \,,\, (V^l \Delta x)_d\,]\delta^{il} \,. \tag{68}$$

If $|U\text{-}U_g|\delta t > (V^l \Delta x)_d$, then we take the next material in line according to

$$\Delta^i = \text{sign}\,(U\text{-}U_g)\,\min[\,|U\text{-}U_g|\delta t - \Delta^l \,,\, (V^{l'} \Delta x)_d\,]\delta^{il'} \,, \tag{69}$$

and so on until everything $|U\text{-}U_g|\delta t$ behind the flow has been taken. In this case, the donor cell may become clean and the acceptor cell may become mixed.

case 3: mixed to mixed

Referring to Fig. 7d, we define aperatures α^i through which material i may pass as follows:

$$\alpha^i = (V^i_L + V^i_R)/2, \tag{70}$$

Then we have simply

$$\Delta^i = \alpha^i \, (U\text{-}U_g)\delta t \,. \tag{71}$$

Notice that $\Sigma\, \alpha^i = 1$, so that the scheme is conservative.

Total densities for the zone are found by summing the fractional densities. If the interface is in a region of the flow with a velocity gradient normal to its surface, then in general $\Sigma V^{i,n+1} \neq 1$, in which case the fractional volumes are renormalized so that they sum to unity. Finally, integrated mass fluxes are computed for use in the linear momentum transport calculation (cf. Sec. 4.4):

$$F^{\rho} = (\sum_{i} \mathfrak{I}^{i})\Sigma/\delta t. \tag{72}$$

5.3 Properties of the interface method

The interface method just described works best on, and in fact was developed for, isolated contact discontinuities in flows with little velocity shear both normal and tangential to the discontinuity's surface. This will not be the case if the discontinuity is interacting with a strong shock or rarefaction wave, nor if it is a *slip* discontinuity.This can be seen by noticing that only one set of velocity components are used to describe both mixed zones and clean zones alike. Indeed, incorporating "fractional velocities" into a such a technique would be difficult because the orientation of the interface is only loosely defined, and one would naturally want to work in terms of discontinuities in the normal and tangential velocity components. By definition, the normal component of velocity is continuous at a contact discontinuity, and therefore in such problems as material boundaries moving normal to their surface, as arise in Rayleigh-Taylor instabilities, one velocity per zone is adequate to give an accurate representation of the interface's motion. In problems with a large amount of slip across the discontinuity, as arise in Kelvin-Helmholtz instabilities, the interface dynamics is driven by the mean flow in which it is embedded. In both cases, the primary function of the fluid interface is to act as a *material separator* , as they were termed originally, preventing numerical diffusion from artificially broadening the discontinuity into several zone-widths.

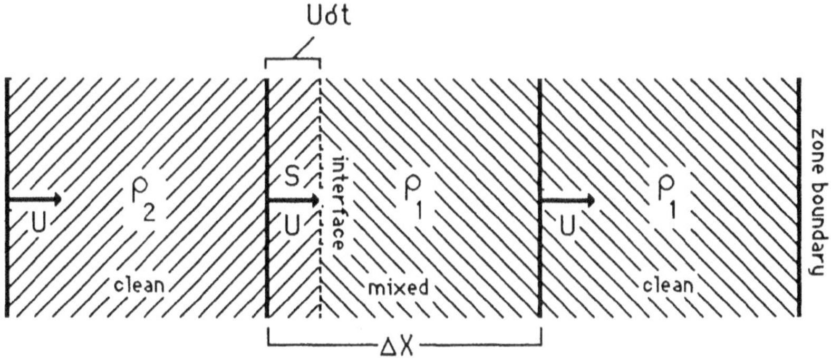

Fig. 8 One-dimensional interface advection in a uniform velocity field U. Discontinuity between $\rho=\rho_1$ and $\rho=\rho_2$ initially coincides with the second-from-the-left zone boundary. Consistent momentum advection (cf. Sec. 4.4) insures that U remains constant despite a large jump in mass and momentum densities.

We shall now demonstrate that the LeBlanc interface method is exact when applied to the uniform advection in 1-D of a discontinuity normal to its surface. Consider three zones as shown in Fig. 8 with a density discontinuity initially coinciding with the zone boundary second from the left. Let the density to the left and right of the discontinuity be ρ_1 and ρ_2, respectively, and let there be a uniform velocity field U pointing to the right. In time δt, the interface will move to the right a distance $U\delta t$, and the new fractional volumes in the middle zone will be $V^1 = 1 - U\delta t/\Delta x$ and $V^2 = U\delta t/\Delta x$. This is precisely what eqs. (64) and (67) yield if we set $U_g=0$, $\Sigma=1$ and $\tau=\Delta x$.

In addition, we can show that consistent advection of momentum [cf. Sec. 4.4 and eq. (72)] insures that the uniform velocity field willl be unaltered by the interface treatement. Summing eq. (65) over material index, we find

$$\rho^{n+1} = (1-\sigma)\rho_1 + \sigma\rho_2, \tag{73}$$

where $\sigma \equiv U\delta t/\Delta x$. Updating the average momentum in the zone centererd about S in Fig. 8, we have from eq. (54)

$$(S^{n+1} - \langle\rho\rangle U)/\delta t = -(\langle\rho\rangle - \rho_2)U^2/\Delta x, \tag{74}$$

or

$$S^{n+1} = [(1-\sigma)\langle\rho\rangle + \sigma\rho_2]U\ , \tag{75}$$

where $\langle\rho\rangle \equiv (\rho_1 + \rho_2)/2$. It is then easy to show from eqs. (73,74) that

$$U^{n+1} \equiv 2S^{n+1}/\ (\rho^{n+1} + \rho_2) = U. \tag{76}$$

Thus, we have passive advection of the interface with no modification of the background velocity field.

6. SELF-GRAVITY

Sections 3-5 describe the hydrodynamic part of the calculation whereby the fluid variables are advanced from timelevel n to n+1. In problems where the self-gravitational forces of the fluid are important, we must also solve the Poisson equation (4) subject to appropriate boundary conditions in order to determine the gravitational potential at the new timelevel $\Phi^{n+1}_{k,j}$ which enters in eqs. (19,20).

The boundary values of gravitational potential are computed from a multipole expansion,

$$\Phi_B = -G\sum_l P_l(\mu_B)\, r_B^{-(l+1)} M_l\ , \tag{77}$$

where the multipole moments are given by

$$M_l = \int d\tau\, \rho(\mathbf{r})\, r^l P_l(\mu)\ . \tag{78}$$

Here **r** is the position vector from the center of the self-gravitating structure, usually at Z=0, R=0; r=|**r**|; μ is the cosine of the angle between the rotation axis and **r**; and P_l are the Legendre polynomials. The subscript B means that these quantities are to be evaluated on the boundary of the computational domain. With assumed equatorial symmetry, the odd moments vanish, and the boundary value at Z=0 becomes $\partial\Phi/\partial Z=0$. In practice only the l=0, 2 and 4 terms are used, which has proven to be adequate if the outer boundary is sufficiently removed from the structure. The boundary value at the axis is, of course, $\partial\Phi/\partial R=0$.

As we generally deal with nonuniform meshes in both coordinate directions which are not spatially periodic, Fourier transform methods to solve eq. (4) are ruled out. Also, direct methods such as Gaussian elimination would be too time-consuming, and hence we must consider iterative techniques. The solution technique for the Poisson equation we use has been described by Black and Bodenheimer (1975), but will be repeated here. The plan is to find the steady-state solution to the diffusion equation

$$\partial\Phi/\partial t = \nabla^2\Phi - 4\pi G\rho^{n+1} , \tag{79}$$

using the ADI method (Peaceman and Rachford 1955) for a series of iterative "timesteps". The time appearing in eq. (79) bears no relation to the physical time of the evolution; the timesteps are chosen to speed convergence.

Let Φ^p be the pth estimate for the gravitational potential, and δt^p be the timestep for the pth iteration. Defining

$$\Delta_Z\Phi_{k,j} \equiv \Phi_{k,j} - \Phi_{k-1,j} , \tag{80}$$

$$\Delta_R\Phi_{k,j} \equiv \Phi_{k,j} - \Phi_{k,j-1} , \tag{81}$$

then Φ^p is advanced to Φ^{p+1} by the following two-step ADI procedure:

$$(\Phi^{p+1/2}\text{-}\Phi^p)_{k,j}/\delta t^p = (\Delta_Z\Phi^p{}_{k+1,j}/\Delta Z_{k+1} - \Delta_Z\Phi^p{}_{k,j}/\Delta Z_k)/\Delta Z_{k+1/2}$$
$$+ (R_{j+1}\Delta_R\Phi^{p+1/2}{}_{k,j+1}/\Delta R_{j+1} - R_j\Delta_R\Phi^{p+1/2}{}_{k,j}/\Delta R_j)/(R\Delta R)_{j+1/2} - 4\pi G\rho^{n+1}{}_{k,j} , \tag{82}$$

followed by

$$(\Phi^{p+1}\text{-}\Phi^{p+1/2})_{k,j}/\delta t^p = (\Delta_Z\Phi^{p+1}{}_{k+1,j}/\Delta Z_{k+1} - \Delta_Z\Phi^{p+1}{}_{k,j}/\Delta Z_k)/\Delta Z_{k+1/2}$$
$$+ (R_{j+1}\Delta_R\Phi^{p+1/2}{}_{k,j+1}/\Delta R_{j+1} - R_j\Delta_R\Phi^{p+1/2}{}_{k,j}/\Delta R_j)/(R\Delta R)_{j+1/2} - 4\pi G\rho^{n+1}{}_{k,j} . \tag{83}$$

The implicit sweeps generate a set of tridiagonal matrix equation which are solved directly using the well-known technique of "forward sweep, backward substitution" described by Richtmyer and Morton (1967).

Eqs. (82,83) are solved for $0 \leq p \leq P-1$, where the iterative timesteps δt^p fo m a geometric series

$$\delta t^p = \alpha^p \delta t_{max} \; ; \; p=0,1,\ldots,P-1, \tag{84}$$

with

$$\delta t_{max} = \max(Z_{max}{}^2, R_{max}{}^2)/4, \tag{85}$$

and

$$\alpha = (\delta t_{min}/\delta t_{max})^{1/(P-1)}, \tag{86}$$

$$\delta t_{min} = \min(\Delta Z_{min}{}^2, \Delta R_{min}{}^2)/4 \, . \tag{87}$$

The new potential is given by $\Phi^{n+1}=\Phi^P$. This timestep prescription is constructed in analog to a treatment by Peaceman and Rachford (1955), who solved diffusion in a square, rectangular mesh. The timesteps are chosen to reduce the amplification factors of eqs. (82,83) for modes of wavelength comparable to $(\delta t^p)^{1/2}$.

Convergence is checked by monitoring

$$\chi = \max_{k,j} |\nabla^2 \Phi - 4\pi G\rho| \, / \, 4\pi G\rho, \tag{88}$$

which should be $\leq 10^{-5}$ to achieve a potential gradient accuracy of about a percent. We find typically that $10^{-6} < \chi < 10^{-10}$ if P is approximately half the number of zones in one dimension.

7. TIMESTEP CONTROL

The final operation in the problem cycle is the calculation of a new timestep to be used in the next cycle. Explicit hydrodynamics requires the timestep to satisfy the Courant condition for stability, which for a one-dimensional problem is

$$\delta t \leq \min \Delta x/(C+|U|), \tag{89}$$

Where C and U are the local sound speed and flow speed, and the minimum is taken over the entire domain. A simple and effective prescription for multidimensional calculations which we use is

$$\delta t^{n+1} = b \left[\max_{k,j}\left(\sum_{i=1}^{3} \delta t_i{}^{-2}\right) + \delta t_4{}^{-2}\right]^{-1/2}, \tag{90}$$

where b is the safety factor, usually ≈ 0.5, and the δt_i are defined for each zone k,j as follows:

$$\delta t_1 = \min(\Delta Z, \Delta R)/C, \tag{91}$$

$$\delta t_2 = \Delta Z/|U-U_g| \, , \tag{92}$$

$$\delta t_3 = \Delta R/|V-V_g| \, . \tag{93}$$

The artificial viscosity also limits the timestep, since Q^{ZZ} and Q^{RR} are used to form a momentum diffusion problem. For an explicit diffusion scheme the timestep is limited by

$$\delta t \leq \Delta x^2 / 4\nu \tag{94}$$

where ν is the kinematic viscosity. A comparison of eqs. (21,22,24 & 25) with C_1=0 to the Navier-Stokes equation shows the numerical kinematic viscosity to be

$$\nu^{ZZ} = \Delta Z(C_2 Q^{ZZ}/\rho)^{1/2} = C_2|\Delta U|\Delta Z \,, \tag{95}$$

$$\nu^{RR} = \Delta R(C_2 Q^{RR}/\rho)^{1/2} = C_2|\Delta V|\Delta R \,, \tag{96}$$

thus we define a fourth timestep for zones with nonzero Q,

$$\delta t_4 = \min\,(\Delta Z/4C_2|\Delta U|, \Delta R/4C_2|\Delta V|) \,. \tag{97}$$

Finally, the timestep is limited to a 30 percent increase per cycle to maintain accuracy when the system makes abrupt dynamical changes, yet may decrease by an arbitrary amount in order to maintain numerical stability.

8. NUMERICAL EXAMPLES

In this section we illustrate the use of our numerical techniques on a number of problems of astrophysical interest involving static and dynamic fluid interfaces.

8.1 Self-gravitating isothermal clouds

The picture of a cold, dense isothermal cloud in pressure equilibrium with a hot intercloud medium is a paradigm for the interstellar medium that is often used as initial conditions for calculations of gravitational collapse and star formation. Unlike self-gravitating equilibria with polytropic equations of state, isothermal equilibria are infinite in spatial extent unless truncated at some finite radius with a finite boundary pressure, such as would be provided by a hot intercloud medium. Such truncated self-gravitating isothermal equilibria possessing zero angular momentum are called Bonner-Ebert spheres, named after the men who first determined their structure (Bonner 1956; Ebert 1955). The rotating analogs to the Bonner-Ebert spheres were first investigated by Norman (1980) using the hydrodynamic techniques described above, and subsequently by Stahler (1983) and Hachisu and Eriguchi (1984) using hydrostatic codes.

In the hydrodynamic approach, a fluid interface was used to delineate the cloud-intercloud boundary, which is a free boundary. As initial conditions for the calculation, Norman assumed a constant density and temperature sphere with a specified rotation law embedded in a constant pressure background. The initial

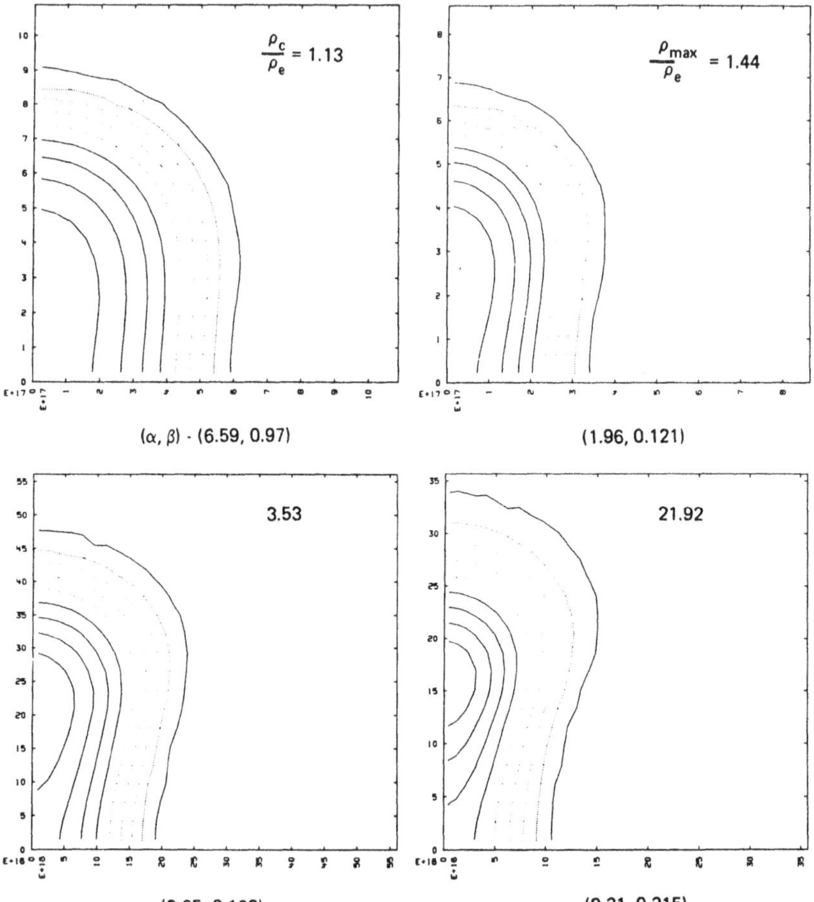

Fig. 9 Equidensity contours of rotating, isothermal equilibria of constant mass, angular momentum and distribution of angular momentum for several values of α=|internal energy / gravitational energy| and β=|rotational energy / gravitational energy| (in parenthesis). One quadrant is displayed and the rotation axis is horizontal. The lower, right-hand model is near criticality to gravitational collapse. The cloud boundary (outermost solid line) is described by a fluid interface. From Norman (1980).

conditions were then evolved to equilibrium hydrodynamically with a velocity-dependent damping term added to the momentum equation in order to hasten the approach to equilibrium. Once equilibrium was reached, the cloud temperature was slowly decreased, generating a "cooling sequence" of quasi-static isothermal spheroids of constant mass, angular momentum and its distribution. During the initial relaxation phase and cooling phases, the intercloud medium was not evolved hydrodynamically, but rather was kept at constant density and pressure.

Fig. 9 shows four equilibria from a cooling sequence initiated wtih a sphere whose angular velocity was ten times higher at its center than at its edge. The fluid interface is indicated by the outermost solid line. A consequence of this angular momentum distribution is toroidal equilibria for $\alpha<1$, where α is the ratio of the cloud's internal energy to its gravitational self-energy. As the temperature is further decreased, the minimum α for stable equilibrium α_c is encountered. Below α_c the cloud is dynamically unstable to gravitational collapse. Collapse is computed numerically at constant cloud temperature without the damping term in the equation of motion. Fig. 10 shows the cloud structure well into the collapse phase when the peak density on the toroidal axis exceeds 10^3 times the edge value. For a complete discussion of the collapse dynamics and its dependence on angular momentum distribution, the reader is referred to Norman (1980).

8.2 Supersonic jets

Calculations of supersonic jets of the sort displayed in Plate 1 have been performed in connection with radio galaxy studies and their associated radio jets (Norman et al. 1982; Norman, Winkler and Smarr 1983,1985; Norman, Smarr and Winkler 1984; Smarr, Norman and Winkler 1984; Smith et al. 1985). The calculations are performed in 2-D axisymmetry neglecting self-gravity and rotation. Initially, the computational domain is filled with a uniforn, static background gas which is to represent the intergalactic medium surrounding the radio galaxy. Subsequently, a perfectly collimated supersonic beam of gas is continuously injected through an area on the domain boundary, and its interaction with the ambient gas is computed. The beam's incident pressure is chosen to match the undisturbed ambient pressure, whereas its incident density and velocity are varied from evolution to evolution. A fluid interface (shown in black) is used to track the contact discontinuity seperating the jet gas from the ambient gas.

Plate 1a shows the establishment of a Mach 3 jet with an input density of 10% the background density. A characteristic of low density jets is that as they propagate, they enshroud themselves in a cocoon of gas that has "splashed back" from the leading end of the jet. The cocoon is generally less dense and hotter than the beam gas because of shock-heating at the terminal shock front. This can be seen as a difference in colors between the central beam (green) and cocoon (blue) in Plate 1, where different densities have been assigned different colors according to the scale accompanying Plate 2. As can bee seen in Plates 1b-d, the jet boundary is subject to Kelvin-Helmholtz instabilities which lead to turnover and mixing of the jet and ambient gases. The fluid interface allows one to follow these interfacial instabilities into the nonlinear regime with a minimum of numerical diffusion. A wealth of hydrodynamical detail can be extractred from

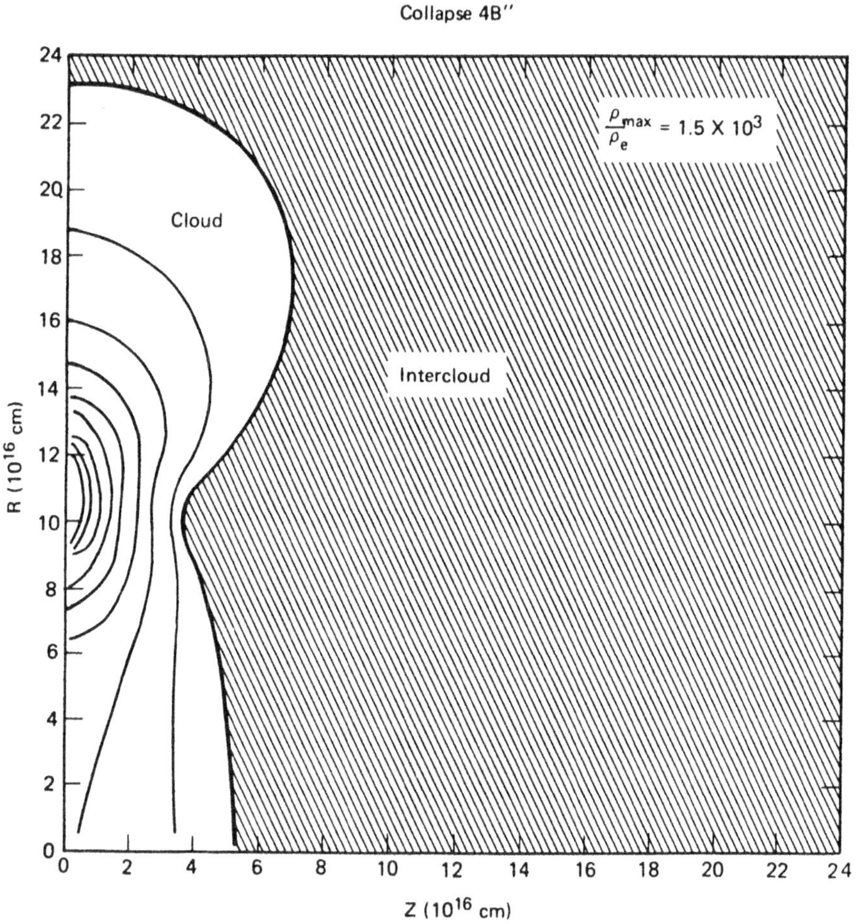

Fig. 10 Gravitational collapse from the equilibrium toroidal cloud shown in Fig. 9d. The calculation was halted at a density contrast of 1.5×10^3, when the zoning resolution became inadequate to follow the subsequent condensation. Isodensity contours are plotted. From Norman (1980).

the calculations using the color imaging techniques described by Winkler and Norman in these proceedings. A comprehensive overview of the key physical results is given in Smarr, Norman and Winkler (1984).

Plate 2 illustrates a second application of our numerical techniques to the propagation of supersonic gas jets. In this calculation, 2-D cartesian geometry is assumed so that nonaxisymmetric "kink" instabilities can be studied. The numerical procedure is idendical to the axisymmetric jet calculations described above, except now the jet is admitted with a transverse velocity component which varies sinusoidally in time according to

$$v_y(t) = 0.05\ v_x \sin\omega_r t,$$

where ω_r satisfies Woodward's resonance condition (Woodward, these proceedings) for the fastest-growing unstable mode

$$\omega_r = \pi v_x (M^2-1)^{-1/2}/W$$

where v_x is the beam velocity, M is its internal Mach number and W is the slab width. Plates 2a-d show the rapid growth of the kink instability and its disruptive effect on the directed bulk flow.

8.3 Blastwaves in exponential atmospheres

A strong point explosion in a plane-stratified exponential atmosphere presents some interesting phenomena not found in the Taylor-Sedov type blastwaves produced in a uniform atmosphere. Plate 3 illustrates these phenomena. At t=0, the background gas is distributed according to

$$\rho(Z) = \rho_0 \exp(-Z/h(Z))$$

$$T(Z) = T_0$$

and in hydrostatic equilibrium with the gravitational potential given by

$$\Phi(Z) = \Phi_0 - RT_0 \ln(\rho/\rho_0).$$

h(Z) is the local scale height given by

$$h(Z) = h(Z_c) + a(Z-Z_c),$$

where Z_c is the height of the explosion above the midplane. Although the problem is scale-free, the following astrophysical numbers were used: $\rho_0=1.67\times10^{-24}$ g cm^{-3}, T_0=500K, $Z_c=1.4\times10^{20}$ cm, $h(Z_c)=4.21\times10^{19}$ cm, a=-0.125 cm^{-1}. A sphere of high temperature ($T_s=4\times10^{10}$K) and low density ($\rho_s=6\times10^{-26}$ g cm^{-3}) is emplaced at $Z=Z_c$ with initial radius $r_s=10^{19}$ cm. A fluid interface seperates the high pressure "driver gas" from the constant temperature background gas. The computational domain spans 6×10^{19} cm $\leq Z \leq 3.9\times10^{20}$ cm, $0 \leq R \leq 1.1\times10^{20}$ cm. The subsequent evolution is computed assuming a γ=5/3 adiabatic equation of state in both gases.

Plate 3a shows the color-coded entropy distribution shortly after the explosion begins. The distribution of colors in the atmosphere indicates a stable entropy stratification. The red circular region is the high entropy explosion gas that has been shocked by the expanding blastwave. As the blastwave barely extends over one atmospheric scale-height at this time, it is still circular. Plate 3b shows how the blastwave distorts and becomes egg-shaped as it samples different regions of the stratified pressure distribution. The upper apex of the blastwave propagates the fastest since it is following the steepest pressure gradient. Plate 3c and 3d show blastwave "breakout" as first predicted by Kompaneets (1960), and the subsequent buoyant rise of the hot bubble. Note the growth of Rayleigh-Taylor instabilities on the leading edge of the bubble, indicating the necessity of using a fluid interface in this calculation.

8.4 Twin-exhaust jets

As a final example, we consider the production of jets via the Blandford-Rees (1974) Twin-Exhaust mechanism. This mechanism was first proposed to account for the production of twin jets in the nuclei of radio galaxies, and is currently being applied to jet production by protostars embedded in molecular clouds in our own galaxy (Königl 1982). The model holds that if a continuous source of hot, buoyant gas is established in a relatively colder, denser background gas that is gravitationally confined, then the buoyant gas will preferentially escape along the path or paths of least resistance; i.e., parallel to the steepest pressure gradient, which in a radio galaxy nucleus could be taken to be along the minor axis of a rotationally-flattened central gas cloud. According to this model, the boundary between the cold confining gas and the buoyant outflowing gas would naturally assume the shape of a deLaval nozzle, which would accelerate the outflow to supersonic speeds and collimate it into jets. This is manifestly a two-fluid problem requiring a dynamic fluid interface to study the formation and stability of the flow channel boundary.

Plate 4 illustrates the nozzle formation process. The confining atmosphere is initially isothermal, plane-stratified and in hydrostatic equilibrium. The density and hence pressure distribution is a power-law with a central plateau given by

$$\rho(Z) = \rho_0 / [1+(Z/h)^2]$$

$$P(Z) = (\gamma-1)\rho(Z)\varepsilon_0,$$

where ρ_0 and ε_0 are the midplane density and specific internal energy, respectively. Hot gas is continuously created in a spherical source region of radius h/10 with zero velocity at a mass rate m with specific internal energy ε_j. Both fluids were assumed to obey $\gamma=5/3$ ideal gas equations of state. The following dimensionless quantities define the evolution:

$$\lambda = m\varepsilon_j / \rho_0\varepsilon_0^{3/2}h^2$$

$$\theta = \varepsilon_j / \varepsilon_0.$$

This example illustrates an energetic (λ=2) source of hot (θ=100) gas, which we had previously determined to be susceptible to Rayleigh-Taylor instabilities (Norman et al. 1981). The computational domain spans $0 \leq Z \leq 10h$, $0 \leq R \leq 10h$.

Plate 4a shows the initial bubble of hot gas inflated by the central source. Gas temperature is color-coded such that high temperatures are red and low temperatures are blue. The bubble is elongated in the direction of the pressure gradient as its size exceeds the plateau scale-length h. Since the interface between the bubble gas and the background gas is Rayleigh-Taylor unstable, any kinks or ripples on the bubble surface will be amplified by the instability. The growth of these instabilities in subsequent frames is tracked with our numerical fluid interface, shown in black.

Plates 4b and 4c show the establishment of the cavity-nozzle-jet structure. The throat of the nozzle forms as dense Rayleigh-Taylor "fingers" penetrate the bubble from the side and converge toward the axis. As the throat necks down, tha cavity inflates with subsonic gas (Plate 4d), and now the top of the cavity develops the characteristic Rayleigh-Taylor "spike and bubble" structure. The dense spikes merge on axis in Plate 4e forcing the jet gas to flow out in an annular region. The annular jet breaks through the layer of dense gas seen in blue in Plate 4e to form the continuous diverging jet of Plate 4f. The jet has an embedded spindle of dense gas along its axis of symmetry, which is slowly being blown downstream by the jet ram pressure.

Further numerical evolutions of this sort are described in Norman et al. (1981), and an analytic discussion of the flow stability is given in Smith et al. (1983). The relevance of these calculations to jet formation in active galactic nuclei is discussed in Smith et al. (1981).

REFERENCES

Black, D.C. & Bodenheimer, P. 1975, Ap. J. 199, p619.
Blandford, R.D. & Rees, M.J. 1974, Mon. Not. Roy. Astron. Soc. 169, p395.
Bonner, W.B. 1956, Mon. Not. Roy. Astron. Soc. 116, p351.
deBar, R. 1974, *Fundamentals of the KRAKEN code*, Lawrence Livermore National Laboratory Internal Report UCIR-760.
Ebert, R. 1955, Zs. f. Astrophys. 37, p217.
Glimm, J. 1984, Courant Institute Preprint.
Hachisu, I. & Eriguchi, Y. 1984, Max-Planck-Institut f. Astrophysik preprint No. 149.
Kompaneets, A.S. 1960, Soviet Phys. Doklady 5, p46.
Königl, A. 1982, Ap. J. 261, p115.
Noh, W.H. & Woodward, P.R. 1976, in *Proceedings of the 5th International Conference on Numerical Methods in Fluid Dynamics* , (Springer Verlag: New York).
Norman, M.L. 1980, Ph.D. dissertation, University of California, Davis; Lawrence Livermore National Laboratory report UCRL-52946.
Norman, M.L., Smarr, L., Wilson, J.R. & Smith, M.D. 1981, Ap. J. 247, p52.

Norman, M.L., Smarr, L., Winkler, K.-H.A. & Smith, M.D. 1982, Astron. Astrophys. 113, p285.
Norman, M.L., Smarr, L. & Winkler, K.-H.A. 1984, in *Numerical Astrophysics*, ed. J. Centrella, J. LeBlanc & R. Bowers, (Jones and Bartlett: Portola Valley, CA.).
Norman, M.L., Wilson, J.R. & Barton, R. 1980, Ap. J. 239, p968.
Norman, M.L., Winkler, K.-H.A. & Smarr, L. 1983, in *Astrophysical Jets*, ed. A. Ferrari & A.G. Pacholczyk, (Reidel:Dordrecht).
Norman, M.L., Winkler, K.-H.A. & Smarr, L. 1985, in *Physics of Energy Transport in Extragalactic Radio Sources*, ed. A. Bridle & J. Eilek, NRAO Conference Proceedings No. 9.
Richtmyer, R.D. & Morton, K.W. 1967, *Difference Methods for Initial-Value Problems*, (Interscience:New York), p198.
Smarr, L., Norman, M.L. & Winkler, K.-H.A. 1984, Physica D 12, p83.
Smith, M.D., Smarr, L., Norman, M.L. & Wilson, J.R. 1981, Nature 293, p277.
Smith, M.D., Smarr, L., Norman, M.L. & Wilson, J.R. 1983, Ap. J. 264, p432.
Stahler, S. (1983), Ap. J. 268, p165.
vanLeer, B. 1977, J. Comp. Phys. 23, p276.

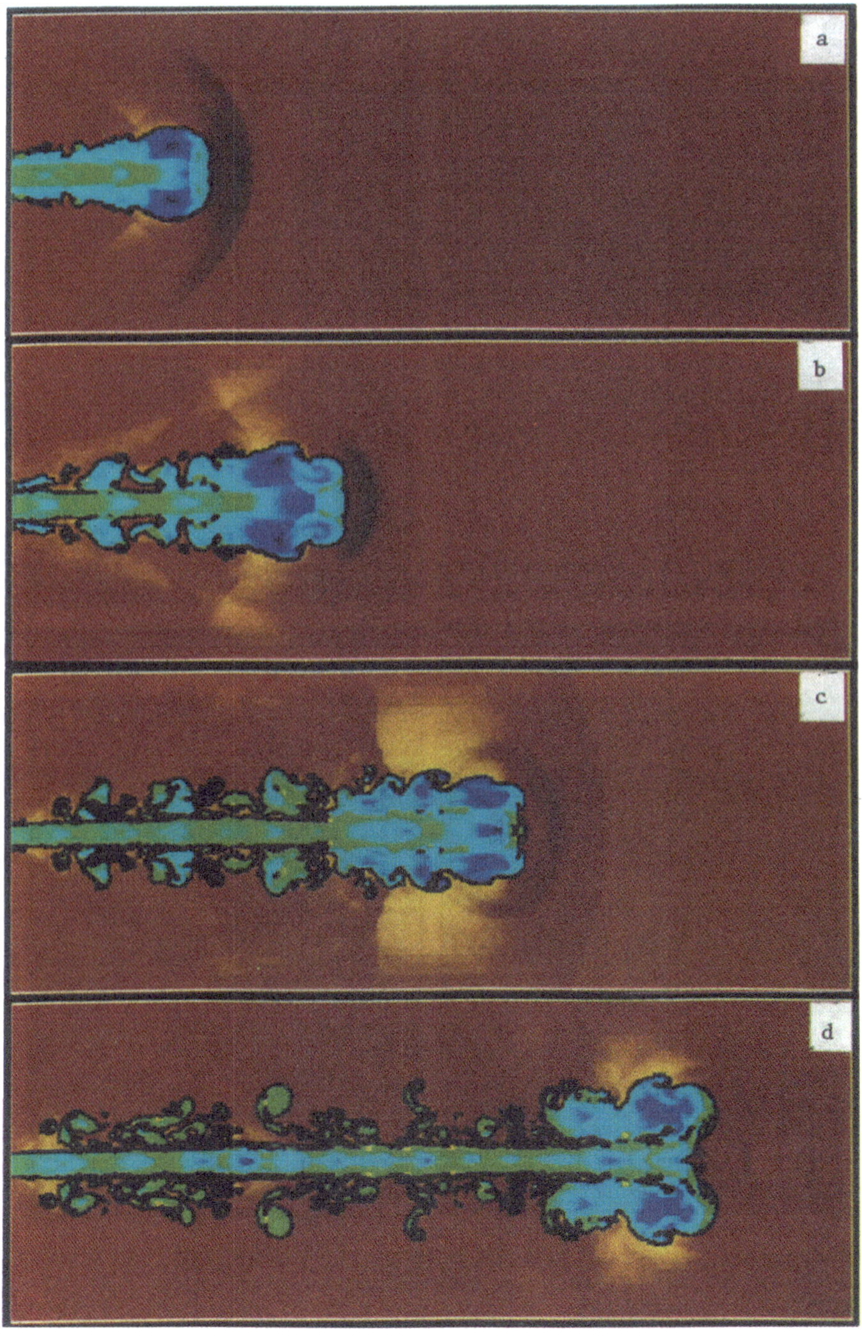

Plate 1

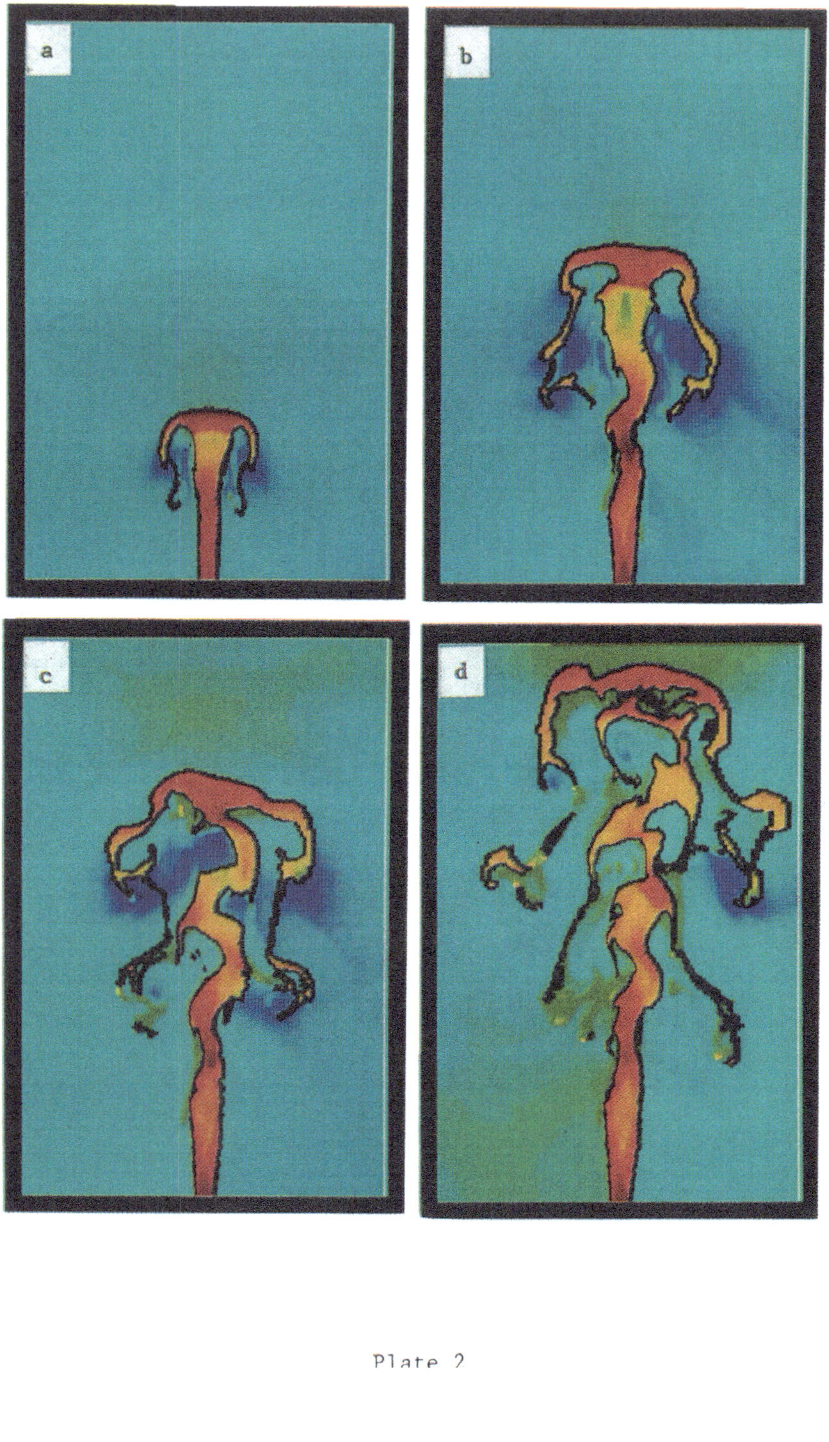

Plate 2

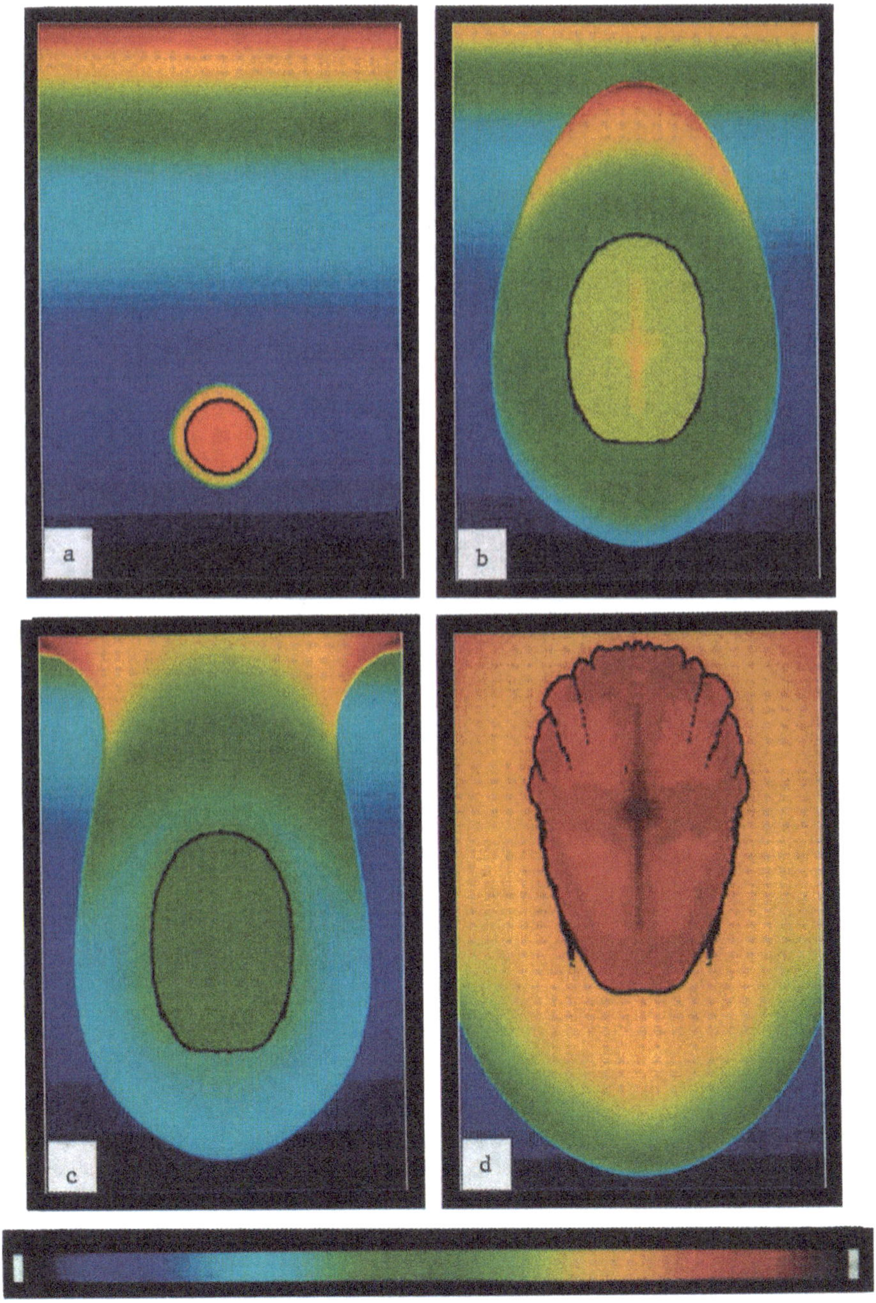

Plate 3

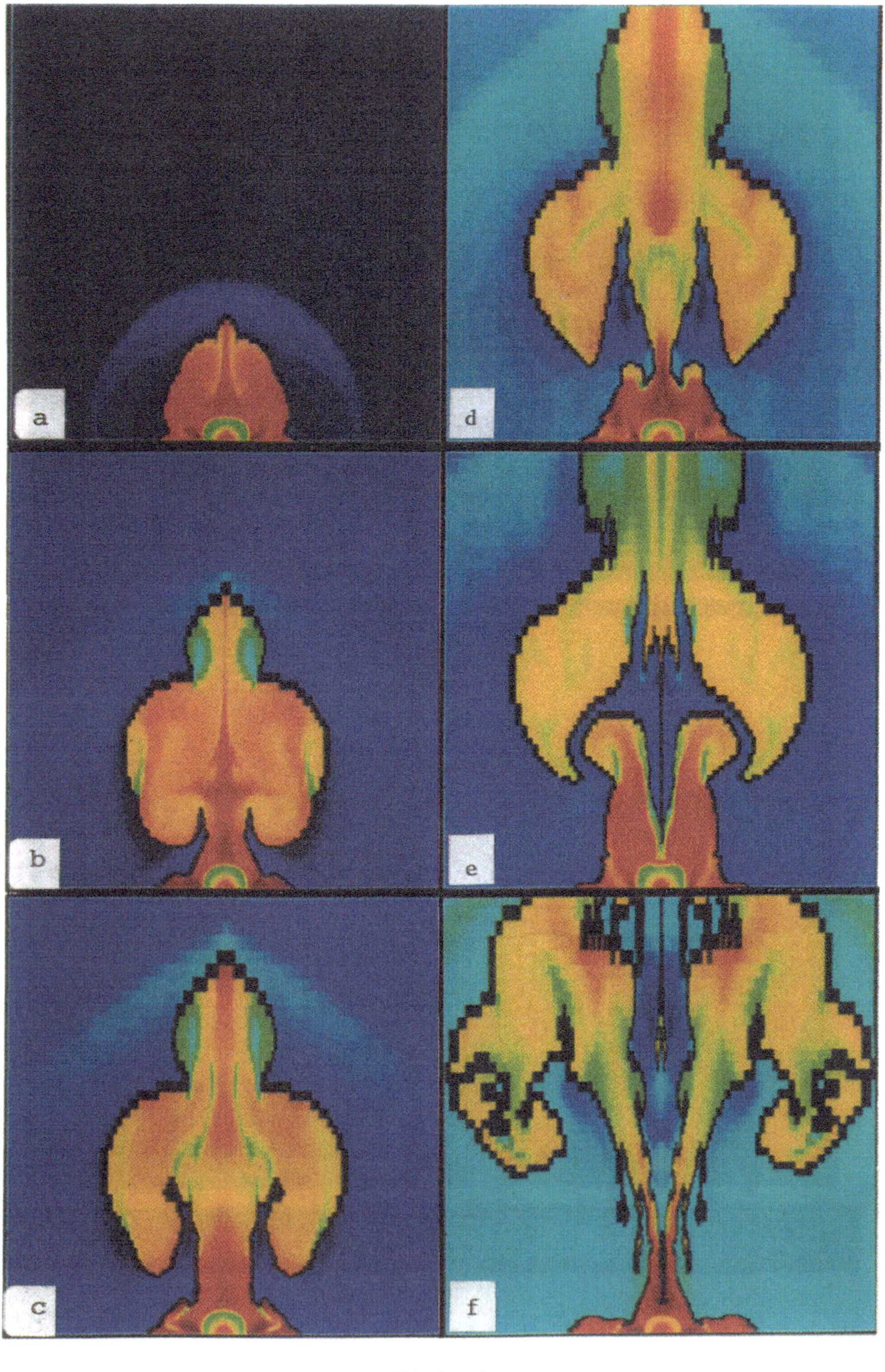

Plate 4

Captions to color plates

Plate 1. The time evolution of an axisymmetric supersonic jet. The plane of the picture contains the symmetry axis of the jet. Gas is continuously introduced from a circular inlet at left, with an internal Mach number of 3, a density 0.1 times the ambient density, and a pressure equal to the ambient pressure. Gas density is represented in 73 shades of color, each color representing an equal logarithmic interval between the maximum and minimum value of the density. The colors are ordered according to the color scale accompanyingPlate 2. Dark blue is minimum and ruddy red is maximum. The jet is divided into a forward moving supersonic beam (green) and a surrounding backward moving cocoon (blue). The boundary between the cocoon and the ambient medium is subject to nonlinear Kelvin-Helmholtz instabilities, which lead to turnover and mixing of the two gases. The computational half-plane comprises 640 equidistant zones in the axial direction and 60 equidistant zones in the radial direction out to 7.5 beam radii, with an additional 15 ratioed zones out to 15 beam radii.

Plate 2. Nonlinear kink instability in a 2-dimensional "slab" jet. Gas is continuously introduced from a slit at the bottom, with an internal Mach number of 3, a density 10 times the ambient density, and a pressure equalt to the ambient pressure. The instability is excited by applying a time-varying transverse velocity equal to 5% of the longitudinal velocity, with a frequency corresponding to the fastest-growing unstable mode. Gas density is displayed in color as described in the caption to Plate 1. The instability grows to nonlinear amplitude after convecting a few wavelengths downstream, effectively disrupting the directed bulk flow. Computation performed in Cartesian geometry. The computational plane comprises 300 equidistant zones in the longitudinal direction, 20 equidistant zones across the slab width, and 70 ratioed zones on either side of the midplane extending out to $\pm$ 10 slab widths.

Plate 3. Strong point explosion in a plane-stratified exponential atmosphere. Entropy is displayed in color as described in the caption to Plate 1. **a**) Initial spherical expansion of the high entropy "driver gas" (red) and blastwave-heated ambient gas (yellow). Blastwave radius $\approx$ 1 scale height. **b**) Nonspherical evolution of blastwave and hot bubble as it encompasses many scale heights. **c**) "Breakout" of the blastwave apex as predicted by Kompaneets (1960). **d**) Bouyant rise of the hot bubble and growth of the Rayleigh-Taylor instability on the leading surface of the bubble. Computation performed in cylindrical geometry assuming axisymmetry, with axis running vertically through the center of each plot. Computational half-plane comprises 360 axial by 120 radial equidistant, square zones.

Plate 4. Time evolution of jet formation via the Blandford-Rees (1974) Twin-Exhaust mechanism. Gas temperature is displayed in color as described in the caption to Plate 1. **a**) A bubble of hot gas is initially inflated by the central source. **b-c**) The nozzle forms as the bubble rises due to bouyancy. **d**) Nozzle constricts leading to inflation of the subsonic cavity surrounding the central source. **e-f**) A global Rayleigh-Taylor instability introduces dense ambient gas into the newly-formed jet. Computation performed in cylindrical geometry assuming axisymmetry and equatorial symmetry. Computational domain comprises 100 x 100 ratioed zones in the radial and axial directions spanning $0 \leq Z \leq 10h$, $0 \leq R \leq 10h$, with a central zone size of h/100, where h is the plateau scale height.

MUNACOLOR: UNDERSTANDING HIGH-RESOLUTION GAS DYNAMICAL SIMULATIONS THROUGH COLOR GRAPHICS

Karl-Heinz A. Winkler and Michael L. Norman

Los Alamos National Laboratory and
Max-Planck Institut fuer Physik und Astrophysik

I. INTRODUCTION

The bottom line is: once you have a billion numbers, it doesn't matter any more how you got them - whether from observation, the laboratory or a numerical experiment. You have to analyze them in order to understand their physical significance. One way of dealing with huge amounts of data, often employed by observers, is to use image processing techniques. In this method, the original data are carefully screened and processed with enhancement algorithms until the various structures hidden in the image become clearly visible and can be displayed graphically.

Numerical fluid dynamicists, on the other hand, have always relied heavily on the use of contour and vector plots for coping with the enormous amounts of data generated by multidimensional high resolution gas dynamical simulations. However, with the increased complexity of the flows under investigation, new display tools are needed to adequately display the simulation visually and to extract the essential flow features. Black and white contour plots alone are not enough any more.

Therefore, the use of color graphics and image processing methods is required to complement contour and vector plots in the concise presentation of numerical simulations. Furthermore, we will show how a new dimension of penetrating analysis of the numerical simulation becomes possible through the use of image processing. Finally, we outline our concept of a Numerical Fluid Dynamics Simulator (NFDS). Here, through a combination of raw processing speed, massive storage capacity, and realtime inter-

K.-H. A. Winkler and M. L. Norman (eds.), Astrophysical Radiation Hydrodynamics, 223–243.

active image analysis and display capability, the productivity of the individual research can be increased by several orders of magnitude, compared to current working conditions.

II. MUNACOLOR

Figure 1 and Plate 1 show the same four snapshots of the density evolution of a supersonic jet. Here a gas with density $\rho_b = .01$ (compared to the density of $\rho_a = 1.0$ of the ambient medium) in pressure equilibrium with the surroundings is continuously shot with Mach 12 (v_b / c_b = 12) into the ambient medium from the left side of the box. Because the beam gas is less dense than the ambient gas a cocoon consisting of supersonically backflowing gas surrounds the beam. The gas in the beam is processed by various shock systems at the head of the jet - called the working surface - and is redirected into the cocoon. A bow shock is racing ahead of the jet in the ambient medium. Furthermore, we can see the onset of Kelvin-Helmholtz shear instabilities along the boundary between the cocoon and the ambient medium. The simulation of this 2-D axially-symmetric jet was performed with the method described by Norman and Winkler elsewhere in this book on a grid of 640 equidistant gridpoints in the axial direction and with 75 gridpoints in the radial direction of the mesh. The radius of the beam itself is resolved with 8 equidistant gridpoints. The remaining 67 radial gridpoints have an equidistant distribution out to 7.5 beam radii, and a logarithmic distribution beyond out to 15 beam radii. What we now would like to discuss in detail is how the various graphical displays are made and what they tell us about the flow.

Figure 1 is a combined density-contour, velocity-vector plot. Contour-levels are equally spaced in the logarithm of density with a ratio of 1.2. Vectors are shown at every 4th zone if the absolute value of velocity is larger than .01 of the maximum value. The length of the vectors is scaled by the absolute value of velocity.

Figure 1. Evolution of a supersonic jet in black and white. Density-contour, velocity-vector plot of a supersonic jet propagating at Mach 12 through a 100 times denser medium. Regions of high gradients are highlighted by this display technique.

Plate 1. Evolution of the supersonic jet of Figure 1 in color. Density is shown in 73 spectral colors, (see color bar of Plate 2) where blue, green, and red indicate increasing levels of density.

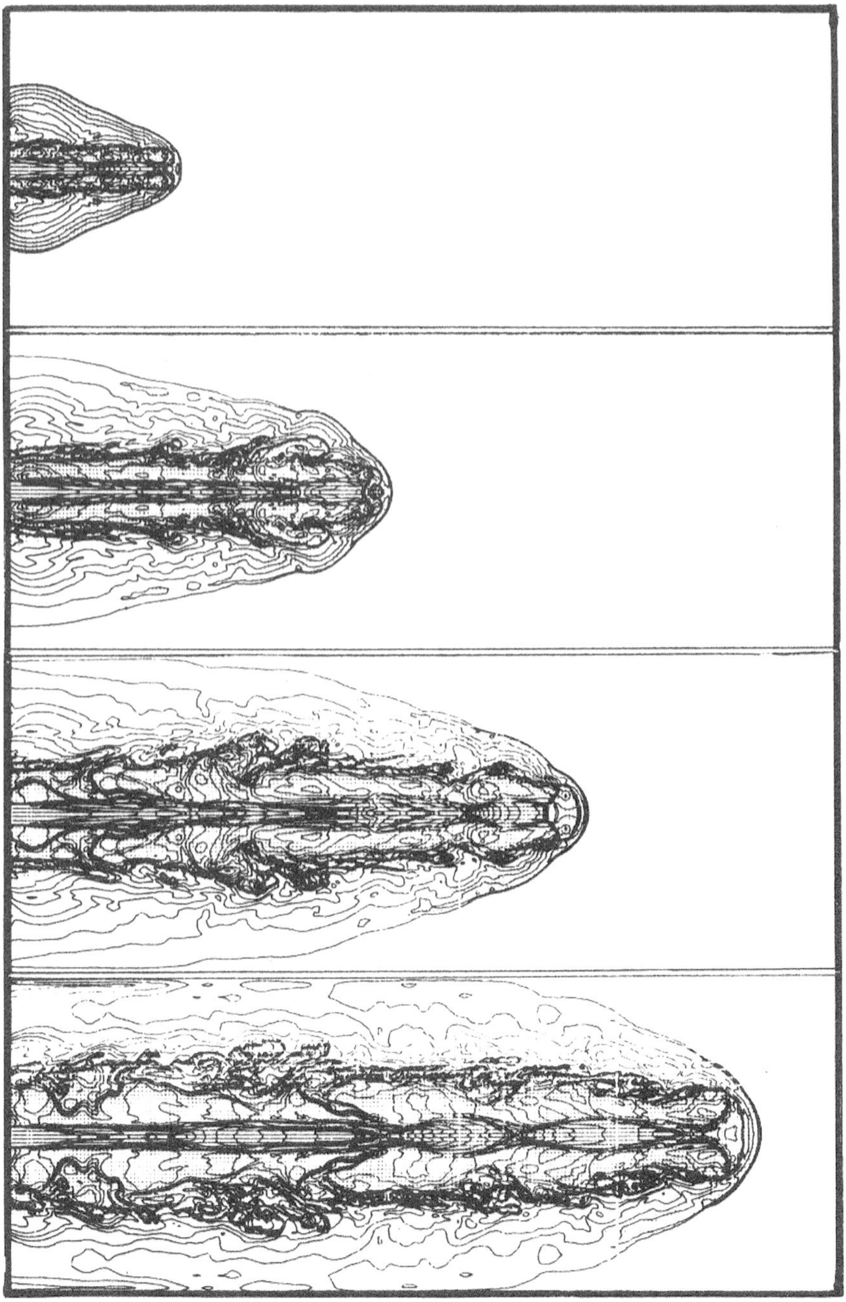

Figure 1

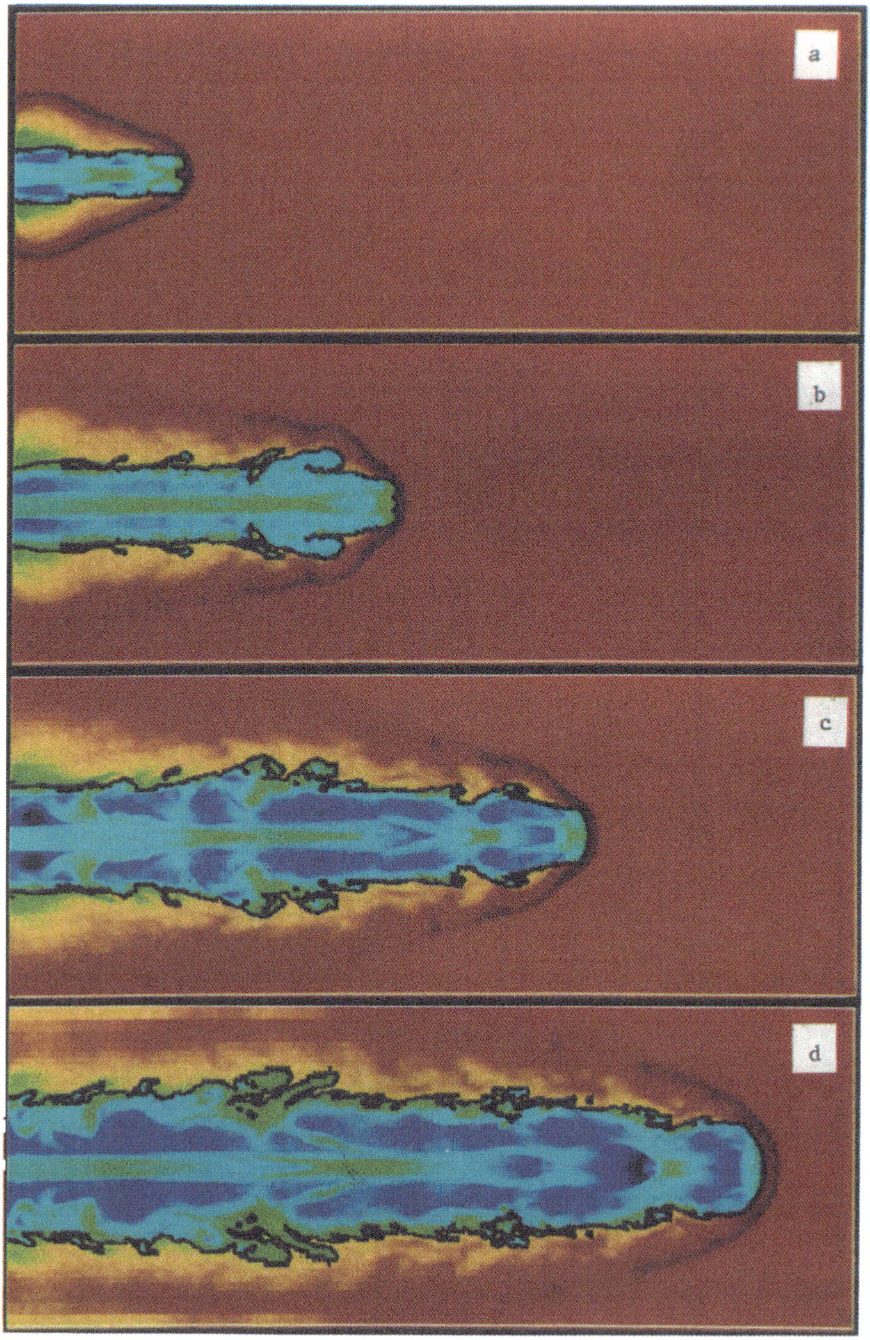

Plate 1

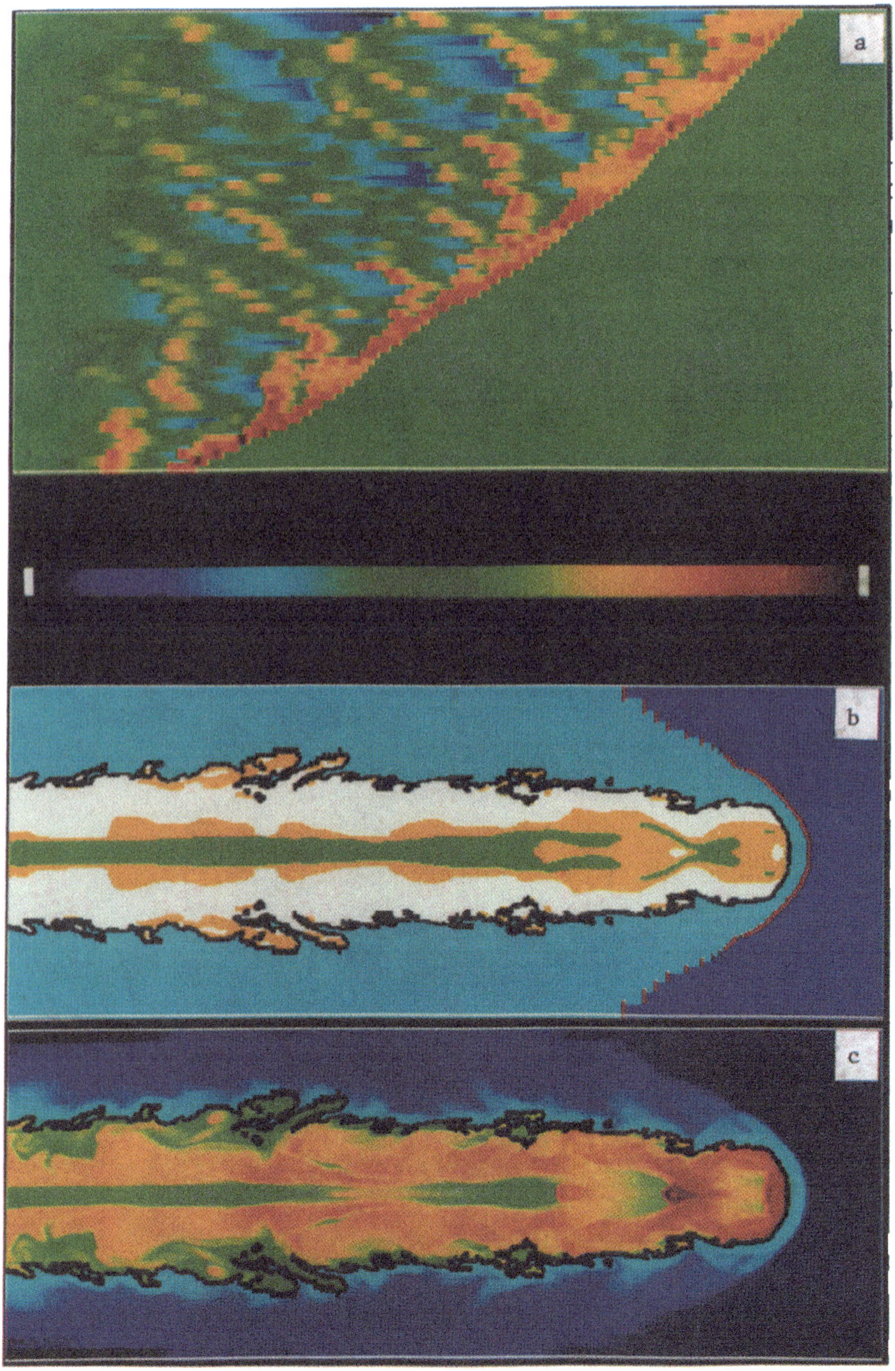

Plate 2

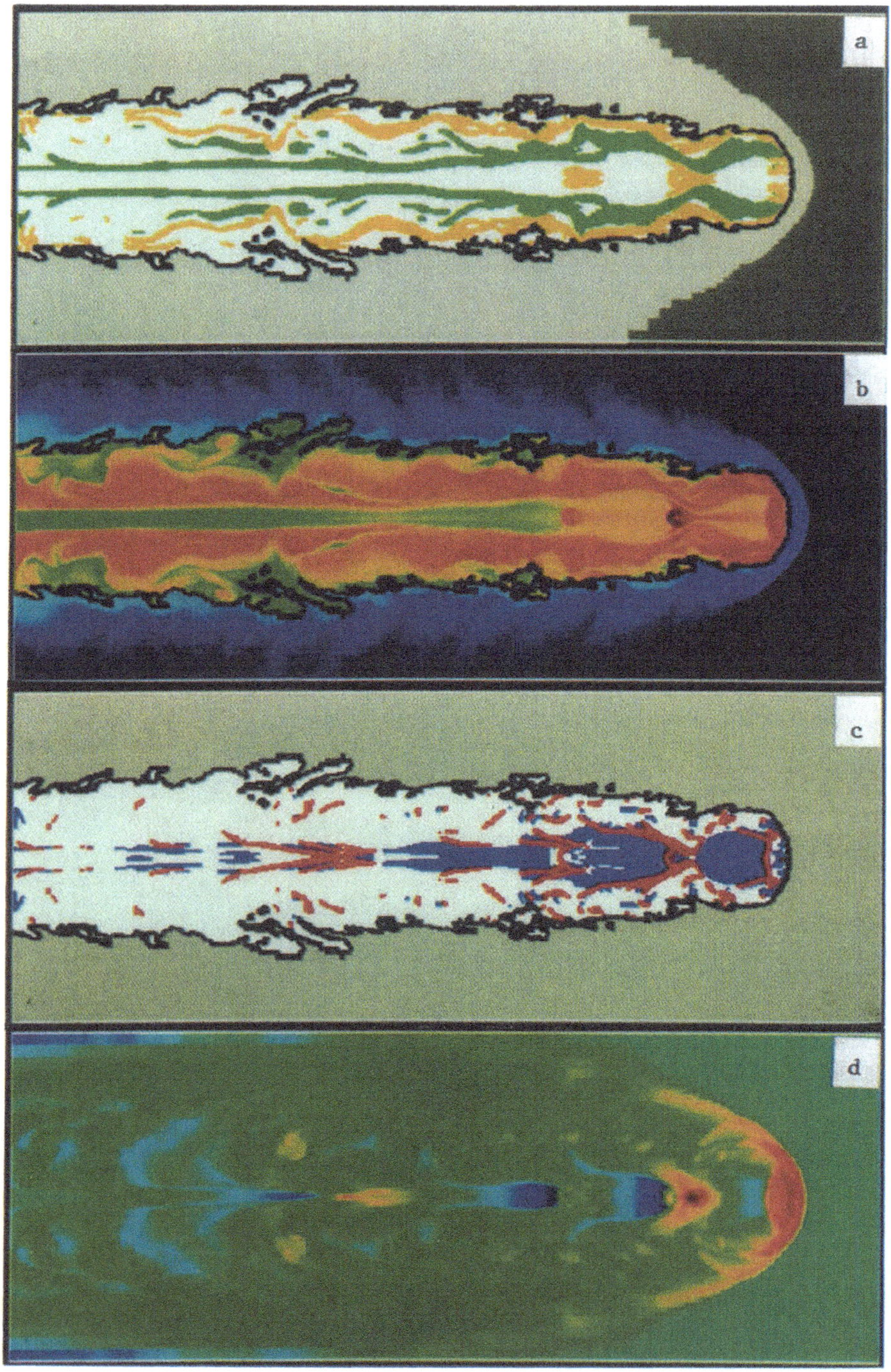

Plate 3

Plate 2. a) Space-time diagram of the supersonic jet of Figure 1 for the on-axis pressure sampled every 500th time step of the computation. Time increases from bottom to top. Low, intermediate, and high pressure levels are indicated by the spectral colors of the color bar.
b) Kinematic structure of the supersonic jet. Colors indicate: blue - undisturbed medium, red - bow shock, black - interface separating the two gases, white - backflowing cocoon, yellow - forward flowing cocoon, and green - beam.
c) Temperature shown in the 73 spectral colors of the color bar, same orientation.

Plate 3. a) Large absolute values of azimuthal component of vorticity $(\underline{\nabla}\times\underline{v})_\theta$. Positive levels are shown in green, negative levels are shown in yellow.
b) Entropy shown in 73 spectral colors.
c) Shock structure (red) and rarefaction (blue) plot. Notice the Mach disk configuration at the head (working surface) of the jet
d) Pressure distribution shown in 73 colors.

Plate 1 shows the same density distribution, now color-coded. We have mapped the density distribution from computational space into the pixel space of an image device (see Appendix for details). The logarithm of density is divided into 73 equidistant steps between its minimum and maximum value and assigned a spectral color according to the color bar shown in Plate 2. Blue indicates low densities; turquoise, green, and yellow are intermediate densities; and red denotes high densities. The lowest and highest values are shown in dark blue and dark red, respectively. As the boundary or contact discontinuity dividing the ambient medium from the cocoon is numerically treated by the LeBlanc interface tracking method, we have indicated all zones containing a piece of the interface in black. The Ramtek Image device on which the color plot was made has a resolution of 512 × 512 pixels. Therefore, our simulation, containing 640 gridpoints in the z-direction of the mesh, is slightly underresolved by the color image.

Nevertheless, the color image does contain additional important information compared to the contour plot. Whereas the contour plot shows narrow lines and focuses on steep gradients in the flow, the color image is area filling and puts emphasis on the gross structures in the flow. With the help of the color bar, one can immediately extract quantitative information from the color plot. In contrast to the contour plot, we can instantly decide, for example, where regions of low or high density are. In the contour plot a labeling of the individual contour levels indicating their density values is simply not feasible because of the complexity of the structures in the flow and the large number of different contour lines needed to describe them.

On the other hand, in the contour plot one can easily recognize shock systems and contact discontinuities through a clustering of several contour levels, indicating a steep gradient in the particular variable displayed, in this case density. An important detail of Fig. 1 is the vector plot, which gives us information about the direction and magnitude of the flow. This information can easily be included in the contour plot, which typically is done on a device with 200 or more points per inch resolution. In contrast, an image device with 512 × 512 pixel resolution is too coarse; one with 1280 × 1024 pixel resolution is only marginally useful for showing a vector plot.

Summarizing what we have learned so far from these two truly complementary modes of display, we can state the following: the gas is propagating forward in a narrow channel - the beam - and encounters several shock systems before it is turned around in the working surface. Although the backflowing gas inflates an extensive cocoon, the primary backflow takes place inside a narrow, modulated cylinder, close to the outer edge of the cocoon. In between this cylinder and the beam we can recognize several vortices, which appear to impose shock patterns in the beam whenever they touch it. Density is highest at the bow shock in the ambient medium, and drops considerably below the ambient value in rarefaction zones separated by shocks in the beam.

To investigate the flow further we would first like to isolate the essential gross features of the flow in a broad-brush way, showing no details. This is done in Plate 2b, which grew out of necessity at a time when the computer system used put a prohibitive limit on the amount of data that could be transferred into the image device. In such a case, algorithm (1) described in the Appendix works best.

In Plate 2b, the undisturbed ambient medium is shown in blue and the part of the medium disturbed by the bow shock (shown in red) is shown in turquoise. The interface separating the ambient gas from the cocoon is black. The beam, defined to be that volume where the kinetic luminosity $\rho u_z^3 > 5\%$ of its largest value, is shown in green. Forward flowing cocoon material is shown in yellow; backward flowing cocoon material is shown in white. Again, we recognize backward drifting vortices, i.e., regions where the beam (green) is surrounded by temporarily forward flowing bits of cocoon (yellow) which cause the beam to undulate.

Remarkable features in Plate 2b are the regions of backflowing gas (white) at and near the axis of symmetry. These regions are just downstream from very strong shock systems (Mach disk configurations) in which the flow is considerably slowed down in such a way that a temporarily backflowing situation can exist. This phenomena becomes much clearer when one examines the distribution of vorticity, Plate 3a, and looks at the shock systems and rarefaction waves shown in Plate 3c.

Plate 3a shows the azimuthal component of vorticity, i.e., curl of velocity $(\underline{\nabla} \times \underline{v})_{\theta}$. Positive values of vorticity from the maximum down to 5% of the maximum are shown in green; negative values from the minimum up to 5% of the minimum are shown in yellow. The beam and cocoon are shown in white, the ambient medium in light grey and dark grey.

In Plate 3c, the divergence of velocity is used to locate strong pressure waves and shock systems, as well as strong rarefaction zones. The former waves (compression zones) are indicated in red, and show negative regions of velocity divergence $\underline{\nabla} \cdot \underline{v}$ from the minimum value up to 5% of the minimum. Rarefaction (expansion) waves are shown in blue, and show positive regions of velocity divergence from the maximum value down to 5% of the maximum. The ambient medium is shown in grey.

The vortex sheet (green) originating at the inlet of the beam in Plate 3a extends all the way to the head of the jet, where it bends around and collapses onto itself all the way from the working surface back to the inlet. By comparison of Plate 3a and 3c, we see that vorticity (in this case negative vorticity, yellow) is generated in curved shock systems. The triplepoint of the Mach disk configuration at the working surface is an extreme case of a curved shock. A closer look at the structure of the working surface will reveal the source of the temporary backflow on-axis in Plate 2b. The backflow is caused by the vorticity generated in the Mach disk configuration.

Plate 2c, showing temperature, and Plate 3b, showing entropy, are generated in the same way as the density plots of Plate 1. We see that temperature varies more strongly than entropy. As we have solved the ideal (Euler) equations of gas dynamics, entropy can be changed only at shock fronts or by mixing materials of different entropy. In evaluating changes of entropy, we always must be aware of the fact that we are looking at a highly time-dependent phenomenon and not at a stationary flow situation. Thus, entropy may decrease along a streamline, reflecting the thermodynamic history of the fluid. By comparing Plates 3a, 3b, 3c, and 2b with Fig. 1 and Plate 1, we confirm that indeed the gas in the beam goes through an alternating sequence of rarefaction and compression waves before it escapes into the cocoon and flows efficiently backwards in a modulated cylinder close to the outer edge of the cocoon. The latter flow pattern leaves a clear mark in the entropy, density and vorticity plots. In the vorticity plot, the primary backflow channel lies between the green and yellow vorticity regions in the cocoon.

Entropy is almost like an open book, which tells us the whole story of the simulation by keeping a detailed account of the thermodynamic history of the flow. Consider, for example, the green regions of entropy in the cocoon close to the outer boundary. They are outside the main backflow channel and cannot have been produced by entropy changes of the beam/cocoon material alone. The real news in Plate 3b is that the material in the green regions of the cocoon is a turbulent boundary layer, which consists of mixed ambient and cocoon material. Why then is the interface separating the cocoon from the ambient material located so far outside? The explanation can be found in how we apply interface tracking. Interface tracking is a reliable method so long as enough gridpoints (10 to 20) are available to resolve the wavelength of an instability. However, one characteristic feature of the Kelvin-Helmholtz instability is that it generates smaller and smaller wavelengths until they cannot be resolved any more by any fixed number of meshpoints. If we encounter isolated regions of space where the instability can no longer be resolved adequately, we dispense with interface-tracking altogether and simply mix the materials according to their actual values. Thus, we see that the boundary between the cocoon and the ambient gas is not a simple vortex sheet but is instead a complicated turbulent boundary layer, a feature which one would have difficulty recognizing in a contour plot. "And now you know the rest of the story." (Paul Harvey)

From the entropy plot, we also learn that the speed of the bow shock is not steady, but instead alternately decreases and increases. This behavior is closely related to quasiperiodic vortex-shedding at the working surface. As the vortex surrounding the working surface grows in size, the dynamic drag associated with it slows down the jet somewhat. Then, the vortex is finally shed, the jet head itself experiences less dynamic drag and accordingly accelerates, resulting in an increased entropy production by the now-stronger bow shock. Evidence for about a dozen such vortex-shedding events can be identified in the entropy plot as ripples of higher entropy in the ambient medium. They are also visible in the density distributions of Plate 1. For completeness, we would like to mention that the rundown of entropy corresponding to the last 3 density snapshots in Plate 1 has been published on the cover of Physics Today in July 1984.

Coming back to Plate 3c, we discover that the bow shock is missing in that particular plot. Why? Perhaps an oversight? Answer: no. According to our prescription for finding compression regions (red), we require that such regions have at least 5% of the maximum compression in the flow. However, this is not the case for the bow shock. So we make the discovery that internal shocks in the beam are at least 20 times stronger than the bow shock as measured by the divergence of velocity. The explanation for this behavior can be found in the strong rarefaction zones separating different shock systems in the beam. In order for the beam to propagate at all, the pressure at the head of the working surface just inside the contact discontinuity must be larger than, or at least comparable to, the pressure in the ambient medium. So we see that the shock systems in the beam simply have to be a lot stronger than the bow shock to make up for the pressure loss in the rarefaction zones.

That the rarefaction zones are indeed related to low pressure regions follows from a comparison of Plates 3c and 3d. In Plate 3d, pressure is shown in 73 rainbow colors in the same way as density in Plate 1, or temperature and entropy in Plates 2c and 3b, respectively. We see that pressure in the rarefaction zones drops way below the equilibrium pressure and obtains its highest values just downstream of shock waves in the beam.

For the interpretation of extragalactic radio jets, the pressure- and shock system-plots are the most important ones. What one observes is synchrotron radiation which supposedly has been produced by in situ acceleration of relativistic electrons in shock fronts, which radiate as they spiral around a local magnetic field. In fact, many of the observed morphologies of radio emission from hot spots (i.e., working surfaces) of extragalactic jets can be reproduced by gas-dynamical simulations similar to the one presented here (see Smith et al., 1984).

If we monitor pressure variations along the jet axis as a function of time, we obtain the space-time diagram shown in Plate 2a. Again, 73 spectral colors are used to indicate the pressure distribution. However, because we have sampled the pressure variation along the axis for only every 500th time step, the plot has a ragged appearance. Nevertheless, we can track the space-time paths of about a dozen high-pressure ridges, originating near the bow shock and propagating to the left towards the jet source. These pressure ridges give us the location in space-time of the shock systems in the beam that are generated by the slowly backward drifting vortices in the cocoon.

This concludes our discussion of the MUNACOLOR graphics presented here. More examples of B/W and color presentations can be found in Norman and Winkler, this book, and in Norman et al., 1982, 1983, 1984a,b, Smarr et al., 1984, and Smith et al., 1984.

Other very useful plots we have tried thus far are pressure-contour and momentum-vector plots superposed, contour plots of the z- and r-components of velocity, specific vorticity (i.e., curl velocity divided by density) plots, and plots of a combination of interfaces and kinetic luminosity. These all focus on additional aspects of the flows under consideration. What one really would like is to have all these various aspects of the same evolution available in a highly interactive environment, coupled with a variety of additional software tools which would allow one not only to visualize the flows, but also to scrutinize the simulation and to extract hard quantitative information of all kinds from it. What such an environment would look like is described next.

III. IMAGE PROCESSING AND NUMERICAL EXPERIMENTS

So far, we have discussed only a selection of different plots and images. But what really will make a difference in analyzing a numerical simulation is to process the outcome of the gas dynamical simulation, that is, the entire set of models, which are dumped frequently. It is at this point we introduce the powerful tools of image processing methods into our concept of a NFDS. For example, one could Fourier transform the distribution of entropy in the ambient gas and compute its power spectrum. The same could be done for the entire space-time diagram of pressure distribution along the axis of symmetry. In this way, one could immediately find out, at the touch of a button, whether there is some quasiperiodicity in the flow, and if so, what its dominant wavelengths are. Another way of measuring this parameter, known as the Strouhal number, is to introduce a rake of probes at several locations perpendicular to the direction of the flow and monitor, at each rakepoint, the velocity components as a function of time (Zabusky, private communication). Then one feeds these newly extracted data into a Fast Fourier Transform (FFT) routine and measures the power spectrum, and hence, the Strouhal number directly. Incidentally, this is precisely the procedure followed in actual laboratory experiments. The important point to emphasize here is that in this way one can build up a set of diagnostic tools which allow one to analyze the experimental data, be they from a laboratory experiment or a numerical experiment. In other words, we really would like to get away from the level of producing merely pretty and colorful pictures and instead move to the level where we extract hard scientific information from the original data in an objective, automatic way, without human interference.

Another important issue is pattern recognition. One would like to know whether there are any coherent structures in the flow and what they look like. For pattern recognition, the human eye and mind are undoubtedly still the best, but they can be greatly aided by image processing software. For example, we could take the density images and instead of spreading out the full range of spectral colors over the whole density range, just use all the colors in a relatively small density window. Then, in an interactive session, we could move the window over the entire density range for a given snapshot, or use the same window and move it over many consecutive frames, always watching alertly for new patterns to pop up, which then could be analyzed subsequently with objective methods. We could also treat the vorticity and shock system plot similarly. But now in contrast to the examples given in the previous section, we could dispense with the arbitrary cutoff at 5% and monitor the functions smoothly over all windows of interest.

In the same vein, one would like to track features of the flow in time and measure their mass or energy output (or whatever). The best way to do this would be to lock onto a certain feature in an interactive session and just follow its temporal evolution automatically and objectively.

Another very successful way of finding patterns in any given data, be they of observational or numerical origin, is to treat them first with various kinds of filters, and then look for patterns. For example, one could use an edge-enhancement filter and thus suppress all low contrast structures while highlighting all high contrast structures in the flow. This would be an ideal tool for finding and isolating all sorts of discontinuities and nonlinear waves in the flow.

Yet another useful application of image processing methods for gas dynamical simulations is to subtract a stationary or equilibrium solution out of a time dependent or nonequilibrium solution and study the remainder in detail. Similarly, one can filter numerical solutions of time-dependent problems and study the most significant or the least significant bits. An illustrative application of this procedure would be to get a better handle on the numerical noise in any simulation. One also could do the same simulation on 2 different mesh resolutions, and image process the difference to find the actual "error bars" of the numerical experiment. Further, when we go from 2-D to 3-D simulations, we will need to exploit image-processing technology to perform the frequent coordinate transformations required to allow visualization of the numerical solution by projection onto a view plane at various viewing angles.

We could go on with this list of different applications of image processing techniques in the analysis of numerical simulations for quite a while, but we think that the usefulness of image processing in the present context should be clear by now. Instead, we would like to discuss some parameters and constraints the computer environment must satisfy in order to allow the individual researcher to utilize fully the ideas described above.

IV. NUMERICAL FLUID DYNAMICS SIMULATOR

A typical session with such a system ideally would look like this. Suppose one does a 2-D gas dynamical simulation with 10^6 grid points and up to 10^5 timesteps. While the computation is proceeding, one would like to monitor it and display in real time a collection of the various plots and images described above on a blackboard-size, high-resolution, flat panel display. One also would like to do the various analyses (checks) of the numerical solution periodically in order to make a well founded decision whether to proceed with the simulation or to terminate it, either because one got the desired result, or because something went wrong. One also would like to sample the numerical solution from time to time and apply the diagnostic tools.

Instant replay capability, and the possibility of careful and thorough post-processing, would require the taking of frequent model dumps, and put a large strain not only on the processing speed of the computer environment, but also on the communication channels and storage capacities.

However, this mode of operation, i.e. performing an actual numerical experiment, had already been foreseen by John von Neumann in 1946. According to him, one defining characteristic of a numerical experiment is the ability of the human researcher "to exercise his intuitive judgement as the calculation develops". This also necessitates the continuous read out of all essential information "while the calculation is in progress". The user "can then intervene whenever he sees fit". Of course, von Neumann was thinking in terms of the technological base of his time, and therefore discussed the use of oscilloscopes as the primary graphical output medium. Nevertheless he placed enormous emphasis on what he called the input-output organ.

The question therefore is what performance characteristics have to be fulfilled by the computer environment to allow the user to carry out his real-time numerical experiments productively. The answer will depend on the complexity of the problem under consideration and the baudrate with which humans can absorb information visually. By the latter condition an impedance match of information flow between man and machine is established, see Winkler et. al., 1985. This means that the information flow of the numerical experiment has to match the human needs, not the other way around, in order to avoid excessive waiting times at the human end.

Let us think of (Winkler et al., 1985) the simulation processors as an information pump generating 0 [word/s] of raw output data, where 0 is given by

$$0 = \frac{X}{\bar{C}} \tag{1}$$

Here X [operation/s] is the processors' aggregate execution speed and $\bar{C}$ [operation/word (word = updated field variable)] the application algorithm's average complexity. We would like to route this output data to an image device for display purposes as fast as it is being generated. Taking into account the redundancy in the raw output data we find for the inter-system throughput baudrate T_{is} [baud = bit/s]

$$T_{is} = 0\ W\ f_c\ f_d\ f_s\ f_r\ \min\left(\frac{P}{S},1\right) \tag{2}$$

where W [bit] is the word length, and the f's are redundancy factors as follows: f_c is the data compression factor representing the truncation of a full machine word of length W to of order 24 bits of color information; thus, $f_c = 24/W$; f_d is the dump frequency factor equal to the inverse of the interval in cycles between dumps; f_s is the selection factor, reflecting the fraction of the total number of field variables you would like to display at once; and f_r is the redundancy factor reflecting fractional change of the image in time. The last factor in Eqn. (2) reduces T_{is} by a factor (P/S) if the problem size S [number of computational grid zones] exceeds P [number of pixels on the image device dedicated to display a single variable on the computational grid].

Our principle of impedance matching man and machine now implies that the internal throughput of the image device T_{id} must equal the effective human throughput T_h^{eff}:

$$T_{id} = \frac{T_{is}}{f_r} = T_h^{eff} = \min(\frac{S}{P},1)T_h \tag{3}$$

T_h^{eff} is less than the human throughput T_h by a factor S/P for $S < P$ simply because a deficit in information due to low resolution or restricted field of view cannot be compensated for by a higher display rate. T_{id} must be greater than T_{is} by a factor of $1/f_r$ because the nonredundant data must be expanded to fill the screen. What remains to be determined is the human throughput T_h for a particular device. An upper limit for the maximum human throughput T_h^{max} is given by the following relation:

$$T_h^{max} = \frac{FV}{AR} \, CR \, DR, \tag{4}$$

where $FV = 60^o \times 60^o$ is the field of view of the human eye, $AR = 1' \times 1'$ its angular resolution, $CR \leqslant 24$ bits its color resolution, and $DR \leqslant 24$ Hz its display rate (Winkler et. al., 1985). Inserting these numbers we find

$$T_h^{max} \leqslant 7.5 \text{ gigabaud} \tag{5}$$

Using more realistic values for $CR \leqslant 18$ bits and $DR \sim 12$ Hz we find $T_h^{max} \leqslant 3$ gigabaud. However, current image devices typically don't operate that fast. Taking this fact into account we arrive at

$$\begin{aligned} T_h &= P_{dev} \quad CR_{dev} \quad DR_{dev} \\ &= \left(\frac{P_{dev}}{\frac{FV}{AR}}\right) \left(\frac{CR_{dev}}{CR}\right) \left(\frac{DR_{dev}}{DR}\right) T_h^{max} \end{aligned} \tag{6}$$

where P_{dev} is the number of pixels on the screen, CR_{dev} is the depth [bit per pixel], and DR_{dev} [Hz] is the rate with which the device can accept images. Typical ratios for commercially available equipment are

$$\frac{P_{dev}}{\frac{FR}{AR}} = \frac{1280 \times 1024}{3600 \times 3600} \simeq 0.1 \tag{7}$$

$$\frac{CR_{dev}}{CR} = \frac{8}{24} \simeq 0.3 \tag{8}$$

$$\frac{DR_{dev}}{DR} = \frac{1}{24} \simeq 0.04 \tag{9}$$

So we find that the baudrate T_h with which humans can view images on commercially available equipment is often up to a factor of 1000 lower than T_h^{max}.

Combining Eqns. (1), (2), and (3) we find

$$X\ W\ f_c\ f_d\ f_s = \left(\frac{S}{P}\right) \bar{C}\ T_h \tag{10}$$

relating the processors' aggregate execution speed X to the ratio (S/P) of problem size over the number of pixels, the algorithm's complexity $\bar{C}$, and the human throughput T_h. This relation can now be used in several ways to study system requirements. We could for example determine what baudrate results from a simulation with S gridpoints and complexity $\bar{C}$ running on several processors with the aggregate execution speed X, provided we want to monitor the numerical experiment in real time. Or for example, we may want to derive how large X must be in order that a 2-D gas dynamical simulation (carried out with the code used for all examples above) with $S = 1280 \times 1024$ gridpoints and complexity $\bar{C} = 50$, yields a movie-like display rate of 24 Hz for 24 bits per pixel on a screen with P = S pixels. We find

$$X = \frac{\left(\frac{S}{P}\right) \bar{C}\, T_h}{W\, f_c\, f_d\, f_s} = \frac{1 \quad 50 \quad 750 \times 10^6}{64 \quad \frac{24}{64} \quad 1 \quad 1} = 1.6 \text{ gigaflops}, \tag{11}$$

where gigaflops stands for 10^9 floating point operations per second. If we only looked at every tenth time step, in order to get a dynamical impression of the evolution from an explicit hydrocode, we arrive at 16 gigaflops. These numbers are not as excessive as they appear and probably can be obtained within a few years with then state of the art supercomputers. The assumed baudrate of 750 megabaud could be brought down to 250 megabaud by cutting down the color resolution to 8 bits, allowing the simultaneous display of 256 different colors. Such baudrates should certainly be obtainable with present day fiber optics technology.

Finally let's give an estimate of what kind of storage capacity will be needed for such a run with 5 different variables and 10^5 time steps if every 10th time step is recorded with 8 bits/pixel accuracy for viewing purposes. We get

$$\begin{aligned} \text{Storage} &= 1280 \times 1024 \times 10^4 \times 5 \times 8 \\ &= 5.2 \times 10^{11} \text{ bits} \end{aligned} \tag{12}$$

This number will go up to 15×10^{11} bits if we retain 24 bits per pixel for real image processing purposes. Such large storage capacities will hopefully become available soon with the arrival of the laser disk storage technology.

In our opinion, the conditions described above can be met with relatively modest extensions of existing technology. They would put the scientific user of the NFDS into a position to ask specific questions relating to the physics of the flows under investigation, and <u>actually have them answered in short enough time to keep a lively dialog between the scientist and the system going</u>. A quantum jump in productivity of the individual researcher would certainly result as a consequence of such a facility.

APPENDIX

In principle, one has two options to generate a color-image on an image-device. First, one may use the graphical features of the device, and plot in the normal mode with area fill on, and leave it to image processing software to generate an image field. This is done in the first method described in detail below. Alternatively, one may generate directly on a computer a file which already contains all the necessary information in the appropriate format, and which can then be read directly into the memory of the image device and displayed on the system. This is the second method described below.

1) 2-D Pattern Search (Motto: Divide and Conquer)

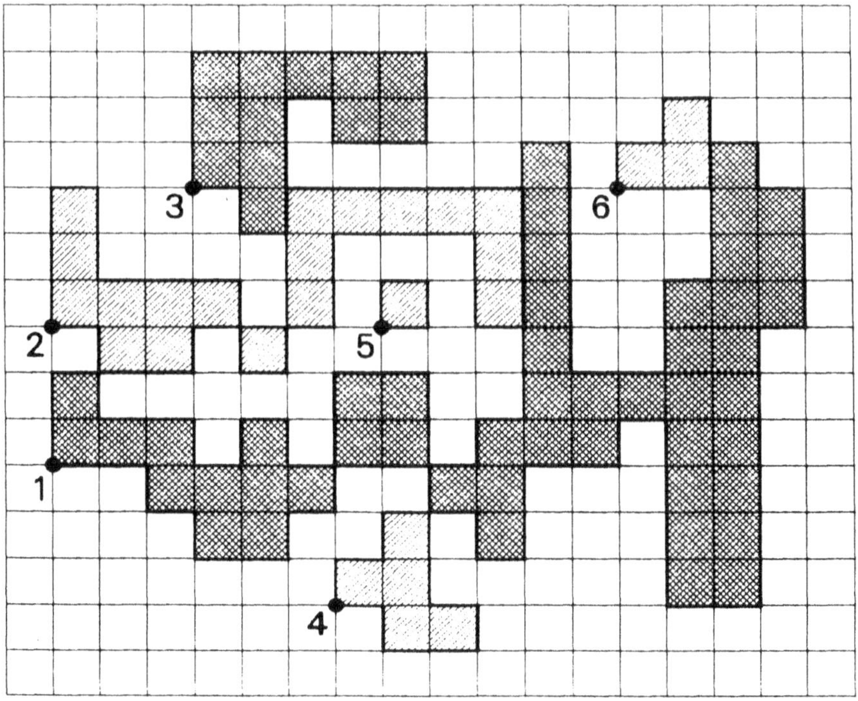

Given a matrix of color indices we proceed in the following way. Scan, from left to right, each column, from bottom to top, for the first appearance of the nth color. Then follow the lower boundary of the color area by going only right, or up and down (never go left) until you run out of that color. Then go from right to left, and scan each column from the bottom of a color area for the top of that color area until you hit the beginning of the color area. Plot the polygon with area fill on. Set the area just plotted to a fake color value in order to avoid plotting the same color area twice. Go back to 1) and proceed as described there for the next patch of the same color, and so on. Finish by resetting the fake color value to the correct original color value. Continue this procedure for the next color.

What we have then done is to divide a complex and multiply connected area into several simple convex areas (i.e. six such areas in the example above), and plot them individually. This procedure works for arbitrary topologies.

In detail, we decide which line to track as follows:

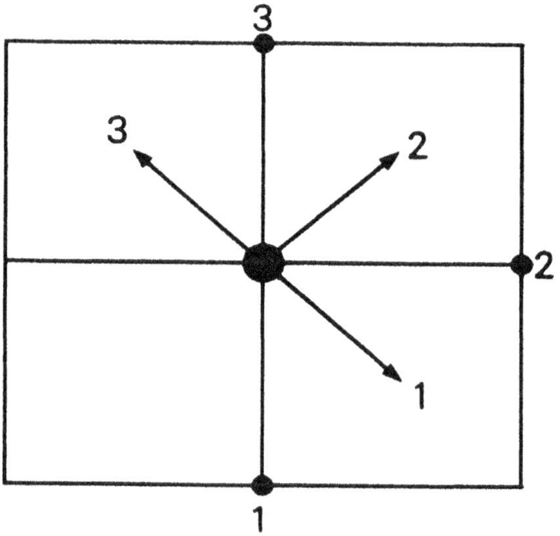

If Area 1 is same color, next point is 1; if area 2 is same color, next point is 2; if area 3 is same color, next point is 3. If none of the above, terminate; you have hit the end of the trail. In order to minimize the I-O further, we drop all non-corner points from the polygon; that is, all the interior points of a linear boundary piece.

The above procedure works well if one is limited by I-O and has a hardware polygon fill on the image device. The disadvantage of this approach is that the random pixel writing speed of the device is typically much slower than the sequential pixel writing speed.

2) If I-O is not a problem, one can map the matrix of color indices into pixel space, and write it directly (sequentially) in image file format onto the image device. The disadvantage of this approach is that one must then transfer larger amounts of data.

REFERENCES

1. Norman, M. L., Smarr, L. L., Winkler, K.-H. A., and Smith, M. D., 1982, Astron. & Astrophys., 113, p. 285.

2. Norman, M. L., Winkler, K.-H. A., and Smarr, L. L., 1983, in "Astrophysical Jets", ed. A. Ferrari and A. G. Pacholczyk, (Dordrecht: Reidel).

3. Norman, M. L., Smarr, L. L., Winkler, K.-H. A., 1984a, "Numerical Astrophysics: A Festschrift in Honor of James R. Wilson", ed. R. Bowers, J. Cantrella, J. LeBlanc, (Jones and Bartlett: Boston).

4. Norman, M. L., Winkler K.-H. A., and Smarr, L. L., 1984b, "Proceedings of NRAO Workshop No. 9 on Physics of Energy Transport in Extragalactic Radio Sources", ed. A. H. Bridle and J. A. Eilek

5. Smarr, L. L., Norman, M. L., and Winkler, K.-H. A., 1984, Physica 12D, p. 83.

6. Smith, M. D., Norman, M. L., Winkler, K.-H. A., and Smarr, L. L., 1984, submitted M.N.R.A.S.

7. von Neumann, J., 1946, in "John von Neumann, Collected Works", Vol. 5, p. 1, Pergamon Press, 1963.

8. Winkler, K.-H. A., Norman, M. L., and Norton, J. L., 1985, "On the Characteristics of a Numerical Fluid Dynamics Simulator", in Austin Symposium on Algorithms, Architectures, and the Future of Scientific Computation, in press.

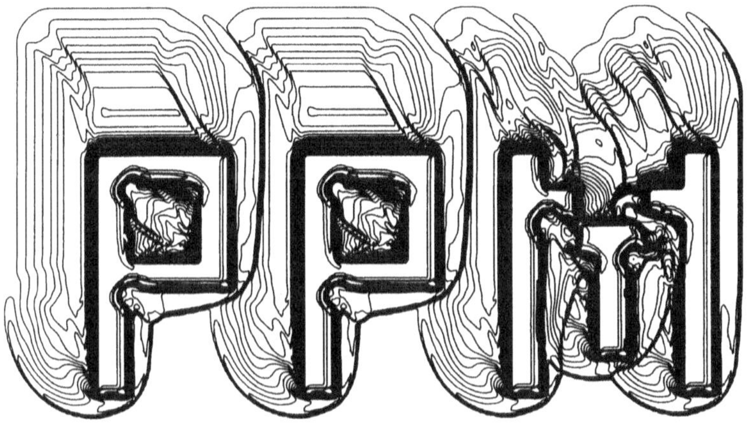

PIECEWISE-PARABOLIC METHODS FOR ASTROPHYSICAL FLUID DYNAMICS

Paul R. Woodward
Lawrence Livermore National Laboratory
Livermore, California 94550

ABSTRACT

A general description of some modern numerical techniques for the simulation of astrophysical fluid flow is presented. The methods are introduced with a thorough discussion of the especially simple case of advection. Attention is focussed on the piecewise-parabolic method (PPM). A description of the SLIC method for treating multifluid problems is also given. The discussion is illustrated by a number of advection and hydrodynamics test problems. Finally, a study of Kelvin-Helmholtz instability of supersonic jets using PPM with SLIC fluid interfaces is presented.

K.-H. A. Winkler and M. L. Norman (eds.), Astrophysical Radiation Hydrodynamics, 245–326.

INTRODUCTION

There is a long history of the use of computer simulations in astrophysics because of the difficulty or impossibility of obtaining useful observational data. It should come as no surprise, for example, that nearly all our understanding of stellar structure and evolution has come from computer calculations. In other areas such as star formation, supernova explosions, formation of interstellar clouds, and the evolution of galaxies some observational data exists. However, computer simulations offer the potential for controlled experiments in which the various parameters of a theoretical model may be varied independently and unequivocal observations of the results may be made. Scant astronomical observational data often allows several theoretical interpretations. Computer simulations have been used to choose between such interpretations by means of checking for a rigorous consistency of theoretical arguments with known physical laws whose workings are too complex for the unaided mind to fathom.

For these reasons computer simulations have played an important role in astrophysics since the early 1960's, and there is every indication that their role will continue to grow along with the power of the computing machines they exploit. Consequently it is advisable for modern astrophysicists to develop some familiarity with the methods by which computer simulations are performed. Without such a familiarity the astrophysicist is forced to take all computational results at face value, and this can lead to misunderstandings. It is the purpose of this article to acquaint the uninitiated with some fundamental ideas behind a group of modern techniques for simulating fluid flow on computers. There is no intent to set down detailed equations, to give any more than an indication of the historical development of this subject, or to maintain any semblance of objectivity in choosing which techniques merit discussion. This editorial attitude will simplify the discussion so that the essential ideas of the various techniques which receive attention may be more clearly understood. The reader who desires a fuller account will be referred to a number of articles where he may find it.

THE RELATIONSHIP BETWEEN SIMPLE ADVECTION AND ASTROPHYSICAL FLUID DYNAMICS

Every informed astrophysicist should know that the foundation of fluid dynamics is the Boltzmann equation. Apart from the collision integral on the right-hand side, this equation describes simple advection, an incompressible motion of fluid in phase space. We write the collisionless Boltzmann equation in a two-dimensional phase space as

$$\frac{\partial f}{\partial t} + v \frac{\partial f}{\partial x} + g \frac{\partial f}{\partial v} = 0 \quad . \tag{1}$$

Here f is the density of the fluid in phase space, x is the distance coordinate, v the speed, and g the acceleration. When g is independent of v the motion in phase space is obviously incompressible. Inappropriate as it may seem, this equation with the last term dropped (i.e., g = 0) has been the conceptual basis for the design of numerical methods for treating the much more complicated, nonlinear equations of fluid dynamics. The success of this approach rests on the fact that the fluid equations are derived from the <u>collisional</u> Boltzmann equation, and hence at a fundamental level they involve simple advection. To see this, consider the equations for the one-dimensional, isentropic flow of a polytropic fluid. In Lagrangian coordinates moving with the fluid the governing equations are:

$$\frac{\partial x}{\partial t} = u \quad , \tag{2}$$

$$\frac{\partial V}{\partial t} - \frac{\partial u}{\partial m} = 0 \quad , \tag{3}$$

$$\frac{\partial u}{\partial t} + \frac{\partial p}{\partial m} = 0 \quad , \tag{4}$$

$$dm = \rho \, dx \quad , \tag{5}$$

$$V = 1/\rho \quad , \tag{6}$$

$$p = Ap^{\gamma} \quad , \tag{7}$$

Here m is a mass coordinate related to the distance coordinate through the density ρ as in Eq. (5), A is a constant, and γ is the constant polytropic index. The Eulerian sound speed c and the Lagrangian sound speed C are defined by

$$c^2 = \gamma \, p \, V \quad , \tag{8}$$

$$C = \rho c \quad . \tag{9}$$

If we define two Riemann invariants R_+ and R_- by

$$R_{\pm} = u \pm \frac{2}{\gamma - 1} c \quad , \tag{10}$$

it is a trivial matter to verify that these quantities satisfy the advection equations

$$\frac{\partial R_\pm}{\partial t} \pm C \frac{\partial R_\pm}{\partial m} = 0 \quad . \tag{11}$$

This implies that the two Riemann invariants are simply advected in opposite directions at the speed C in the Lagrangian coordinate system (R_+ goes to the right and R_- to the left). Thus the hydrodynamical problem has been reduced to simple advection.

Under more general circumstances fluid dynamics is not so simple. For large amplitude sound signals the two Riemann invariants interact and entropy is produced (A increases). In this case formulae like Eq. (10) for the Riemann invariants cannot be written; however, equations for the differentials $dR_\pm$ are still useful. Now A in Eq. (7) is no longer constant and we must augment our nonlinear system of equations with an energy equation

$$\frac{\partial E}{\partial t} + \frac{\partial up}{\partial m} = 0 \quad . \tag{12}$$

Here E is the total energy per unit mass, which is related to the specific internal energy ε by

$$E = \varepsilon + \frac{1}{2} u^2 \quad . \tag{13}$$

The equation of state which replaces Eq. (7) is now

$$p = (\gamma-1) \rho\varepsilon \quad . \tag{14}$$

The characteristic equations for changes in Riemann invariants now become

$$dR_\pm = du \pm \frac{dp}{C} = 0 \quad . \tag{15a}$$

Equations (15a) hold only along characteristic paths defined by

$$dm = \pm C \, dt \quad . \tag{15b}$$

In addition, a third characteristic equation

$$dR_o = dp - c^2 d\rho = 0 \tag{15c}$$

applies along characteristic paths defined by $dm = 0$. Equations (15) hold only in smooth flow; if shocks or contact discontinuities are encountered on the characteristic paths, $dR_{\pm}$ and dR_0 are nonzero across these discontinuities. Even when no simple formulae for quantitites $R_{\pm}$ can be written, the concept of advection of Riemann invariants along characteristic paths still proves useful.

Advection also enters the study of fluid dynamics more directly when Eulerian calculations are performed. In this article we will consider an Eulerian calculation to consist of two steps, a Lagrangian calculation followed by a remapping of the results of that calculation onto the Eulerian grid. The remap step of such a composite calculation requires the solution of the following nonlinear system of equations:

$$\frac{\partial \rho}{\partial t} + u \frac{\partial \rho}{\partial x} = 0 \quad , \qquad (16)$$

$$\frac{\partial u}{\partial t} + \rho u \frac{\partial u}{\partial m} = 0 \quad , \qquad (17)$$

$$\frac{\partial E}{\partial t} + \rho u \frac{\partial E}{\partial m} = 0 \quad . \qquad (18)$$

We have written the last two equations in this set in an unusual manner which gives the best guide to the consistent numerical solution of the set of equations as a whole. Clearly, all three of these equations are just Eq. (1) with $g = 0$ and with different definitions of the density f, the coordinate x, and the advection speed v.

It should be noted that Eq. (1) has other uses in astrophysics. When g is determined from Poisson's equation, it becomes the equation of stellar dynamics. A similar equation with a source term on the right is the equation of radiative transfer. Equations like Eq. (1), often with source terms on the right, must be solved whenever a constituent of the fluid must be treated which has a long mean free path.

NUMERICAL TECHNIQUES FOR EXPLICIT CALCULATIONS OF SIMPLE ADVECTION

a. The PPM and PPB advection schemes

We will discuss two general approaches to solving the simple advection equation in one dimension

$$\frac{\partial f}{\partial t} + v \frac{\partial f}{\partial x} = 0 \quad . \tag{19}$$

For a fuller discussion of this subject and for historical references the reader is referred to van Leer (1977), Woodward and White (1983), and Colella and Woodward (1983). Of course, there is really only one way to solve Eq. (19) numerically; one takes a finite amount of data about f(x), from it he infers an approximation to f(x) everywhere, and then he uses this approximation to update his finite data in some straightforward way. This simple truth stands in relation to numerical analysis of partial differential equations as the following summary of algebra does to that subject: "You just let something be x and then solve for it" (E. J. Woodward 1959, private communication). Needless to say, many mathematicians have based careers on finding out just how the approximation to f(x) everywhere should be constructed.

We will consider here only difference methods in conservation form. For these methods the finite data about f(x) must in part consist of or be equivalent to the locations x_i of zone interfaces and the average values $\langle f \rangle_i$ of f(x) within the intervals from x_i to x_{i+1}. Here the zone number i runs from 1 to some finite number N. For one method we will discuss, PPM, this is all the data provided for the calculation. For another method, PPB, additional moments of f(x) within the zones must be provided. These moments will be written as $\langle f\chi^k \rangle$, where k = 0, 1, 2 and where χ is a zone-centered coordinate defined by

$$\chi = [x - \tfrac{1}{2} (x_L + x_R)] \,/\, \Delta x \quad . \tag{20}$$

Here x_L and x_R denote the locations of the left- and right-hand interfaces of the zone, that is $x_L = x_i$ and $x_R = x_{i+1}$ for zone i. The zone width is Δx, and therefore χ ranges from - 1/2 to 1/2 across the zone. The moments $\langle f\chi^k \rangle$ are defined by

$$\langle f\chi^k \rangle = \int_{-1/2}^{1/2} f(\chi)\, \chi^k \, d\chi \quad . \tag{21}$$

In the first approach to solving Eq. (19), here represented by the difference scheme PPM (the Piecewise-Parabolic Method; Woodward and Colella 1980, 1983; Colella and Woodward 1983), only the masses of the zones are used as data. These are given by Eq. (21) with k = 0. In order to approximate f(x) everywhere PPM constructs a parabola to represent $f(\chi)$ within each zone:

$$f(\chi) = f_0 + f_1\chi + f_2\chi^2 \quad . \tag{22}$$

To compute the coefficients f_k, PPM first interpolates values f_L and f_R of f at the left- and right-hand interfaces of the zones. This is done by assuming that f(x) is smooth, so that it can be approximated near x_L by the unique cubic curve which has the average values in the four zones nearest x_L which are prescribed by the data for the scheme. This cubic curve is then evaluated at x_L to give the interpolated density f_L. For the special case of uniform zone widths the formula for f_L, or $f_{i-1/2}$, has the following simple form:

$$f_{i-1/2} = \frac{7}{12}\left(\langle f\rangle_{i-1} + \langle f\rangle_i\right) - \frac{1}{12}\left(\langle f\rangle_{1-2} + \langle f\rangle_{i+1}\right) . \qquad (23)$$

Once values for f_L and f_R have thus been obtained, the parabola in Eq. (22) is determined by demanding that it pass through f_L and f_R at x_L and x_R and that its integral over the zone should yield the prescribed zone mass $\langle f\rangle \Delta x$. The coefficients f_k must therefore be

$$f_0 = \frac{3}{2} \langle f\rangle - \frac{1}{4} (f_L + f_R) , \qquad (24a)$$

$$f_1 = f_R - f_L , \qquad (24b)$$

$$f_2 = 3 (f_L + f_R) - 6 \langle f\rangle . \qquad (24c)$$

We will later describe important modifications to the above procedure which improve its performance near discontinuities in f(x).

In the second approach, represented by the difference scheme PPB (the Piecewise-Parabolic Boltzmann scheme; van Leer 1977, Woodward and White 1983), the three moments $\langle f\rangle$, $\langle f\chi\rangle$, and $\langle f\chi^2\rangle$ are used as data. This information is sufficient to uniquely determine the coefficients of the interpolation parabola in Eq. (22):

$$f_0 = \langle f\rangle - 15 \langle f\chi^2\rangle_2 , \qquad (25a)$$

$$f_1 = 12 \langle f\chi\rangle , \qquad (25b)$$

$$f_2 = 180 \langle f\chi^2\rangle_2 , \qquad (25c)$$

$$\langle f\chi^2\rangle_2 = \langle f\chi^2\rangle - \langle f\rangle / 12 . \qquad (25d)$$

When the advection velocity, v, in Eq. (19) is constant, the derivation of formulae to update the original data is straightforward, once the interpolation parabolae have been determined at the original time t = 0. In this special case Eq. (19) has a trivial solution:

$$f(x,t) = f(x - vt, 0) \quad . \tag{26}$$

Thus the interpolation parabolae for f(x,0) also give f(x,t), and this function need only be integrated over the zones to give the new data (see Fig. 1). For the PPB scheme care must be taken in evaluating the higher moment integrals in Eq. (21) to avoid confusing the χ's referring to time 0 and to time Δt. For details the reader is referred to Woodward and White (1983). To keep the logic in the program simple it is customary to demand that the timestep Δt be limited so that no zone interface moves out of the interval defined by the two zones adjacent to it.

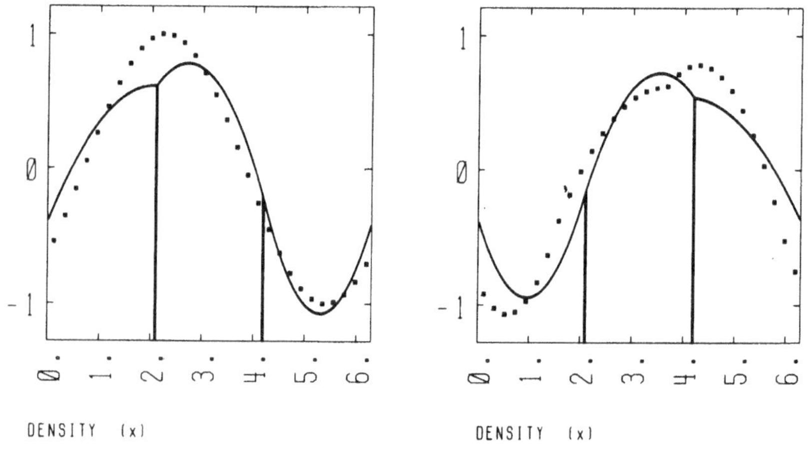

Fig. 1 The PPM advection scheme moves a sine wave resolved with 3 zones a distance of 3/4 zone widths to the right in a single timestep. (a) The zone-averaged values for sine wave (shown dotted) are used to construct a parabolic representation of the curve within each zone (solid lines). (b) These original interpolation parabolae (dotted) have been translated 3/4 zone widths to the right. Integrating these translated curves over the original zone intervals gives the new zone averages. These have been used to construct new interpolation parabolae (solid lines), the final PPM representation of the translated sine wave.

When the advection velocity v depends upon x, as is usually the case in the remap step of a hydrodynamics calculation, interpolation parabolae are constructed directly for the zones at time Δt after they have translated, expanded, or contracted. The necessary data referring to these zones is provided by the output of the Lagrangian step of the calculation. Straightforward integrations over the interpolated structures again yield data referring to the original Eulerian mesh. As is indicated by the form of Eqs. (16), (17), and (18), the density should be interpolated as a function of a volume coordinate. The integrations over these density structures then yield masses advected from one zone to another. These should be used to define the advection velocities relative to the mass coordinate. The fluid velocity and specific total energy are then interpolated as functions of the mass coordinate, and integration of these structures over the advected masses gives the momenta and total energies advected from one zone to another.

b. Comparison of PPM and PPB on a Gaussian Advection Problem

The performance of the two schemes PPM and PPB is illustrated by the Gaussian advection tests shown in Figs. 2 and 3 and tabulated in Table I. The initial data for these runs is constructed using Eq. (21) with f set to f_0, where

$$f_0(x) = \exp\,[(x - 1/2)^2 / 0.01125] \quad . \tag{27}$$

Thus $f_0(x)$ is a Gaussian of height 1 and standard deviation 0.075 centered at x = 1/2. Periodic boundary conditions are applied at x = 0 and x = 1. Using grids with 8, 16, 32, 64, and 128 zones, Eq. (19) has been solved with v = 1. Solutions were obtained with two values of the Courant number, σ, defined by

$$\sigma = v\,\Delta t \,/\, \Delta x \quad . \tag{28}$$

The values of σ used, 0.08 and 0.8, represent approximate worst and best cases, with the errors at σ = 0.08 as much as four times larger than those at σ = 0.8. The results of the PPM calculations are shown in Fig. 2, while those of the PPB calculations are shown in Fig. 3. In generating these figures the schemes themselves have been used to determine interpolation parabolae within the zones, so that f(x) is defined everywhere for plotting. In addition integrated errors ε_0, ε_1, ε_{10}, and ε_{100} have been computed from the results at times 0, 1, 10, and 100 using the interpolated f(x) obtained by the scheme:

$$\varepsilon = \int_0^1 | f(x) - f_0(x) | dx \Big/ \int_0^1 f_0(x) dx . \qquad (29)$$

These errors are tabulated in Table I.

A glance at Figs. 2 and 3 reveals a striking difference in accuracy between the PPM and PPB schemes. Apparently, it makes a great deal of difference which interpolation parabola is used to describe the internal structure of a zone. The PPM scheme requires grids of 16, 32, 64, or 128 zones if the Gaussian pulse is to be well represented after 0, 1, 10, or 100 transits of the grid, respectively. In contrast to this, the PPB scheme requires only 8, 8, 16, or 32 zones for the same purpose.

Thus the PPB scheme uses three times as much data per zone but needs only one fourth as many zones as PPM. One might have thought that the smoothness of the Gaussian would allow the higher moment data used by PPB to be effectively regenerated from the information provided by $\langle f \rangle$ in neighboring zones. The PPM results indicate that if such a reconstruction of the data is ever possible, it must require a much finer grid than is needed by PPB to adequately resolve the density structure. (This result holds true even when seventh-order curves are used in place of cubics to interpolate the interface values f_L and f_R in PPM.) Because PPB requires only 40% more time to update a zone than PPM (on the Cray-I PPB updates 1.2×10^6 zones in a second), it is therefore much more efficient to use PPB on the coarser grid.

A close look at the errors on the fine grids in Table I shows that both schemes are converging faster than we might have expected. Generally, if a polynomial of order n is used to describe the internal zone structure, the associated advection scheme is $\underline{n+1}^{th}$ order accurate. The interpolation polynomial cannot account for the $\underline{n+1}^{th}$ derivative of f within the zone, so the advected mass calculated at each interface contains an error proportional to this derivative, which is assumed to be of order 1, multiplied by Δx^{n+1}. However, a similar error is introduced in the advected mass at both interfaces of the zone, and these tend to cancel when the two advected masses are differenced to obtain the net change in the zone's mass. Thus the error in this net change is of order Δx^{n+2}. If we hold the Courant number σ fixed as Δx is reduced, then the number of timesteps needed to reach a specified time increases as $1/\Delta x$. This effect implies that the order of the error in computing to a fixed time, that is the order of the difference scheme, is one less than the order of the error incurred in a single timestep. Hence a scheme employing an $\underline{n}$th-order polynomial to describe the internal structure of a zone should be $\underline{n+1}^{th}$ order accurate.

The errors in Table I for the fine grid calculations indicate that PPM converges nearly as fast as a fourth-order scheme and that PPB converges nearly as fast as a fifth-order scheme. The argument given above implies that both should be third-order schemes in a certain strict sense, but obviously as a practical matter both are more accurate than this. The accelerated convergence of PPM comes from its "higher-order spatial

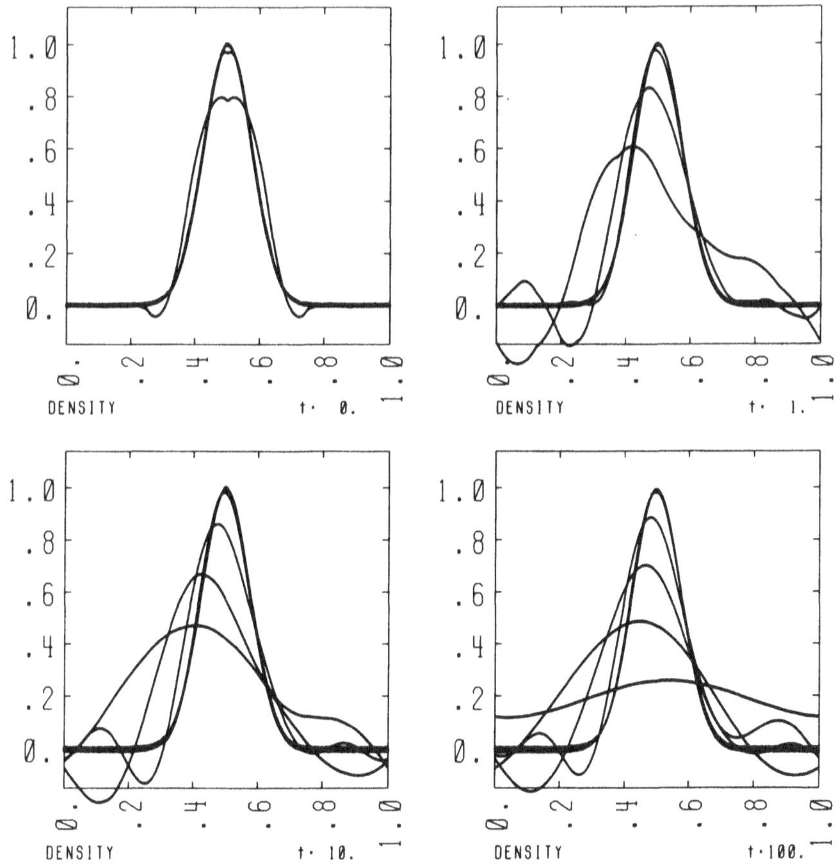

Fig. 2a A Gaussian of height 1 and standard deviation 0.075 is advected by the PPM scheme at a Courant number of 0.08. Results for grids of 8, 16, 32, 64, and 128 zones are plotted against the exact solution at times 0, 1, 10, and 100. The parabolae used by PPM to interpolate within zones are used here for plotting. After the Gaussian has traversed the grid 100 times, only the finest grid gives adequate results.

accuracy". Because the density values at zone interfaces are interpolated using cubic curves, and because in the limit of small Courant numbers (or of σ near unity) these interface values alone give the advected masses, in this same limit the scheme must be fourth-order accurate. Thus the accelerated convergence of PPM is more noticeable in the runs with $\sigma = 0.08$, but it is still effective even when $\sigma = 0.8$.

The accelerated convergence of PPB is due to its exact conservation of the moments $\langle f\chi \rangle$ and $\langle f\chi^2 \rangle$. Consider the

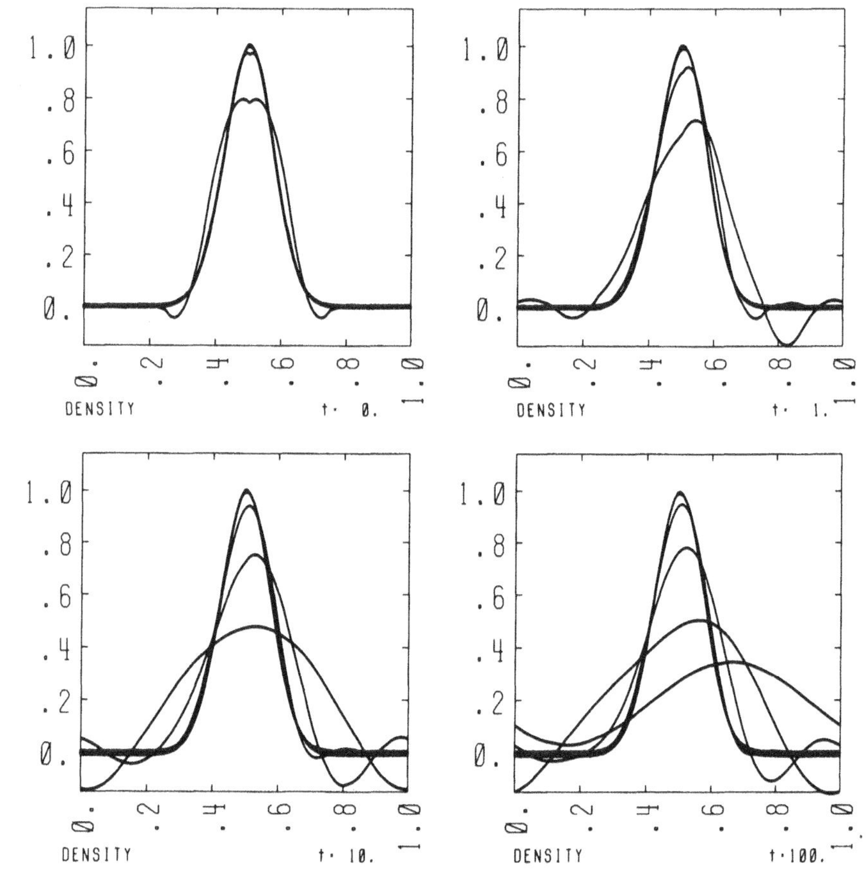

Fig. 2b Same as Fig. 2a except that a Courant number of 0.8 was used. This larger timestep improves the results substantially, so that the 64-zone grid now gives good results at time 100.

initial replacement of the exact distribution $f_0(x)$ with the PPB representation $f(x)$. If we define the local error $\varepsilon(x)$ by

$$\varepsilon(x) = f(x) - f_0(x) \quad , \tag{30}$$

then the preservation of the moments $<f_0\chi^k>$ by the PPB

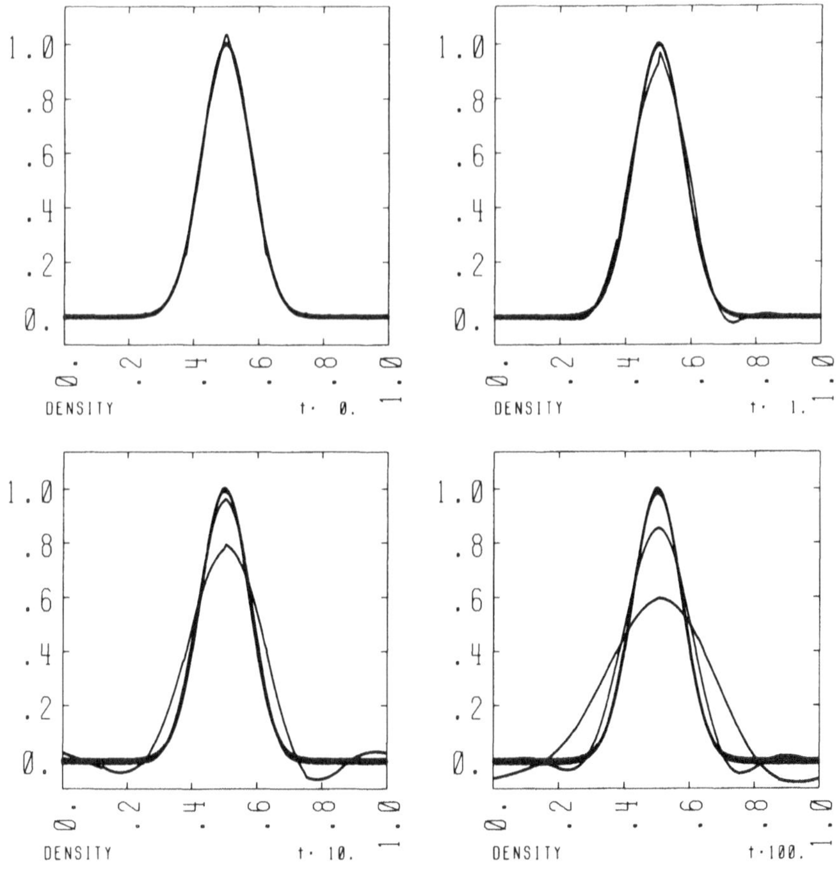

Fig. 3a The same advection experiment shown in Fig. 2a is here carried out using the PPB scheme. The interpolation parabolae used by the scheme to represent the distributions within zones are used here for plotting. Note the quality of this representation on the 8-zone grid at times 0 and 1. The three finest grids give very good results at time 100. Note the low level of the negative densities generated on the coarsest grids, particularly in contrast to the PPM results in Fig. 2a.

representation guarantees that

$$\langle \varepsilon \chi^k \rangle = 0 \quad , \qquad \text{for } k = 0, 1, 2 \quad . \tag{31}$$

This very special property implies that any meaningful definition of global error which involves integrals over $\varepsilon(x)$ rather than its absolute value will yield a sixth-order rather than a third-order error. For example, consider the following error definition:

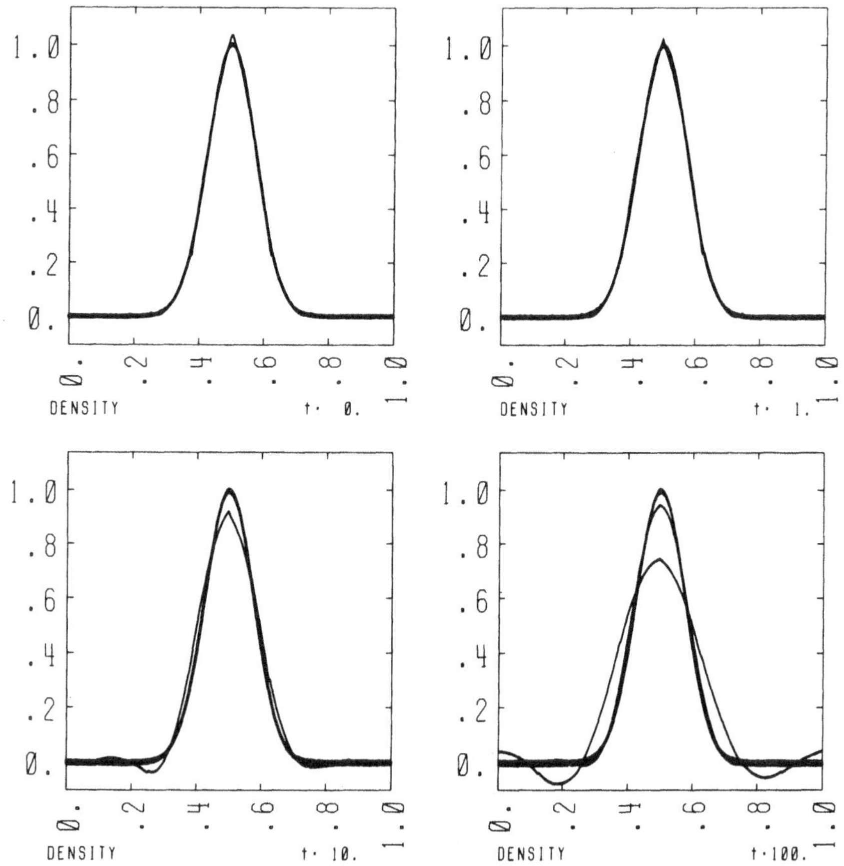

Fig. 3b Same as Fig. 3a except that a Courant number of 0.8 was used. The results are improved by the use of this larger timestep, so that now only 16 zones are needed to produce good results at time 100.

$$e_{f0} = \int_0^1 \varepsilon(x)\, f_0(x)\, dx \;\Big/\; \int_0^1 [f_0(x)]^2\, dx \;, \tag{32}$$

The numerator of this expression can be broken up into a sum of terms like

$$\begin{aligned}\Delta x \int_{-1/2}^{1/2} \varepsilon(\chi)\, f_0(\chi)\, d\chi \;\simeq\; & \Delta x \,\langle\varepsilon\rangle\, f_0(0) + (\Delta x)^2 \,\langle\varepsilon\chi\rangle \left.\frac{\partial f_0}{\partial x}\right|_{\chi=0} \\ & + \frac{1}{2}(\Delta x)^3 \langle\varepsilon\chi^2\rangle \left.\frac{\partial^2 f_0}{\partial x^2}\right|_{\chi=0} \\ & + \frac{1}{6}(\Delta x)^4 \langle\varepsilon\chi^3\rangle \left.\frac{\partial^3 f_0}{\partial x^3}\right|_{\chi=0}\end{aligned} \tag{33}$$

By virtue of Eq. (31), the first three terms on the right in Eq. (33) vanish. Therefore the numerator in Eq. (32) can be expressed as a sum of terms each of which is of order $(\Delta x)^4 \langle\varepsilon\chi^3\rangle$. Now $\varepsilon(\chi)$ is of order $(\Delta x)^3$, and the number of terms is of order $1/\Delta x$, so the integrated error ε_{f0} is of order $(\Delta x)^6$. An advection timestep for a constant advection velocity consists of an exact "Lagrangian step" in which the zones all move over a constant distance followed by a remap which preserves three moments. If we regard the piecewise-parabolic representation in the zones which have shifted over a distance $v\,\Delta t$ as the "exact" distribution $f_0(x)$, then our argument above shows that ε_{f0} is increased by an amount of order $(\Delta x)^6$ when we use the exact moments of this $f_0(x)$ in the original Eulerian zones to construct a new piecewise-parabolic representation $f(x)$. Consequently, using the error ε_{f0} to measure the performance of the PPB scheme, we discover that this scheme is fifth-order accurate.

van Leer (1977) found another explanation for the unexpected accuracy of moment-conserving difference schemes. His work applied to the MUSCL scheme, which conserves only $\langle f\rangle$ and $\langle f\chi\rangle$, but similar arguments apply to PPB. The initial data constructed from $f_0(x)$ by finding the exact moments $\langle f_0\chi^k\rangle$ can be broken into two parts, one which will be advected by PPB with fifth-order accuracy and one which will be damped by a factor of order unity in each timestep. The second, rapidly damped part decreases in size as $(\Delta x)^3$. It therefore contributes an error of this size which, after several timesteps, does not continue to grow. This error makes the scheme third-order accurate, but after many timesteps this error is overwhelmed by the growing fifth-order error which arises from the advection of

TABLE I

Numerical Errors Generated in Gaussian Advection Tests

Scheme	Grid	ε_0	ε_1	ε_{10}	ε_{100}
PPM	8	2.28e-1	1.04	9.82e-1	1.15
σ = 0.08	16	1.92e-2	4.12e-1	9.00e-1	8.89e-1
	32	1.49e-3	5.87e-2	3.51e-1	6.54e-1
	64	1.17e-4	4.05e-3	3.96e-2	2.66e-1
	128	1.09e-5	2.76e-4	2.74e-3	2.65e-2
PPB	8	2.54e-2	1.02e-1	3.50e-1	6.66e-1
σ = 0.08	16	2.53e-3	6.99e-3	5.08e-2	2.14e-1
	32	3.00e-4	4.50e-4	2.37e-3	2.06e-2
	64	3.70e-5	4.74e-5	9.02e-5	7.66e-4
	128	4.62e-6	5.80e-6	6.36e-6	2.49e-5
PPM	8	2.28e-1	5.28e-1	9.52e-1	1.13
σ = 0.8	16	1.92e-2	1.29e-1	4.82e-1	9.57e-1
	32	1.49e-3	1.20e-2	9.67e-2	3.98e-1
	64	1.17e-4	8.59e-4	8.32e-3	7.13e-2
	128	1.09e-5	7.23e-5	7.14e-4	6.99e-3
PPB	8	2.54e-2	3.67e-2	1.50e-1	4.30e-1
σ = 0.8	16	2.53e-3	2.95e-3	1.11e-2	7.51e-2
	32	3.00e-4	3.26e-4	5.04e-4	3.77e-3
	64	3.70e-5	3.98e-5	4.20e-5	1.29e-4
	128	4.62e-6	4.96e-6	4.99e-6	6.33e-6

the first component of the initial data. This behavior can be clearly seen in the final entries in Table I for the PPB run with 128 zones at a Courant number of 0.8. These numbers show a relatively rapid initial growth of the error followed by a much slower phase of error accumulation.

In Fig. 4 the initial data for a pure sine wave on a grid of 16 uniform zones is decomposed into parts which are damped at different rates by the PPB scheme. In Fig. 4a the most slowly damped component is the one which closely resembles a sine wave. The other component shown is very rapidly damped. In order to make it visible, it has been amplified for plotting by the factor $100/(\Delta x)^3$. This component itself consists of two parts which are damped by PPB at different rates. These are shown in Fig. 4b. In fact, these three components of the sine wave are eigenfunctions of the PPB scheme; they are reduced in amplitude and shifted in phase by the scheme but their shapes are preserved. The larger one in Fig. 4b is damped by a factor of 7/8 in each timestep at a Courant number of 1/2, while the smaller one is damped only by a factor of 1/2. The larger eigenfunction contributes an amount of order $(\Delta x)^3$ to the initial data but

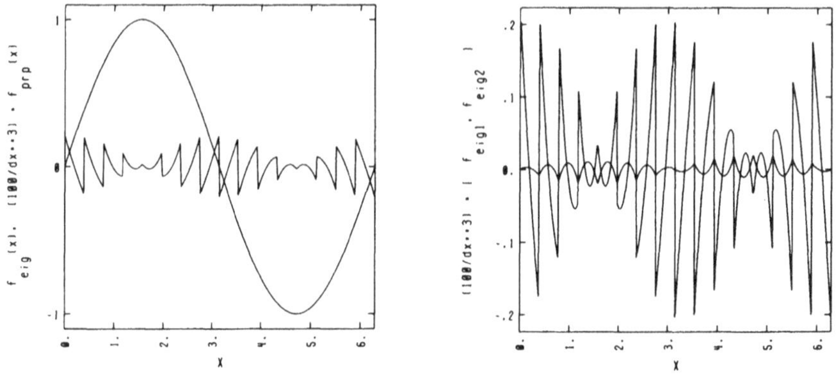

Fig. 4 The initial representation of a pure sine wave by the PPB scheme on a 16-zone grid is here decomposed into parts which are damped at different rates by the scheme. On the left, the largest eigenfunction component, which closely resembles the sine wave, is shown. This eigenfunction is advected by PPB with fifth-order accuracy. The saw-toothed curve plotted with this eigenfunction is the difference between that eigenfunction and the sine wave. It has been miltiplied by a factor $(100/(\Delta x)^3)$ to make it visible. This component, which is third-order in Δx, consists of two eigenfunctions of the PPB scheme. These are shown on the right. Both are damped strongly in a single timestep.

its contribution to the fundamental mode in the Fourier spectrum of this data is of order $(\Delta x)^6$. This is consistent with our previous discussion of the error measure ε_{f0}. These contributions for the smaller eigenfunction scale as $(\Delta x)^4$ and $(\Delta x)^8$.

PRACTICAL LIMITS TO THE ACCURACY OF A DIFFERENCE SCHEME FOR ADVECTION

From the above comparison of PPM and PPB one might be tempted to conclude that the best way to solve advection problems numerically would be to use a very high-order moment conserving method. This conclusion is false for a number of reasons. The most important of these reasons is that the advection problems which arise in computational physics are usually either multidimensional or nonlinear or both. For multidimensional problems, the amount of work for both the programmer and the computer rises rapidly as the accuracy increases beyond the second order. This can most easily be seen in two dimensions, with Cartesian coordinates x and y. Strang (1968) showed that a second-order, two-dimensional difference operator D_{xy} could be constructed from a symmetrized product $D_xD_yD_yD_x$ of second-order, one-dimensional difference operators. The second-order error generated by using D_yD_x in place of $D_{xy} = D_x + D_y$ is cancelled by the error in using D_xD_y subsequently. This error in D_yD_x arises for nonlinear advection equations, such as the characteristic equations of isentropic gas dynamics [Eqs. (6)-(11)]. In that case the advection speeds are altered by D_x and D_y, so that D_yD_x and D_xD_y must give different results. For such nonlinear advection equations even third-order methods seem impractically complex.

For linear advection equations multidimensional calculations can be performed in a sequence of "passes" in which the fluid is permitted to move in only one dimension. However, if more than second-order accuracy is required these passes must be two dimensional in nature. The reason for this is illustrated in Fig. 5. If all gradients in the y direction are ignored in the x-pass, then the difference of the masses of the upper shaded triangles in the figure is being effectively equated to that of the lower shaded triangles. The error this introduces in the net change in the zone mass is proportional to

$$\Delta x \, (\Delta y)^2 \, \frac{\partial}{\partial x} \left(\frac{\partial \rho}{\partial y} \, \frac{\partial v_x}{\partial y} \right) \quad .$$

Therefore a third-order difference scheme must account for this term. Even though an advection equation to be solved is nonlinear, so that strict third-order accuracy is unattainable by means of 1-D passes, it may still make sense to use a scheme as elaborate as PPB or PPM and to account for the above error term.

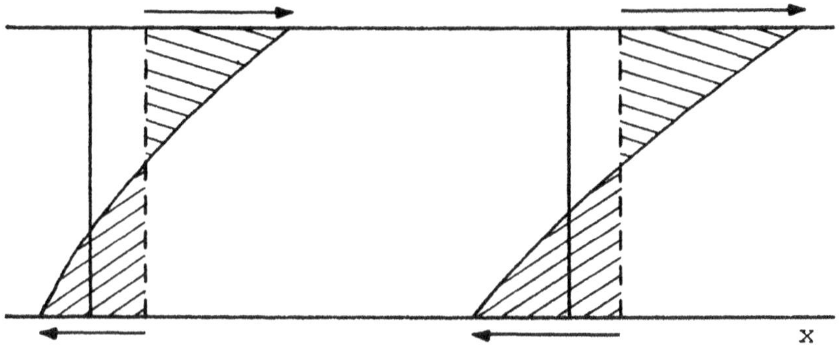

Fig. 5 The x-pass of an operator-split advection calculation in which a purely one-dimensional operator D_x is used. Ignoring gradients of ρ and v_x in the y-direction effectively causes the mass difference of the upper shaded triangles to be equated to that of the lower ones.

Such an application is stellar dynamics, in which the Boltzmann Eq. (1) must be solved. The equation is nonlinear by virtue of the coupling of the acceleration g to the phase space density through Poisson's equation. However, this coupling is weak, as only the velocity-integrated density enters Poisson's equation. Another reason why high-order advection techniques are appropriate in this application is that the high dimensionality of the problem means that only very coarse grids can be afforded. Therefore one is unlikely to find the calculation in the asymptotic regime where third- or fifth-order convergence is achieved. Actually, in such problems in 4 or 6 phase space dimensions one is lucky to resolve even the gross features of the problem adequately. Then schemes like PPB are especially attractive because of their good behavior when the flow is badly underresolved.

The usefulness of the PPB scheme in such stellar dynamical problems is illustrated by the gravitational two-stream instability problem discussed in detail by Woodward and White (1983). The time development of this flow is shown in Fig. 6. The problem begins with the imposition of a sinusoidal perturbation in the x-direction on the equilibrium density f_0 of fluid in phase space. This equilibrium consists of a uniform distribution of gravitating fluid in physical space. In velocity space the distribution in the v_y- and v_z-directions is a delta function at velocity zero. In the v_x-direction the distribution consists of the superposition of two Gaussians centered at $v_x = \pm 4$ and with standard deviations of unity. Each Gaussian contains a velocity-integrated density of 1/2. The sinusoidal perturbation imposed on this equilibrium at time zero has a wavelength of 12

Figure 6a

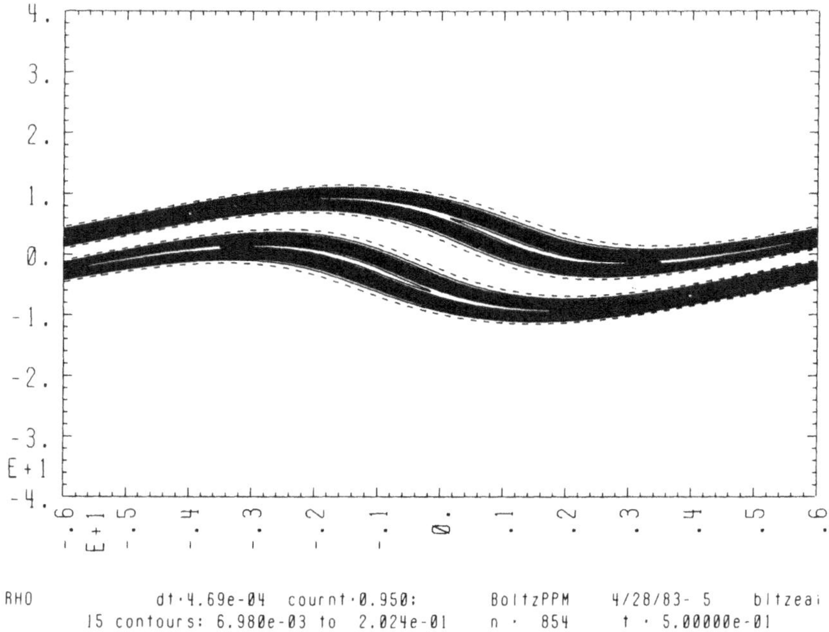
4.
3.
2.
1.
0.
-1.
-2.
-3.
E+1
-4.
-.6
E+1
-.5
-.4
-.3
-.2
-.1
0.
.1
.2
.3
.4
.5
.6
RHO dt·4.69e-04 cournt·0.950: BoltzPPM 4/28/83- 5 bltzeai
15 contours: 6.980e-03 to 2.024e-01 n · 854 t · 5.00000e-01

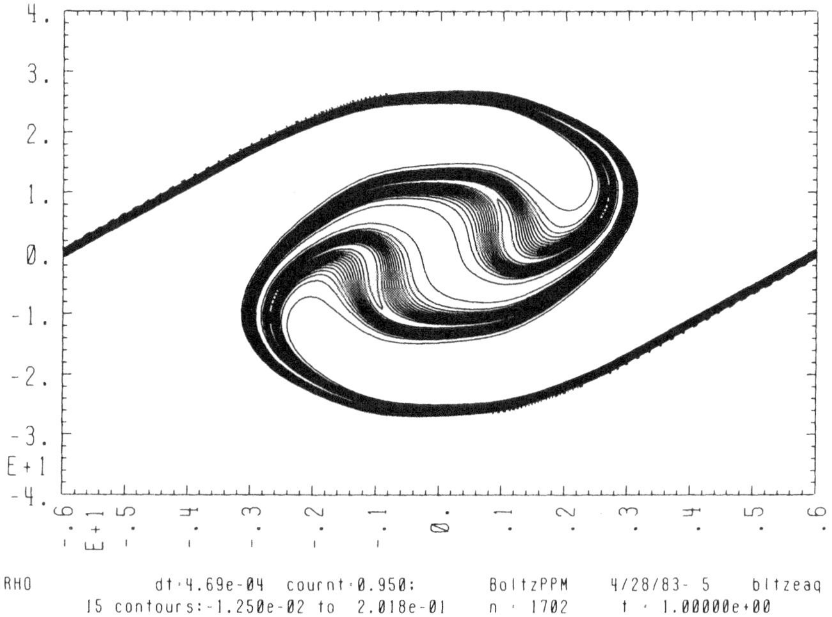
4.
3.
2.
1.
0.
-1.
-2.
-3.
E+1
-4.
-.6
E+1
-.5
-.4
-.3
-.2
-.1
0.
.1
.2
.3
.4
.5
.6
RHO dt·4.69e-04 cournt·0.950: BoltzPPM 4/28/83- 5 bltzeaq
15 contours:-1.250e-02 to 2.018e-01 n · 1702 t · 1.00000e+00

Figure 6b

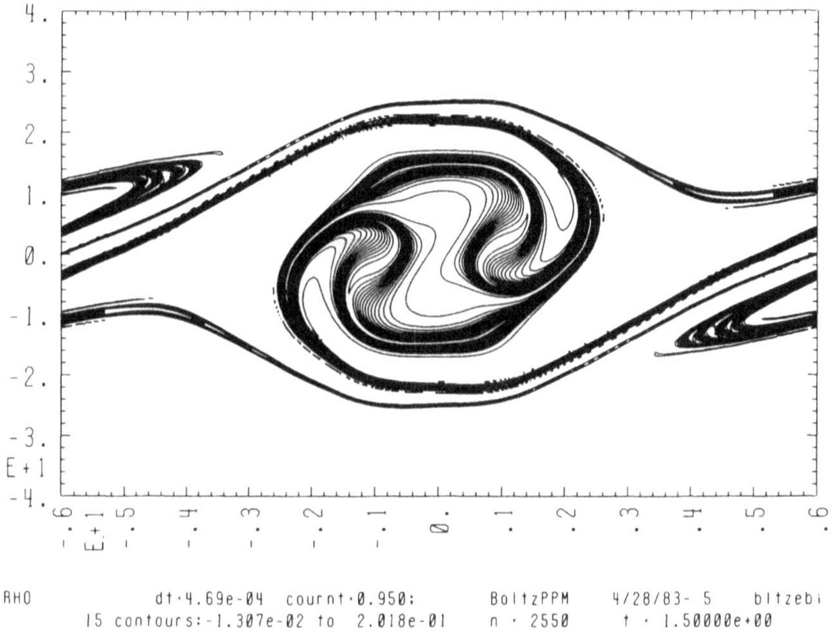

RHO dt·4.69e-04 cournt·0.950; BoltzPPM 4/28/83- 5 bltzebi
15 contours:-1.307e-02 to 2.018e-01 n · 2550 t · 1.50000e+00

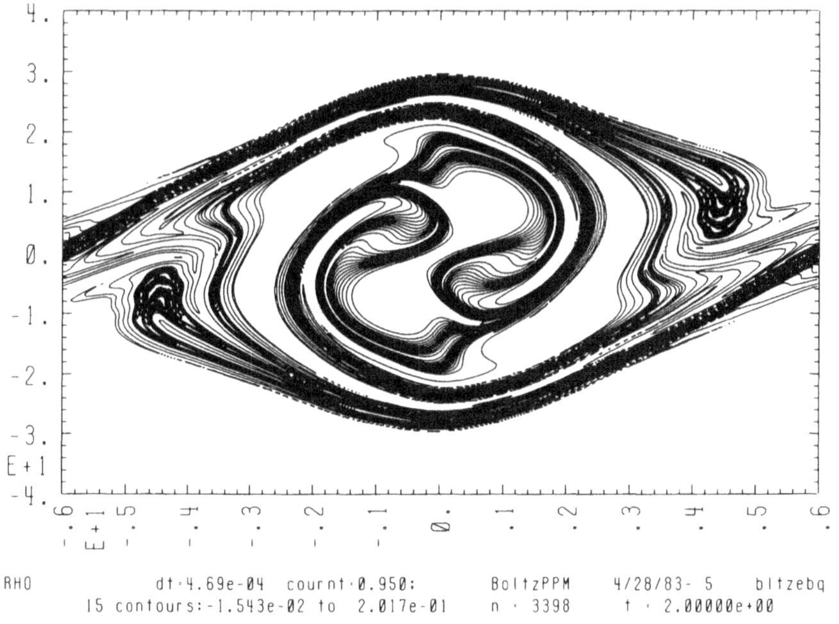

RHO dt·4.69e-04 cournt·0.950; BoltzPPM 4/28/83- 5 bltzebq
15 contours:-1.543e-02 to 2.017e-01 n · 3398 t · 2.00000e+00

Figure 6c

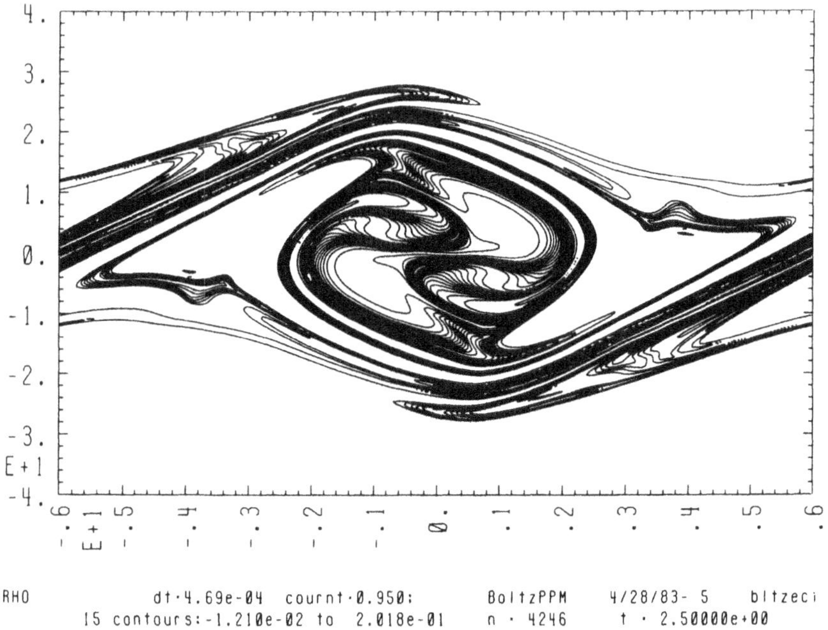

Figure Captions

Fig. 6 Contours of density in phase space for the gravitational two-stream instability problem described in the text. These results were obtained with the PPB scheme using a fine grid of 480x480 uniform zones. The flow begins with two Gaussian velocity distributions at $v = \pm 4$ and with unit standard deviation. A 5% sinusoidal perturbation of the velocity-integrated density is imposed. The flow is shown at time intervals of 1/2. The spiralling motion about the location of the initial density maximum causes ever thinner cords of phase space density to develop which eventually can no longer be resolved.

Fig. 7 PPB results at time 2.5 are displayed for two coarse grid calculations of the gravitational two-stream instability problem shown in Fig. 6. Grids of 30x30 (top) and 120x120 (bottom) uniform zones were used. As the grid is coarsened, fine-scale features are blended, but very little numerical noise is generated. The large-scale features near the center of the plots are adequately described even on the coarser grid. The PPB interpolation parabolae have been used to generate a 5x5 grid of density values per zone for plotting.

Figure 7

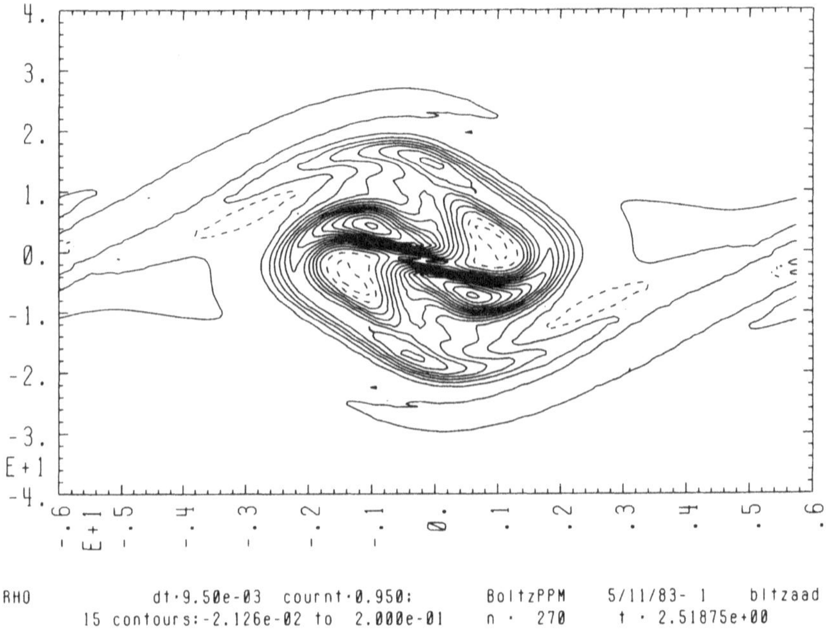
4.
3.
2.
1.
Ø.
-1.
-2.
-3.
E+1
-4.
-.6
E+1
-.5
-.4
-.3
-.2
-.1
Ø.
.1
.2
.3
.4
.5
.6
RHO dt·9.5Øe-Ø3 cournt·Ø.95Ø; BoltzPPM 5/11/83- 1 bltzaad
15 contours:-2.126e-Ø2 to 2.ØØØe-Ø1 n · 27Ø t · 2.51875e+ØØ

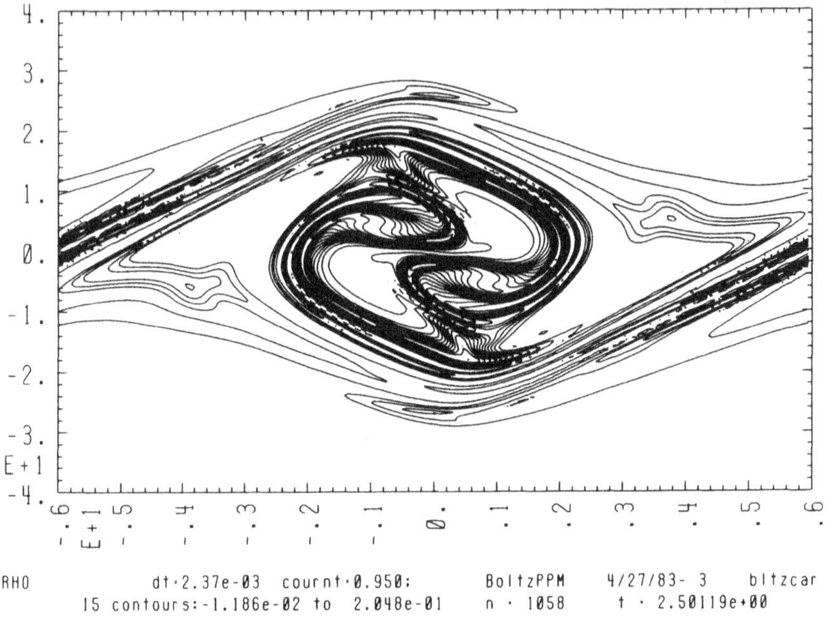
4.
3.
2.
1.
Ø.
-1.
-2.
-3.
E+1
-4.
-.6
E+1
-.5
-.4
-.3
-.2
-.1
Ø.
.1
.2
.3
.4
.5
.6
RHO dt·2.37e-Ø3 cournt·Ø.95Ø; BoltzPPM 4/27/83- 3 bltzcar
15 contours:-1.186e-Ø2 to 2.Ø48e-Ø1 n · 1Ø58 t · 2.5Ø119e+ØØ

and an amplitude of 5%. The specification of a value for the gravitational constant G in Poisson's equation determines the relationship of these mass, length, and time units to the cgs system. We have set $G = \pi$ for this problem. The Jeans length for a collisional gas with a sound speed of unity would then be unity. Our perturbation, with a wavelength of 12 is therefore quite unstable.

In Fig. 6 the phase space density is represented at time intervals of 1/2. Each snapshot displays 15 evenly spaced contours of phase space density with the lowest contour dotted. The calculation employed a very fine grid of 480x480 zones (only half of these were computed and the others were generated for plotting by symmetry). The standard deviations of the initial Gaussians were well resolved on this grid by six zones. The nature of the flow is a spiraling of most of the fluid about the origin and a translation of small parts of the fluid acorss the entire grid. Because the problem is periodic this translating envelope of fluid goes from one gravitationally bound clump to another, lingering between each pair. As the flow continues to wind up, ever thinner cords of phase space density are created, so that ultimately the flow must become unresolved. At this point PPB behaves relatively well. Instead of generating large amounts of numerical noise, as would be the case for most high-order methods, PPB averages over the individual thin structures to obtain a smoother structure. In this averaging process information is lost and entropy is created, but six moments of the local phase space distribution are nevertheless conserved. This averaging process is illustrated by the results in Fig. 7. These results were obtained with PPB using grids of 30x30 (Fig. 7a) and 120x120 zones (Fig. 7b). For plotting, the PPB interpolation parabolae were used to generate data values on a 5x5 grid within each zone. This figure shows how detailed features are smeared out and blended as the grid is coarsened, but the important larger structures near the origin are not lost. In large part this blending constitutes a numerically accelerated representation of the physical relaxation process of phase mixing.

The gravitational two-stream instability problem just discussed is a good example of the type of problem for which PPB is well-suited. The advection equation is nonlinear, so that still higher order methods are probably unwarranted; nevertheless the continuous generation of smooth but fine-scaled structures demands a scheme which breaks down in a very graceful way. The PPB scheme is complicated, but the simplicity of Eq. (1) allows this complicated, moment conserving approach to be applied without undue programming difficulty. Problems of similar nature which are encountered in astrophysics and where the PPB approach would be useful are radiation transport and the treatment of nonthermal particles such as cosmic rays or neutrinos (in supernovae) which are loosely coupled to the thermal gas.

PRACTICAL LIMITS TO THE ACCURACY OF A DIFFERENCE SCHEME FOR HYDRODYNAMICS

For hydrodynamical simulations the practical limits placed upon a difference scheme are even stronger than for the case of advection. As noted earlier, the nonlinearity of the equations limits the accuracy to second order if the calculation is performed in a sequence of 1-D passes. However, if strong pressure waves are present the nonlinearity of the equations can effectively limit the accuracy to first order. This limitation is caused by discontinuities in the flow. Strong pressure waves can steepen into shocks, and these can interact to produce contact discontinuities. At these discontinuities the differential equations describing smooth flow are not obeyed, and the results of difference approximations to them are meaningless. The best one can expect from a difference scheme is a smeared out representation of a discontinuity which has a width of a zone (or a few zones) and which describes approximately the correct jumps in the flow variables across the discontinuity. Much labor can be expended to make the numerical representation of a discontinuity thin, but the width of the smeared out jump must scale linearly with Δx. Therefore, in so far as this width affects the overall accuracy of the calculation, that accuracy is limited to first order. In many flows of interest, the representation of discontinuities is crucial; obviously for these flows it would be a waste of time to build a difference scheme along the lines of PPB. It is just for such flows that the PPM hydroynamics scheme was designed.

An excellent example of the first-order errors which can be caused by the numerical treatment of shocks is the simple problem of the propagation of a strong shock through a uniform gas described by a nonuniform grid. For simplicity, we will assume that the calculation is performed in Lagrangian coordinates and that the masses Δm_i of the 3N zones obey the following relations:

$$R = 100^{1/N-1} , \qquad \Delta m_1 = (1-R^{-1})/(1-R^{-N}) ,$$

$$\Delta m_i = \Delta m_{i-1}/R , \qquad \text{for} \quad i \leq N ,$$

$$\Delta m_i = \Delta m_{2N+1-i} , \qquad \text{for} \quad N < i \leq 2N ,$$

$$\Delta m_i = 1/N , \qquad \text{for} \quad N > 2N .$$

We will assume that the left-hand boundary of the grid at $x = 0$ is a reflecting wall and that all gradients vanish at the right-hand boundary at $x = 3$. The initial conditions are uniform flow with $u = -1$, $\rho = 1$, $p = 10^{-6}$, and the equation of state is $p = (2/3)\rho\varepsilon$ (thus $\gamma = 5/3$). The solution to this flow problem consists of the motion of a shock from left to right across the grid at a velocity of 4/3 with respect to the fluid and with a post-shock velocity of zero, density of 4, and pressure of 4/3. The gas is brought to rest by this shock and all its kinetic energy is transformed into internal energy.

The numerical solution to this problem can be gravely in error when N is 20 or less. Solutions obtained with PPM in Lagrangian coordinates are displayed in Fig. 8. For these solutions, the values of N are 10, 20, 40, and 80. Zone-averaged values at time 2 are connected by straight lines in Fig. 8, and both the density and the total energy per unit mass are plotted. The exact solution is $\rho = 4$ for $m < 8/3$, $\rho = 1$ for $m > 8/3$, and $E = 1/2$ everywhere. All four calculations tend to give correct answers where the grid is uniform, but large errors in the post-shock entropy are generated where the grid is not uniform. Where the zones are gradually becoming smaller in the direction of shock propagation, not enough heat is generated in the shock, and the gas is overcompressed. Where the zones become larger, too much heat is generated in the shock, and the gas is undercompressed. At $m = 2$ the zones suddenly become smaller, and a large overcompression is produced. The large undercompression at $m = 0$ is caused by the reflecting wall there. Neither of these latter two errors is diminished when the grid is refined.

The peculiar behavior of the PPM scheme which is shown in Fig. 8 can be understood without a knowledge of the detailed workings of that difference scheme. It is only necessary to realize that Eqs. (3), (4), and (12) are differenced in conservation form. That is to say that at each interface L between two zones numerical approximations to the time-averaged fluxes $\bar{u}_L$, $\bar{p}_L$, and $(\overline{up})_L$ of specific volume, momentum, and total energy are computed. Thus for example

$$\bar{u}_L = \frac{1}{\Delta t} \int_0^{\Delta t} u_L(t)dt \quad , \tag{34}$$

$$(\overline{up})_L = \frac{1}{\Delta t} \int_0^{\Delta t} u_L(t)p_L(t)\, dt \quad . \tag{35}$$

If the differential equations (3), (4), and (12) are integrated over a rectangle in space-time with sides Δm and Δt, exact difference equations in conservation form are obtained:

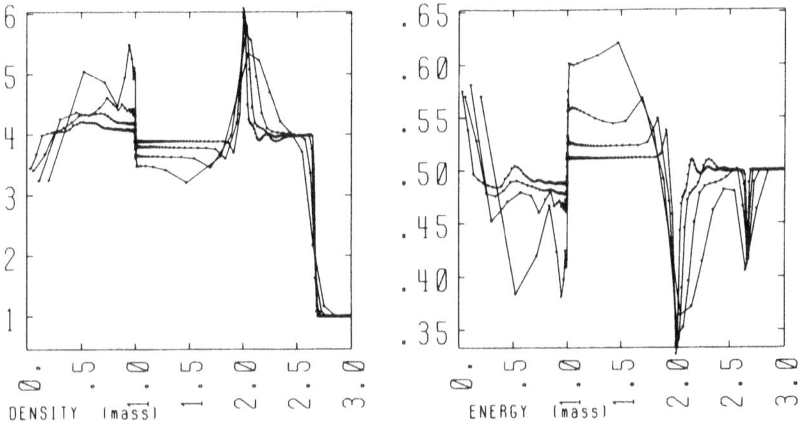

Fig. 8 The propagation of a strong shock through a nonuniform grid using the PPM scheme in Lagrangian coordinates. The exact cancellation of large errors which normally guarantees that correct shock jumps are computed is vitiated by the nonuniformity of the grid, and incorrect post-shock entropies are therefore computed. Results for grids of 30, 60, 120, and 240 zones, as described in the text, are displayed.

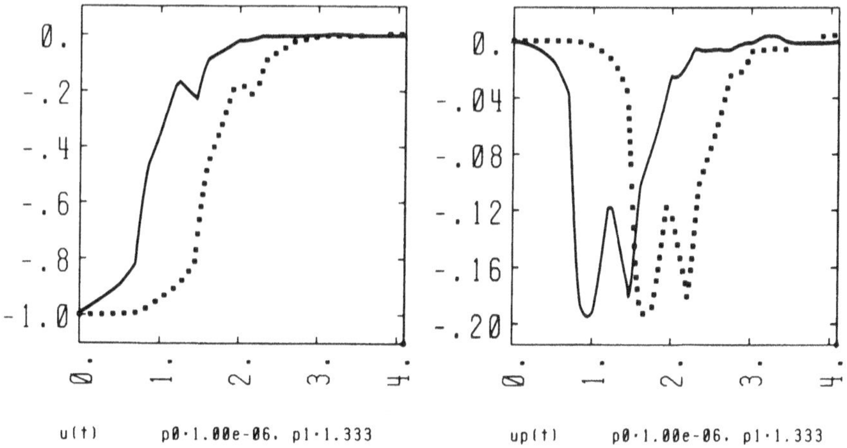

Fig. 9 Time histories of the fluxes $\bar{u}_L$ and $(\overline{up})_L$ (solid lines) and the fluxes $\bar{u}_R$ and $(\overline{up})_R$ (dotted lines) as computed by the PPM scheme on a uniform Lagrangian grid for the strong shock in Fig. 8.

$$\langle V\rangle_{new} = \langle V\rangle + (\bar{u}_R - \bar{u}_L)\,\Delta t/\Delta m \quad , \tag{36}$$

$$\langle u\rangle_{new} = \langle u\rangle - (\bar{p}_R - \bar{p}_L)\,\Delta t/\Delta m \quad , \tag{37}$$

$$\langle E\rangle_{new} = \langle E\rangle - [(\overline{up})_R - (\overline{up})_L]\,\Delta t/\Delta m \quad . \tag{38}$$

Here the subscripts L and R refer to the left- and right-hand interfaces of the zone, while the subscript "new" refers to the new time level. For a conservative difference scheme the numerical approximation consists entirely of a means of estimating the time-averaged fluxes $\bar{u}_L$, $\bar{p}_L$, and $(\overline{up})_L$. The advantage of conservative differencing is that Eqs. (36)-(38) are valid across discontinuities in the flow, when the differential equations break down.

If a difference scheme is dissipative, as it must be to convert kinetic to internal energy in a shock, and if it is conservative, then it must give precisely the correct jumps for a shock propagating through a uniform medium on a uniform computational mesh. The reason for this is that the translational symmetry of the problem guarantess that the same fluxes will be computed at both interfaces of a zone, but they will be computed a time interval $\Delta m/(\rho_0 v_s)$ later at one interface, say interface R. Here ρ_0 is the preshock density and v_s is the shock speed. Now to find the value of the jumps in V, u, and E for the zone as the shock passes we will integrate Eqs. (36)-(38) numerically using very small values for Δt. This procedure will give nearly continuous, smooth curves for the fluxes $\bar{u}_L$, $\bar{u}_R$, $\bar{p}_L$, $\bar{p}_R$, $(\overline{up})_L$, and $(\overline{up})_R$ as functions of time. Such curves generated by the PPM scheme are shown in Fig. 9. These curves were constructed in the uniform section of the grid for the shock problem shown in Fig. 8. The dissipative character of the scheme guarantees that constant post-shock values are eventually attained. Integrating Eqs. (36)-(38) over the time interval 0 to t, shown in the figure and realizing that V, u, and E have their preshock values V_0, u_0, and E_0 at t = 0 and their post-shock values V_1, u_1, and E_1 at t_1, we find

$$V_1 - V_0 = -(u_1 - u_0)\,/\,\rho_0 v_s \quad , \tag{39a}$$

$$u_1 - u_0 = (p_1 - p_0)\,/\,\rho_0 v_s \quad , \tag{39b}$$

$$E_1 - E_0 = (u_1 p_1 - u_0 p_0)\,/\,\rho_0 v_s \quad . \tag{39c}$$

On the right we have replaced $\bar{u}(0)$ by u_0, $\bar{u}(t_1)$ by u_1, and similarly for $\bar{p}$ and $(\overline{up})$. Augmenting these equations with the equation of state gives a full set, and all the shock jumps and the shock speed v_s may be determined. Because Eqs. (36)-(38) are exact, and because the detailed structure of the fluxes $\bar{u}(t)$, $\bar{p}(t)$, and $\overline{up}(t)$ in Fig. 9 cancelled out upon integration due to translational symmetry, Eq. (39) and the other equations determined in this way must be exact. This is an amazing fact; it implies that crude difference approximations to the derivatives in Eqs. (3), (4), and (12) may be used, and with a little care one can obtain the right answer under conditions where those differential equations themselves break down.

The above arguments depend critically upon the symmetry which guarantees that the curves $(\overline{up})_L(t)$ and $(\overline{up})_R(t)$ represented in Fig. 9 are identical except for a phase shift. This symmetry guarantees that the gross error which must be incorporated in the time integral of $(\overline{up})_L(t)$ for any difference scheme will be exactly cancelled by a compensating gross error in the integral of $(\overline{up})_R(t)$. Anything which causes this symmetry to be broken will upset this cancellation of errors, and the computed shock jumps will be wrong. This has happened in the runs shown in Fig. 8. In this case, the symmetry breaking at the reflecting wall at $m = 0$ is most easily grasped. At this wall $\bar{u}_L$ is always set to zero, the exact value. Thus $(\overline{up})_L$ is also zero and exact. However, $\bar{u}_R$ and $(\overline{up})_R$ for the zone next to the wall are very inexact, and all of this error must appear in the post-shock values of V and E, hence ε, for this zone. As the grid is refined, the errors in $\bar{u}_R$ and $(\overline{up})_R$ are spread over shorter time intervals, but then so is the integration to obtain V_1 and E_1. Thus the errors in V_1 and E_1 are large and independent of zone size. These errors do obviously depend upon the thickness of the numerical shock structure relative to the zone size, so this structure should be made as thin as possible. A glance at Fig. 8 shows that for PPM this structure is indeed close to the limit imposed by the width of a single zone.

The overheating at the reflecting wall in Fig. 8 can be understood because $\bar{u}_R$ must always be negative, while $\bar{p}_R$ must always be positive if the shock profile is monotone (i.e. there is no ringing). Then $(\overline{up})_R$ is always negative, while it should actually vanish. Consequently $(\overline{up})_R$ in Eq. (38) contributes a spurious positive energy to the first zone near the wall. The time integral of up must always be negative and should be roughly proportional to the numerical shock width, because up vanishes outside the numerical shock structure. Now, the dissipative mechanisms in the PPM difference scheme, which cause shocks to be spread out on the mesh, scale with the zone width. The result is a shock structure which is always about a zone wide. Therefore in the first region of the grid in Fig. 8, where the zones are decreasing in size as m increases, the time integrals of up at successive zone interfaces must also be decreasing in magnitude. Equation (38) then implies that the computed post-shock energy should be too low, as indeed it is. Where the zones increase in size the sign of this effect is flipped, and the computed energy is too high. As the grid is refined the ratio of neighboring zone widths is reduced toward unity in a nearly linear fashion. This causes the noncancellation of errors in the time integrated energy fluxes at the different sides of each zone to be diminished in a nearly linear fashion. This brings about the roughly linear convergence of the solution which is shown in Fig. 8 away from the wall and the discontinuous jump in zone width at $m = 2$.

The above long discussion shows how a difference scheme which is formally second-order accurate may nevertheless converge only linearly to the solution of a problem involving discontinuities. The problem in Fig. 8 is not as artificial as it may seem. Zoning irregularities are unavoidable in many practical problems. An extreme case which appears frequently is the central region of a uniform grid in a cylindrical or spherical radial coordinate. Then factors of r or r^2 appear in the flux integrals and destroy the cancellation of errors. Error cancellation is also destroyed when two discontinuities collide. Then glitches like those near $m = 2$ in Fig. 8 can be generated. Finally, it should also be noted that PPM is a good difference scheme. Results of, say, the standard von Neumann-Richtmyer scheme on a problem like that in Fig. 8 are much worse (see for example Noh et al., 1979).

A somewhat extreme one-dimensional, hydrodynamical test problem which brings these considerations into better focus is the collision of two strong blast waves discussed in Woodward (1982) and exhaustively studied in Woodward and Colella (1983). A wave diagram for this problem is shown in Fig. 10. In this diagram the bunching together of many density contours marks the trajectory of a shock or contact discontinuity, while the spreading of density contours marks a rarefaction fan. At time zero the density is unity and the velocity vanishes everywhere on the unit spatial interval of the problem. For $x < 0.1$, $p = 1000$; for $x > 0.9$, $p = 100$; and for $0.1 < x < 0.9$, $p = 0.01$. Reflecting walls are located at $x = 0$ and $x = 1$. The gas has a gamma-law equation of state (Eq. 14) with $\gamma = 1.4$. The hot regions near the walls expand, driving strong shocks into the cold central gas. The hot gas is separated from the shells of dense, shocked gas by two contact discontinuities. At about $t = 0.028$ the shocks collide and cause a new contact discontinuity to be created. This contact discontinuity is soon strongly accelerated to the right by a rarefaction created by one of the strong shocks breaking out of the dense, previously shocked gas into the near vacuum next to the right-hand wall. Much of the wave interaction takes place in an extremely small region of space-time which is enlarged in the lower portion of Fig. 10.

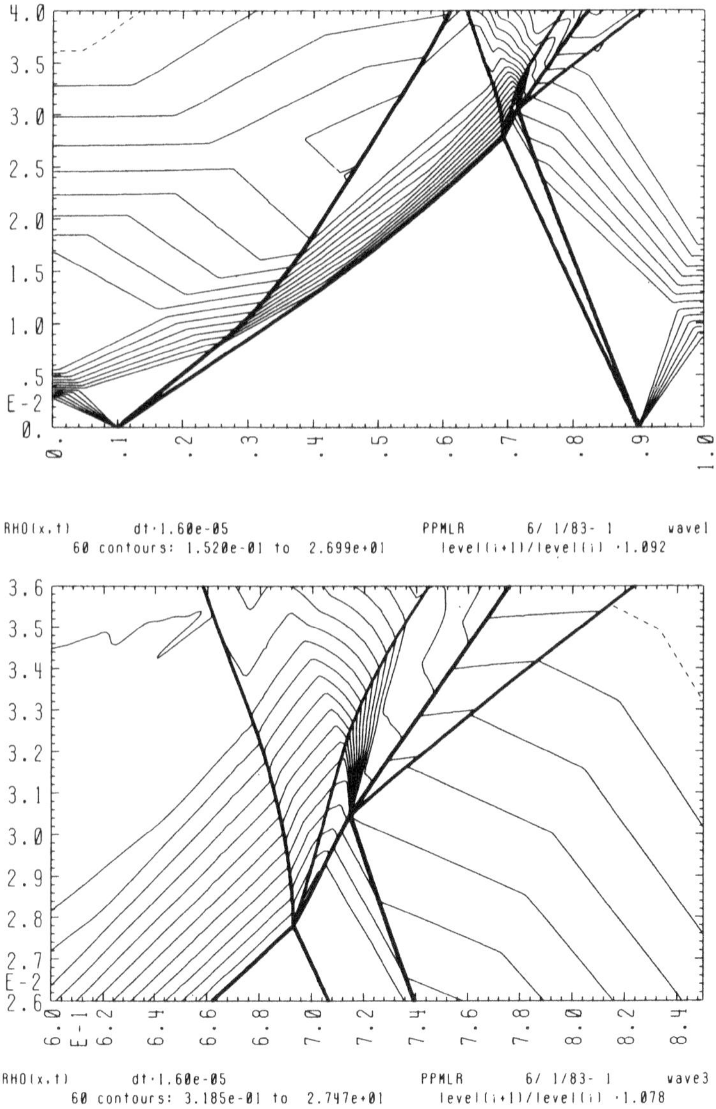

Fig. 10. Contours of density on the space-time plane for the interacting blast wave problem of Woodward (1982) and Woodward and Colella (1983). Two strong shocks approach each other closely followed by contact discontinuities. Rarefaction fans reflected from the walls interact with these waves and with each other in a complicated fashion. The region of this interaction is enlarged in the lower part of the figure. The creation of a new contact discontinuity by the collision of the two shocks can be clearly seen in this enlargement.

The results in Fig. 10 may be regarded as exact. They have been generated on a nonuniform grid of 3096 zones with the smallest zone width set to 1/9600 in the region of the shock collision. In addition, a five-zone section of the grid around each discontinuity is refined by an additional factor of 8, and the two principle contact discontinuities are treated as specially tracked interfaces between distinct fluids. The resulting solution is good to 1% everywhere, and is much better in most places. More detailed displays of this calculation are given in Woodward and Colella (1983). This interacting blast wave problem has been designed as a worst case to challenge Eulerian difference schemes to the utmost. The solution demands accurate treatment of the resolution of the jumps in the initial data into separate discontinuous waves and of the interaction of these waves with other discontinuous and strong continuous waves. A method which can solve this problem well should be able to handle just about anything which can arise in one-dimensional pure hydrodynamic flow.

PPM is such a scheme. The comparison with many other methods in Woodward and Colella (1983) makes this abundantly clear. However, the rate of convergence of PPM on this problem is linear. PPM contains many second-order features, many third-order features, and even a fourth-order feature, but for problems dominated by the interactions of discontinuous waves, as is this one, these high-order features serve mainly to keep the discontinuous waves sharp; they do not give high-order convergence. In fact, the main purpose for the use of parabolic interpolation in PPM is the sharp resolution of contact discontinuities. In Fig. 11 results of PPM solutions of the blast wave problem for uniform Eulerian grids of 200 and 1200 zones are presented at time 0.038. In this figure the computed zone-averaged densities are shown as dots while the "exact" solution is drawn as a solid line for comparison. These results show that the shocks and contact discontinuities are only about one zone wide. Nevertheless, the contact discontinuity formed by the shock collision contains 8 zones, and the jump at this discontinuity is not correctly computed on the 200-zone grid. A variety of solutions of lower quality are presented for this problem, along with quantitative measures of the errors and convergence rates in Woodward and Colella (1983). Here it suffices to say that PPM is the only one of six schemes studied in that article which converges as rapidly as linearly on this extremely difficult problem.

The results for this interacting blast wave problem are intended to discourage any overzealous pursuit of additional terms in Taylor expansions or additional moments to be conserved within zones. The intent is not to dampen the enthusiasm behind such pursuits but instead to redirect it in more profitable avenues, so far as hydrodynamical problems are concerned. One such avenue has already been mentioned in passing; that is the technique of adaptive local mesh refinement. This technique has been quietly pursued by many people for many years, but recently it has been more systematically developed and more openly and systematically discussed by Gropp (1980), Berger (1982), and others. The simplest application of the technique is to refine the grid in 4 or 5 zones around each discontinuity in the flow. The grid should be refined in both space and time in order to keep the calculation both efficient and explicit. For PPM a mesh refinement factor of 8 is recommended, while for less powerful methods larger factors should be used. In the interacting blast wave problem discussed above, this procedure plus treating the problem as involving three distinct fluids brings the error for a 100-zone uniform Eulerian reference grid down below that of the unalloyed scheme using 1200 uniform Eulerian zones. Obviously this is a powerful technique, and it has the very attractive feature that it can be carried out in two dimensions quite easily.

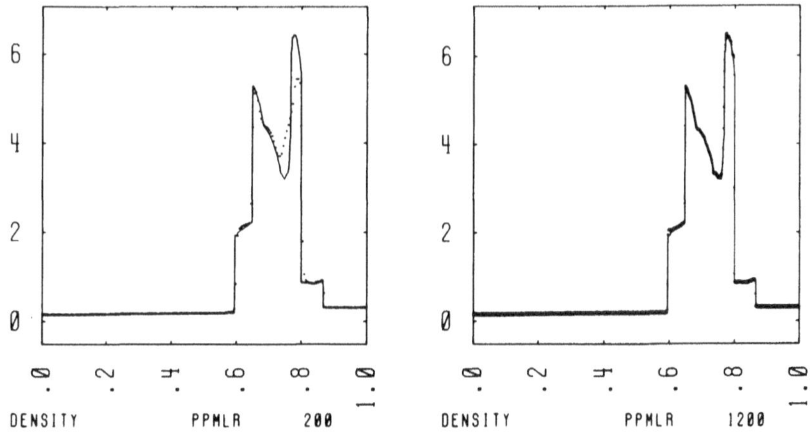

Fig. 11 PPM results at time 0.038 for the interacting blast wave problem shown in Fig. 10. Results for uniform Eulerian grids of 200 to 1200 zones are plotted as dots against an extremely accurate solution (solid lines). Note especially the sharpness of the contact discontinuities. A linear convergence of the integrated error for this problem is observed, despite the formal second-order accuracy of the PPM scheme.

THE TREATMENT OF DISCONTINUITIES IN ADVECTION SCHEMES -- MONOTONICITY CONSTRAINTS AND CONTACT DISCONTINUITY STEEPENERS

We have seen that the most important limitation to the accuracy of hydrodynamical difference schemes is the necessity to represent discontinuities in the flow. One of the principal means of doing so may be discussed in the simpler context of the advection equation (19). This discussion will explain the treatment of flow discontinuities in the remap step of an Eulerian hydrodynamics calculation, their treatment in the Lagrangian step will be discussed later.

In Figs. 12 and 13 PPM and PPB advection calculations are presented which show the generation of oscillations near sharp jumps in the advected distribution. These oscillations always appear when high-order interpolation polynomials are used to interpolate between undersampled data. The oscillations are much less severe for the PPB scheme because the additional moment data provided for this scheme allows the smooth structure of the jumps to be nearly resolved on the finer grids. The format for Figs. 12 and 13 is similar to that of Figs. 2 and 3. Periodic boundary conditions are applied on the unit interval, the advection velocity is unity, and the results are displayed at times 0, 1, 10, and 100. The internal zone structures generated by the PPM and PPB schemes for grids of 8, 16, 32, 64, and 128 zones are plotted; and a Courant number of 0.08 was used so that these results represent a worst case. The advected function is

$$f_o(x) = 1 \,/\, \left\{1 + \exp\left[80(|x - 1/2| - 0.15)\right]\right\} \quad . \tag{40}$$

The two jumps in this function are quite sharp; 95% of each jump is spread over a distance of only 0.092, and 50% is spread over only 0.028. On the finest grid these intervals are described by about 12 and 3.6 zones, respectively. This resolution proves insufficient to avoid oscillatory behavior for either the PPM or the PPB schemes, although the oscillations for the PPB scheme on the finest grid are very slight and appear only at the latest time shown in Fig. 13. Other tests reveal that with twice this resolution the PPB scheme will propagate these sharp features properly. However, if we are willing to give up the formal high-order accuracy of such a scheme, excellent results may be obtained at much less cost. The trick is to know when to give up formal accuracy and to what degree.

For the simple case of advection the source of oscillations near sharp jumps can be readily identified. The plots in Figs. 12 and 13 for time 0 reveal interpolated structures which are not monotone increasing (decreasing) even though they have been constructed from monotone data. These over- and undershoots in the interpolated internal zone structures eventually give rise to over- and undershoots in the zone-averaged data. It is clear that if the interpolated function f(x) is monotone, then regardless of which bins are used to construct new zone-averaged data (that is, regardless of the value of the Courant number) this new data must also be monotone. It is also clear that the first-order advection scheme which sets f(x) constant within each zone must always preserve the monotonicity of its initial data. van Leer (1977) realized that an advection scheme may be made to preserve the monotonicity of its initial data if any non-monotone interpolated zone structures are flattened so that they become monotone. This flattening process can be viewed as a local

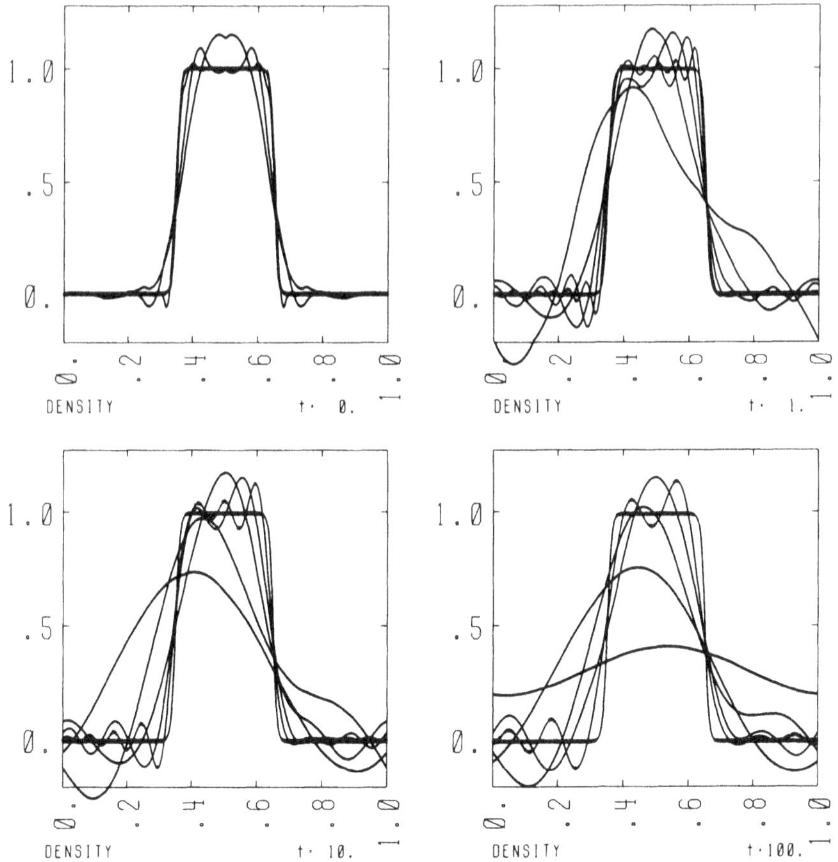

Fig. 12 Square wave advection tests using PPM without monotonicity constraints.

blending of the high-order scheme with the first-order scheme which uses constant zone structures. In order to treat the structure of each zone independently, the structures in neighboring zones are ignored when the flattening factor for the zone of interest is determined. This leads to van Leer's monotonicity constraint: no value interpolated within a zone shall lie outside the range defined by the zone averages for this zone and its two neighbors. If one wants to take the trouble, this constraint may be relaxed slightly to read as follows: the average density within the section of a zone to be advected into a neighboring zone and in the section which remains in the original zone for this timestep must lie within the range defined by the zone averages for this zone and its

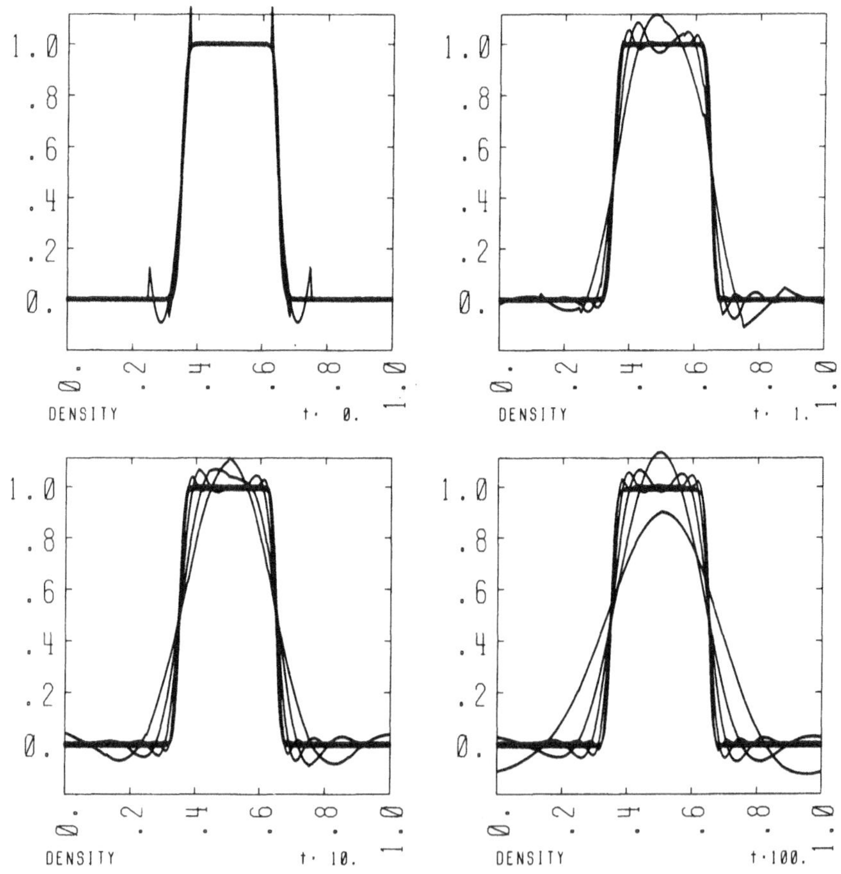

Fig. 13 Square wave advection tests using PPB without monotonicity constraints.

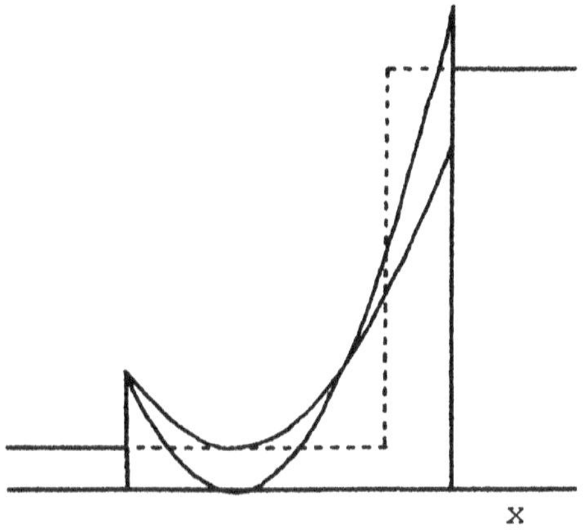

Fig. 14a Interpolated and flattened zone structures appropriate for the representation of a sudden jump. This flattened structure will give monotone advection regardless of the timestep.

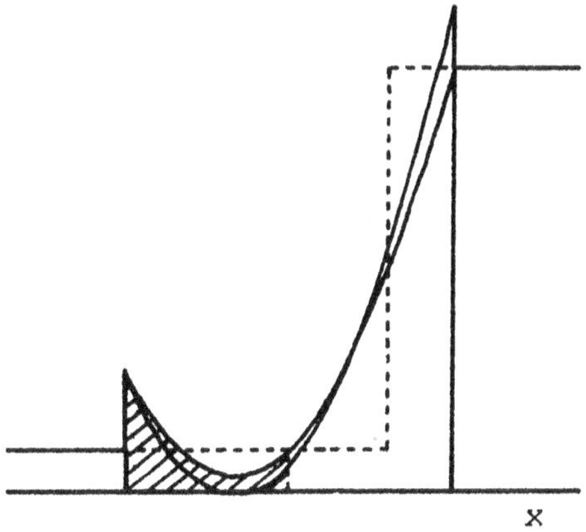

Fig. 14b Interpolated and flattened zone structures appropriate for monotone advection to the right at a Courant number of 0.5. The shaded region of the zone which is not advected into the zone on the right has the same average density as the zone on the left.

two neighbors. These monotonicity criteria are illustrated in Fig. 14. A similar monotonicity constraint for advection schemes was developed earlier by Boris and Book (1973), but that algorithm is not so easily understood and it has been much less successful when applied to other problems such as hydrodynamics.

An unfortunate characteristic of these monotonicity constraints is that they tend to alter a waveform dramatically once the decision is made to switch them on. This is illustrated by the Gaussian advection results in Fig. 15. These may be directly compared with the results in Fig. 2b, from which they differ only by the use of monotonicity constraints and contact discontinuity steepeners to be described shortly. The results in Fig. 15 are definitely an improvement over those in Fig. 2b, but the shape of the waveform has been completely lost in the process of keeping it monotone. One might well prefer the non-monotone results of the PPB scheme which are shown in Fig. 3b. In fact, the PPB results are so good that if any monotonicity constraints are to be applied to this scheme at all they must be applied with great care in order not to lose the advantage of this scheme's accuracy in smooth flow. In particular, the monotonicity constraints of van Leer (1977) and of Boris and Book (1973) cause zones near local extrema to be completely flattened. For the PPB scheme this is a disaster. The advantage of PPB is that its extra moment data allows the accurate treatment of very short wavelength Fourier modes. For these modes flattening zone structures near the frequent extrema results in gross wave distortion and damping.

For the PPB scheme it is worthwhile to expend some computer time to make sure that monotonicity constraints are really needed before they are applied. This procedure may be useful for other schemes as well, but for PPB it is essential if the constraints are to be applied at all. The entire procedure more than doubles the running time, so if the unconstrained scheme will do, it should be used. The monotonicity constraints described above modify the advection algorithm in regions where the second derivative of the advected function is large compared to the first derivative -- more precisely, when

$$\Delta x \left| \frac{\partial^2 f}{\partial x^2} \right| > \left| \frac{\partial f}{\partial x} \right| .$$

For schemes employing parabolic interpolation, large second derivatives of f do not necessarily pose any special problem. Therefore the monotonicity constraint may often be inappropriate, especially near extrema, where $\partial f/\partial x$ vanishes.

The key to modifying the monotonicity constraint is not to measure the monotonicity of the interpolated zone structure but instead to measure the ability of that structure when extrapolated into the neighboring zones to predict the correct zone-averaged values there. This will be a more direct measure of the extent to which the density distribution is undersampled by the grid than is the monotonicity criterion described above. In fact, both criteria will locate regions in which short wavelength Fourier modes play a large role, but the monotonicity criterion will also locate perfectly innocent extrema as well.

To measure the quality of the interpolation parabola given by Eqs. (22) and (25) (or by Eqs. (22)-(24) for PPM) we proceed as

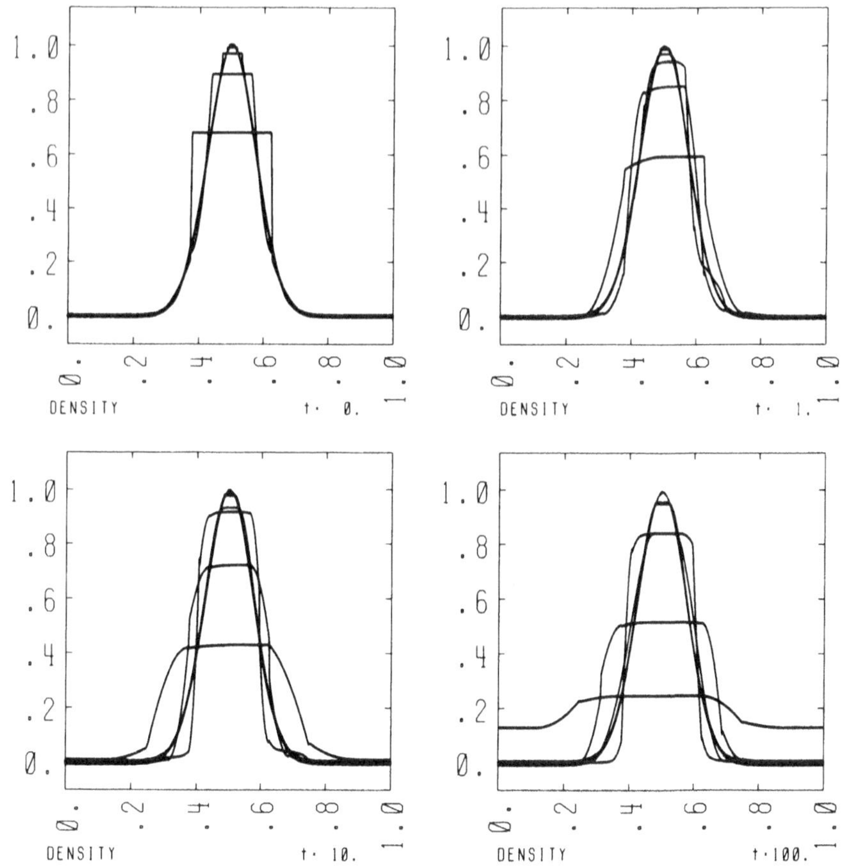

Fig. 15 Gaussian advection tests using PPM with monotonicity constraints and contact discontinuity steepeners.

follows. First we check the extrapolation f(x) into the zone on the left (subscript ZL). This extrapolation gives a guess $\langle f \rangle_{XZL}$ at the neighboring zone average $\langle f \rangle_{ZL}$. For a uniform mesh we have

$$\langle f \rangle_{XZL} = f_0 - f_1 + \frac{13}{12} f_2 \quad . \tag{41}$$

The quality Q_{ZL} of this extrapolation is then defined as

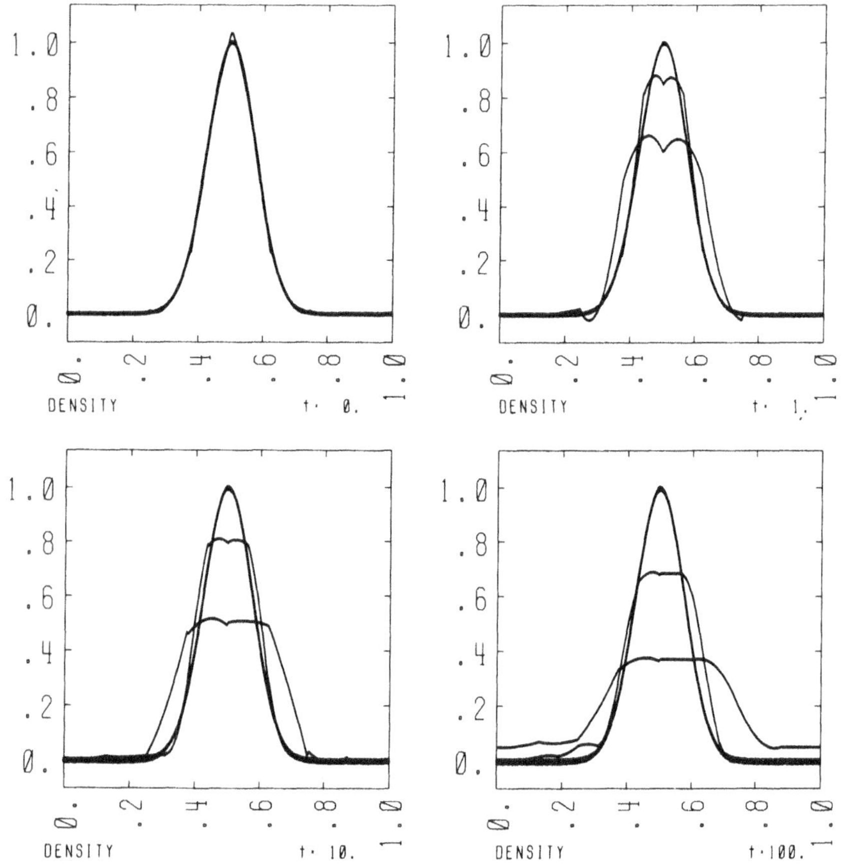

Fig. 16 Gaussian advection tests using PPB with monotonicity constraints under the control of a "quality" criterion. The Courant number is 0.8.

$$Q_{ZL} = \frac{|\langle f\rangle_{XZL} - \langle f\rangle_{ZL}|}{|\langle f\rangle_{XZL} - \langle f\rangle| + 0.005\ (|\langle f\rangle_{ZL}| + |\langle f\rangle|)} \quad . \qquad (42)$$

The extrapolated average $\langle f\rangle_{XZR}$ and its quality Q_{ZR} are then computed similarly for the zone on the right. The overall quality Q of the interpolated zone structure f(x) is then taken to be

$$Q = \max \{Q_{ZL} , Q_{ZR}\} \quad . \qquad (43)$$

If Q exceeds a threshold value of 0.4 (a more conservative figure is 0.2) in this zone or in either of its neighbors, then the interpolated zone structure may be modified according to the less restrictive of the two monotonicity constraints described above. Gaussian advection tests with PPB using this quality criterion and monotonicity constraint are shown in Fig. 16. These results are to be compared with those in Figures 3b and 15. The comparison with Fig. 15 shows that the quality criterion is smart enough to leave the results on the three fine grids essentially unchanged. Whether or not the coarse grid results are superior to those in Fig. 3b is a matter of taste and would depend upon the particular application as well. If the threshold for Q is raised above 0.4, only the 8-zone run will be affected by the monotonicity constraint. The threshold of 0.4 causes PPB to clip a pure sine wave on a grid of 3 zones or less, while a threshold of 0.2 clips the wave on grids of 4 zones or less.

Treatment of Discontinuities in PPM

The technique for treating discontinuities in PPM is most easily explained in terms of a specific example. Formulae for more general cases may be found in Colella and Woodward (1983). We will follow the scheme as it constructs a representation f(x) from zone-averaged data $\langle f\rangle_i$ corresponding to an $f_0(x)$ which is abstracted from the square wave advection tests in Figs. 12 and 13:

$$f_0(x) = \tanh (x) \quad . \qquad (44)$$

In Fig. 17 the six zones near the origin are shown, with $\Delta x = 3$ and the leftmost zone interface at $x = -10$. The dotted lines in Fig. 17a show the interpolation parabolae constructed by the PPM scheme described earlier and used in the Gaussian advection tests in Fig. 2. The values of f at zone interfaces, f_L and f_R, have been generated from Eq. (23). These interface values are attained by the unique cubic curves which have the same average values as $f_0(x)$ in the four zones closest to each interface. The parabolae within zones are constructed to have these interface values and the same zone-averaged values as $f_0(x)$. Note that the resulting interpolated f(x) is not monotone. In Fig. 17b a construction of zone-averaged slopes Δf_P is illustrated. The dotted lines show the zone averaged slopes for the parabolae which have the same zone-averaged values as $f_0(x)$ in each zone and its two neighbors. For a uniform grid Δf_P is given by

$$\Delta f_P = \frac{1}{2} \left(\langle f \rangle_{ZR} - \langle f \rangle_{ZL} \right) \quad . \tag{45}$$

The solid lines in Fig. 17b show monotonized slopes Δf_M. These are constructed by reducing Δf_P toward zero if necessary to keep the interpolated values implied by these slopes within the range of the three zone averages $\langle f \rangle_{ZL}$, $\langle f \rangle$, and $\langle f \rangle_{ZR}$.

The solid lines in Fig. 17a illustrate the interpolation parabolae implied by a provisional set of interface values computed by the full PPM scheme. Like the dotted lines in Fig. 17a these are also obtained by evaluating cubic polynomials at the zone interfaces. These interface values are determined by rewriting the interpolation formula in Eq. (23) in terms of $\langle f \rangle_{ZL}$, $\langle f \rangle$, Δf_{PZL}, and Δf_P and then substituting Δf_{MZL} and Δf_M for the two slopes:

$$f_L = \frac{1}{2} \left(\langle f \rangle_{ZL} + \langle f \rangle \right) - \frac{1}{6} \left(\Delta f_M - \Delta f_{MZL} \right) \quad . \tag{46}$$

Thus if the original slopes Δf_{PZL} and Δf_P are unchanged by the application of the monotonicity constraint, Eq. (46) gives the same result as Eq. (23). In our example only the slope for the zone nearest the origin is unaffected by the monotonicity constraint. The use of Eq. (46) gives values of f_L which are more reasonable than those obtained from Eq. (23). As the solid lines in Fig. 16a demonstrate, each value f_L lies between the neighboring zone averages $\langle f \rangle_{ZL}$ and $\langle f \rangle$. This monotonicity property is guaranteed by the use of monotonized slopes in obtaining f_L. A further improvement resulting from the use of Eq. (46) is the slight steepening of the implied internal structure for the zone nearest the origin. This is indeed a small effect, but such effects accumulate over thousands of timesteps and can eventually mean the difference between a contact discontinuity which is three zones wide and one which is five zones wide.

The full PPM algorithm includes a contact discontinuity detector and steepener which will cause the implied structures for the two zones between x = -4 and x = 2 to be steepened further. This part of the PPM algorithm is responsible for the very thin contact discontinuities in Fig. 11. It is also the cause of the very steep, incorrect structures in Fig. 15. The idea behind this part of the algorithm is to detect regions of the flow in which the density jumps so sharply that these regions may be considered to be contact discontinuities (for hydrodynamics, we demand that the density jump not be accompanied by a pressure jump). When these regions are detected, the PPM

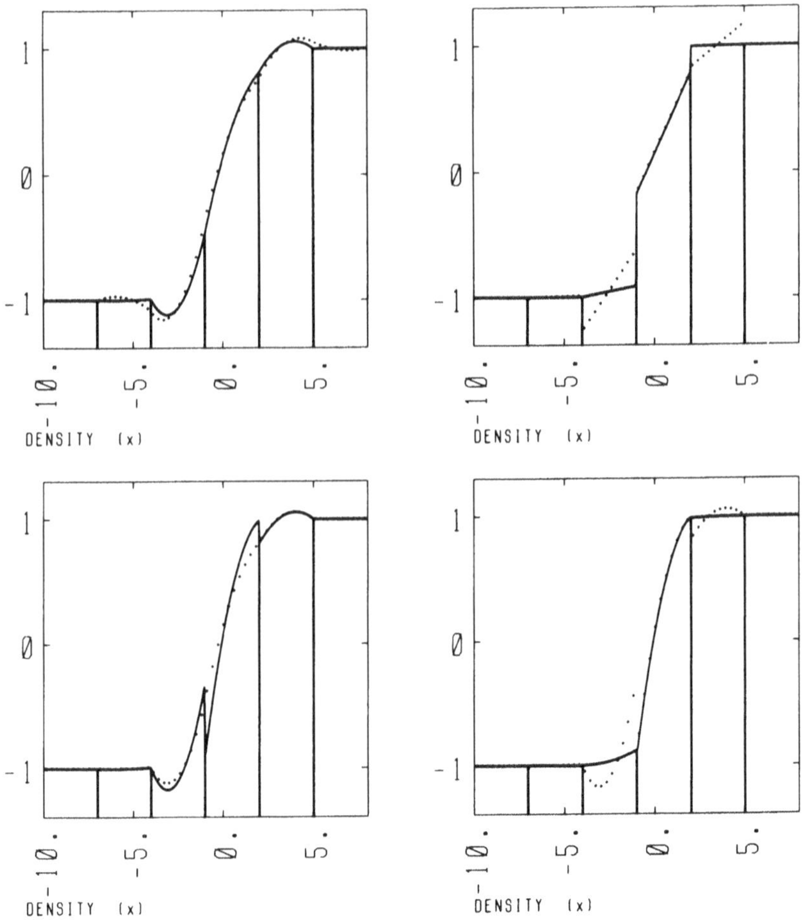

Fig. 17 (a) Interpolation parabolae for the unconstrained scheme used for the tests in Figs. (2) and (12) are shown as dots. The solid lines are generated from interface values constructed using monotonized zone-averaged slopes. (b) Unconstrained (dotted) and monotonized (solid) zone-averaged slopes. (c) Unsteepened (dotted) and steepened (solid) interpolation parabolae. (d) Unmonotonized, steepened parabolae (dotted) and montonized, steepened parabolae (solid).

algorithm causes them to be continually steepened, whether they are actually contact discontinuities or not. In doubtful cases some choice must be made, and it may be wrong, but we are careful to guarantee that nothing involved in the treatment of detected discontinuities will vitiate the formal second-order accuracy of the scheme in well-resolved, smooth flow. The behavior in the Gaussian advection tests in Fig. 15 results from the clipping of the peak of the Gaussian due to the monotonicity constraint. This introduces sudden jumps near the clipped peak which are detected as contact discontinuities and thus steepened still further. This behavior ceases when the grid is sufficiently fine. The point at which this transition occurs may be altered by adjusting the dimensionless constants in the contact discontinuity detector.

The action of the contact discontinuity steepener is illustrated in Fig. 17c. Here the dotted lines are the unsteepened interpolated structures which were drawn as solid lines in Fig. 17a. The solid lines in Fig. 17c are the steepened interpolated structures. Steepeneing occurs only for the two zones nearest the origin. For this case of uniform zone widths the discontinuity detection algorithm is as follows (more general formulae are given in Colella and Woodward 1983). First, an estimate of the second derivative of f is computed for each zone:

$$\Delta_2 f = \langle f\rangle_{ZR} - 2\langle f\rangle + \langle f\rangle_{ZL} \quad . \tag{47}$$

Then a provisional steepness parameter S_0 is computed which compares the magnitudes of the first and third derivatives in each zone:

$$S_0 = [(\Delta_2 f)_{ZR} - (\Delta_2 f)_{ZL}] \;/\; [6\,(\langle f\rangle_{ZL} - \langle f\rangle_{ZR})\,] \quad . \tag{48}$$

Note that S_0 is positive for sudden jumps in f, but it is negative for small plateaus in f. To locate these sudden jumps we must also demand that $\Delta_2 f$ change sign in crossing the region of the jump. Therefore we define a revised steepness parameter S_1 by

$$S_1 = S_0 \;, \quad \text{if } (\Delta_2 f)_{ZL}\,(\Delta_2 f)_{ZR} < 0 \tag{49}$$

$$= 0 \;, \quad \text{otherwise.}$$

We also demand that S_1 be zero unless some reasonable measure of the fractional change of f exceeds a small threshold of, say, 1%. This will avoid the steepening of tiny glitches of numerical origin. The final step in the detection procedure is to make some quantitative decision as to the steepness which warrants special treatment as a discontinuity. The quantity S_1 is scaled from 0 to 1 via:

$$S_2 = \max\{0 , \quad \min\{1 , \quad 20\,(S_1 - .05)\}\} \quad . \tag{50}$$

The dimensionless constants in this formula have been chosen to guarantee that discontinuities one zone in width will be steepened fully (i.e., $S_2 = 1$ for these cases) regardless of their phase with respect to the zone boundaries. In our specific example in Fig. 17c, this criterion causes steepening to be applied only to the two zones nearest the origin. For the zone nearest the origin this steepening is substantial.

When the steepening parameter S_2 is nonzero, the previously interpolated interface values f_L and f_R are blended with values f_{SL} and f_{SR} which will generally give a steeper structure within the zone. The blending factor is just S_2. The values f_{SL} and f_{SR} are obtained by extrapolating the monotonized zone-averaged slopes in neighboring zones (see Fig. 17b) to the zone interfaces. Thus

$$f_{SL} = \langle f\rangle_{ZL} + \frac{1}{2}\,\Delta f_{MZL} \quad , \tag{51a}$$

$$f_{SR} = \langle f\rangle_{ZR} - \frac{1}{2}\,\Delta f_{MZR} \quad . \tag{51b}$$

Note that these interface values are formally second-order accurate, although they are intended for use in underresolved regions where order of accuracy is a concept of limited value.

The final step of the PPM construction is shown in Fig. 17d. This figure illustrates the application of a monotonicity constraint, first used by the PPM scheme in Woodward and Colella (1981). The zone averages of f, $\langle f\rangle$, and the interface values f_L and f_R determine parabolic internal structures for the zones of the form given in Eq. (22) through the relations in Eq. (24). These parabolae will be monotone so long as the absolute value of f_2 is less than that of f_1. For the marginal cases where these absolute values are equal we have either

$$f_L = f_{ML} = 3\,\langle f\rangle - 2f_R \tag{52a}$$

or

$$f_R = f_{MR} = 3 \langle f \rangle - 2f_L \quad . \tag{52b}$$

When one of these marginal values is exceeded, either f_L or f_R is reset to that marginal value so that the implied interpolation parabola is monotone. Also, at extrema in $\langle f \rangle$ both f_L and f_R are set to $\langle f \rangle$. This flattening at extrema is done first. Then the difference of f across the zone is computed

$$\Delta f = f_R - f_L \quad . \tag{53}$$

The criteria for resetting f_L and f_R can be written

$$f_L \rightarrow f_{ML} , \quad \text{if} \quad \Delta f \, (f_L - f_{ML}) < 0 \quad ; \tag{54a}$$

$$f_R \rightarrow f_{MR} , \quad \text{if} \quad \Delta f \, (f_R - f_{MR}) > 0 \quad . \tag{54b}$$

This monotonicity constraint has a dramatic effect on the steepened interpolation parabolae shown as solid lines in Fig. 17c and as dotted lines in Fig. 17d. The final result, the solid lines in Fig. 17d is a very sharp representation of the jump in $f_0(x)$. It is considerably sharper than the profile in Fig. 17a which is produced by the completely unconstrained PPM scheme. In fact the profile in Fig. 17d is much closer to $f_0(x)$ than is that in Fig. 17a. It is an empirical fact that the PPM interpolation procedure described above is sufficient to cause a square pulse 10 zones in width to be advected essentially without distortion. Unfortunately, as Fig. 15 demonstrates, other smoother pulses may be turned into square waves as they are advected if they are not well resolved. The philosophy here is that if such a pulse is not sufficiently resolved on the grid to be advected properly its shape may be greatly distorted but at least its position and height will be fairly well represented.

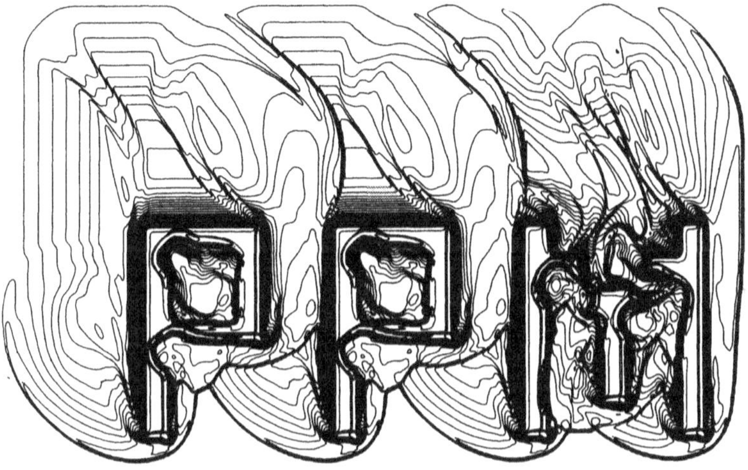

THE PPM METHOD FOR HYDRODYNAMICS

It is easiest, both conceptually and computationally, to perform hydrodynamic calculations with PPM if these are split into a sequence of one-dimensional calculations which in turn are split into Lagrangian hydrodynamics calculations followed by remap operations. Therefore we need only describe here the one-dimensional Lagrangian formulation of PPM hydrodynamics (for other formulations see Colella and Woodward 1983). The PPM hydrodynamics scheme is built upon the PPM advection scheme described above. The basic idea is to exploit the relationship between hydrodynamics and the advection of Riemann invariants which was discussed earlier. One would like to use the PPM advection scheme to update each family of Riemann invariants separately, but unfortunately hydrodynamics is not so simple. In general the Riemann invariants are defined only differentially through Eqs. (15), and they are not advected independently of one another. However, it is still true that the time dependence of the flow at any point in the fluid is determined by two separate domains of dependence bounded in space-time by the trajectory of that point and by two characteristic paths corresponding to pressure waves travelling toward the point from either side (see Fig. 18a). The information contained in each of these domains of dependence which affects the time evolution of the point in question still basically refers to only one of the two Riemann invariants. The approach in PPM is to let a nonlinear Riemann solver handle any difficulties in sorting out what aspects of the information in each domain of dependence correspond to which Riemann invariant and to what extent the information in the two separate domains should interact. Such use of a nonlinear Riemann solver in hydrodynamic calculations was first introduced by Godunov (1959) and was first adapted for higher-order difference schemes by van Leer (1979).

A nonlinear Riemann solver is an essential ingredient in PPM. It computes the solution to Riemann's problem, that is the nonlinear interaction of two constant states of the fluid. The initial conditions for Riemann's problem are two constant states separated by a discontinuous jump. This sort of initial condition is found in the blast wave problem discussed earlier

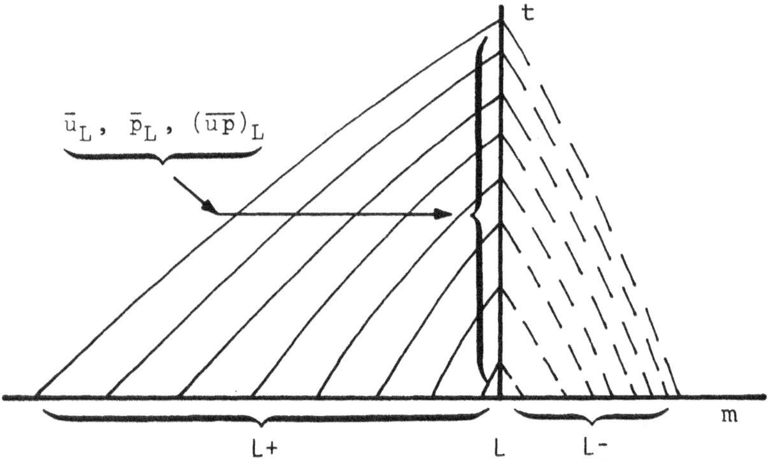

Fig. 18a A space-time diagram in Lagrangian coordinates showing the domains of dependence for zone interface L. Paths of sound waves travelling to the right (left) are drawn as solid (dotted) lines. These trace out the domain of dependence L+ (L−). The characteristic equations (15a) hold along these paths if the flow is smooth. These paths have slopes equal to the local sound speed. This is approximated by PPM as $\langle C\rangle_{ZL}$ or $\langle C\rangle$ for the solid or dotted paths, respectively. The interaction of these sound waves produces the time evolution of u_L and p_L. The averages of these over the time interval of Δt shown here are the fluxes $\bar{u}_L$ and $\bar{p}_L$ used by PPM in the conservation laws in Eqs. (36) and (37).

and displayed in Fig. 10. The early evolution of that problem illustrates the character of the general solution to Riemann's problem (see also Fig. 18b). In general two nonlinear sound waves, either shocks or rarefactions, emerge from the initial jump, leaving a contact discontinuity behind there. The solution is self-similar, and in particular the state at the location of the initial jump is constant in time. A space-time diagram for such a Riemann problem is shown in Fig. 18b. The sound wave paths, or characteristic curves in Fig. 18b are straight lines, but they kink as they cross the shock and rarefaction waves. The algorithm for solving Riemann's problem involves an iteration, as will be described shortly.

We have already noted that the PPM hydrodynamics scheme is cast in strict conservation form, so that it will compute shock

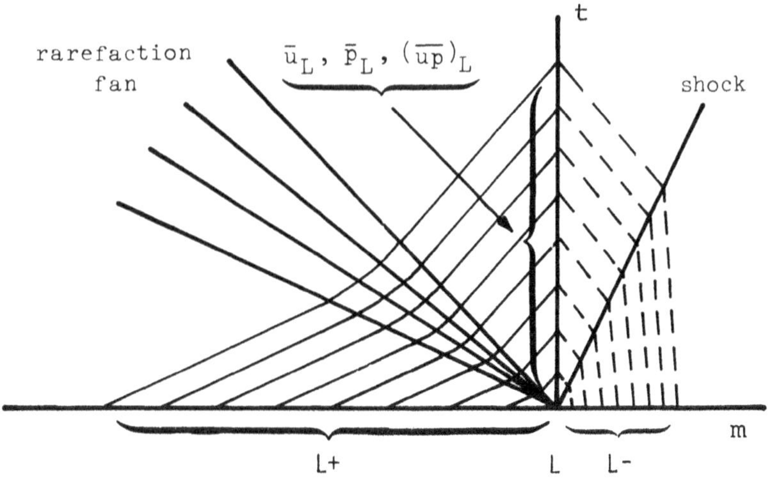

Fig. 18b A space-time diagram showing the use of a Riemann solver in PPM to estimate the time-averaged fluxes $\bar{u}_L$, $\bar{p}_L$, and $(\overline{up})_L$ at zone interface L. A case similar to that occuring in the blast wave problem in Fig. 10 is shown. Constant states for the Riemann problem are obtained by averaging the interpolated initial data over the domains L+ and L-. The nonlinear interaction of these constant states produces a shock and a rarefaction wave and a contact discontinuity at interface L. The velocity and pressure $\bar{u}_L$ and $\bar{p}_L$ at the interface are constant in time and serve as estimates of the true time-averaged fluxes.

jumps correctly on uniform grids in otherwise smooth, well resolved flow. This means that PPM uses Eqs. (36)-(38) to update the zone-averaged specific volume V, flow velocity u, and specific total energy E. The remainder of the PPM scheme thus boils down to an algorithm for computing time-averaged fluxes $\bar{u}_L$, $\bar{p}_L$, and $(\overline{up})_L$ of zone volume, momentum, and total energy at the zone interfaces. This computation of the fluxes is illustrated in Fig. 18b. First, the interpolation algorithm from the PPM advection scheme described earlier is used to generate monotone internal zone structures for the variables V, u, and p. These variables are chosen for interpolation because of their close relationship to the variables appearing in the characteristic equations (15). The average Lagrangian sound speeds $\langle C \rangle_{ZL}$ and $\langle C \rangle$ in the zones adjacent to the interface L are then used to estimate the extents $\langle C \rangle_{ZL} \Delta t$ and $\langle C \rangle \Delta t$ of the two domains of dependence L+ and L- indicated in Fig. 18b. The interpolation parabolae are then used to compute average values $\langle V \rangle_{L\pm}$, $\langle u \rangle_{L\pm}$, and $\langle p \rangle_{L\pm}$ of the variables within these spatial domains of dependence. These average values will be used together with the equation of state to determine the two constant states for a Riemann problem. The solution to this Riemann problem is illustrated in Fig. 18b. It yields constant values for u_L and p_L, which will serve as estimates for the true time-averages $\bar{u}_L$ and $\bar{p}_L$. The time average of the product up is then simply

$$(\overline{up})_L = \bar{u}_L \, \bar{p}_L \quad . \tag{55}$$

This procedure effectively replaces the time average of the sound wave interactions at the zone interface by the interaction of the spatial averages over the information carried by the sound waves. It is the job of the Riemann solver to correctly compute this nonlinear interaction and to sort out from the spatially averaged states just that information which the sound waves actually transport to the zone interface.

Before describing the technique used to solve Riemann's problem numerically, we first describe the means of treating a complicated equation of state. A more complicated treatment is described by Colella (1983) with examples of its use for strongly nonlinear flows in air (see also Colella and Glaz 1983). The simpler method here appears to work equally well. It is based upon a model equation of state suggested by J. LeBlanc:

$$p = p_{00} + (\gamma - 1)\, \rho\varepsilon \quad . \tag{56}$$

Using this model equation of state we may derive formulae for the Eulerian and Lagrangian sound speeds c and C:

$$c^2 = (\gamma - 1)(\varepsilon + pV) \quad , \tag{57a}$$

$$C = \rho c \quad . \tag{57b}$$

Here we have used the fact that an adiabatic change $d\varepsilon$ in the specific internal energy is given by $-p\,dV$. The model equation of state (56) is meant to apply only to one zone and only for one timestep. The parameters p_{00} and γ are determined at the beginning of each timestep for each zone by reference to the true, presumably more complicated, equation of state. This true equation of state is asked to compute zone-averaged pressures $\langle p \rangle$ and sound speeds $\langle c \rangle$ from $\langle V \rangle$ and $\langle \varepsilon \rangle = \langle E \rangle - \langle u \rangle^2/2$. Equations (56) and (57a) are then used to determine p_{00} and γ:

$$p_{00} = \langle p \rangle - \frac{\langle \varepsilon \rangle \langle c \rangle^2}{\langle V \rangle (\langle e \rangle + \langle p \rangle \langle V \rangle)} \quad . \tag{58a}$$

$$\gamma = 1 + [\, \langle c \rangle^2 / (\langle \varepsilon \rangle + \langle p \rangle \langle V \rangle)] \quad . \tag{58b}$$

Methods for solving Riemann's problem may be found in fluid dynamics text books such as Courant and Friedrichs (1948). We will only summarize a method here. Positive-definite nonlinear wave speeds $W_\pm$ for discontinuous waves travelling to the right (+) or to the left (-) may be found from the model equation of state and the jump conditions in Eqs. (39), with

$$W_\pm = \pm \rho_0 v_s \quad . \tag{59}$$

These jump conditions also imply a useful relation for the jump in the internal energy:

$$\varepsilon_1 - \varepsilon_0 = -\frac{1}{2}(p_0 + p_1)(V_1 - V_0) \quad . \tag{60}$$

If we approximate rarefaction waves by discontinuous jumps, the solution $\bar{u}_L$ and $\bar{p}_L$ to the Riemann problem illustrated in Fig. 18b satisfies

$$\left(\bar{u}_L - \langle u \rangle_{L\pm}\right) \pm \left(\bar{p}_L - \langle p \rangle_{L\pm}\right) / W_\mp = 0 \quad . \tag{61}$$

The similarity of Eqs. (61) to the characteristic equations (15a) for smooth flow is striking. In fact, in the limit of small jumps between the states L+ and L−, the waves in the solution to Riemann's problem are weak and $W_{\mp}$ approaches $\langle C\rangle_{L\pm}$. In this limit of weak waves Eqs. (61) and (15a) become identical.

The Riemann problem can be solved iteratively by beginning with the assumption of weak waves and replacing $W_{\mp}$ in Eq. (61) with $\langle C\rangle_{L\pm}$. These sound speeds are computed from $\langle V\rangle_{L\pm}$ and $\langle p\rangle_{L\pm}$ by using the equation of state (56) together with Eqs. (57). We may then solve for a first guess $\bar{p}_{L1}$ at $\bar{p}_L$:

$$\bar{p}_{L1} = \langle p\rangle_{L+} + [(\langle p\rangle_{L-} - \langle p\rangle_{L+}) - \langle C\rangle_{L-} (\langle u\rangle_{L-} - \langle u\rangle_{L+})]$$

$$\times \langle C\rangle_{L+} / (\langle C\rangle_{L-} + \langle C\rangle_{L+}) \quad . \tag{62}$$

Using the jump conditions (39a), (39b), and (60) together with the equation of state (56) and Eq. (59), we may derive a formula for the wave speeds $W_{\mp}$ in terms of the pressure $\bar{p}_L$:

$$W_{\mp}^2 = \frac{1}{2\langle V\rangle_{L\pm}} [(\gamma_{L\pm} - 1)(\langle p\rangle_{L\pm} + \bar{p}_L)$$

$$+ 2(\bar{p}_L - p_{00L\pm})] \quad . \tag{63}$$

Substituting our first guess $\bar{p}_{L1}$ for $\bar{p}_L$ in these equations gives first guesses $W_{\mp 1}$ at the wave speeds. These may now be inserted in Eq. (61) in place of $W_{\mp}$ to give two estimates, $\bar{u}_{L\pm 1}$, for $\bar{u}_L$. We will use Newton's method to find the value of $\bar{p}_L$ for which the difference between $\bar{u}_{L\pm}$ vanishes. Usually only a single iteration need be performed to achieve sufficient accuracy for our purposes. Using only the first guess for $\bar{p}_L$ is usually not sufficient.

The Newton iteration for $\bar{p}_L$ is illustrated in Fig. 19. It requires a formula for $\partial\bar{p}_L / \partial\bar{u}_{L\pm}$, as can be seen in the figure. This formula is obtained by differentiating the jump conditions for discontinuous waves:

$$\frac{\partial\bar{p}_L}{\partial\bar{u}_{L\pm}} = \frac{\mp 4\langle V\rangle_{L\pm} W_{\mp}^3}{4\langle V\rangle_{L\pm} W_{\mp}^2 - (\gamma_{L\pm} + 1)(\bar{p}_L - \langle p\rangle_{L\pm})} \quad . \tag{64}$$

Inserting the first guesses $\bar{p}_{L1}$ and $W_{\mp 1}$ in this formula yields $(\partial \bar{p}_L / \partial \bar{u}_{L\pm})_1$. These derivatives give a second guess $\bar{p}_{L2}$ at $\bar{p}_L$, and the iteration can then either be continued or terminated as desired:

$$\bar{p}_{L2} = \bar{p}_{L1} + (\bar{u}_{L-1} - \bar{u}_{L+1}) \left[\frac{(\partial \bar{p}_L / \partial \bar{u}_{L-})_1 \, (\partial \bar{p}_L / \partial \bar{u}_{L+})_1}{(\partial \bar{p}_L / \partial \bar{u}_{L-})_1 - (\partial \bar{p}_L / \partial \bar{u}_{L+})_1} \right] \quad (65)$$

If the iteration is terminated at this point, we estimate $\bar{u}_L$ by

$$\bar{u}_{L2} = \bar{u}_{L+1} + \frac{(\bar{u}_{L-1} - \bar{u}_{L+1}) \, (\partial \bar{p}_L / \partial \bar{u}_{L-})_1}{(\partial \bar{p}_L / \partial \bar{u}_{L-})_1 - (\partial \bar{p}_L / \partial \bar{u}_{L+})_1} \, . \quad (66)$$

The manner in which the PPM scheme treats a strong shock is illustrated in Fig. 20 from Woodward and Colella (1983). In this figure five zones near a strong shock are shown at various stages of the calculation. This calculation uses a uniform Eulerian

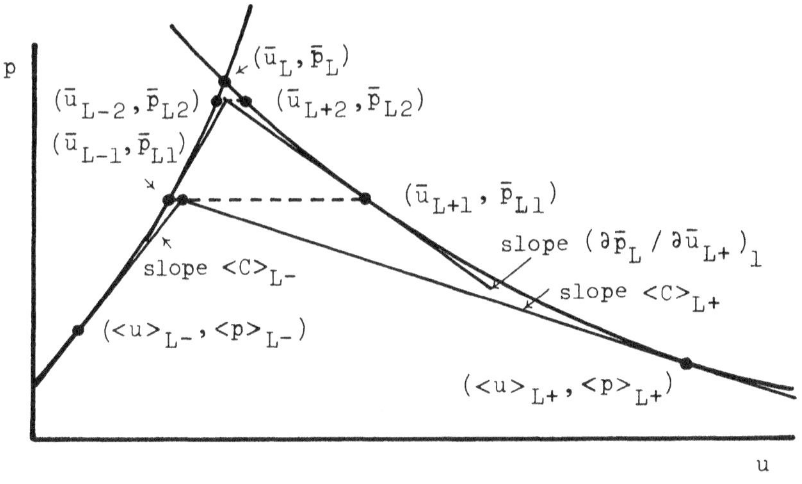

Fig. 19 The Newton iteration for the solution of Riemann's problem in the case involving two strong shocks. The curved lines through the initial states represent all the states which can be connected to the initial states through discontinuous waves.

grid and the shock moves three quarters of a zone width to the right in a timestep. The gas has a gamma law equation of state (Eq. (14)) with $\gamma = 1.4$. Ahead of the shock $u = 0$, $\rho = p = 1$, and behind the shock $u = 3.4056$, $\rho = 4.4091$, and $p = 16$. In Fig. 20a the initial zone-averaged velocities $\langle u \rangle$ are indicated by dotted lines, while the monotonized PPM interpolation parabolae are represented by the solid lines. The shock profile is very steep, with essentially only one zone describing the internal structure of the shock. Figure 20b illustrates the averaging over the domains of dependence for each zone interface in order to generate appropriate Riemann problems. The dotted lines again show the interpolation parabolae, while the solid lines indicate the averages in the domains of dependence. Figure 20c shows the Riemann problems generated in Fig. 20b as dotted lines. The solutions $\bar{u}_L$ to these Riemann problems determine the effective velocity gradients $\partial u/\partial m$ which compress the zones in the Lagrangian step of the calculation. These effective velocity gradients are shown by the solid lines in Fig. 20c. In computing $\bar{u}_L$ the Riemann solver is clearly doing its job of taking values of the Riemann invariant transported by this wave only from the "upstream" domains of dependence. In each case $\bar{u}_L$ is nearly equal to the "upstream" value $\langle u \rangle_{L+}$. This suggests that we might have computed this simple result much more cheaply. In this trivial instance that is indeed true, but the Riemann solver pays for itself when strong shocks interact with other continuous or discontinuous waves. Then reliable short cuts are very difficult to devise. In Fig. 20d the remap step is displayed. The monotonized interpolation parabolae are shown by the solid lines in the Lagrangian zones, which have been compressed and translated to the right. The dotted lines represent the average values of the velocity in the original Eulerian zones. Notice that the shock is once again confined essentially to a single zone.

The ability of the PPM scheme to compute such thin numerical shock structures without introducing post-shock oscillations is critically dependent upon the use of nonotonicity constraints coupled with a nonlinear Riemann solver. Consider the case shown in Fig. 20. In order to avoid overcompressing the second zone from the left in that figure, the interface velocity computed at the right-hand side of that zone must be very nearly equal to the post-shock velocity. This is indeed the case in the calculation, because the monotonicity constraint has generated an interpolation parabola for that zone which is almost flat and also the Riemann solver has essentially chosen the "upstream" velocity from the two values presented to it. As important as avoiding overcompressing the second zone from the left in Fig. 20 is the need to begin compressing the fourth zone. Again, the monotonicity constraint keeps the average velocity large in the right-hand domain of dependence of the third zone, and the Riemann solver

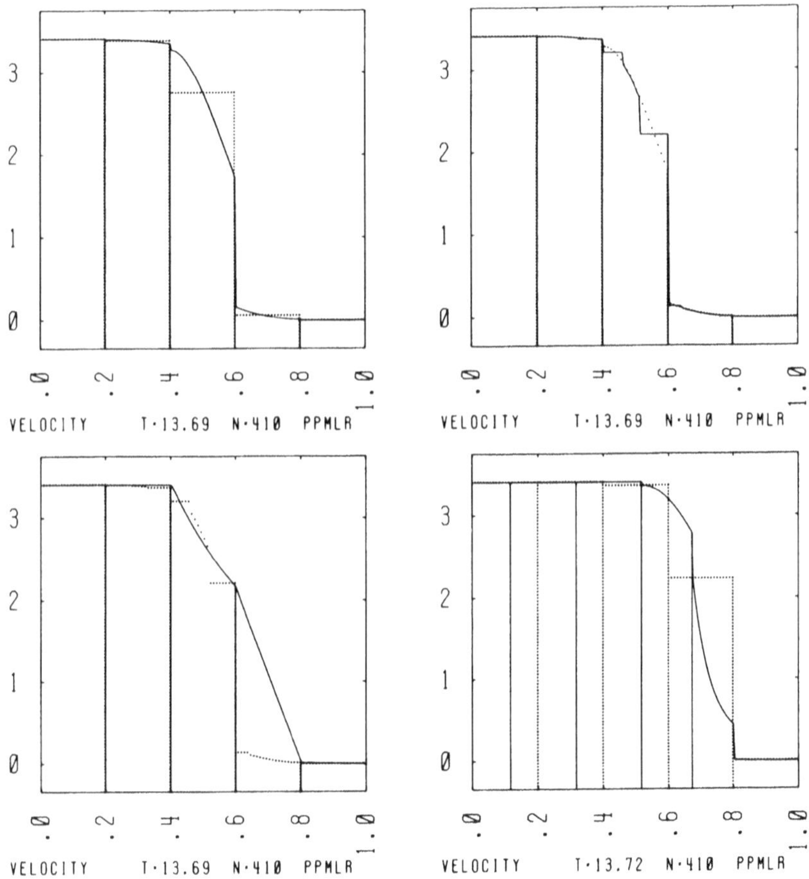

Fig. 20 The propagation of a isolated strong shock on a uniform grid by the PPM scheme. (a) Original zone-averaged data (dotted lines) is used to generate monotone interpolation parabolae (solid lines) within the five zones shown. (b) By averaging over the interpolation parabolae (dotted lines) within the domains of dependence of the zone interfaces, constant states (solid lines) are obtained for Riemann problems. (c) Solution of the Riemann problems (dotted lines) yields interface velocities which imply compressional velocity gradients (solid lines) for the Lagrangian zones. (d) Interpolation parabolae (solid lines) are generated for the compressed Lagrangian zones which yield upon integration the new average velocities in the original Eulerian zones (dotted lines). The shock is again one zone wide after moving 3/4 of a zone width to the right.

chooses this velocity to be assigned to the zone interface. This gives a relatively strong compressional velocity gradient for the fourth zone.

The results in Fig. 20 demonstrate that with the shock moving 3/4 Δx in a single timestep there is very little room for error. The third Eulerian zone is accelerated very nearly to the post-shock velocity in this time step; the slightest readjustment of the fluxes at the zone interfaces could result in an overshoot. One might well ask how other difference methods without monotonicity constraints and Riemann solvers accomplish this feat. Of course, the answer is that they do not accomplish it; but they may nevertheless propagate such strong shocks without oscillations. The means by which this is achieved is an artificial viscosity. The type of artificial viscosity which is most easily described is that proposed by Lapidus (1967). It amounts to adding to each interface flux $\bar{u}_L$, $\bar{p}_L$, and $(\overline{up})_L$ a small diffusive flux $\bar{u}_{DL}$, $\bar{p}_{DL}$, and $(\overline{up})_{DL}$:

$$\bar{u}_{DL} = C_{DL} \left(\langle V \rangle - \langle V \rangle_{ZL} \right) \quad , \tag{67a}$$

$$\bar{p}_{DL} = C_{DL} \left(\langle u \rangle_{ZL} - \langle u \rangle \right) \quad , \tag{67b}$$

$$(\overline{up})_{DL} = C_{DL} \left(\langle E \rangle_{ZL} - \langle E \rangle \right) \quad , \tag{67c}$$

where C_{DL} is a Lagrangian diffusion speed given by

$$C_{DL} = C_{D0} \min \left\{ \langle \rho \rangle_{ZL} \, , \; \langle \rho \rangle \right\} \; X \; \max \left\{ \left(\langle u \rangle_{ZL} - \langle u \rangle \right) \; , \; 0 \right\} \quad . \tag{67d}$$

Here C_{D0} is a constant of order unity. For the example in Fig. 20, the effect of the diffusive flux $\bar{u}_{DL}$ would be to increase $\bar{u}_L$. In this example the monotonicity constraints and the Riemann solver act in this same way. In effect they behave here like a very complicated prescription for computing C_{DL} which gives precisely the desired amount of diffusive flux to prevent osicllations near the shock. It is clear that no simple formula like Eq. (67) with a constant value of C_{D0} will match this performance. Therefore the value of C_{D0} must be chosen high enough to be safe in all situations. This leads to too much diffusion most of the time and much broader shock structures than PPM delivers.

It is generally true that one designs a difference scheme to give shock structures which are as narrow as possible. However, one can overdo this. As is illustrated in Woodward and Colella (1983) and discussed further in Colella and Woodward (1983), difference schemes which produce shocks which are too narrow may be plagued by low level noise behind the shocks, particularly in the entropy distribution. These difficulties show up most strongly in two-dimensional calculations. For PPM this problem arises when a strong shock is nearly aligned with one of the coordinate directions (in 2-D) and when it is nearly stationary with respect to the grid. For such nearly stationary shocks PPM generates extremely narrow structures. Because the fundamental frequency of any numerical noise generated in such narrow shocks is just the frequency with which the shock crosses zone boundaries, nearly stationary shocks tend to generate noise with wavelengths of several zones. The PPM scheme is so accurate in smooth flow that this noise is not effectively damped. In order to eliminate this noise the shock structures must somehow be broadened slightly.

This problem and its solution are illustrated in Fig. 21. The dotted line in the figure displays the interpolated internal zone structures of the density computed by the PPM scheme described above. Twenty zones in the neighborhood of a very strong, nearly stationary shock are shown. The shock is extremely narrow, and the density oscillations behind it are partly due to spurious sound waves travelling away from the shock and partly due to spurious oscillations in the post-shock entropy. In an earlier section we argued that a difference scheme in conservation form must compute proper shock jumps. However, the argument of that section does not apply to the case in Fig. 21 because there is not sufficient dissipation in the PPM scheme to damp the relatively long wavelength oscillations behind the shock in a reasonable distance. The cure for this situation is to prevent their generation within the shock region by broadening the shock slightly. The oscillations are generated by a periodic expansion and contraction of the numerical shock structure as the shock crosses the zones. The shock is broadest when it is centered at the middle of a zone and narrowest when at a zone interface. The oscillations are drastically reduced when the shock is spread by additional dissipative mechanisms so that its shape varies only slightly with its phase relative to the zone boundaries. Such a PPM shock profile is shown by the solid line in Fig. 21.

The extra dissipation needed to give the solid line in Fig. 21 is added to PPM by flattening the interpolated zone structures in the shock region at the outset of both the Lagrangian step and the remap step. A more complicated and more effective dissipation involving jiggling of the computational grid near stationary shocks is discussed in Woodward and Colella (1983) and in Colella and Woodward (1983). This viscosity is "smart"; it recognizes which shock structures may remain thin and which must be broadened. It is also effective; it accomplishes the greatest damping of post-shock noise with the least spreading of the shock structure. However, it is too complicated to be described here.

The dissipation mechanism we will describe broadens all shocks, whether they need it or not, to the point where post-shock oscillations are strongly damped in the worst cases. To accomplish this, the method relies on a measure of the shock steepness, S_{shk}. This measure must be sensitive to the pressure profile, which controls the generation of spurious sound waves behind the shock. The steepness of the pressure profile is measured by comparing narrow- and wide-based differences centered on a particular zone i. Thus we define a steepness parameter S_p by:

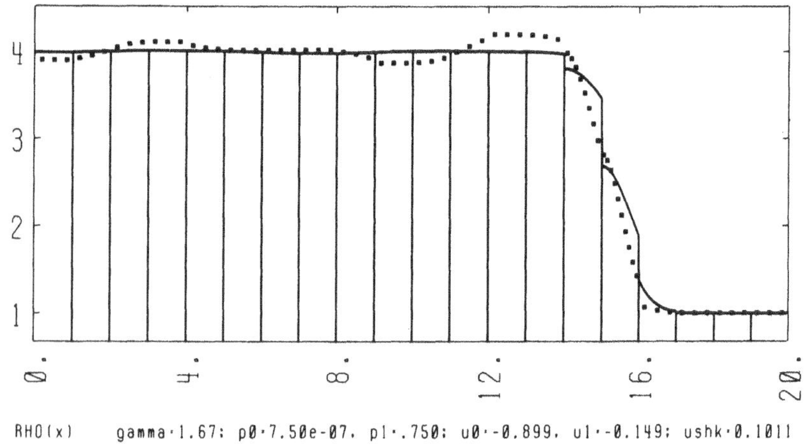

Fig. 21. Nearly stationary structures computed with PPM for a very strong shock on a uniform grid. The preshock (subscript 0) and post-shock (subscript 1) states of this gas with $\gamma = 5/3$ are: $\rho_0 = 1$, $\rho_1 = 4$; $p_0 = 7.5 \times 10^{-7}$, $p_1 = 0.75$; $u_0 = -0.899$, $u_1 = -0.149$. The dotted line shows the interpolated density structures in 20 zones near the shock for the PPM scheme without special shock broadening. The solid line shows these structures as computed by the PPM scheme with the shock broadening mechanism described in the text. This worst-case oscillation has been reduced in amplitude by a factor of 20 by slightly broadening the numerical shock structure as shown.

$$S_p = (\langle p\rangle_{i+1} - \langle p\rangle_{i-1}) \,/\, (\langle p\rangle_{i+2} - \langle p\rangle_{i-2}) \quad . \tag{68a}$$

This parameter is scaled from 0 to 1/2 via

$$S_{01} = \max\{0, \ \min\{0.5, \ 5\,(S_p - 0.75)\}\} \quad . \tag{68b}$$

In well-resolved, smooth flow S_p will have a value very close to 1/2, so it is clear that S_{01} will be nonzero only near sudden jumps in p. The choice of the dimensionless constants in the definition of S_{01} indicates that we are aiming at broadened shock structures with values of S_p around 0.8 to 0.85, values at which PPM gives quite smooth post-shock regions in the worst of cases.

We wish to apply extra dissipation only near shocks. Therefore we multiply S_{01} by a parameter A_{shk} which is unity near shocks and vanishes elsewhere. We wish to ignore shocks with fractional pressure jumps dp_f which are less than 1/3, where

$$dp_f = (p_1 - p_0) \,/\, p_0 \quad . \tag{69}$$

Here the subscripts 0 and 1 refer to the pre- and post-shock states, respectively. Large pressure gradients can be set up without causing shocks if for example gravitational or centrifugal forces are at work. Therefore we examine instead the size of any compressional velocity jump relative to the sound speed. The shock jump conditions allow a threshold in d_{pf} to be related to one in the quantity dM^2 defined by

$$dM^2 = (u_1 - u_0)^2 \,/\, c_0^2 \quad . \tag{70}$$

For a gamma-law gas a threshold dp_{f0} corresponds to a threshold dM_0^2 given by

$$dM_0^2 = 2\,(dp_{f0})^2 \,/\, \{\ \gamma[2\gamma + (\gamma+1)\ dp_{fo}]\ \} \tag{71}$$

For each zone we estimate dM^2 by

$$\langle dM^2\rangle = (\langle u\rangle_{ZR} - \langle u\rangle_{ZL})^2 \,/\, \min\{\ \langle c^2\rangle_{ZL}\ ,\ \ \langle c^2\rangle_{ZR}\ \} \ . \tag{72}$$

Then the parameter A_{shk} is given by

$$A_{shk} = 1 \quad , \quad \text{if} \quad \langle dM^2 \rangle \; > \; dM_0^2$$

$$\text{and} \quad \text{if} \quad (\langle u \rangle_{ZL} - \langle u \rangle_{ZR}) \; > \; 0 \quad , \tag{73}$$

$$= 0 \quad , \quad \text{otherwise.}$$

A value of 1/3 for dp_{f0} in Eq. (71) is recommended. The shock steepness parameter is then

$$S_{shk} = A_{shk} \, S_{01} \quad . \tag{74}$$

To achieve sufficient dissipation it is necessary to take for S_{shk} not this value, but the maximum of this value as computed for the zone in question and for its neighbor on the higher pressure side.

Extra dissipation is achieved in the Lagrangian step by flattening the montonized interpolated zone structures of all variables by a factor $(1 - S_{shk})$. That is, the new structure is a blend of $(1 - S_{shk})$ times the monotonized interpolated structure and S_{shk} times a totally flat structure. This has the effect of blending the PPM scheme with Godunov's first-order scheme near shocks which are too thin. In the remap step the same flattening procedure is used. For purely Lagrangian calculations best results are obtained if the value of S_{shk} is doubled. Some care must be taken when this technique is generalized to 2-D computations. One must be sure to sense the presence of shocks which compress mainly in the y-direction. Thus differences of the transverse component of velocity, u_y, must figure in the definition of $\langle dM^2 \rangle$ in Eq. (72). Also, a numerical approximation of the velocity divergence should replace the velocity difference in Eq. (73). Basically, the flattening factor for a zone should take the larger of the values which would be obtained for the zone in a 1-D x-sweep and a 1-D y-sweep. This will give much needed dissipation along the direction of a shock nearly aligned with the mesh in 2-D. An example of the consequences of omitting such dissipation is given in Fig. 8 of Woodward and Colella (1983).

MULTIFLUID CALCULATIONS WITH PPM

The Multifluid Lagrangian Step

When Eulerian hydrodynamics calculations are performed in a sequence of 1-D passes each of which consists of a Lagrangian step followed by a remap, an especially simple and powerful approach to multifluid hydrodynamics is easily employed. This approach was first used by DeBar in his KRAKEN code in the late 1960's and it is described in DeBar (1974). The idea behind DeBar's technique is to introduce a variable f_j for each fluid j in the problem. This variable is the fraction of the zone volume occupied by fluid j. In the Lagrangian step of the calculation the fluids are treated as well mixed within the zones, while in the remap step interfaces between the fluids are constructed within each zone. An essential feature of this method is its use of only the fractional volumes in the surrounding zones to construct the configuration for the fluids within each zone. This use of only local data makes the method both simple and robust. In particular, the method does not break down when a simply-connected region of fluid becomes multiply connected. Thus the method can follow complicated flows such as the breaking of water waves.

The Lagrangian step of a multifluid calculation in which the fluids are treated as well mixed can be performed in a number of ways. In the original KRAKEN code, the individual equations of state of the fluids were used to compute partial pressures which were summed to give the pressure of the zone. This pressure was used to accelerate the zones, and all fluids in the zone were assigned the same velocity. In KRAKEN an equation for the internal rather than the total energy was solved. This equation was solved separately for each fluid under the assumption that all fluids experienced the same fractional change in volume. Because of the differing compressibilities of the fluids, this last assumption led to inaccuracies in certain situations, and means of equilibrating the partial pressures within multifluid zones were introduced in other, later codes employing the same basic approach as KRAKEN.

We now describe a possible way of performing PPM Lagrangian hydrodynamics for a zone containing two fluids. If we assume that two fluids in a zone are always well mixed and in pressure equilibrium, then for small changes in zone volume this mixed fluid obeys an equation of state which can be derived from those of the individual fluids. If we call the two fluids A and B, with fractional volumes f_A and $f_B = 1 - f_A$, then pressure equilibrium implies that

$$p_A = p_B \quad , \tag{75a}$$

$$dp_A = dp_B \quad . \tag{75b}$$

For small adiabatic changes we must then have

$$c_A^2 \, d\rho_A = c_B^2 \, d\rho_B \quad . \tag{76a}$$

The constancy of the mass fractions of the fluids implies that for small changes

$$\frac{d\rho}{\rho} = \frac{d\rho_A}{\rho_A} + \frac{df_A}{f_A} \quad , \tag{76b}$$

where ρ, the density of the composite fluid, is given by

$$\rho = f_A \rho_A + f_B \rho_B \quad . \tag{77}$$

Differentiating this equation for ρ and using

$$df_B = - \, df_A \quad , \tag{78a}$$

we obtain

$$d\rho = \left(f_A + f_B \, \frac{c_A^2}{c_B^2} \right) d\rho_A + (\rho_A - \rho_B) \, df_A \quad . \tag{78b}$$

We now find

$$\frac{1}{f_B} \, \frac{df_A}{f_A} = \left(\frac{\rho_A c_A^2 - \rho_B c_B^2}{\rho_B c_B^2} \right) \frac{d\rho_A}{\rho_A} \quad . \tag{78c}$$

The definition of the composite sound speed, c, is

$$c^2 = \frac{\partial p}{\partial \rho}\Big|_S \quad , \tag{79}$$

where the derivative is taken at constant entropy. Using the definition of the composite pressure, p,

$$p = f_A p_A + f_B p_B \quad , \tag{80}$$

we find

$$c^2 = f_A\, c_A^2\, \frac{\partial \rho_A}{\partial \rho}\Big|_S + f_B\, c_B^2\, \frac{\partial \rho_B}{\partial \rho}\Big|_S \quad , \tag{81}$$

which leads to

$$c^2 = \frac{\rho_A\, \rho_B\, c_A^2\, c_B^2}{[\rho_A \rho_B\, (f_A^2\, c_B^2 + f_B^2\, c_A^2) + f_A f_B\, (\rho_A^2\, c_A^2 + \rho_B^2\, c_B^2)]} \, . \tag{82}$$

Equations (80) and (82) allow both p and c^2 to be computed, and our model equation of state discussed earlier may be fitted to this data. Lagrangian PPM hydrodynamics may then be computed as for a single fluid, so long as the fractional volumes are adjusted after the computation to achieve precise pressure balance of the two fluids within the zone.

In practice, the method of treating fluids as well mixed within zones for the purpose of Lagrangian hydrodynamical calculations works very well. The reason for this must be that the main falsification introduced appears in the sound speed rather than in the pressure or velocity, which appear directly in the fluxes of specific volume, momentum, and total energy. The pressure and velocity are not greatly affected by assuming that the fluids are well mixed, because these variables are continuous across a fluid interface. In two dimensions, demanding that both fluids share a common velocity makes the treatment of slip along the fluid interface difficult. Any such slip must be smeared over a thin region about two zones wide.

The Multifluid Remap Step

In the remap step of a multifluid Eulerian calculation interfaces separating the various fluids within a zone must be drawn. The SLIC method of Noh and Woodward (1977) is a particularly simple and effective means of drawing these interfaces. It is a simplification of the original method of DeBar (1974), and it is especially well behaved when many fluids are present in a single zone. For the special case of only two fluids in a zone SLIC is just about the simplest possible procedure for defining a fluid interface, and it is remarkable how well it works in practice. There are only five possible fluid configurations, and these are illustrated in Fig. 22.

The SLIC algorithm is specifically designed to work in 1-D passes using a minimum of data. The fluid interface within a zone is constructed using only the fractional volumes of the fluids in that zone and a knowledge of the presence or absence of each fluid in each of the two neighboring zones. Representing presence of the fluid by a 1 and absence by a 0, we may characterize this data in neighboring zones by an ordered pair of numbers for each fluid. The pair (1,0), for example, would mean that the fluid was present only in the neighboring zone on the left. SLIC chooses the horizontal fluid configuration on the left in Fig. 22 whenever these ordered pairs of numbers are the same for both fluids in the zone. If we call the fluids A and B, then the vertical configuration with A on the left will be chosen if we have either (1,0) for fluid A or (0,1) for fluid B. The remaining vertical configuration is the same except that the names of the fluids are interchanged. One of the two center-slab configurations on the right in Fig. 22 will be chosen if the ordered pairs for the two fluids are (0,0) and (1,1). These center-slab configurations occur only rarely in practice. They generally arise when there is a tendency to form spray at a badly distorted fluid interface. They will not be properly advected, and the presence of many such zones in a problem is an indication that the calculation should be terminated. Disregarding these zones, SLIC simply chooses to represent the fluid interface within a zone as horizontal or vertical. This gives a first-order accurate description of the fluid interface. This description is extremely robust; it makes it quite difficult for a fluid region to break up or to form thin structures. This

Fig. 22 The five permitted two-fluid configurations in SLIC for the x-pass of an "operator-split" calculation.

robustness is amply demonstrated by the 10-fluid problem presented in Noh and Woodward (1977).

After the lengthy discussion of advection algorithms at the beginning of this article it should seem natural to apply the PPM advection scheme to the treatment of the fluid interface. If this is done, it should be done only in regions where a single, isolated fluid interface is to be described over a large enough number of zone widths to make the high-order interpolations in PPM meaningful. A means for applying PPM to the problem of fluid interface definition is illustrated in Fig. 23. To construct the representation of the fluid interface which appears at the left in Fig. 23, the fractional volumes of the two fluids have been summed over five-zone strips. The sums over vertical strips have been chosen to define the fluid interface, because viewed from this direction the interface appears more horizontal. The figure illustrates how the fluid interface might appear if the PPM interpolation algorithm with monotonicity constraints were used to construct it from data summed over strips. In the central portion of Fig. 23 this form for the fluid interface has been used within each strip to construct interfaces within zones which match the fractional volumes prescribed there. This representation of the interface would be appropriate for the y-pass of the calculation, because the monotonicity constraint would act to damp strongly any short wavelength ripples on the fluid interface. Such damping serves to prevent the creation of spray, which cannot be adequately treated by the method unless it is well resolved on the computational mesh.

For an x-pass of the calculation the representation in the center of Fig. 23 is inappropriate. The portions of the fluid interface which have been flattened by the monotonicity constraint would now cause an undesirable proliferation of multifluid zones. In these zones the fluid interfaces are drawn nearly horizontal, while the SLIC algorithm would draw them vertical. A possible means of modifying the fluid interface to deal with this problem is illustrated at the right in Fig. 23. In constructing this part of the figure an additional monotonicity constraint has been applied. In each multifluid zone the usual monotonicity criterion has been applied in the y-direction. That is, the interface has been steepened if necessary to give values of the fractional volume in the x-volume coordinate which lies between those for the zones just above and below the zone in question. Note that this constraint is not applied to the data summed over strips but instead to that for individual zones.

The ideas illustrated in Fig. 23 are provisional and are now under development. They do not provide a tested means of improving the SLIC algorithm, but Fig. 23 should warn the ambitious reader that improving significantly upon SLIC is not as easy as at first it seems. It also remains to be shown that improvements in drawing the fluid interface make sense in the absence of a better treament of the hydrodynamics in the Lagrangian step and a better treatment of slip along the interface. A true treatment of fluid slip requires that the calculation not be split into separate x- and y-passes. However, significant improvements can be made by accounting for the shear term, formally third-order small, which was discussed earlier and illustrated in Fig. 5 (J. LeBlanc, private communication).

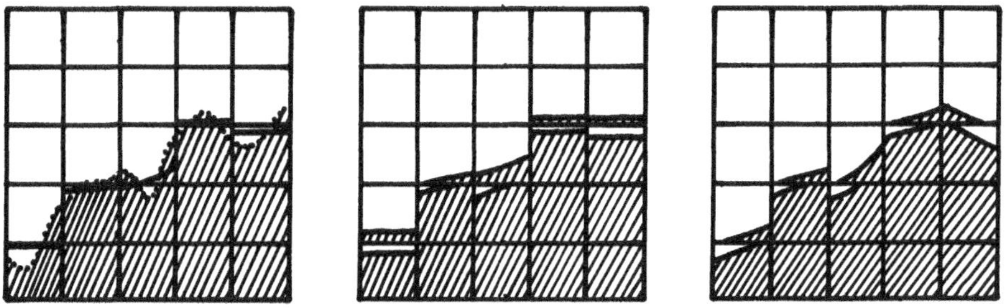

Fig. 23 A proposed piecewise-parabolic embellishment of the SLIC algorithm for defining a fluid interface. (a) Monotonized PPM interpolation parabolae are generated from fractional volume data summed in five-zone strips. The direction of summation is determined so that the fluid interface appears most nearly horizontal. (b) The interface shapes constructed in (a) are adjusted vertically to correspond to fractional volumes in individual zones. This representation is appropriate for advection in the direction of summation in strips (vertical in this example). (c) For advection in the other direction (horizontal in this example) the individual fluid interfaces are steepened if necessary to satisfy a monotonicity constraint in the direction of the strips (vertical). This will prevent excessive proliferation of multifluid zones. The block of 25 zones shown here would be used only to generate the fluid interface within the central zone. Generation of interfaces in other zones is indicated here to give an idea of the variety of situations which can arise.

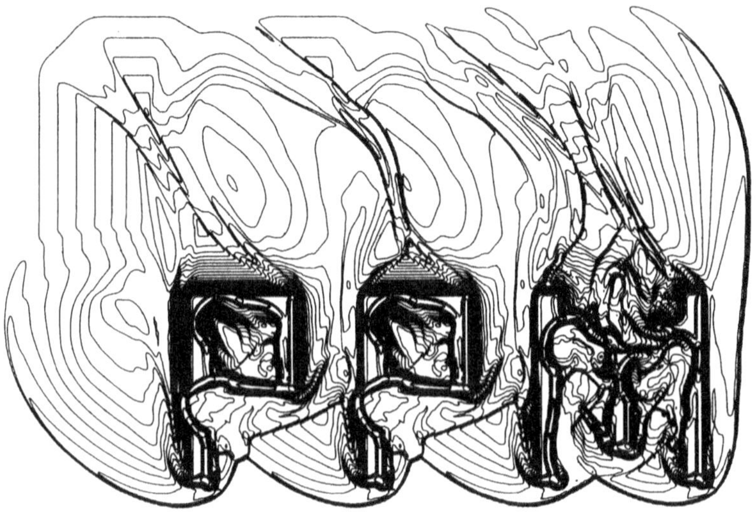

SIMULATION OF UNSTABLE SUPERSONIC JETS WITH PPM

As an example of an astrophysical application of multifluid PPM hydrodynamics with SLIC tracking of fluid interfaces, we present here some results from a study of Kelvin-Helmholtz instabilities in supersonic jets. A detailed discussion of this work may be found in Woodward (1983). An earlier example of the use of SLIC imbedded in a different hydrodynamics code may be found in Woodward (1979, 1980). The motivation for the work on jet instability is the need to establish whether or not simple hydrodynamical models are appropriate for the study of jets emitted from the nuclei of active galaxies. A realistic simulation of a hydrodynamical jet must be three-dimensional; however, such calculations are beyond the capabilities of present computers. Consequently, two-dimensional simulations have been performed. In the very detailed work of Norman et al. (1982, 1983, see also this volume), cylindrical symmetry was assumed, and no catastrophic instabilities were observed. To complement this work, I have performed two-dimensional simulations of supersonic fluid sheets in Cartesian geometry. These simulations, unlike those in cylindrical geometry, permit the "jet" to meander. The meandering, or "firehose" instability is potentially more disruptive than those Kelvin-Helmholtz modes allowed in cylindrical geometry. Therefore it is important to see if this instability grows rapidly for popular hydrodynamic models of jets from active galactic nuclei. It is hoped that the behavior of fluid sheets in two dimensions will be a good indicator of that of roughly cylindrical jets in three dimensions.

It is well known that an isolated slip surface between two uniform fluids is stable when the motion of both fluids relative to a disturbance of the slip surface exeeds a critical Mach number slightly greater than unity. A similar result does not hold for a fluid sheet (or jet) because of the possibility of symbiotic interactions between perturbations at either side of the sheet. Such interactions, which result in a violent instability of the fluid sheet, are illustrated in Fig. 24. The calculation in Fig. 24 began with a sheet of gas with density 0.1, pressure 0.6, and x-velocity 6.32 (Mach 2) located between $y = -0.5$ and $y = 0.5$. This gaseous sheet was initially in pressure equilibrium with a surrounding ambient gas of density 1, pressure 0.6, sound speed 1, and x-velocity 0. Both gases obey a gamma-law equation of state with $\gamma = 5/3$. This system was perturbed by introducing a small vertical component to the velocity within the "jet" (we will use this term loosely from now on):

$$u_y = [0.05 \sin (2\pi x/3) + 0.025 \sin (4\pi x/3)]\, u_{xo}. \qquad (83)$$

The subsequent flow, computed with PPM using a SLIC fluid interface, is shown at times 0.5, 1, 1.5, and 2 in Fig. 24. At each time 30 density contours are plotted with each contour level a fixed factor higher than the preceding one. The lowest and highest levels are given in the figure caption, and the lowest contour is dotted. The fluid boundary as well as the shocks in Fig. 24 are made evident by the bunching together of several density contours. Centered rarefaction fans are marked by regions in which several density contours appear to fan out from a common center. The calculation simulates only a small section of the jet, and periodic boundary conditions are applied at $x = 0$ and $x = 3$. The grid is very fine. There are 180x240 square zones in the region $0 < x < 3$ and $-2 < y < 2$ which is shown. Eighty additional zones above and 80 below this region grow steadily larger so that the y-boundaries are at ±12. This many additional y-grid lines proved to be unnecessary, because the jet was completely disrupted before sound singals could reach these distant boundaries. To demonstrate the correctness of the general flow pattern in Fig. 24, an identical calculation using precisely half the grid resolution in each direction is presented for comparison in Fig. 25. Note that even the amount of entrainment of ambient gas into the jet is the same in these two calculations. Only the fine details of the flow near the jet boundaries require the finer grid for a more accurate description.

Although the initial disturbance given in Eq. (83) has no y-dependence within the jet, the disturbance at time 0.5, in Fig. 24a has quite a different character. It tends to be similar along Mach lines rather than along vertical lines. Mach lines are the sound wave fronts which are generated in supersonic flow by signals emitted at specific points. The oblique features within the jet in Fig. 24a are weak shocks aligned at the Mach angle in the frame of reference moving with the small kinks in the jet boundaries which generate these disturbances. The lack of y-dependence in the initial state was inconsistent with causality, and the flow has rearranged itself accordingly in short order. By time 1, in Fig. 24b, a characteristic and rather special structure has emerged. Nearly all the various weak shocks inside the jet have merged into two principal shocks of greater strength. These shocks, which form a zig-zag pattern inside the jet, are the means by which one jet boundary influences and destabilizes the other.

The flow pattern in Fig. 24b is fundamental to an understanding of supersonic jet instability. Beginning at the left in Fig. 24b, a shock is generated as the supersonic jet gas strikes a kink in the upper wall of the channel. This shock then propagates at roughly the Mach angle to the opposite wall of the channel. Here the shock strikes the wall and generates a dent in

Figure Captions

Fig. 24 The fundamental odd mode of instability of a Mach 2 jet. The jet density is initially 0.1 and the ambient gas has a density of 1. Both are gamma-law gases with $\gamma = 5/3$. They are initially in pressure equilibrium, and the ambient sound speed is unity. Thirty density contours, each larger by a fixed factor, are plotted at times 0.5, 1, 1.5, and 2. A perturbation of the transverse velocity of the jet causes a characteristic zig-zag pattern of shocks within the jet to develop. Eventually some of the jet material is reversed in direction and ambient gas is entrained in the jet. By time 2.5 the jet is completely disrupted. The section of the grid which is shown is uniform, with 60 zones per jet width. The calculation was performed with PPM using a SLIC fluid interface.

Fig. 25 Results of a calculation identical to that in Fig. 24 except for the use of precisely half the grid resolution in both x- and y-directions and half the number of timesteps. The general morphology of the flow in this calculation -- the amplitude of the jet oscillation, the locations of shock fronts, and the amount of entrainment of ambient gas into the jet -- agrees very well with the results for the finer grid in Fig. 24. Only fine details of the flow near the jet boundaries are not well represented here.

Fig. 24

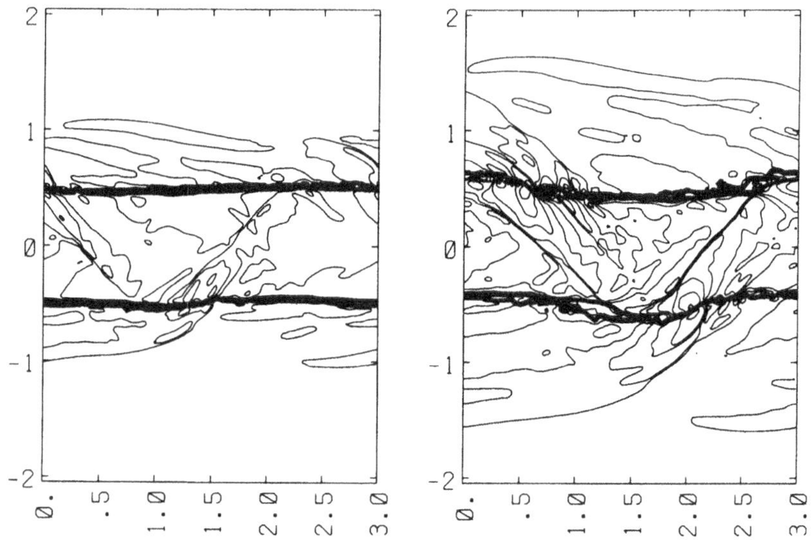

RHO, RHO t=5.02e-01, 1.00e+00; PPMLR 9/27/82- 1 mach2eac, mach2eag
n= 292, 586; 30 contours: 3.87e-02 to 1.54e+00, 3.05e-02 to 2.14e+00

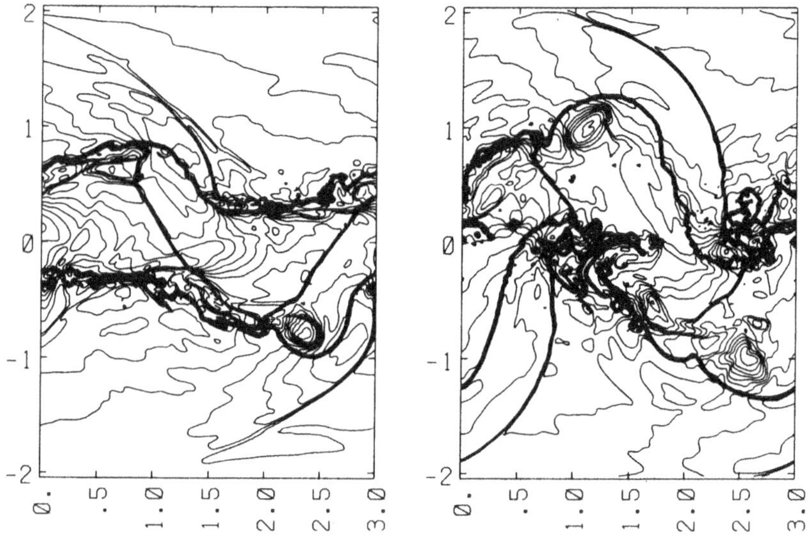

RHO, RHO t=1.50e+00, 2.00e+00; PPMLR 9/27/82- 1 mach2eak, mach2eao
n= 894, 1210; 30 contours: 2.75e-02 to 2.07e+00, 1.44e-02 to 2.20e+00

Fig. 25

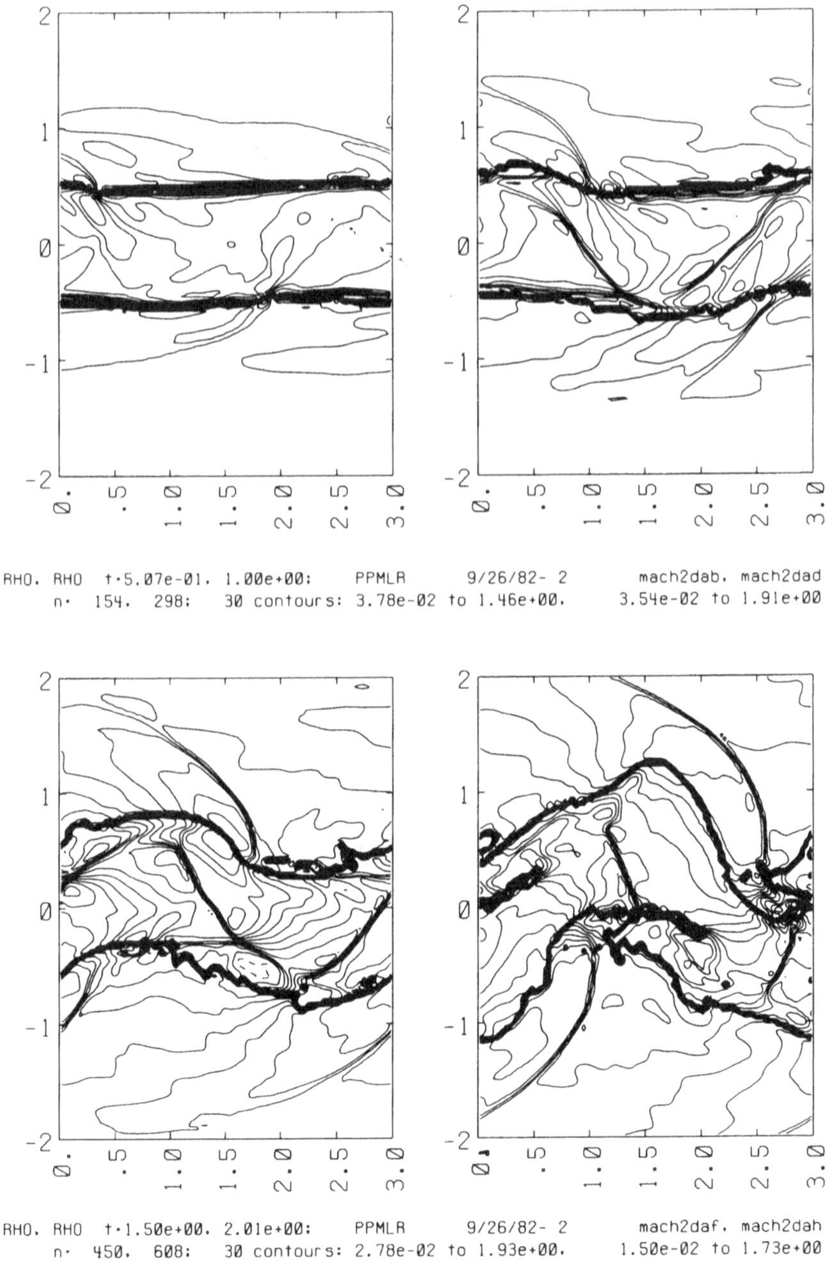
RHO. RHO t·5.07e-01. 1.00e+00: PPMLR 9/26/82- 2 mach2dab. mach2dad
n· 154. 298: 30 contours: 3.78e-02 to 1.46e+00. 3.54e-02 to 1.91e+00
RHO. RHO t·1.50e+00. 2.01e+00: PPMLR 9/26/82- 2 mach2daf. mach2dah
n· 450. 608: 30 contours: 2.78e-02 to 1.93e+00. 1.50e-02 to 1.73e+00

it. The obliquity of the shock is such that the jet gas is turned downward here into the channel wall. This turning is aided by a small centered rarefaction fan which returns the post-shock pressure to the value in the adjacent ambient gas. The jet gas which has been turned into a collison course with the channel wall again causes a shock to form. When this shock reaches the opposite wall, the cycle is completed. The kink on each wall generates a shock which turns the flow into the kink on the opposite wall. When this turned gas strikes the wall, the pressure generated both sustains a shock and further excavates the kink in the wall. The effect of this interaction between the perturbations on the two channel walls is a rapid growth of the amplitude of the wobbling of the jet and of the strength of the imbedded shocks. By time 1.5, in Fig. 24c, some of the jet gas strikes the wall head on and is turned around to flow backwards along the jet boundary. In Fig. 24d, at time 2, the two vortices associated with this reverse flow are very large. Also the shocks inside the jet are nearly orthogonal to the flow. They therefore decelerate and heat the gas considerably. The propagation of the kinks in the jet is supersonic with respect to the ambient gas, and by time 1.5 fairly strong shocks are generated in that gas. In addition, chunks of ambient gas are entrained in the jet so that by time 2.5 the jet is completely disrupted.

Because the instability of the jet in Fig. 24 is driven by a reinforcement of the perturbation on one boundary of the jet by the perturbation on the other, it is natural to expect resonant effects. In fact, if the two perturbations are to reinforce each other the shocks set up on one side of the jet must strike the other side in proper phase with the perturbation there. Because the shocks must be inclined at nearly the Mach angle, there must therefore be a resonance condition involving the Mach number of the jet, the jet width, and the wavelength of the perturbation. These parameters are near resonance for the jet in Fig. 24. This particular resonant mode corresponds to a single zig-zag of shocks within the jet which are inclinded at roughly the Mach angle. This is the fundamental odd mode, the resonant mode which has the longest wavelength, the fastest growth rate, and the most destructive effect upon the jet. It is a wobbling mode with alternating symmetrical regions of positive and negative velocity in the y-direction.

A whole series of modes can be constructed by drawing additional shock zig-zags spaced evenly in phase. Odd numbers of zig-zags give odd, wobbling modes, while even numbers of zig-zags give even, pinching modes. Even modes have even symmetry for the density distribution about the jet center line, and they are therefore the modes which appear when this line is treated as a reflecting boundary (as is the case when cylindrical symmetry is imposed). The fundamental even mode is shown in Fig. 26. The initial, unperturbed jet for this calculation was identical to that used in the run in Fig. 24. However, in this case reflection symmetry about the jet center line at $y = 0$ was assumed. Also a sinusoidal perturbation of u_y within the jet was introduced with only half the wavelength of the fundamental odd mode:

$$u_y = 0.1 \quad y \quad u_{xo} \quad \sin(4\pi x/3) \quad . \tag{84}$$

The double zig-zag pattern which is characteristic of the fundamental even mode of jet oscillation is modified in Fig. 26 by the Mach reflections of the shocks at the jet center. For higher Mach numbers these become regular shock reflections, and a simple double zig-zag pattern of shocks within the jet appears. The grid used for the calculation in Fig. 26 is matched to that in Fig. 24. There are 90x120 square zones in the region $0 < x < 1.5$ and $0 < y < 2$, and 40 additional grid lines of increasing separation carry the computational domain out to $y = 5$. The flow is shown in Fig. 26 at time intervals of 0.5 beginning at $t = 0.5$. This mode rapidly causes breaking waves to form along the jet boundary and by time 2 a considerable amount of ambient gas has been entrained in the jet; however, this mode is still less disruptive than the fundamental odd mode in Fig. 24.

As can be seen at the earlier times displayed in Fig. 26, it is possible to draw a tighter shock zig-zig pattern within the jet at the Mach angle. Thus the fundamental even mode could be generated at higher frequencies; however, at higher frequencies its growth is less rapid and its effects are less disruptive of the jet. This is demonstrated in Fig. 27a. In this calculation the same even perturbation of the jet as in Eq. (84) but with half the wavelength was introduced. The flow is shown at time 1.5 in Fig. 27a, and the characteristic double zig-zig pattern of internal shocks is evident. During the evolution of this run the fundamental even mode shown in Fig. 27a competes with its first harmonic. As the instability grows and the effective jet width is reduced slightly by the entrainment of ambient gas, it becomes possible to fit the double zig-zag within the imposed wavelength of 3/4 of the jet width. Then this fundamental even mode grows in strength to dominate the flow. To obtain the second even mode, with a quadruple zig-zag pattern of internal shocks, we must decrease the wavelength of the perturbation further. In the calculation shown at time 1.5 in Fig. 27b, a wavelength of 0.5 has been imposed:

$$u_y = 0.1 \quad y \quad u_{xo} \quad \sin(4\pi x) \quad . \tag{85}$$

Fig. 26

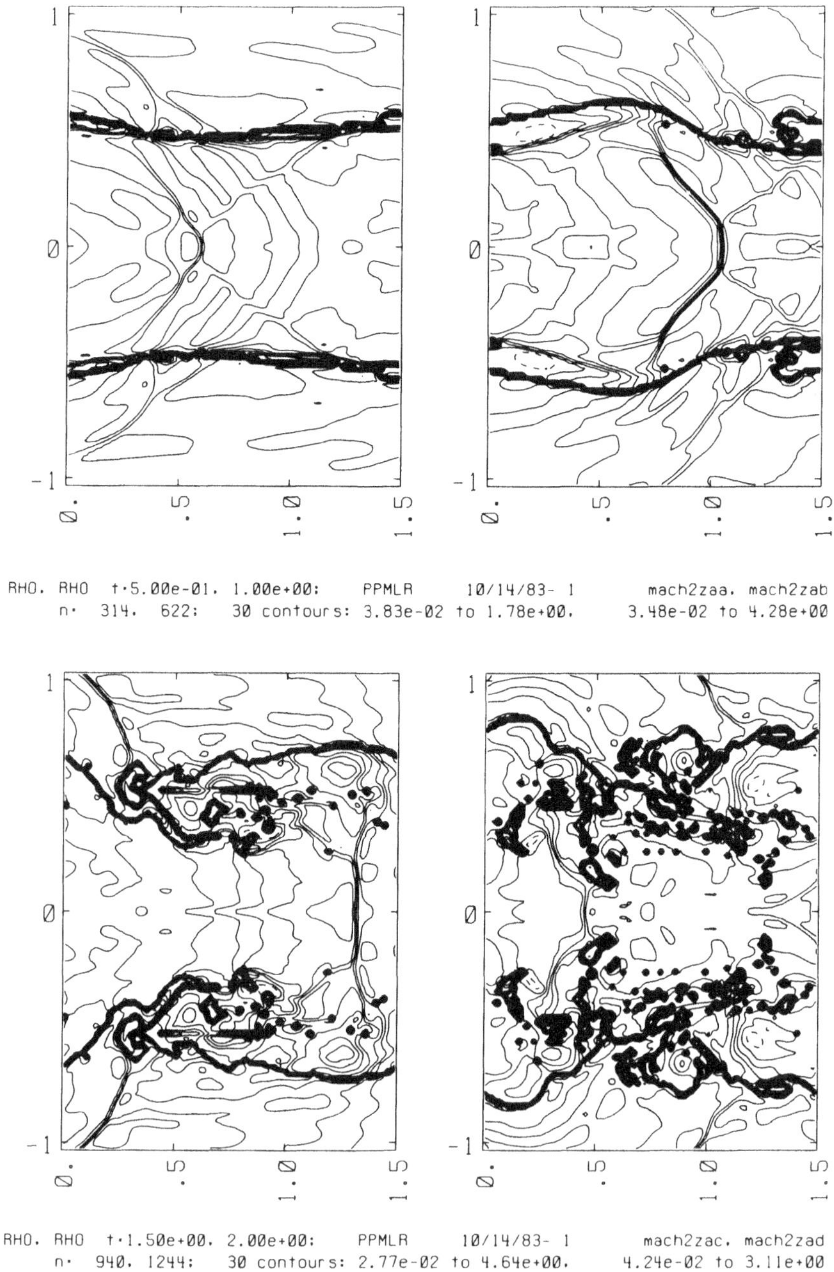
1
0
-1
0.
.5
1.0
1.5
RHO, RHO t=5.00e-01, 1.00e+00: PPMLR 10/14/83- 1 mach2zaa, mach2zab
n= 314, 622; 30 contours: 3.83e-02 to 1.78e+00, 3.48e-02 to 4.28e+00
RHO, RHO t=1.50e+00, 2.00e+00: PPMLR 10/14/83- 1 mach2zac, mach2zad
n= 940, 1244; 30 contours: 2.77e-02 to 4.64e+00, 4.24e-02 to 3.11e+00

Figure Captions

Fig. 26 The fundamental even mode of instability of the Mach 2 jet in Fig. 24. This PPM computation uses a uniform grid within the region shown, with 60 zones per jet width as in Fig. 24. Density contours are shown at times 0.5, 1, 1.5, and 2. The growth of this resonant pinching mode is very rapid. The characteristic x-pattern of internal shocks appears by time 0.5, and the wave disturbances at the jet boundary break by time 1. Nevertheless, this mode is less disruptive than the "firehose" mode shown in Fig. 24. Note the Mach reflection of the internal shocks at the jet center line.

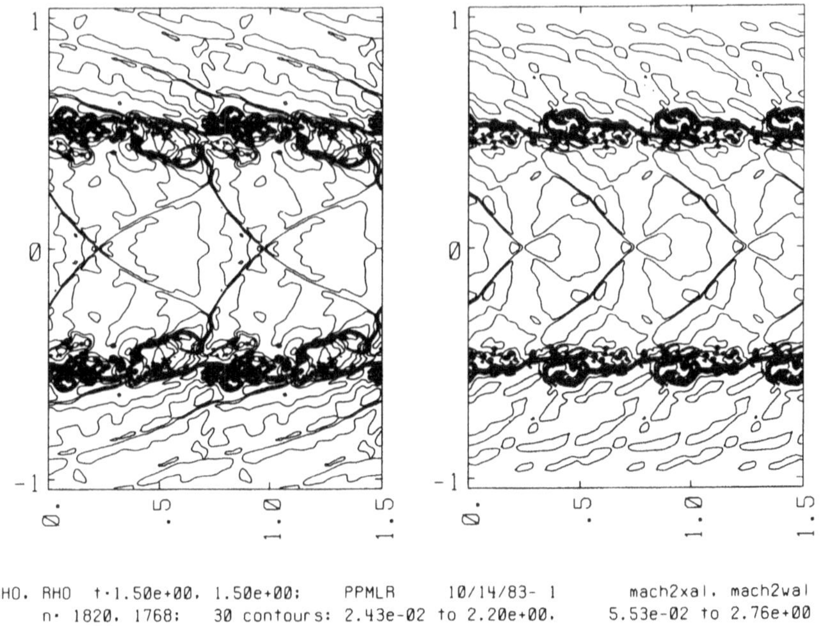

Fig. 27 The effects of short wavelength perturbations of even symmetry about the jet center line are indicated by these two PPM simulations. The unperturbed jets here are the same as in Fig. 24. In both simulations the grids are uniform in the region shown, with 120 zones per jet width, and both flows are shown at time 1.5. At the left the perturbation wavelength is 0.75 and at the right it is 0.5 jet widths. At the left a competition between the fundamental even mode of Fig. 26 and its first harmonic is eventually won by the former mode; at the right, the latter, harmonic mode is dominant.

The second even mode now appears clearly, but despite its rapid ruffling of the jet boundary it is less disruptive than the fundamental even mode shown in Fig. 27a. The relatively minor effect of this mode on the flow near the center of the jet can be gauged by the weakness of the shocks which it sets up there.

In Fig. 28 the second odd mode of jet instability can be seen. The calculation in Fig. 28 is identical to that in Fig. 24 except in two respects. First, the grid is somewhat coarser. There are 120x160 square zones in the region $0 < x < 3$ and $-2 < y < 2$. Twenty y-grid lines of increasing separation on either side of this region carry the computational domain out to $y = \pm 4$. The second difference between this calculation and that in Fig. 24 is of course the initial perturbation:

$$u_{yo} = 0.05 \, [\sin(2\pi x) + \sin(8\pi x/3)] \, u_{xo} \quad . \tag{86}$$

Initially the third resonant mode represented by the first term in u_{yo} grows to produce at time 1.25 the shock lattice structure shown in Fig. 28a. However, nonlinear coupling of this mode with the fourth one (the second term in u_{yo}) soon causes one of the zig-zag shocks to overwhelm the others, and the fundamental odd mode emerges at time 2.25 in Fig. 28b. By time 3 entrainment of ambient gas in the jet disrupts it considerably.

The experiment in mode-mode coupling represented in Fig. 28 demonstrates the dominance of the fundamental odd mode. In this case it has arisen as the difference frequency oscillation corresponding to the two shorter wavelength initial perturbations. That this behavior is not just a general tendency for ever longer wavelength oscillations to grow is demonstrated by two further experiments. The flow in Fig. 29 began with the same initial conditions as that in Fig. 28, except that the initial perturbations had longer wavelengths by a factor of 4:

$$u_{yo} = 0.05 \, [\sin(\pi x/2) + \sin(2\pi x/3)] \, u_{xo} \quad . \tag{87}$$

The grid has 360x120 square zones in the region $0 < x < 12$ and $-2 < y < 2$, with 10 additional y-grid lines of increasing separation expanding the computational domain to $y = \pm 2.5$. The flow is displayed at time 2, and the mode in Fig. 24 is clearly dominant. There is barely any hint of power at the difference frequency, corresponding to a wavelength of 12. In Fig. 30 a similar run is shown at time 3. Here the initial perturbation contained a wavelength of 12:

$$u_{yo} = [0.06 \sin(\pi x/6) + 0.02 \sin(\pi x/2)]\, u_{xo} \quad . \tag{88}$$

A somewhat coarser grid in x of 240 evenly spaced x-lines was used, while a somewhat finer grid in y of 160 evenly spaced y-lines was used from y = -2 to y = 2. On either side of this domain forty additional y-lines of increasing separation carry the grid to y = ±6. In the flow at time 3 the long wavelength initial perturbation has only modulated a pattern dominated by the shorter wavelength.

These simulations suggest that if a meandering supersonic jet has a principal wavelength discernable, this wavelength can be related to the other parameters of the jet by assuming that the

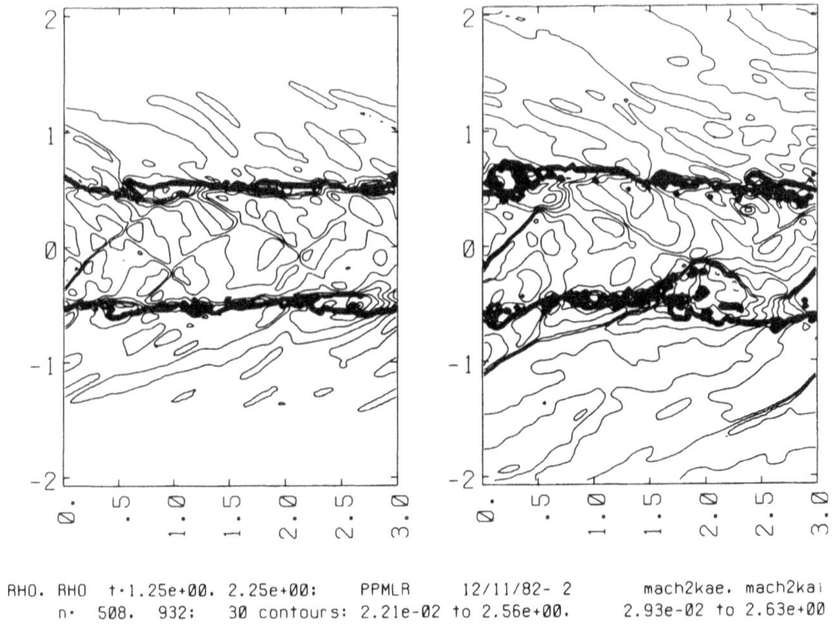

Fig. 28 The second resonant odd mode of instability of a Mach 2 jet. The same jet as in Fig. 24 has here been perturbed with disturbances with 3 and 4 wavelengths over the length of the grid. At first, the longer wavelength sets up an unstable resonant oscillation characterized by the triple zig-zag of shocks within the jet which is seen at the left (time 1.25). Later (time 2.25) at the right, nonlinear coupling between the initial disturbances has excited the dominant, fundamental odd mode.

fundamental odd resonant mode has been observed. This mode is so violently unstable that if it is not seen, the jet is unlikely to be a simple hydrodynamic jet of the type studied here. Instead, it may contain a stabilizing internal magnetic field. Alternatively it may be stabilized by a strong pressure gradient in the external ambient gas which gives the jet a strongly preferred direction of propagation and resists meandering (rivers flowing down hills do not meander). Another possibility is that the jet Mach number is so large (of the order of 100) that it gets where it's going before the instability can develop. Remember that the resonant wavelength scales linearly with the jet Mach number when this is large (because of its relation to the Mach angle).

In the supersonic exhausts of jet engines the fundamental even mode is usually dominant for some time, but the fundamental odd mode, a helical mode for a cylindrical jet, eventually takes over (cf. Yu and Steiner 1983). The reason for the initial dominance of the even mode is an enormous even perturbation of the jet caused by its sudden emergence into a pressure reservoir to which it must adjust by either a lateral expansion or contraction. Norman et al. (1982) have suggested that hot spots of synchrotron radiation emitted from astrophysical jets may be due to

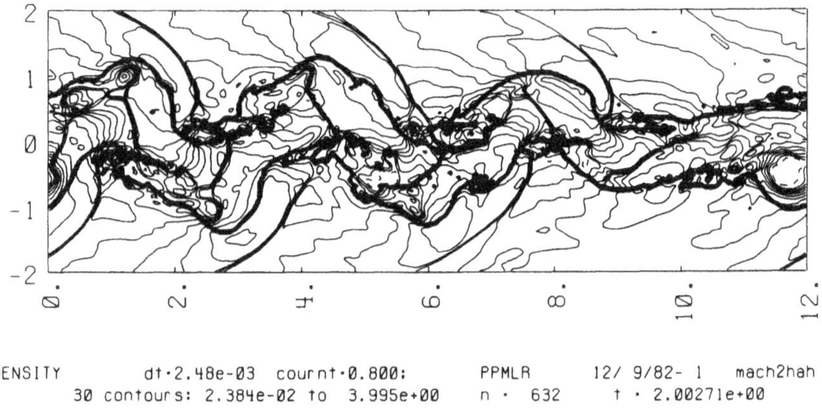

Fig. 29 The same initial jet as in Fig. 24 here has been perturbed by 5% sinusoidal oscillations of the transverse velocity with wavelengths of 3 and 4 jet widths. Because the shorter wavelength initial perturbation is near the fundamental resonant mode shown in Fig. 24, this mode grows so rapidly that it disrupts the jet before nonlinear mode-mode coupling can introduce any significant oscillation with a wavelength of 12 jet widths. Density contours are shown here at time 2, shortly before the jet is disrupted.

internal shocks of the type they observe in their calculations. If such hot spots correspond to the even modes in their calculations, perhaps the astrophysical jets, like their terrestrial counterparts, have suddenly emerged from a galaxy into an intergalactic pressure reservoir to which they adjust by exciting strong even modes of oscillation.

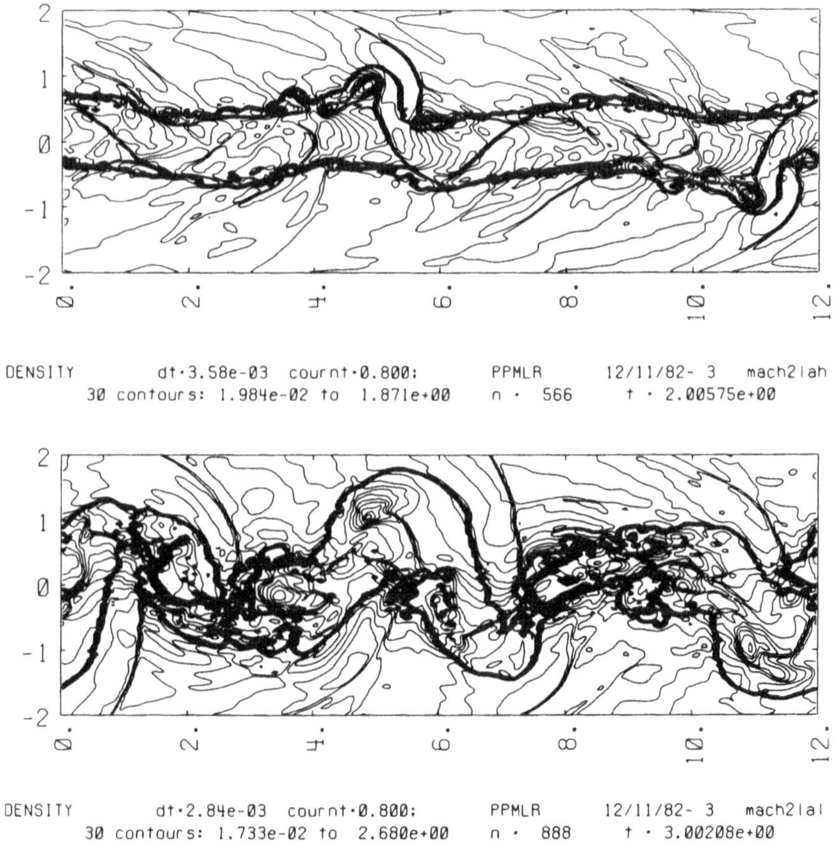

Fig. 30 The same initial jet as in Fig. 24 here has been perturbed by 6% and 2% sinusoidal oscillations in the transverse velocity with wavelengths of 12 and 4 jet widths. Despite its smaller initial amplitude, the shorter wavelength perturbation soon dominates the flow. Density contours are shown here at times 2 and 3. By time 3 the jet is nearly disrupted by a short wavelength oscillation. Although it is not immediately apparent, nonlinear effects have generated a large amount of the fundamental resonant mode with a wavelength of 3 jet widths.

The PPM hydrodynamics scheme with SLIC fluid interfaces is a powerful tool for investigating fluid flows. As the above investigation of jet instability demonstrates, nonlinear flow phenomena are easily and accurately computed which can then be understood in relatively simple terms. Although only one case may be computed at a time, the calculations are cheap enough that many can be afforded. This allows similarities and trends to be identified which may be used to predict the outcome of further simulations in order to test the correctness of their interpretation. From this process a phenomenological understanding emerges, which can be made quantitative as desired by performing additional simulations. This procedure allows theoretical work to be extended into regimes which are intractible for analytic methods.

The contrast between this numerical approach and analytic techniques is illustrated by the problem of jet instability. The numerical work sketched above and discussed more fully in Woodward (1983) should be compared with the analytic work of Ferrari et al. (1978) and of Hardee (1982, 1983). In that work long formulae involving special functions and graphs of growth rates are presented, but the role of resonant effects mediated by shocks at the Mach angle in defining the dominant mode in a given situation is obscured. Undoubtedly this is largely because such shocks, even when they are weak, are inherently nonlinear phenomena. Another reason that the Mach angle does not emerge from the linear analysis is that the modes studied for supersonic jets are not constrained to be causally meaningful. Flows which may actually be realized in meaningful experiments must be obtained by a superposition of a great many of the modes studied analytically. Thus although one may conclude from the analytic work that a supersonic jet is unstable, he may not learn what that jet will actually do in nature.

The numerical and analytic approaches can reinforce one another. The phenomenological understanding which emerges from a series of numerical simulations can be used to guide analytic work in profitable directions. As every student knows, it is helpful to know the answer before beginning the analysis of a problem. Numerical simulations, like physical experiments, make this possible. Consider the example of supersonic jet instability. In principle, we could find suitable trial perturbations for an analytic treatment by the following method. We could idealize the results of the simulations at the point where a principal mode has been established but where the shocks involved are still weak. This idealization would serve to remove unavoidable small numerical noise from the computed result and also to yield a perturbation simple enough to treat analytically. If growth rates for such perturbations could be computed analytically as functions of the jet parameters, the results would be most useful. As the power of computing machinery and techniques continues to grow we may hope to see an accompanying growth of such interplay of the numerical and analytical approaches to theoretical physics.

Summary

Piecewise-parabolic methods for advection and for hydrodynamics have been discussed thoroughly here outside of the context of applied mathematics from which they have grown. This is not because other methods are without merit, but instead it is the result of the author's personal preference. Nevertheless, many of the fundamental issues discussed -- construction of interpolation functions, monotonicity constraints, contact discontinuity steepeners, conservation form, operator splitting, multifluid techniques, etc. -- are more universal than the limited context in which they have been discussed here. The author hopes that readers who do not intend to use piecewise-parabolic methods may still find much to interest them in these pages.

This work was performed under the auspices of the U. S. Department of Energy by the Lawrence Livermore National Laboratory under contract No. W-7405-ENG-48.

REFERENCES

Berger, M.: 1982, "Adaptive Mesh Refinement for Hyperbolic Partial Differential Equations," Thesis, Stanford University.

Boris, J. P., and Book, D. L.: 1973, J. Comp. Phys. 40, pp. 202.

Colella, P.: 1983, "Approximate Solution of the Riemann Problem for Real Gases," Lawrence Berkeley Laboratory Report LBL-14442, to appear in J. Comp. Phys.

Colella, P., and Woodward, P. R.: 1983, "The Piecewise-Parabolic Method (PPM) for Gas Dynamical Simulations," J. Comp. Phys., in press.

Courant, R., and Friedrichs, K. O.: 1948, Supersonic Flow and Shock Waves, New York: Interscience.

DeBar, R.: 1974, "Fundamentals of the KRAKEN code," Lawrence Livermore National Laboratory Report UCIR-760.

Ferrari, A., Trussoni, E., and Zaninetti, L.: 1978, Astron. Astrophys. 64, pp. 43.

Godunov, S. K.: 1959, Nat. Sb. 47, pp. 271.

Gropp, W. D.: 1980, SIAM J. Sci. Stat. Computing 1, pp. 191.

Hardee, P. E.: 1982, Ap. J. 257, pp. 509.

Hardee, P. E.: 1983, Ap. J. 269, pp. 94.

Noh, W. F., Gee, M., and Kramer, G.: 1979, Lawrence Livermore National Laboratory Report UCID-18515.

Noh, W. F., and Woodward, P. R.: 1977, Lecture Notes in Physics 59, pp. 330.

Norman, M. L., Smarr, L., and Winkler, K.-H. A.: 1982, Astron. Astrophys. 113, pp. 285.

Norman, M. L., Winkler, K.-H. A., and Smarr, L.: 1983, to appear in the proceedings of the conference on astrophysical jets, Torino, Italy, October, 1982.

Strang, G.: 1968, SIAM J. Num. Anal 5, pp. 506.

van Leer, B.: 1977, J. Comp. Phys. 23, pp. 276.

van Leer, B.: 1979, J. Comp. Phys. 32, pp. 101.

Woodward, P. R.: 1979, in The Large-Scale Characteristics of the Galaxy, ed. W. B. Burton, I.A.U. Symp. No. 84, pp. 159.

Woodward, P. R.: 1980, in Early Solar System Processes and the Present Solar System, ed. D. Lal, Proc. Int. Sch. Phys. "Enrico Fermi," Course 73, pp. 1.

Woodward, P. R.: 1982, in Parallel Computations, ed. G. Rodrigue, Academic Press, pp. 153.

Woodward, P. R.: 1983, "Kelvin-Helmholtz Instability of Pressure-Confined, Supersonic Jets," Lawrence Livermore National Laboratory report, in preparation.

Woodward, P. R., and Colella, P.: 1981, Lecture Notes in Physics 141, pp. 434.

Woodward, P. R., and Colella, P.: 1983, J. Comp. Phys., in press.

Woodward, P. R., and White, R. L.: 1983, "The Piecewise-Parabolic Boltzmann Scheme (PPB)," Lawrence Livermore National Laboratory report, in preparation.

Yu, J. C., and Steiner, J. M.: 1983, Proc. A.I.A.A. Eighth Aero-acoustics Conf., Atlanta, April, 1983.

FINITE ELEMENT METHODS FOR TIME DEPENDENT PROBLEMS

David F. Griffiths

Department of Mathematical Sciences
University of Dundee, Scotland

1. INTRODUCTION

Finite element methods have, from their origins in structural engineering in the mid 1950's, grown to form a large and important part of the toolkit of methods for solving partial differential equations. Their range of application has expanded inexorably so that they now pervade almost all areas. In certain applications, such as time dependent and fluids problems, finite elements have been in use for only a relatively short period of time and one can therefore expect further developments to take place to improve their overall efficiency.

By the term finite element method (FEM) we shall mean the Galerkin formulation or its extension, the Petrov-Galerkin (PG-) FEM though several other variations are well established, for example the Ritz, Mixed and Hybrid FEMs. Our aim is to give a description of the basic method along with some of its pertinent properties. The main attributes of the FEM are its optimality in energy (though this has often to be sacrificed) its ability to treat very general regions and the relative ease with which the order of approximation may be changed. The mathematical theory of FEMs is also much better developed than for finite difference methods, though we will not really touch on this aspect.

The essential idea underlying the FEM for elliptic equations is a development of the classical Galerkin method. For example, to solve the Poisson equation

$$-\nabla^2 u = f \quad \text{in} \quad \Omega \tag{1.1}$$

K.-H. A. Winkler and M. L. Norman (eds.), Astrophysical Radiation Hydrodynamics, 327–357.

with u = 0 on the boundary $\partial\Omega$ by the Galerkin method, let $\{\phi_1,\phi_2,..,\phi_N\}$ denote a set of trial functions each of which vanish on the boundary. The solution u is approximated by the function

$$U = \sum_{j=1}^{N} a_j\phi_j \tag{1.2}$$

and the coefficients $\{a_j\}$ computed by insisting that the residual function $r = -\nabla^2 U - f$ be orthogonal to each of the trial functions:

$$\int_\Omega \phi_i r \, d\Omega = 0, \qquad i = 1,2,..,N. \tag{1.3}$$

Now provided that the trial functions are sufficiently smooth, (1.3) provides an N×N system of equations

$$S\underline{a} = \underline{F} \tag{1.4}$$

where the entries of the stiffness matrix S and source term $\underline{F}$ are given by

$$S_{ij} = -\int_\Omega \phi_i \nabla^2 \phi_j \, d\Omega \qquad \text{and} \qquad F_i = \int_\Omega \phi_i f \, d\Omega.$$

The matrix S will be positive definite so long as the trial functions are linearly independent. It will also be full thus precluding the use of specialised solution techniques. The success of the Galerkin FEM depends on the following modifications.

i) Instead of using trial functions which are generally non-vanishing on the whole of Ω, the region is divided into a large number of small non-overlapping subregions - called elements - and each trial function is defined in such a way as to vanish on all but a small number of elements. The (i,j) entry of S will therefore be zero unless ϕ_i and ϕ_j do not vanish simultaneously on at least one element. In this way the matrix S will contain a large proportion of zeros and allow the use of sparse matrix techniques.

ii) Employing Greens Theorem in the definition of S_{ij} leads to

$$S_{ij} = \int_\Omega \underline{\nabla}\phi_i . \underline{\nabla}\phi_j \, d\Omega \quad . \tag{1.5}$$

Consequently only first derivatives of the trial functions need be computed. Of course these functions must be smooth enough to ensure that the integrals in (1.5) are finite. In this case it is sufficient that the trial functions be continuous on Ω.

If V is any other combination of trial functions as in (1.2), then the GFEM solution U satisfies

$$\int_\Omega \{\underline{\nabla}(U-u)\}^2 \, d\Omega \leq \int_\Omega \{\underline{\nabla}(V-u)\}^2 \, d\Omega$$

leading to the assertion that U is the best fit to u in the 'energy' norm

$$\|v\|_E = \{\int_\Omega (\underline{\nabla} v)^2 \, d\Omega\}^{\frac{1}{2}} . \tag{1.6}$$

The extension of these ideas to time dependent problems is straightforward. For the heat equation

$$u_t = \nabla^2 u + f \quad \text{in} \quad \Omega\times\{t>0\} \tag{1.7}$$

the coefficients $\{a_i\}$ in (1.2) are allowed to vary with time. The residual now becomes $r = U_t - \nabla^2 U + f$ and this leads, via (1.3), to the system of ordinary differential equations

$$\underline{M}\dot{\underline{a}} + \underline{S}\underline{a} = \underline{F} \tag{1.8}$$

where the dot denotes differentiation with respect to t, S remains as before and the mass matrix M has

$$M_{ij} = \int_\Omega \phi_i \phi_j \, d\Omega. \tag{1.9}$$

In practice the element shapes are either triangles or quadrilaterals and the trial functions designed to be polynomials on each element. The trial functions must have a basic approximation property: it must be possible, for each t, to determine coefficients so that $U \to u$ ans $N \to \infty$, where convergence is measured in the energy norm (1.6) (the limit process $N \to \infty$ is associated with $h \to 0$ where h is the diameter of the largest element). In most instances the trial functions are constructed so that the coefficients in (1.2) have a physical interpretation. Typically, a_i would refer to the value of U or one of its derivatives at a specific point(node).

In the next two sections we describe, respectively, the spatial discretisation of parabolic equations in one and two space dimensions. The discretisation process is completed in section 4 with a finite difference method for time integration. Section 5 is devoted to application of the FEM to first order hyperbolic equations. The main feature here is the introduction of the Petrov-Galerkin formulation which, unlike the GFEM, can allow a certain amount of diffusion to be incorporated into the approximation. Finally, in section 6, we describe the so-called Moving Finite Element (MFE), proposed in (9), in which the grid is moved in time to track the principal features of the solution.

For a more comprehensive introduction to the FEM the reader should consult one of the numerous texts on the subject (see for instance (25), (31)), whilst for specific application to the Navier-Stokes equations, see (10), (32) (which are fairly mathematical in nature) or (33).

2. THE FEM IN ONE SPACE DIMENSION

In this section we shall look in greater detail into the steps involved in constructing the discrete equations (1.8) by considering the simple model of diffusion - convection

$$u_t + \lambda u_x = \varepsilon u_{xx} + f, \quad 0 < x < 1, \ (\varepsilon > 0) \tag{2.1}$$

with boundary conditions,

$$u(0,t) = \alpha, \quad u_x(1,t) + \beta u(1,t) = \gamma \tag{2.2}$$

and initial condition

$$u(x,0) = u_0(x), \qquad 0 < x < 1 . \tag{2.3}$$

The coefficients λ and ε will be assumed constant though little additional complexity would be introduced if they were allowed to vary or, indeed, depend on u.

The first stage is to transform the problem into the more convenient weak form. Let v(x) be any smooth function (called a test function). Equation (2.1) is multiplied by v and the result integrated with respect to x. The term involving second derivatives is integrated by parts to give

$$\int_0^1 \{v(u_t + \lambda u_x) + \varepsilon v_x u_x - vf\} dx - \varepsilon v u_x \Big|_0^1 = 0 . \tag{2.4}$$

Since u satisfies what is known as an essential boundary condition at x=0, we insist that v(0) = 0 (for a problem involving derivatives of order 2m, m partial integration would be necessary. Any boundary condition specifying the jth derivative (j<m) of u in terms of lower derivatives would be an essential boundary condition and would necessitate the vanishing of the jth derivative of v at the point). Now, with the definitions of inner product

$$(v,u) = \int_0^1 vu \, dx, \tag{2.5}$$

bilinear form

$$a(v,u) = \int_0^1 \{\lambda v u_x + \varepsilon v_x u_x\} \, dx + \beta \varepsilon v(1) u(1,t) \tag{2.6}$$

and source

$$F(v) = \int_0^1 vf\,dx + \varepsilon\gamma v(1) \ , \tag{2.7}$$

the weak form of (2.1)-(2.3) reads:
Find u, with $u(0,t) = \alpha$, satisfying

$$\frac{\partial}{\partial t}(v,u) + a(v,u) = F(v) \tag{2.8}$$

for all sufficiently smooth functions v with $v(0) = 0$.

The main purpose of introducing the weak form is to reduce the order of the highest spatial derivatives appearing in the statement of the problem. For this reason the data might be such that a solution would exist to (2.8) (called a weak solution) where one might not exist to (2.1). However, whenever a solution exists to (2.1) it must coincide with the weak solution.

For the integrals involved in (2.8) to exist, the functions u and v must satisfy the 'admissibility' conditions

$$\int_0^1 (u_x)^2 dx < \infty \qquad \int_0^1 (v_x)^2 dx < \infty, \tag{2.9}$$

i.e. they must have finite 'energy'. In one dimension these conditions are equivalent to the requirement that u and v belong to the class of continuous functions on the interval (0,1).

The next stage in the procedure is to replace the infinite set of ordinary differential equations (2.8) by a finite set. First we divide the interval (0,1) into N elements $e_j \equiv (x_{j-1}, x_j)$, $j = 1,2,..,N$ by the points

$$0 = x_0 < x_1 < \ . \ . \ . < x_N = 1.$$

For each time t, we will approximate the functions v and u by finite element functions V and U which reduce to polynomials over each element. Enforcing the essential boundary conditions gives $U(0,t) = \alpha$, $V(0) = 0$ and, in accordance with (2.9), both U and V must be continuous.

2.1 Piecewise Linear Approximation

The lowest degree polynomials consistent with the admissibility conditions are linear so we begin with an account of piecewise linear finite element approximation. The global nature of U is shown in fig. 1 - the behaviour of V is similar except that $V(0) = 0$. Introducing the trial functions $\{\phi_j\}$ defined by (see fig. 1)

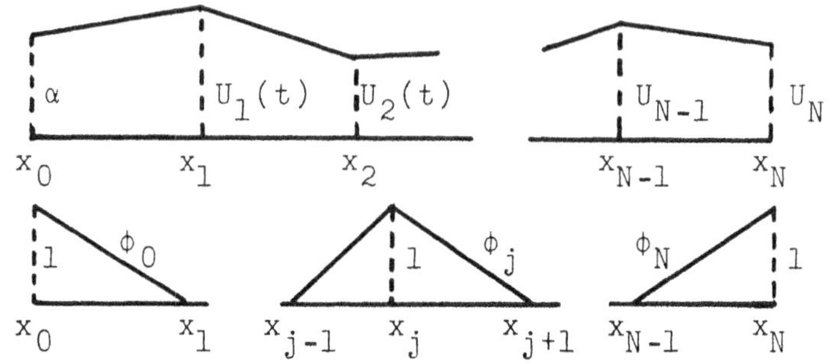

Figure 1. The global approximant U (upper) and, below, the trial functions for piecewise linear approximation.

$$\phi_0(x) = \begin{cases} (x_1-x)/h_1, & x_0 \le x \le x_1 \\ 0, \text{ otherwise} & \end{cases}$$

$$\phi_j(x) = \begin{cases} (x-x_{j-1})/h_j, & x_{j-1} \le x \le x_j \\ (x_{j+1}-x)/h_{j+1}, & x_j \le x \le x_{j+1} \\ 0, \text{ otherwise} & \end{cases} \tag{2.10}$$

$$\phi_N(x) = \begin{cases} (x-x_{N-1})/h_N, & x_{N-1} \le x \le x_N \\ 0, \text{ otherwise,} & \end{cases}$$

where $h_j = x_{j+1}-x_j$, we have the global representations

$$U(x,t) = \alpha\phi_0(x) + \sum_{j=1}^{N} U_j(t)\phi_j(x), \quad 0 \le x \le 1 \tag{2.11}$$

$$V(x) = \sum_{j=1}^{N} V_j\phi_j(x) \tag{2.12}$$

and $U_j(t) \equiv U(x_j,t)$, $V_j \equiv V(x_j)$ are the nodal values of U and V. The approximation of the weak form (2.8) is:
Find U, of the form (2.11), satisfying

$$\frac{\partial}{\partial t}(V,U) + a(V,U) = F(V) \tag{2.13}$$

for all functions V of the form (2.12). The system (2.13) is equivalent to the N differential equations

$$\frac{\partial}{\partial t}(\phi_i,U) + a(\phi_i,U) = F(\phi_i), \quad i = 1,2,..,N. \tag{2.14}$$

Strictly speaking the initial data should be determined by choosing the nodal parameters so that $U(x,0)$ is a best least squares approximation to $u_0(x)$ (see 2.3). However, since this process would involve solving an $N \times N$ system of linear equations, it is more common to assign the initial data by interpolation:

$$U_j(0) = u_0(x_j), \qquad j = 1,2,..,N.$$

The evaluation of the various terms in (2.14) is straightforward. For instance, with $0 < i < N$ and making use of (2.10) and (2.11) we find

$$(\phi_i, U) = \int_{x_{i-1}}^{x_{i+1}} \phi_i U(x,t)dx$$

$$= \int_{x_{i-1}}^{x_i} h_i^{-1}(x-x_{i-1})U\,dx + \int_{x_i}^{x_{i+1}} h_{i+1}^{-1}(x_{i+1}-x)U\,dx$$

$$= \frac{1}{6}\{h_i(U_{i-1}+2U_i) + h_{i+1}(2U_i+U_{i+1})\}\,.$$

Similarly,

$$a(\phi_i, U) = \frac{1}{2}\lambda(U_{i+1}-U_{i-1}) + \varepsilon\{h_i^{-1}(U_i-U_{i-1}) - h_{i+1}^{-1}(U_{i+1}-U_i)\}.$$

Assuming now that $f \equiv 0$ and that the elements are all of equal length (h), the complete set of equations may be written as

$$\frac{h}{6}(\dot{U}_{i-1}+4\dot{U}_i+\dot{U}_{i+1}) + \frac{\lambda}{2}(U_{i+1}-U_{i-1}) =$$

$$= \frac{\varepsilon}{h}(U_{i-1}-2U_i+U_{i+1}), \qquad 0 < i < N, \tag{2.15}$$

with $U_0(t) = \alpha$ and, for $i = N$,

$$\frac{h}{6}(\dot{U}_{N-1}+2\dot{U}_N) + \lambda(U_N-U_{N-1}) = \frac{\varepsilon}{h}(U_N-U_{N-1}) - \varepsilon(\beta U_N-\gamma)$$

where the dot denotes differentiation with respect to t. Note that, for $0 < i < N$, the spatial derivatives are replaced by standard central difference approximations. This method is, even for irregular grids, convergent at a rate $O(h^2)$ in the usual finite difference sense provided that the ratio of maximum to minimum element size remains bounded as the grid is refined.

Whilst the use of the global representation (2.11) is perfectly suitable for both analysis and hand derivation of the discrete equations, it is not at all convenient for automatic computer generation of these equations. The alternative is to use local representations of U and V which, when pieced together, are equivalent to (2.11) and (2.12). Let $x = x(p)$ denote the linear transformation of the jth element $e_j \equiv (x_{j-1}, x_j)$ to the so-called standard element $\hat{e} \equiv \{0 \leq p \leq 1\}$. A linear function on e may be written as

$$\begin{aligned} U &= U_{j-1}\sigma_1(p) + U_j\sigma_2(p), \qquad 0 \leq p \leq 1 \\ &= \underline{U}_j^T\underline{\sigma}(p) \end{aligned} \tag{2.16}$$

where σ_1 and σ_2 are the shape functions for linear interpolation:

$$\sigma_1 = 1 - p, \quad \sigma_2 = p , \tag{2.17}$$

$\underline{\sigma} = (\sigma_1, \sigma_2)$ and $\underline{U}_j = (U_{j-1}, U_j)^T$ is the vector of nodal parameters on e_j. The transformation from e_j to e may also be expressed in terms of these shape functions,

$$x = \underline{X}_j^T\underline{\sigma}(p), \quad \underline{X}_j = (x_{j-1}, x_j)^T$$

from which we compute the Jacobian $J_j = h_j$. Transformations which make use of the same shape functions as the approximant are known as isoparametric transformations and the elements as isoparametric elements.

The constituent terms in the approximate weak equation (2.13) may be conveniently computed in terms of these local representations. For example,

$$(V,U) = \int_0^1 VU\, dx = \sum_{j=1}^{N} \int_0^1 VUJ_j\, dp.$$

Now, using (2.16) and a similar expression for V, given

$$(V,U) = \sum_{j=1}^{N} \underline{V}_j^T M_j \underline{U}_j$$

where M_j is the 2×2 element mass matrix

$$M_j = \int_{\hat{e}} J_j\, \underline{\sigma}\, \underline{\sigma}^T\, dp.$$

Note that all integrations now take place over the standard element ê. In an enitrely similar way we find

$$a(V,U) = \sum_{j=1}^{N} \underline{V}_j^T S_j \underline{U}_j, \quad F(V) = \sum_{j=1}^{N} \underline{V}_j^T \underline{F}_j$$

where the element stiffness matrix and element source term are given by, respectively,

$$S_j = \int_e \{\lambda \underline{\sigma\sigma}'^T + \varepsilon J_j^{-1} \sigma'\sigma' \; dp, \quad \underline{F}_j = \int_e J_j f\underline{\sigma} \; dp$$

where the prime denotes differentiation with respect to p. When $j = N$ the terms

$$\begin{pmatrix} 0 & 0 \\ 0 & \beta\varepsilon \end{pmatrix}, \quad \begin{pmatrix} 0 \\ \varepsilon\gamma \end{pmatrix}$$

have to be added, respectively, to S_N and $\underline{F}_N$ so as to include the effect of the boundary terms.

In general the integrals involved must be evaluated by numerical means - Gauss quadrature is universally used for this purpose - but in the present case we easily compute

$$M_j = \frac{1}{6} h_j \begin{pmatrix} 2 & 1 \\ 1 & 2 \end{pmatrix}, \quad S_j = \lambda \begin{pmatrix} -1 & 1 \\ -1 & 1 \end{pmatrix} + \frac{\varepsilon}{h} \begin{pmatrix} 1 & -1 \\ -1 & 1 \end{pmatrix}.$$

The final stage in the construction process is the assembly of the element level quantities M_j, S_j and $\underline{F}_j$ into global arrays $\hat{M}$, $\hat{S}$ and $\underline{\hat{F}}$. Since $\underline{U}_j$ contains the (j-1)st and jth nodal variables, the entries of element matrices are added to the (j-1,j-1), (j-1,j), (j,j-1) and (j,j) positions of the respective global matrices whilst the entries of $\underline{F}_j$ are added to the (j-1)st and jth positions of $\underline{F}$. This process is shown schematically in fig.2. Defining $\underline{\hat{U}} = (U_0, \overline{U}_1, .., U_N)^T$ with a similar expression for $\underline{V}$, we may write

$$\sum_{j=1}^{N} \underline{V}_j^T M_j \underline{U}_j = \underline{\hat{V}}^T \hat{M} \underline{\hat{U}}, \quad \sum_{j=1}^{N} \underline{V}_j^T \underline{F}_j = \underline{\hat{V}}^T \underline{\hat{F}}$$

so that

$$(V,U) = \underline{V}^T M \underline{U}, \quad a(V,U) = \underline{V}^T S \underline{U}, \quad F(V) = \underline{V}^T \underline{F} \tag{2.18}$$

which give an algebraic description of the terms appearing in (2.13). Next, by virtue of the essential boundary condition at $x = 0$, $V(0) = 0$ so that $V_0 = 0$. Therefore, since all entries in the first rows of the assembled arrays are multiplied by zero, they may be deleted. Furthermore the entries in the first column

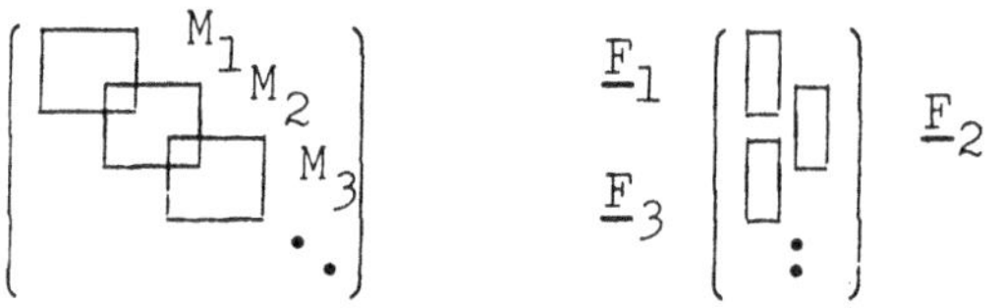

Figure 2. Schematic illustration of the assembly of local matrices and vectors into global arrays.

of both $\hat{M}$ and $\hat{S}$ multiply $U_0(=\alpha)$ and may also be deleted provided that the known terms are transposed to become part of the source term. We shall denote the respective modified mass and stiffness matrices by M and S and the new source term by $\underline{F}$. Thus, with $\underline{U} = (U_1, U_2, .., U_N)^T$ and $\underline{V} = (V_1, V_2, .., V_N)^T$, (2.13) becomes

$$\underline{V}^T(M\,\dot{\underline{U}} + S\,\underline{U} - \underline{F}) = 0$$

which must hold for all N-vectors $\underline{V}$. It therefore follows that $\underline{U}$ satisfies

$$M\,\dot{\underline{U}} + S\,\underline{U} = \underline{F}\,, \tag{2.19}$$

the matrix equivalent of equations (2.15).

In summary, from the weak form, the steps leading up to (2.19) are:
a) Specification of elements by their shape functions,
b) Evaluation of element level arrays,
c) Assembly of the global arrays, and
d) Imposition of essential boundary conditions.
In practice these steps are amalgamated so that each element needs to be referenced only once during the entire process. It is important to appreciate that the calculations for any one element are entirely separate (and unaffected by) those for the remaining elements.

2.2 Piecewise Quadratic Approximation

To improve the approximating ability of the finite element process, the degree of polynomial may be increased so that U is a quadratic function on each element. As three points are necessary to uniquely determine a quadratic function, so three nodal variables are required. On the jth element these are usually taken to be U_{j-1}, $U_{j-\frac{1}{2}}$ and U_j, the values of U at the nodes $x = x_{j-1}$, $x_{j-\frac{1}{2}}$ and x_j, respectively. Consequently, the local representation of the element may be written in the form (2.16) where $\underline{U}_j =$ $= (U_{j-1}, U_{j-\frac{1}{2}}, U_j)^T$ and the corresponding shape functions are (see fig. 3)

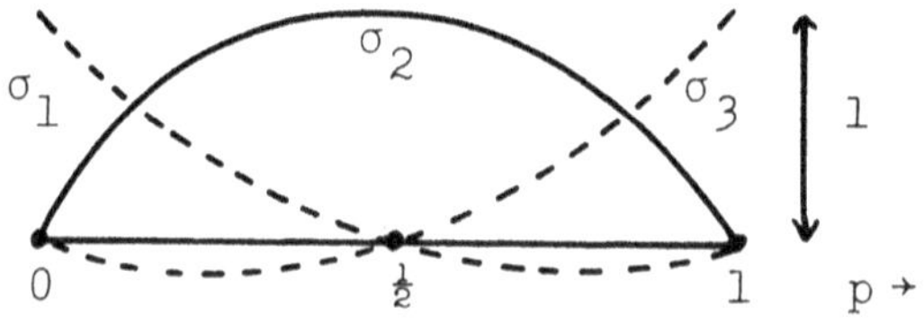

Figure 3. The shape function for a quadratic element.

$$\sigma_1 = (1-p)(1-2p), \quad \sigma_2 = 4p(1-p), \sigma_3 = p(2p-1). \qquad (2.20)$$

The evaluation of the element level arrays follows the same procedure as for linear elements except that they now have dimensions 3×3. In the assembly process, again depicted in fig. 2, successive submatrices overlap by just one entry and the global matrices have, alternatively, three and five non-zero entries in each row.

2.3 Hermite Cubic Approximation

The two previous examples have used Lagrange-type representations where the nodal parameters are values of U at specific points. By including four nodes on each element we could, in a like manner define a piecewise cubic approximation. Global continuity of U in Lagrange schemes is ensured by including the extremities of elements as nodes, for the local representations on adjacent elements share the same value at their point of contact. Perhaps the most popular cubic element is that of Hermite type which has, as its four nodal parameters, the values of U and its derivative U_x at the extremities of the element. This nodal arrangement ensures that the global approximant is continuously differentiable, a property that is necessary for solving fourth order problems in one space dimension. The local form of the Hermite cubic approximation is given by (2.16) with

$$\underline{U}_j = (U_{j-1}, h_j U_{x,j-1}, U_j, h_j U_{x,j})^T$$

and the shape functions

$$\sigma_1 = (1+2p)(1-p)^2, \quad \sigma_3 = (3-2p)p^2$$

$$\sigma_2 = p(1-p)^2, \quad \sigma_4 = -p^2(1-p).$$

The factors h_j are necessary in the definition of $\underline{U}_j$ in order to take account of the fact that $d/dx = h_j/dp$. With 4×4 element matrices and adjacent elements sharing two nodal parameters, there is a 2×2 overlap between successive submatrices in the assembly shown in fig 2.

Under mild conditions, the Galerkin finite element methods using piecewise polynomials of degree r converge at a rate $O(h^r)$ in the energy norm (the one-dimensional analogue of 1.6) and at $O(h^{r+1})$ in the L_2 (least squares) sense.

3. THE FEM IN TWO SPACE DIMENSIONS

We now turn our attention to the two dimensional analogue of (2.1), namely,

$$u_t + \underline{\lambda}\cdot\underline{\nabla}u = \varepsilon\nabla^2 u + f \qquad \text{in} \qquad \Omega\times\{t > 0\} \tag{3.1}$$

where ε is a positive constant and $\underline{\lambda}$ a constant 2-vector. The initial condition is

$$u = u_0 \qquad \text{on} \qquad \Omega\times\{t = 0\} \tag{3.2}$$

and boundary condition

$$u = g \qquad \text{on} \qquad \partial\Omega\times\{t > 0\} \tag{3.3}$$

($\partial\Omega$ denotes the boundary of the domain Ω). We restrict ourselves to the Dirichlet type boundary condition (3.3) only for the sake of brevity. The implementation of more general conditions follows the procedure outlined for the one dimensional problem.

To reduce (3.1) to its weak form, we multiply the equation by a smooth test function v and use Green's theorem on the term involving the Laplacian operator:

$$\int_\Omega v\nabla^2 u \, d\Omega = - \int_\Omega \underline{\nabla}v\cdot\underline{\nabla}u \, d\Omega + \int_{\partial\Omega} v \frac{\partial u}{\partial n} \, ds \, .$$

The boundary integral vanishes because $v = 0$ on $\partial\Omega$ by virtue of the essential boundary condition (3.3). Therefore, if we define

$$(u,v) = \int_\Omega uv \, d\Omega,$$

$$a(u,v) = \int_\Omega (v\underline{\lambda}\cdot\underline{\nabla}u + \varepsilon\underline{\nabla}v\cdot\underline{\nabla}u)d\Omega$$

and

$$F(v) = \int_\Omega v \, f \, d\Omega$$

the weak form of (3.1) reads (cf. 2.8):
Find u, with $u = g$ on $\partial\Omega$, satisfying

$$\frac{\partial}{\partial t}(v,u) + a(v,u) = F(v) \tag{3.4}$$

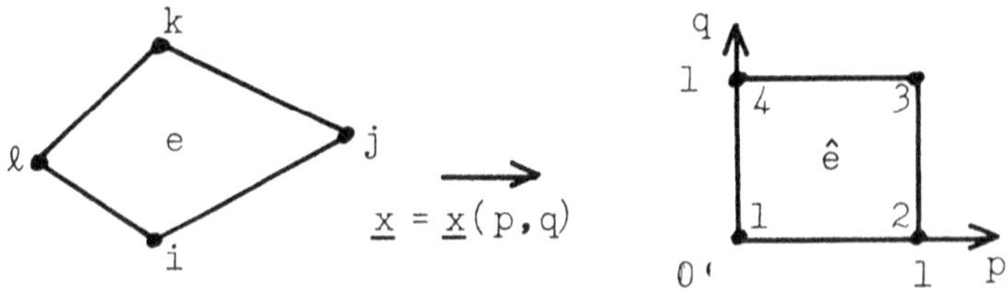

Figure 4. A typical 4-node element, e, and its transformation to the standard element e.

for all sufficiently smooth functions v with $v = 0$ on $\partial\Omega$.

Most finite element approximations in two dimensions begin by dividing the domain Ω into a number of non-overlapping elements which are either quadrilaterals or triangles (or a mixture of both). We shall describe briefly examples of teach type.

3.1 The 4-Node Bilinear Element

We divide the domain into a number of quadrilateral elements and isolate a typical member e which is mapped to the standard element $e = \{0 \le p \le 1,\ 0 \le q \le 1\}$ (see fig. 4). It is natural to take the values of U at the vertices of e as the nodal parameters, i.e. $\underline{U}_e = (U_i, U_j, U_k, U_\ell)^T$ where i,j,k and ℓ refer to the sequence numbers of the nodes on e (taken counterclockwise). The four shape functions are constructed from products of the linear one dimensional shape functions defined in (2.17). Using an asterisk to denote these latter functions, we have

$$\sigma_1(p,q) = \sigma_1^*(p)\sigma_1^*(q), \quad \sigma_2(p,q) = \sigma_2^*(p)\sigma_1^*(q),$$

$$\sigma_3(p,q) = \sigma_2^*(p)\sigma_2^*(q), \quad \sigma_4(p,q) = \sigma_1^*(p)\sigma_2^*(q)$$

so that $\sigma_1(p,q) = (1-p)(1-q)$, etc. The local approximant U now takes the form $U(p,q) = \underline{U}_e^T\underline{\sigma}(p,q)$ and the required transformation

$$x = \underline{X}_e^T\underline{\sigma}(p,q), \quad \underline{X}_e = (x_i,x_j,x_k,x_\ell)^T$$

$$y = \underline{Y}_e^T\underline{\sigma}(p,q), \quad \underline{Y}_e = (y_i,y_j,y_k,y_\ell)^T,$$

i.e. the vectors $\underline{X}_e$, $\underline{Y}_e$ contain, respectively, the x and y coordinates of the nodes on e. The nodes with indices i,j,k and ℓ are transformed to those with numbers 1 - 4 respectively on $\hat{e}$. Though the transformation cannot be inverted explicitly to give (p,q) in terms of (x,y), it is a straightforward matter to verify that U reduces to a linear function along each edge of the element e in physical space. It therefore follows that the local representations, when pieced together, combine to give a continuous global approximant on Ω.

The entire region Ω cannot generally be covered by elements of this type - the true boundary $\partial\Omega$ is approximated by a polygon. Moreover the essential boundary condition (3.3) can only be imposed at the nodes lying on $\partial\Omega$. Though these factors do not affect the rate of convergence of this element ($O(h^2)$ in L_2 norm), they may, of course, have an adverse affect on accuracy.

3.2 The 9-Node Biquadratic Element

With the division of Ω into quadrilaterals, we now have the added flexibility that the edges of the elements may be curved (see fig. 5). Each element has nine nodes (one at each vertex, one on each edge and one in the interior) and the associated values of U taken as nodal parameters. The shape functions each have the form $\sigma_i^*(p)\sigma_j^*(q)$ $\{1 \leq i, j \leq 3\}$, where $\{\sigma_i^*\}$ are the quadratic shape functions defined by (2.20). The same shape functions are used to generate the transformation from e to e (i.e. an isoparametric transformation).

Though it is still not possible to exactly match the boundary $\partial\Omega$, the greater accuracy available with this element has made it a popular choice amongst practitioners. A cautionary note, applicable to all elements: they should not be distorted too strongly for this may cause the approximation to break down (the Jacobian of the transformation may, for instance, become singular).

3.3 The Linear Triangle

This is probably the simplest element for second order problems. The domain is divided into triangular elements each of which is transformed, in turn, to the standard triangle $e = \{p \geq 0, q \geq 0, p + q \leq 1\}$ by means of an isoparametric transformation (linear in this case). The nodal variables are taken to be the values of U at the vertices for which the vector of shape functions is $(p,q,1-p-q)^T$.

3.4 The Quadratic Tringle

By increasing the degree of the shape functions so that they are quadratic, the edges of the triangular elements may again be curved (see fig. 6). The arrangement of the nodes is shown in the diagram and the requisite shape functions are

$$p(2p-1),\ q(2q-1),\ r(2r-1),\ 4pq,\ 4pr,\ 4qr$$

where $r = 1-p-q$. This is undoubtedly the most popular of all

Figure 5. A typical 9-node element and its image under the isoparametric mapping.

triangular elements. It may be used in conjunction with the 9-node biquadratic element to fill in 'gaps' left near the boundary.

Before concluding this section it is perhaps worthwhile to make a few remarks regarding the implementation of these ideas. A typical finite element package (if such exists) normally consists of three components: a mesh generator, routines for assembling and solving the discrete equations (including a variety of element types together with the ability to use a blend of those which are compatible) and, finally, a post-processor to produce output in the required form.

In addition to subdividing the domain into elements, the mesh generator must also assign sequence numbers to both the nodes and the elements. This information is passed to the main section together with the coordinates of the nodes and, for each element, the sequence numbers of the nodes belonging to that element. Not infrequently the global sequence numbers are chosen so as to minimise 'fill-in' during the solution stage. The assembly process can now be described quite easily for a general element with r nodes. Suppose that the global sequence numbers of these nodes are $\nu_1^e, \nu_2^e, .., \nu_r^e$ for the eth element. The (i,j) entry of the element matrix M_e (or S_e) is then simply added to the (ν_i^e, ν_j^e) entry of the global matrix M (or S).

For further information on these and other elements the reader is referred to the texts (25, 31, 37) and, for details of implementation, to (1).

Figure 6. A typical 6-node quadratic triangle and its image unter the isoparametric transformation.

4. TIME DISCRETIZATION

Thus far we have described how the Galerkin FEM may be used to discretize the initial-boundary value problem in space and reduce it to a finite system of ordinary differential equations of the form

$$M \dot{\underline{U}} + S \underline{U} = \underline{F} \tag{4.1}$$

where M and S are both N×N sparse (banded) matrices and M is furthermore positive definite. We shall assume for simplicity that neither M nor S depend explicitly on t. It is the presence of the mass matrix M which distinguishes FE approximations from their finite difference counterparts and is the main contributor to the greater accuracy achieved by finite element methods. We have seen (see 2.15) that the replacement of M by a suitable diagonal matrix can reduce the equation (4.1) to a recognisable finite difference approximation.

In principle there is no difficulty in adopting the Galerkin procedure to discretize (4.1) in time by finite elements. However, these methods frequently reduce to standard finite difference methods (though this will depend on the way integrals involving $\underline{F}$ are dealt with) and therefore little, if anything, would be gained by the added complexity.

We shall content ourselves with describing a simple finite difference method for solving (4.1) since this is the course that is most usually adopted. With a time step τ, let $\underline{U}^n$ denote the approximation to $\underline{U}(n\tau)$, then the θ-method $(0 \leq \theta \leq 1)$ applied to (4.1) gives

$$M(\underline{U}^{n+1} - \underline{U}^n)/\tau + S[\theta \underline{U}^{n+1} + (1-\theta)\underline{U}^n] = \underline{F}^{n+\theta}$$

that is

$$(M+\tau\theta S)\underline{U}^{n+1} = (M-(1-\theta)\tau S)\underline{U}^n + \tau \underline{F}^{n+\theta} \tag{4.2}$$

where $\underline{F}^{n+\theta}$ is a suitable approximation to $\underline{F}$ evaluated at $t = (n+\theta)\tau$. This scheme is clearly implicit for all values of θ and is the price that has to be paid for the inclusion of a mass matrix. Nevertheless there may still be an advantage in using $\theta = 0$ - the conditionally stable Euler method - in that the coefficient matrix $M+\tau\theta S$ is then always symmetric. The matrix system is invariably solved by a direct method based on Gaussian elimination or, equivalently, LU factorisation. A great deal of effort has been devoted to the development of specialised technieques and the reader is referred to (19,21,22,30) for details and further references.

The order or accuracy (measured in L_2-norm) of the GFEM applied to (2.1) and (3.1) with piecewise polynomials of degree r in space and the θ-method in time is $O(h^r + \tau^q)$, where $q = 2$ for $\theta = 1/2$ and $q = 1$ otherwise. The method is only conditionally stable for $\theta < 1/2$. Pointwise convergence analyses (maximum norm) are technically more difficult than those in L_2, but, for practical purposes, the same estimates hold.

5. FINITE ELEMENTS FOR HYPERBOLIC EQUATIONS

In view of the important role played by hyperbolic equations in fluid dynamics, we now look at some features peculiar to GFEM solution of first order hyperbolic equations. We begin with the simple advection equation

$$u_t + au_x = 0, \quad t > 0 \tag{5.1}$$

with initial condition

$$u(x,0) = u_0(x) \tag{5.2}$$

where a is a constant and u_0 is assumed to be periodic with period 1: $u_0(x+1) = u_0(x)$, for all x. We may therefore restrict ourselves to the interval $0 \leq x \leq 1$, which is divided into N elements as before. The approximant is written in the form

$$U(x,t) = \sum_j U_j(t)\phi_j(x) \tag{5.3}$$

where, for definiteness, we have chosen a Lagrangian type basis for the FE approximation. The GFEM approximation of (5.1) is therefore given by the system of ordinary differential equations

$$\int_0^1 \phi_j(x)\ (U_t + a\ U_x)dX = 0, \quad \text{for all } j. \tag{5.4}$$

An important property of the Lagrange bases is that the trial functions $\{\phi_j\}$ sum to unity. Consequently, by summing equations (5.4) over j, we find that

$$\frac{d}{dt}\int_0^1 U\ dx = 0 \tag{5.5}$$

showing that 'mass' is automatically conserved. Furthermore, by multiplying (5.4) with $U_j(t)$ and summing the result, we have

$$\frac{d}{dt}\int_0^1 U^2\ dx = 0 \tag{5.6}$$

thus guaranteeing the conservation of 'energy'. Similar results apply for initial-boundary problems except that certain boundary contributions appear on the right of (5.5) and (5.6).

Specialising to the case of piecewise linear elements the trial functions are defined by (2.10) and (5.4) becomes, on a uniform grid of size h,

$$\frac{h}{6}(\dot{U}_{j-1}+4\dot{U}_j+\dot{U}_{j+1})+a(\dot{U}_{j-1}-\dot{U}_{j+1})=0, \tag{5.7}$$

a sheme which conservative and fourth order accurate at the nodes and second order accurate in the L_2 norm. When (5.7) is discretized in time by the Crank-Nicolson method the resulting six point difference scheme is second order accurate and non-dissipative.

Increasing the degree of the finite elements to quadratic functions gives rise to the differential-difference equations (see Ref. 26)

$$\begin{aligned}&\frac{h}{30}\{-\dot{U}_{j-1}+2\dot{U}_{j-\frac{1}{2}}+8\dot{U}_j+2\dot{U}_{j+\frac{1}{2}}-\dot{U}_{j+1}\}+\\&\frac{a}{6}\{U_{j-1}-4U_{j-\frac{1}{2}}+4U_{j+\frac{1}{2}}-U_{j+1}\}=0\end{aligned} \tag{5.8}$$

corresponding to each 'integer' node ($x = jh$) and, for each half-integer node ($x = (j+\frac{1}{2})h$),

$$\frac{h}{15}\{\dot{U}_j+8\dot{U}_{j+\frac{1}{2}}+\dot{U}_{j+1}\}+\frac{2}{3}a\{U_{j+1}-U_j\}=0. \tag{5.9}$$

If these equations are used to solve (5.1) subject to the initial condition

$$u(x,0)=e^{i\omega x}, \qquad \omega \text{ real } (|\omega|\le\pi)$$

so that $u(x,t) = \exp(i\omega(x-at))$, we find, by standard Fourier techniques, that (13,16)

$$U_j(t)=A_1e^{i\omega(jh-a\sigma_1t)}+A_2e^{i\omega(jh-\sigma_2t)}, \tag{5.10}$$

for $j = 0, \pm\frac{1}{2}, \pm1,..$ where the constant amplitudes are usually such that $|A_1| \ll |A_2|$ and the functions σ_1, σ_2 depend on ω. It is found that σ_1 increases from the value 1 at $\omega = 0$ to 1.0066 at $|\omega| = \pi$. Thus the leading term in (5.10) provides an extremely accurate approximation to $u(x_j,t)$ with no dissipation and very little phase error. In contrast, σ_2 has the value -5 at $\omega = 0$ and varies smoothly to -1 at $|\omega| = \pi$. The second term on the right of (5.10) therefore represents a spurious wave travelling in a direction opposite to that of the true solution. Hedstrom (16) has shown how this spurious component may be annihilated by modifying the initial values at the half-integer nodes. It is not known what effect this spurious wave has in more realistic problems. A similar situation exists for Hermite cubic approximations and, most probably, for any FEM which uses more than one type of trial function.

5.1 The Petrov-Galerkin FEM

Galerkin finite element methods, being non-dissipative, are not suitable for solving hyperbolic equations over any but short time intervals. Petrov-Galerkin FEMs on the other hand, by allowing the test and trial functions to differ, admit a controllable amount of dissipation to be introduced. The effect of the process is somewhat akin to the use of 'upwinding' in finite difference methods (3).

We shall describe the PG-FEM with reference to equation (5.1) and suppose U to be given by (5.3) where $\{\phi_j\}$ are the piecewise linear trial functions defined by (2.10). Let $\{\psi_j\}$ denote the set of test functions - one corresponding to each trial function - then the PG-FEM equations are (cf. 5.4)

$$\int_0^1 \psi_j(x)(U_t + aU_x)dx = 0, \quad \text{for all } j. \tag{5.11}$$

In the first schemes of this type (5.34) the test functions were chosen to be

$$\psi_j = \{1 - \frac{1}{2}\gamma h \frac{d}{dx}\}\phi_j(x) \tag{5.12}$$

with the parameter γ having the same sign as the wave speed a and its magnitude controlling the required amount of dissipation. By substituting (5.12) into (5.11) and integrating by parts, it is found that

$$\int_0^1 \phi_j(x)(1 + \frac{1}{2}\gamma h \frac{d}{dx}(U_t + aU_x)dx$$

so that (5.11) is, in this instance, equivalent to the GFEM solution of the modified equation

$$(1 + \frac{1}{2}\gamma h \frac{\partial}{\partial x})(u_t + au_x) = 0.$$

Another approach (8,14,15), based on (5.11), begins with the method generated by the most general test function $\psi_j(x)$ which vanishes outside $x_{j-1} \le x \le x_{j+1}$. On a uniform grid, we may, without loss of generality, express ψ_j in the form

$$\psi_j(x) = \phi_j(x) + \alpha\chi_j^1(x) + \beta\chi_j^2(x) + \gamma\chi_j^3(x) \tag{5.13}$$

where using $s = x/h - j$,

$$\chi_j^1(x) = 3 - 6\phi_j(x),$$

$$\chi_j^2(x) = 3(1 - 2|s|)\operatorname{sgn} s$$

and

$$\chi_j^3(x) = 3s - 2\text{sgn } s$$

for $x_{j-1} \leq x \leq x_{j+1}$ $(-1 \leq s \leq 1)$, all the functions vanishing outside this interval. These functions are shown in fig. 7. It is only the antisymmetric functions χ_j^2 and χ_j^3 which are capable of introducing dissipation into the discretization. With $\alpha = 0$ and $\beta = \gamma/2$, the PG-FEM is equivalent to the Galerkin FEM applied to the modified equation

$$(1 + \frac{1}{2}\gamma h \frac{\partial}{\partial x})(u_t + au_x) + h\alpha u_{xxt} = 0.$$

A similar modified equation (the regularised long wave equation) is taken as the starting point in (6) for a GFEM solution of the Buckley-Leverett equation.

Combining (5.11) and (5.13) with the θ-method in time gives the six point, two level scheme

$$(1-\hat{\beta}\Delta_0 + \frac{1}{2}(\hat{\alpha} + \frac{1}{3})\delta^2)(U_j^{n+1} - U_j^n) + p(\Delta_0 - \frac{1}{2}\gamma\delta^2)U_j^n = 0 \qquad (5.14)$$

where $\hat{\beta} = \beta - p\theta$, $\hat{\alpha} = \alpha - \gamma p\theta$, $p = a\tau/h$ (the Courant number) and

$$\Delta_0 U_j \equiv \frac{1}{2}(U_{j+1} - U_{j-1}),$$

$$\delta^2 U_j \equiv U_{j-1} - 2U_j + U_{j+1}.$$

It will be noticed that the Lax-Wendroff method is recovered when the parameters are chosen so that $\hat{\beta} = 0$, $\alpha = -1/3$ and $\gamma = p$. In fact, when $\gamma = p$, (5.14) may be regarded as the Lax-Wendroff method generalized to include a 'mass matrix'. The von Neumann stability requirements for (5.14) are

$$p^2 \leq 1,$$

$$p\{\beta - \frac{1}{2}\gamma - p(\theta - \frac{1}{2})\} \leq 0, \qquad (5.15)$$

$$\gamma p\{\alpha - \frac{1}{6} - \gamma p(\theta - \frac{1}{3})\} \leq 0.$$

The method is, subject to these conditions, first order accurate and second order dissipative. The order of accuracy may be increased by choosing

$$\beta = \frac{1}{2}\gamma + p(\theta - \frac{1}{2}) \qquad (5.16a)$$

for second order accuracy and fourth order dissipation. If, in addition, we impose that

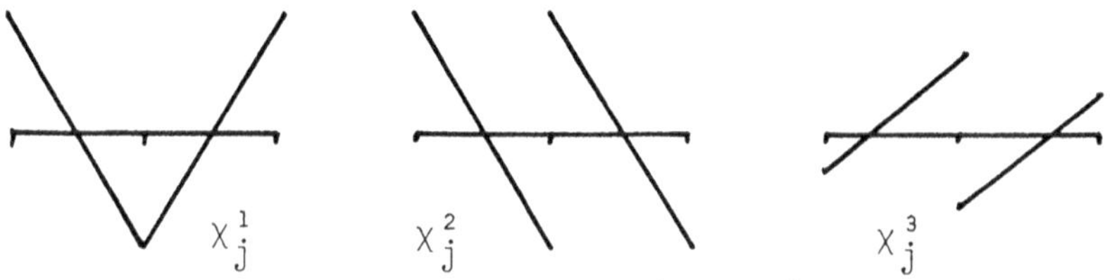

Figure 7. Three functions χ_j^1, χ_j^2 and χ_j^3, shown on the interval (x_{j-1},x_{j+1}), used in the definition (5.13).

$$\alpha = \frac{1}{6}p^2 + \gamma p(\theta - \frac{1}{2}) \tag{5.16b}$$

in order of accuracy become three. Finally, with $\gamma = 0$, we achieve fourth order accuracy and no dissipation. The behaviour of the third order scheme, which contains only one degree of freedom (γ) since θ may be set to zero without loss of generality, has been studied in (8). Writing $\gamma = \mu p$ ($\mu \geq 0$ for stability), it is found that the optimal value for μ is $O(h^5)$ when approximating very smooth solutions and is consequently very small. In general the optimal value for μ lies in (0,1) and increases as the smoothness of the solution diminishes (the value $\mu = 1$ corresponds to the EPGII method of Morton and Parrott (27)). In practice, this method (with μ in the interval (0,1)) gives a dramatic improvement in accuracy over both the GFEM and the Lax-Wendroff method. We refer to (8,27,29) for numerical results.

These ideas generalise in a straightforward manner to systems of hyperbolic equations. Consider, for example, the 2×2 system

$$\underline{u}_t + A\underline{u}_x = 0, \quad 0 \leq x \leq 1, \ t > 0 \tag{5.17}$$

where $\underline{u} = (u,v)^T$ and A is the 2×2 matrix

$$\begin{pmatrix} 0 & 1 \\ 1 & 0 \end{pmatrix}.$$

We supply initial data for $\underline{u}$ and impose the boundary conditions

$$u(0,t) = u(1,t) = 0, \qquad t > 0. \tag{5.18}$$

The approximant is now vector valued:

$$\underline{U}(x,t) = \sum_j \underline{U}_j(t)\phi_j(x)$$

and the test functions $\{\phi_j\}$ regarded as matrix valued functions. That is, for matrix valued parameters α,β and γ, (5.13) is modified to read

$$\psi_j(x) = \phi_j(x)I + \alpha\chi_j^1(x) + \beta\chi_j^2(x) + \gamma\chi_j^3(x)$$

where I is the 2×2 unit matrix. The matrix-vector analogue of (5.11) then gives two discrete equations for each j corresponding to the two components of $\underline{U}$. The analogues of the high order methods defined by (5.16) are obtained by interpreting the Courant number p as $(\tau/h)A$. Finally, to specify the matrix valued parameter γ, we define $\gamma = (\mu\tau/h)A$ where $\mu(\geq 0)$ is a scalar.

At boundary nodes $x = x_J$ $(J = 0,N)$, the first component of $\underline{U}$ is set to zero in accordance with the essential boundary conditions (5.18). The appropriate discrete equation for computing the second component of $\underline{U}$ at the boundary is

$$\underline{Z}^T \int_0^1 \psi_J(x)(\underline{U}_t + A\underline{U}_x)\, dx = 0, \qquad (J = 0 \text{ or } N)$$

where $\underline{Z}$ is any 2-vector not parallel to an eigenvector of A corresponding to an outgoing characteristic. In particular, $\underline{Z}$ may be taken as the (left) eigenvector of A corresponding to an ingoing characteristic. It may be shown that this method of choosing $\underline{Z}$ leads to a numerically stable scheme provided the method is stable at internal nodes. Numerical experiments indicate that the accuracy achievable by this method is comparable to the PG-FEM applied to a periodic initial value problem with the same initial data - that is to say, the accuracy is in no way impaired by the treatment of boundary conditions. It is characteristic of the FEM generally that it provides an approximation of the entire problem, in contrast with the finite difference method where the differential equation and boundary conditions have to be treated separately.

The ideas in this section may also be carried over to the solution of non-linear conservation laws. The quest for improvements in efficiency leads to further approximations, particularly in the treatment of the non-linear terms (see 2,6,8). In this account, we have chosen to interpret the finite element approximations as difference equations in order to place them in a familiar setting. This places great emphasis on nodal values of the approximant and, as argued in (4), this is not necessarily the best interpretation for finite element schemes.

For an alternative approach to PG-FEMs for conservation laws, see (28), where they are combined with characteristic methods. Specific application of the PG-FEM to the compressible Euler equations is made in (20). This reference also contains a comprehensive bibliography of recent work in this area.

Galerkin and Petrov-Galerkin FEMs have also been employed extensively to solve fluid problems governed by the Navier-Stokes equations. Some new difficulties are introduced and we can illustrate one of these by an example presented in (35). Consider the system

$$u_t + p_x = \varepsilon u_{xx}, \qquad 0 < x < 1, \quad t > 0 \qquad (5.19)$$
$$p_t + u_x = 0,$$

with boundary conditions

$$u(0,t) = u(1,t) = 0, \qquad t > 0$$

and prescribed initial values for both u and p. For the Galerkin solution of this problem we adopt continuous piecewise linear approximations to both u and p:

$$U = \sum_{j=1}^{N-1} U_j(t)\phi_j(x), \qquad (5.20)$$

$$P = \sum_{j=0}^{N} P_j(t)\phi_j(x), \qquad (5.21)$$

where $\{\phi_j\}$ are given by (2.10) and we have imposed the boundary conditions in (5.20). The process of using the same trial functions for each component is referred to as equal order interpolation. The Galerkin approximation of the second of equations (5.19) is

$$\int_0^1 \phi_j(x)(P_t + U_x)dx = 0, \quad j = 0,1,..,N, \qquad (5.22)$$

and by summing over j, we obtain

$$\frac{d}{dt}\int_0^1 P(x,t)dx = 0$$

showing that $\int P\,dx$ is conserved during the motion. Now let Z(x) be the continuous piecewise linear function which takes the values $(-1)^j$ at the nodes $x = x_j (j = 0,1,..,N)$. Multiplying (5.22) by $Z(x_j)$ and summing over j leads to

$$\frac{d}{dt}\int_0^1 Z(x)P(x,t)dx = 0$$

so that $\int ZP\,dx$ is also conserved. Therefore, if we write the initial condition for P in the form

$$P(x,0) = A + BZ(x) + W(x,0)$$

where A and B are constants, it follows that

$$P(x,t) = A + BZ(x) + W(x,t). \qquad (5.23)$$

The numerical solution for p therefore contains a standing wave of fixed amplitude, represented by the term BZ(x), which is not present in the exact solution.

The situation is more serious in incompressible Navier-Stokes calculations. The approximate pressure continues to be given by an expression of the form (5.23) where A represents the hydrostatic pressure and, in stationary multi-dimensional problems, the coefficient B is indeterminate in principle. However, in practical calculations, rounding errors contribute to the fixing of B and its value and its value can be (and frequently is) large (10^{10} for example). The solution in this case is to abandon equal order interpolation and adopt different trial functions for u and p - this is called mixed interpolation. Care must still be exercised, for example, using, respectively, quadratic and linear approximations for u and p removes the unwanted standing wave, but using linear and piecewise constant functions does not. For a detailed investigation of this phenomenon see (11,12).

One should not really emphasise the difficulties too strongly; there are a large number of known combinations of velocity and pressure approximations which may be used for Navier-Stokes calculations with no complications. An extensive body of theory exists for these combinations, convergence proofs are given in (10,32).

6. MOVING FINITE ELEMENTS

The techniques we have described thus far are not well suited for certain problems. We have in mind those problems which contain shocks or some form of travelling waves. Resolution of these essentially local phenomena can only be accomplished by using a uniformly fine grid on the entire interval. However, if the nodes of the grid are allowed to move, they could be positioned so as to enable the method to follow the solution accurately even in regions where gradients are large. Before turning to the question of how the node positions may be automatically determined, we first suppose that the spatial grid of (N+1) nodes $\{x_j\}$ is predetermined such that

$$x_0(t) < x_1(t) < \cdots < x_N(t), \quad t \geq 0.$$

For illustrative purposes we consider the conservation law

$$u_t + f(u)_x = 0, \qquad t > 0, \qquad (6.1)$$

with suitable initial data. We shall suppose, for definiteness, that $x = x_0(t)$ is an 'inflow' boundary so that $u(x_0,t)$ must be specified if the problem is to be correctly posed.

There does not appear to be any advantage in using high order interpolation so we shall adopt a piecewise linear FEM where the approximant is given by

$$U(x,t) = \sum_{j=0}^{N} U_j(t)\phi_j(x;t) \tag{6.2}$$

where the trial functions, defined by (2.10), now depend implicitly on t through the movement of the nodes. A straightforward calculation gives

$$\frac{\partial U}{\partial t} = \sum_{j=0}^{N} \{\dot{U}_j(t) - U_x\dot{x}_j\}\phi_j(x;t).$$

Therefore, by defining the continuous piecewise linear functions

$$\dot{U}(x,t) = \sum_{j=0}^{N} \dot{U}_j\phi_j(x;t) \tag{6.3a}$$

and

$$\dot{X}(x,t) = \sum_{j=0}^{N} \dot{x}_j\phi_j(x;t) \tag{6.3b}$$

we can write

$$U_t = \dot{U} - \dot{X}U_x \tag{6.4}$$

where $\dot{U}$ is the total (Lagrangian) time derivative of U and $\dot{X}$ represents the speed of the moving coordinate system ($\dot{X}(x_j,t) = \dot{x}_j(t)$). The Galerkin FEM approximation of (6.1) now gives

$$\int_I \phi_i(U_t + f(U)_x)dx = 0, \qquad i = 1,2,..,N, \tag{6.5}$$

where $I \equiv (x_0,x_N)$ and $i = 0$ is excluded from (6.5) because of the inflow boundary condition at $x = x_0$. Using (6.4), equation (6.5) becomes

$$\int_I \phi_i(\dot{U} - \dot{X}U_x + f(U)_x)dx = 0, \tag{6.6}$$

and, since $\dot{X}$ has been presumed to be a known function, this provides a system of N non-linear ordinary differential equations to determine $\{U_j\}$ and, thereby, $U(x,t)$,

Equations (6.5), (6.6) are capable of a different derivation; we define the residual function $R(x,t)$ by

$$R(x,t) = U_t + f(U)_x \tag{6.7}$$

and, for each t, minimise the expression

$$\int_I R(x,t)^2 dx \tag{6.8}$$

with respect to the parameters $\{\dot{U}_j\}$. This is a standard least squares problem and the minimum is given by the solution of the equations

$$\frac{\partial}{\partial \dot{U}_i} \int_I R^2 dx \equiv 2 \int_I R \frac{\partial R}{\partial \dot{U}_i} dx = 0, \quad i = 1,2,..,N.$$

Equations (6.5) are recovered when one observes that $\partial R/\partial \dot{U}_i = \phi_i$.

One of the main contributions made by Gelinas, Doss and Miller (9) was to recognise that the node positions could be automatically determined if the minimisation process was extended so as to take place with respect to the quantities $\{\dot{x}_i\}$ as well as $\{\dot{U}_i\}$. Performing the necessary minimisation of (6.8) gives rise to the further equations

$$\frac{\partial}{\partial \dot{x}_i} \int_I R^2 dx \equiv 2 \int_I R \frac{\partial}{\partial \dot{x}_i} dx = 0, \quad i = 1,2,..,N$$

which, on using (6.3) and (6.4), reduce to

$$\int_I \phi_i U_x (U_t + f(U)_x) dx = 0, \qquad i = 1,2,..,N. \tag{6.9}$$

Using (6.4), this may be written in the alternative form

$$\int_I \phi_i U_x (\dot{U} - \dot{X} U_x + f(U)_x) dx = 0, \quad i = 1,2,..,N. \tag{6.10}$$

We now have a fully coupled set of 2N ordinary differential equations, (6.6) and (6.10), with which to compute U and X. For future reference, we write these equations in the form

$$\int_I \phi_i R(x,t) dx = 0, \tag{6.11}$$

and

$$\int_I \psi_i R(x,t) dx = 0 \tag{6.12}$$

with $\psi_i = \phi_i U_x$.

There is, however, one immediate problem: ϕ_N, and consequently ψ_N, vanish except on the interval (x_{N-1}, x_N) where U_x is constant. Therefore

$$\int_I \psi_N R\, dx = U_x \big|_{(x_{N-1},x_N)} \int_I \phi_N R\, dx$$

and the equations (6.11), (6.12) with i = N are constant multiples of one another. With only 2N-1 independent equations the simplest course, and the one we shall pursue, is to assume that $x_N(t)$ is known (a similar situation exists at $x = x_0(t)$, but we have already assumed this position to be known).

Linear independence is also lost (degeneracy) when U is a linear (as opposed to piecewise linear) on two adjacent elements, (x_{k-1}, x_k) and (x_k, x_{k+1}), say (see fig. 8). In this case U_x is constant on (x_{k-1}, x_{k+1}), ψ_k is a constant multiple of ϕ_k, and (6.11), (6.12) are dependet for i = k (this is, in fact the only way that linear independence may be lost - see (36)). Now there is effectively no equation with which to determine x_k. The solution proposed in (9) is to minimise the quantity

$$\int_I R(x,t)^2 dx + \sum_j \{\xi_j \dot{h}_j - \sigma_j\}^2 \tag{6.13}$$

Instead of (6.8), where $h_j = x_j - x_{j-1}$,

$$\xi_j = k_x(h_j - k_1)^{-1} + k_3, \quad \sigma_j = k_4(h_j - k_j)^{-1} \tag{6.14}$$

and $k_1,\ldots,k_4$ are small positive constants. The 'penalty' term, represented by the summation in (6.13), is negligible except in the following circumstances:

i) when equations (6.11) and (6.12) are linearly dependent. Minimisation of (6.13) is completely unaffected by the penalty term except that (6.12) now becomes

$$\int_I \psi_i R(x,t)dx + \xi_{i+1}^2 \dot{h}_{i+1} - \xi_{i+1}\sigma_{i+1} - \xi_i^2 \dot{h}_i + \xi_i \sigma_i = 0.$$

Thus, in degenerate cases, the penalty term persists and a singular system of equations prevented.

ii) when $h_\ell \simeq k_1$ for some value of ℓ. The penalty term now dominates and results in

$$\dot{h} \simeq \xi_\ell / \sigma_\ell \simeq k_2/k_4 .$$

The nodes $x_{\ell-1}$ and x_ℓ therefore have a relative speed k_2/k_4 which forces these nodes apart. k_1 is therefore set as the minimum allowable node separation (the functions σ_j are often referred to as spring functions.

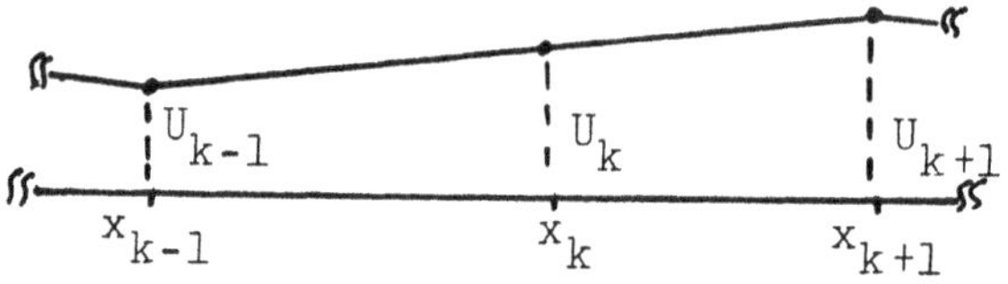

Figure 8. A situation leading to degeneracy in (6.11), (6.12).

Since the constants which feature in the penalty term are small, they only come into play in the two situations mentioned above. The subsequent discussion will therefore ignore the penalty term and concentrate on equations (6.11) and (6.12). Defining $a(u) = \partial f/\partial u$, the residual function becomes

$$R(x,t) = \dot{U} - \dot{X}U_x + a(U)U_x.$$

In those cases that $f(u)$ is a quadratic function of u, $a(u)$ will be linear in u and we shall have $R(x,t) \equiv 0$ when

$$\dot{U} = 0, \quad \dot{X} = a(U)$$

(recall that X is a piecewise linear function). Therefore the moving FE equations (6.11) and (6.12) will be identically satisfied by

$$\dot{U}_j = 0, \quad \dot{X}(x_j,t) \equiv \dot{x}_j(t) = a(U_j)$$

(this property does not depend on the particular test functions used in (6.11), (6.12)). These last relations imply that the nodes exactly follow the characteristics of the problem and the nodal values of U remain constant in time. This occurs, in particular, for the inviscid Burgers equation

$$u_t + uu_x = 0. \tag{6.15}$$

The Moving FEM provides the exact solution to this problem (since u remains constant along characteristics) so long as the solution does not develop shocks.

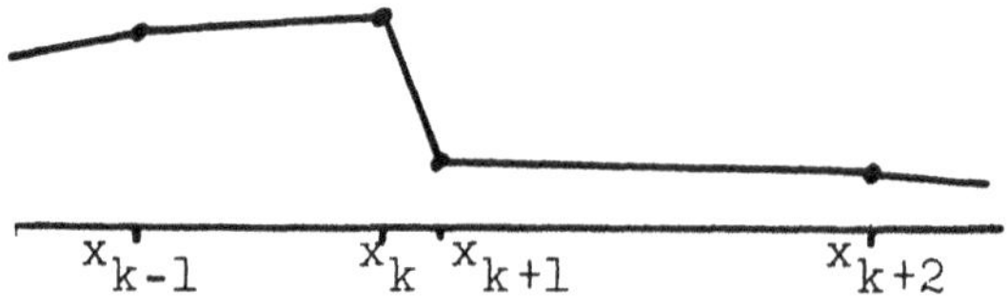

Figure 9. The solution just prior to shock formation.

Furthermore, provided the initial data is such that U(x,0) is not linear over two or more adjacent elements, the equations (6.11) and (6.12) cannot lose their linear independence and, from this point of view, the addition of a penalty term is not necessary. However, in the situation sketched in fig. 9 where the formation of a shock is imminent, it may be shown that

$$\dot{x}_k = a(U_k > \dot{x}_{k+1} = a(U_{k+1})$$

and the node x_k will subsequently overtake x_{k+1}. This possibility is precluded by the penalty term. There are other solutions to this problem of 'zone-tangling' - see (36), for instance, where use is made of the Rankine-Hugoniot conditions.

In order to solve the Moving FE equations in time, recourse to a 'stiff' ordinary differential equation package was made in (9), whilst recognising that this was not necessarily the most efficient possibility. Other workers (18, 36) have found the θ-method, coupled with Newton's method to solve the resulting nonlinear algebraic equations, to be satisfactory. For solutions of (6.15) prior to the formation of shocks, the simple Euler method (θ = 0) continues to return the exact solution at the nodes.

Systems of conservation laws can also be accommodated by the Moving FEM. If we regard u and U as m-vectors (with each component of U using the same trial functions) then the residual R is also an m-vector. Let D denote an m×m positive definite matrix (usually the unit matrix I), then, instead of (6.8), we minimise

$$\int_I R^T DR \, dx$$

with respect to each component of $\{\dot{U}_j\}$ and with respect to $\{\dot{x}_j\}$. The minimisation leads to

$$\int_I \phi_i DR \, dx = 0 \qquad (6.16)$$

and

$$\int_I \phi_i (U_x)^T DR \, dx = 0. \qquad (6.17)$$

Note that, for each i, (6.16) is a vector equation whilst (6.17) is a scalar equation. This approach uses the same grid for all components of U and, at present, appears to be the most practical choice.

For problems involving second derivatives in space, such as (2.1), the residual function will involve δ-functions and as a consequence the quantity (6.8) is not defined. Nevertheless, we may proceed formally which case the appropriate discrete equations again take the form (6.11) and (6.12) with R given by

$$R(x,t) = U_t + \lambda U_x - \varepsilon U_{xx} - f \ .$$

It is understood that the terms containing second derivatives are integrated by parts. The numerical results presented in (9) show that this method continues to accurately approximate the solution u.

Finally, a different approach to the Moving FEM has been presented in (18). For the scalar conservation law (6.1), piecewise linear approximation is again advocated but the method does not make use of a minimisation criterion to determine the discrete equations. Recognising that two discrete equations are required at each node, they borrow from the ideas of Petrov-Galerkin methods and use (6.11) and (6.12) where ϕ_i and ψ_i now represent the two Hermite cubic basis functions at the ith node. In spite of their different origins, the two Moving FEMs share the same properties though there is some evidence to suggest that Petrov-Galerkin form places somewhat less reliance on penalty terms (this is incorporated in the form shown in the equation following (6.14).

REFERENCES

1. Becker, E.B., Carey, G.F., and Oden, J.T.: Finite Elements, An Introduction (Prentice-Hall, Englewood Cliffs) 1981.
2. Christie, I., Griffiths, D.F., Mitchell, A.R. and Sanz-Serna, J.M.: 1981, Inst. Maths. Applics. J. Num. Anal., 1, pp 253-266.
3. Chriestie, I., Griffiths, D.F., Mitchell, A.R. and Zienkiewicz, O.C.: 1976, Int. J. Num. Meth. Eng., 10, pp. 1389-1396.
4. Cullen, M.J.P. and Morton, K.W.: 1980, J. Comp. Phys. 34, pp. 245-267.
5. Dendy, J.E.: 1974, SIAM J. Num. Anal., 11, pp. 637-653.
6. Douglas, J. Jnr., Kendall, R.P. and Wheeler, M.F.: in "Finite Elements for Convection Dominated Flows", (ed. T.J.R. Hughes), A.S.M.E.-AMD Vol. 34 (1979) pp. 201-211.
7. Duff, I.: 1982, Report CSS 125, AERE Harwell.
8. Duncan, D.B. and Griffiths, D.F.: 1982, Report NA/54, Dundee Univ. (to appear: Comp. Meth. Appl. Mech. and Engng.)
9. Gelinas, R.J., Doss, S.K. and Miller, K.: 1981, J. Comp. Phys. 40, pp. 202-249.
10. Girault, V. and Raviart, P.A.: "Finite Element Approximation of the Navier-Stokes Equations", Lect. Notes in Maths. 749 (Springer-Verlag), 1979.
11. Sani, R.L., Gresho, P.M., Griffiths, D.F. and Lee, R.L.: 1981, Int. J. Num. Meth. Fluids, 1, pp. 17-43.
12. Sani, R.L., Gresho, P.M., Lee, R.L., Griffiths, D.F. and Engelman, M.: 1981, Int. J. Num. Meth. Fluids 1, pp. 171-204.
13. Griffiths, D.F.: in "Numerical Analysis of Singular Perturbation Problems", (ed. P.W.Hemker and J.J.H. Miller) Academic Press, 1979.

14. Griffiths, D.F. and Mitchell, A.R.: in "Finite Element Methods for Convection Dominated Flows", (ed. T.J.R. Hughes), A.S.M.E.-AMD Vol. 34, 1979, pp. 91-105.
15. Griffiths, D.F.: in "Mathematics of Finite Elements and Applications IV", (ed. J.R. Whiteman), Academic Press, 1982.
16. Hedstrom, G.W.: 1979, Soc. Ind. Appl. Math. J. Num. Anal. 16, pp. 385-393.
17. Hestrom, G.W. and Rodrigue, G.H.: 1982, Lawrence Livermore Nat. Lab., Report UCRL-87242.
18. Herbst, B.M., Schoombie, S.W. and Mitchell, A.R.: 1982, Int. J. Num. Meth. Engng., 198, pp. 1321-1336.
19. Hood, P.: 1976, Int. J. Num. Meth. Engng., 10, pp. 379-399.
20. Hughes,T.J.R., Tezduyar, T.E. and Brooks, A.N.: in "Numerical Methods in Fluid Dynamics", (ed. K.W. Morton and M.J. Baines), Academic Press, 1982.
21. Irons, B.M.: 1970, Int. J. Num. Meth. Engng., 2, pp. 5-32.
22. Jennings, A.: 1966, Comput. J., 9, pp. 291-285.
23. Lee, R.L., Gresho, P.M., Chan, S.T., Sani, R.L. and Cullen, M.J.P.: 1982, ch. 2 of "Finite Elements in Fluids, Vol. 4", J. Wiley & Sons, N.Y.
24. Miller, K.: 1981, SIAM J. Num. Anal., 18, pp. 1033-1057.
25. Mitchell, A.R. and Wait, R.: 1977, "The Finite Element Method in Partial Differential Equations", J. Wiley & Sons, N.Y.
26. Mitchell, A.R. and Griffiths, D.F.: 1980, "The Finite Difference Method in Partial Differential Equations", J. Wiley & Sons, N.Y.
27. Morton, K.W. and Parrott, A.K.: 1980, J. Comp. Phys., 36, pp. 249-270.
28. Morton, K.W.: 1982, in "Numerical Methods in Fluid Dynamics" (ed. K.W. Morton and M.J. Baines), Academic Press.
29. Raymond, W.H. and Garder, A.: 1976, Month. Weath. Rev., 104, pp. 1593-1590.
30. Reid,J.K.: 1980, Report CSS 86, AERE Harwell.
31. Strang, G. and Fix, G.J.: 1973, "An Analysis of the Finite Element Method", Prentice-Hall, Englewood Cliffs, N.J.
32. Teman, R.: 1977, "Navier-Stokes Equations", North Holland, Amsterdam.
33. Thomasset, F.: 1981, "Implementation of the Finite Element Method for Navier-Stokes Equations", Springer Series in Comp. Physics, Springer-Verlag.
34. Wahlbin, L.B.: 1974, in "Mathematical Aspects of Finite Elements in Partial Differential Equations" (ed. C. de Boor), Academic Press, N.Y., pp. 147-170.
35. Walters, R.A. and Carey, G.F.: 1981, TICOM Report 81-3, Univ. of Texas.
36. Wathen, A.: 1982, Num. Anal. Report 4/82, Univ. of Reading, U.K.
37. Zienkiewicz, O.C.: 1977, "The Finite Element Method" McGraw-Hill.

DESCRIPTION AND PHILOSOPHY OF SPECTRAL METHODS

Philip S. Marcus

Massachusetts Institute of Technology

We describe the use of spectral methods in computational fluid dynamics. Spectral methods are generally more accurate and often faster than finite-differences. For example, the ∇^2 operator in 2 or 3 dimensions is easier to invert with spectral techniques because the spatial dependence of the operator separates in a more natural way. We warn against the use of some of the more common spectral expansions. Bessel series expansions of functions in cylindrical geometries converge poorly. However, other series expansions of the same functions converge quickly. We show how to choose basis functions that give fast convergence and outline the differences between Galerkin, tau, modal, collocation, and pseudo-spectral methods.

INTRODUCTION

After perfecting a numerical code, it is tempting to try and find every physical and astrophysical problem that one can solve with the code. However, in developing a code in the first place, one is often motivated by some particular physical or astrophysical calculation. If motivated by solar convection and rotation, the hydrodynamics appears to be easy: there is no relativity; there are no shocks; the flow is at low Mach number and is smooth. On the other hand, the numerical simulation of these flows is very difficult due to turbulence. In turbulence, many decades of length scales are excited and contribute to the dynamics. To have correct numerical results, it is necessary to resolve or model all of the physically important scales. In choosing a numerical scheme that is useful in computing turbulent flow, we shall need a method that is efficient in three dimensions (since turbulence is inherently

K.-H. A. Winkler and M. L. Norman (eds.), Astrophysical Radiation Hydrodynamics, 359–386.

three-dimensional) and provides the greatest possible spatial resolution. Spectral methods are ideal for these flows.

TWO FAMILIAR EXAMPLES OF SPECTRAL METHODS

Before giving an exact definition of a spectral method, we provide two examples to show that most readers have already used spectral techniques.

Heat Equation

The first example is an analytic calculation similar to a graduate-level electrostatics problem with boundary conditions. Consider the one-dimensional heat equation:

$$\frac{\partial T}{\partial t} = \kappa \frac{\partial^2 T}{\partial x^2} \tag{1}$$

with boundary conditions:

$$T(t,0) = T(t,1) = 0 \tag{2}$$

and initial condition

$$T(0,x) = f(x). \tag{3}$$

One way of solving the heat equation is to expand T in a sine series

$$T(t,x) = \sum_{m=1}^{\infty} a_m(t) \sin(m\pi x) \tag{4}$$

The sines are a natural choice because they are a complete set of basis functions, and each sine individually obeys the boundary conditions. Furthermore, the sines are eigenfunctions of the second derivative operator that appears on the right-hand side of equation (1). Fourier analyzing the initial data, we obtain

$$f(x) = \sum_{m=1}^{\infty} f_m \sin(m\pi x). \tag{5}$$

Substituting T(t,x) from equation (4) into the heat equation and comparing the coefficients of the sines on both sides of the equation [or to be more formal, multiplying both sides of the heat equation by $\sin(m'\pi x)$ and integrating both sides over x from zero to one], we obtain an infinite set of ordinary differential equations for the spectral coefficients $a_m(t)$:

$$\frac{\partial a_m(t)}{\partial t} = -\kappa m^2 a_m(t) \quad m = 1,2,\ldots \infty \tag{6}$$

Equation (6) can be integrated analytically using the initial condition $a_m(0) = f_m$.

$$a_m(t) = f_m \exp(-\kappa m^2 t) \quad m = 1,2,\ldots, \infty \tag{7}$$

Equation (7) is the analytic spectral solution to the heat equation. To find a numerical spectral solution to the heat equation we simply replace $T(t,x)$ by the discrete approximation $T_N(t,x)$,

$$T_N(t,x) \tilde{=} T(t,x), \tag{8}$$

where we have discretized by the simple truncation

$$T_N(x,t) \equiv \sum_{m=1}^{N} a_m(t)\sin(m\pi x) \tag{9}$$

We need to know how well $T_N(t,x)$ approximates the exact solution of the heat equation. One way of measuring the error caused by the approximation is to calculate the mean-square error, L_2, which is defined:

$$L_2 \equiv \left[\int_0^1 |T(t,x) - T_N(t,x)|^2 \, dx\right]^{1/2} \tag{10}$$

Using equations (7) and (9), we find that the mean-square error becomes

$$L_2 = \left[\frac{1}{2} \sum_{m=N+1}^{\infty} |f_m|^2 e^{-2\kappa m^2 t}\right]^{1/2} \tag{11}$$

We see immediately that the error depends solely on the spectral coefficients that we have discarded in the truncation (i.e., those spectral coefficients a_m with $m > N$). Equation (11) shows that for $t > 0$ and $\kappa > 0$, the mean-square error decays exponentially with increasing numerical resolution, N. The hallmark of a good spectral method is that the error decreases exponentially with increasing resolution; whereas, in a second-order accurate finite-difference method, the error (by definition) decreases only as the square of the spatial resolution.

To compare directly the error due to finite-differences with the error due to spectral expansion, we use a spectral analysis to evaluate the finite-difference approximation of the second derivative operator used in the heat equation. Let the initial temperature distribution contain only one Fourier component

$$f(x) = \sin(p\pi x) \tag{12}$$

Using a centered, second-order accurate finite-difference operator, the second derivative of $\sin(p\pi x)$ at x_i is

$$\left.\frac{d^2 \sin(p\pi x)}{d\,x^2}\right|_{x_i} = \frac{\sin[p\pi(x_i+\Delta x)]+\sin[p\pi(x_i-\Delta x)]-2\sin(p\pi x_i)}{(\Delta x)^2}$$

$$= \frac{2[\cos(p\pi\Delta x)-1]}{\Delta x^2}\sin(p\pi x_i) \tag{13}$$

$$\equiv \lambda_{f.-d.} \sin(p\pi x_i)$$

The exact second derivative of $\sin(p\pi x)$ is

$$\left.\frac{d^2}{dx^2}\sin(p\pi x)\right|_{x_i} = -\,p^2\pi^2 \sin(p\pi x_i)$$

$$\equiv \lambda_{exact} \sin(p\,\pi\,x_i) \tag{14}$$

We see that the finite-difference operator gives the exact eigenfunction but an incorrect eigenvalue. If Δx is small, we can Taylor expand the finite-difference eigenvalue and thereby determine the fractional error

$$\left|\frac{\lambda_{f.-d.}-\lambda_{exact}}{\lambda_{exact}}\right| \simeq \frac{\pi^2 p^2 \Delta x^2}{12}\,. \tag{15}$$

If we want the eigenvalue correct to within one percent we require that

$$p\,\Delta x \lesssim 0.11 \tag{16}$$

Since there are $p/2$ wavelengths of $f(x)$ between 0 and 1 and since there are $1/\Delta x$ finite-difference grid points between 0 and 1, equation (16) tells us that we need approximately 20 grid points per wavelength in order to achieve one percent accuracy. This factor of 20 should be compared to the fact that we need only one Fourier mode per wavelength to get perfect accuracy in the numerical second derivative eigenvalue when the derivatives are computed spectrally.

The suspicious reader can argue that the preceding comparison is unfair because our spectral expansion of $f(x)$ uses basis functions that are the exact analytic eigenfunctions of the second derivative operator. However, we shall show soon that expanding $f(x)$ in any other set of suitable basis functions require only three or four modes per wavelength to obtain one percent accuracy; whereas, second-order finite difference methods always need approximately 20 grid points per wavelength to obtain the same accuracy. Furthermore, choosing the basis functions to be the set of analytic eigenfunctions of the second derivative operator is not always wise. We show later that solving the heat equation in cylindrical coordinates with an expansion in the eigenfunctions of the second derivative operator has disastrous consequences.

Quadrature

As a second example of a spectral method, we consider numerical quadrature. One method of numerically integrating a function is to use a Newton-Cotes quadrature formula. These formulae are really just finite-difference methods. Let $f(x)$ be a function that is tabulated at equally spaced intervals x_i, where $x_{i+1} - x_i \equiv \Delta x$. We numerically integrate $f(x)$ from x_{i-1} to x_{i+1} by replacing the function $f(x)$ in the interval $[x_{i-1}, x_{i+1}]$ with the parabola that passes through the three points $(x_{i-1}, f(x_{i-1}))$, $(x_i, f(x_i))$ and $(x_{i+1}, f(x_{i+1}))$. Analytic integration of the parabola gives Simpson's rule:

$$\int_{x_{i-1}}^{x_{i+1}} f(x)\,dx = \frac{\Delta x}{3}\,(f(x_{i+1}) + 4f(x_i) + f(x_{i-1})) + \mathcal{O}(\Delta x^5) \tag{17}$$

To see that Simpson's rule is really a second-order finite difference method, we expand $f(x)$ in a Taylor series about $x = x_i$. Integrating the Taylor series from x_{i-1} to x_{i+1} we obtain

$$\int_{x_{i-1}}^{x_{i+1}} f(x)dx = \int_{x_i-\Delta x}^{x_i+\Delta x} (f(x_i)+xf'(x_i) + \frac{x^2}{2} f''(x_i) + \ldots)dx =$$

$$= 2\Delta x f(x_i) + \frac{1}{3}\Delta x^3 f''(x_i) + \mathcal{O}(\Delta x^5) \qquad (18)$$

In order to evaluate this integral, we need to determine $f''(x_i)$. If we substitute the second-order centered finite difference

$$f''(x_i) = \frac{f(x_{i+1}) + f(x_{i-1}) - 2f(x_i)}{\Delta x^2} + \mathcal{O}(\Delta x^2) \qquad (19)$$

into equation (18), then we recover Simpson's rule.

An alternative approach to quadrature is Gauss's method. In Gaussian quadrature, we abandon the constraint of equally spaced sampling points. The numerical integral of f(x) from -1 to 1 is approximated by a linear sum of weights, w_i, multiplied by f(x) evaluated at the sampling points.

$$\int_{-1}^{1} f(x)dx \cong \sum_{i=0}^{N} f(x_i)w_i \qquad (20)$$

In equation (20) there are 2(N+1) unknown quantities: (N+1) x_i's and (N+1) w_i's. To determine these unknowns we impose 2(N+1) equations: equation (20) must be satisfied exactly for f(x) = 1, f(x) = x, $f(s) = x^2, \ldots, f(x) = x^{2N}$, $f(x) = x^{2N+1}$. Equivalently, the quadrature formula in equation (20) must be exact for all polynomials of order (2N+1) or less. The well-known solution to this problem is that the x_i are roots of the Legendre polynomial of order N+1.

$$P_{N+1}(x_i) = 0 \qquad i = 0,N \qquad (21)$$

What are we really doing when we use Gaussian quadrature? We are actually replacing f(x) with the spectral approximation, $f_N(x)$

$$f_N(x) \equiv \sum_{0}^{N} a_n P_n(x) \qquad (22)$$

where the $P_n(x)$ are Legendre polynomials and where the coefficients a_n, are determined by the method of least squares (cf. Dahlquist, et al., 1979). The quadrature formula in equation (20) is the result of an analytic integration of $f_N(x)$.

DEFINITIONS OF SPECTRAL METHODS - SELECTION OF BASIS FUNCTIONS

Spectral methods are useful in numerical calculations because they allow us to represent a continuous function, $f(x)$, as a discrete approximation, $f_N(x)$. The approximation is written as a finite sum of basis functions multiplied by coefficients.

$$f(x) \cong f_N(x) = \sum_{i=1}^{N} a_i \phi_i(x) \tag{23}$$

The basis functions, $\phi_i(x)$, are arbitrary and do not have to be eigenfunctions or orthogonal. The two tasks of the numericist are to: (1) select a set of basis functions $\phi_i(x)$ and (2) compute the coefficients, a_i. Both the basis functions and the method of computing the coefficients should be chosen so that the boundary and initial conditions are easily satisfied, the spectral sum converges quickly, and both the numericist and computer have a minimal amount of work to perform. The choice of basis functions and the manner of computing the coefficients determines a method's name, such as modal, Galerkin, spectral, tau, collocation, pseudospectral, or Rayleigh-Ritz.

Fourier Series and the 9 % Solution.

We first consider the task of choosing basis functions that make up the spectral series. The simplest choice of a Fourier series with $\phi_k(x) = e^{ikx}$. We remind the reader of an important theorem about Fourier series: if $f(x)$ is a continuous, piecewise function over the domain 0 to 2π, where $f(x)$ is of bounded variation, and if Fourier coefficients, a_k, are defined by

$$a_k \equiv \frac{1}{2\pi} \int_0^{2\pi} f(x)\, e^{-ikx} dx , \tag{24}$$

and if the spectral sum $g(x)$ is defined by

$$g(x) \equiv \sum_{k=-\infty}^{\infty} a_k\, e^{ikx} , \tag{25}$$

then

$$g(x) = 1/2 \left[f(x^+) - f(x^-)\right] . \tag{26}$$

This theorem means that if $f(x)$ is a continuous function, then the Fourier series, $g(x)$, is equal to $f(x)$ at every point. If there is a discontinuity in the function, then $g(x)$ is equal to the arithmetic mean of $f(x)$ at the discontinuity. It is important to realize that for a numericist this theorem has no practical value. In numerical approximations we calculate the partial sums $f_N(x)$

$$f_N(x) \equiv \sum_{k=-N}^{N} a_k \, e^{ikx} \tag{27}$$

The preceding theorem does not guarantee that $f_N(x)$ converges uniformly to $g(x)$ or $f(x)$ as N approaches infinity. In fact, near a discontinuity, $f_N(x)$ never uniformly converges to $f(x)$ or $g(x)$. This lack of convergence is well-known to anyone who has Fourier analyzed a step function and calculated the Gibbs overshoot of approximately 9 % at the discontinuity. If more Fourier modes are included in the partial sum, the 9 % error does not decrease. If the function, $f(x)$, is itself continuous but has a discontinuity in one of its derivatives, then $f_N(x)$ may converge to $f(x)$ but the rate of convergence will be poor. One way of measuring the convergence rate is to examine the mean square error, $L_2(N)$ of the Nth partial sum:

$$L_2(N) \equiv \left[\int_0^{2\pi} |f(x)-f_N(x)|^2 \, dx \right]^{1/2} \tag{28}$$

$$= \left[2\pi \sum_{|k|=N+1}^{\infty} |a_k|^2 \right]^{1/2} \tag{29}$$

To evaluate the error, we must first determine how the Fourier coefficients, a_k, depend on k. Integrating equation (24) by parts we obtain

$$a_k = \left. \frac{if(x)e^{-ikx}}{2\pi k} \right]_0^{2\pi} + \frac{1}{i2\pi k} \int_0^{2\pi} f'(x)e^{-ikx} dx \tag{30}$$

The surface term in equation (30) vanishes if $f(x)$ is continuous and periodic. If all of the derivatives of $f(x)$ are continuous, then further integration by parts produces no surface terms. Integration by parts of equation (24) m times gives

$$a_k = \frac{1}{(ik)^m 2\pi} \int_0^{2\pi} f^{(m)}(x)e^{-ikx} dx \tag{31}$$

From equation (31), we see that a_k falls off as $1/k^m$ for all functions whose first (m-1) derivatives are continuous and periodic. Equation (29 shows that the mean-square error, $L_2(N)$, decreases at least as fast as $1/N^{m-1}$. If f(x) is a c_∞ function, then the Fourier coefficients and mean-square error decrease exponentially with N. Therefore, the convergence of the partial Fourier sums of a c_∞ periodic function is exponential. This exponential rate of convergence should be compared to second-order finite difference methods where (by definition) the rate of convergence is only quadratic.

If f(x) has a discontinuity at x_o, then integration by parts of equation (24) gives

$$a_k = \frac{i}{2\pi k} e^{-ikx_o} [f] + \frac{1}{2\pi ik} \int_0^2 f'(x) e^{-ikx} dx \tag{32}$$

when [f] is the discontinuity in f. The surface term in equation (32) is not zero. No matter how many times we integrate equation (32) by parts, there will always be a contribution to the Fourier coefficient, a_k, that decreases slowly as 1/k. To see how the discontinuity affects the convergence, we write the error in the Fourier partial sum as

$$f(x) - f_N(x) = \sum_{|k|=N+1}^{\infty} a_k e^{ikx} \tag{33}$$

In estimating the sum of the series in equation (33), we can argue that if all the terms have random phases (say, far away from the discontinuity), then the sum is approximately equal to the leading term in the series, a_n, which decreases as 1/N. If the terms in equation (33) are in phase (say, near the discontinuity) then the error in equation (33) is of order unity. The 9 % Gibbs overshoot in the truncated sum made from the Fourier transforms of a square wave is an example of an error of order unity near a discontinuity. Far from the discontinuity the error in the Fourier sum is much smaller and is of order 1/N.

In general, if the lowest order derivative of f(x) that has a discontinuity or non-periodicity is the mth derivative, then the convergence of the partial Fourier sums is of order $(1/N)^{m+1}$ far from the discontinuity and is of order $(1/N)^m$ near the discontinuity. To illustrate this convergence rate, consider the heat equation with an inhomogeneous source term,

$$\frac{\partial f}{\partial t} = \frac{\partial^2 f}{\partial x^2} + 1. \tag{34}$$

Let $f(x)$ be defined over the domain $[0,1]$ with boundary conditions

$$f(0) = f(1) = 0 \tag{35}$$

Expanding both $f(x)$ and 1 in a Fourier sine series,

$$f(t,x) = \sum_{k=1}^{\infty} f_k(t) \sin(k\pi x) \tag{36}$$

$$1 = \sum_{k=1,\ k=\mathrm{odd}}^{\infty} \frac{4}{k\pi} \sin(k\pi x) \tag{37}$$

and substituting into equation (34), we obtain an ordinary differential equation for each Fourier coefficient $f_k(t)$. Integrating these equations we find

$$\begin{aligned} f_k(t) &= f_k(0)\, e^{-k^2\pi^2 t} && \text{for } k = \text{even} \\ f_k(t) &= f_k(0) e^{-k^2\pi^2 t} + \frac{4}{\pi^3 k^3}\,(1-e^{-\pi^2 k^2 t}) && \text{for } k = \text{odd} \end{aligned} \tag{38}$$

We see explicitly that the spectral coefficients do not decrease exponentially with k; instead, we find that they decrease only as $(1/k)^3$. The convergence is not exponential; it is cubic. Why doesn't the series converge any faster? To understand the poor rate of convergence we examine the Fourier coefficients of $f(t,x)$ which are defined

$$f_k(t) \equiv 2 \int_0^1 f(t,x) \sin(k\pi x)dx \tag{39}$$

A first integration by parts yields

$$f_k(t) = \frac{-2f(t,x)}{k\pi} \cos(k\pi x) \Big]_0^1 + \frac{2}{k\pi} \int_0^1 f'(t,x)\cos(k\pi x)dx \tag{40}$$

The surface term in equation (40) vanishes because $f(t,x)$ vanishes at 0 and 1 due to the boundary conditions. A second integration by parts gives

$$f_k(t) = \frac{2f'(t,x)}{k^2\pi^2} \sin(k\pi x) \Big]_0^1 - \frac{2}{k^2\pi^2} \int_0^1 f''(t,x)\sin(k\pi x)dx \tag{41}$$

The surface term vanishes again, not due to any property of $f(t,x)$ or its derivatives, but because the sine vanishes at 0 and 1. A third integration by parts gives

$$f_k(t) = \left.\frac{2f''(t,x)\cos(k\pi x)}{k^3\pi^3}\right]_0^1 - \frac{2}{k^3\pi^3}\int_0^1 f'''(t,x)\cos(k\pi x)dx \quad (42)$$

This time the surface term is non-zero. The surface term and $f_k(t)$ both decrease as $(1/k)^3$. This is the reason that the convergence rate is cubic and not exponential.

Now that we understand why the rate of convergence is slow we can accelerate it. The surface term from the first integration by parts in equation (40) vanishes because $f(x)$ is zero at the boundaries. The surface term vanishes after the second integration by parts because $\sin(k\pi x)$ vanishes at 0 and 1. The surface term from the third integration by parts does not vanish because $f''(0)$ and $f''(1)$ are not equal to zero. In fact, from the inhomogeneous heat equation (34) we see that

$$f''(0) = f''(1) = -1 \quad (43)$$

To improve our rate of convergence we forgo expanding $f(x)$ in a sine series and follow the procedure of Gottlieb and Orszag (1977). Fourier expand a new function $g(t,x)$, where

$$g(t,x) \equiv f(t,x) + \frac{x(x-1)}{2} \quad (44)$$

We have defined $g(t,x)$ so that the boundary conditions on g and g'' are homogeneous:

$$g(t,0) = g(t,1) = g''(t,0) = g''(t,1) = 0 \quad (45)$$

In fact, by using equations (34) and (44) we find

$$g(t,0)^{(2n)} = g(t,1)^{(2n)} = 0 \text{ for } n=0,1,2,\ldots,\infty \quad (46)$$

When we compute the Fourier coefficients of $g(t,x)$ by integration by parts, we find that all of the surface terms vanish. The Fourier sine series expansion of $g(x)$ converges exponentially. Equivalently, the partial sums

$$f_N(t,x) \equiv -\frac{x(x-1)}{2} + \sum_{k=1}^{N} f_k(t)\sin(k\pi x) \quad (47)$$

converge exponentially to the exact solution of the inhomogeneous heat equation which is:

$$f(t,x) = -\frac{x(x-1)}{2} + \sum_{k=1}^{\infty} \hat{f}_k(0)\sin(k\pi x)e^{-\pi^2 k^2 t} \tag{48}$$

where

$$\hat{f}_k(0) \equiv 2 \int_0^{\pi} \left[f(t=0,x) + \frac{x(x-1)}{2}\right] \sin(k\pi x)dx \tag{49}$$

The leading term of the error, $f(x,t)-f_N(t,x)$, is $\hat{f}_{N+1}(0)\sin[(N+1)\pi x]e^{-\pi^2(N+1)^2 t}$, which decays exponentially in N for $t > 0$.

Polynomial Basis Functions

In approximating a non-periodic function with a Fourier sum, we often find that including an appropriate polynomial in the spectral sum improves the rate of convergence. The method of selecting an appropriate polynomial depends on the boundary conditions of the function that is being approximated. We would like to discuss a more general method of finding polynomials to use in spectral approximations so that the rate of convergence does not explicitly depend on boundary conditions. We abandon sines and cosines as basis functions; instead we use the eigenfunction $\{ \phi_n(x) \}$, of a Sturm-Liouville operator:

$$\frac{d}{dx}\left[p(x)\frac{d\phi_n}{dx}\right] + \left[\lambda_n w(x)-q(x)\right]\phi_n(x) = 0 \tag{50}$$

where

$$p(x) \geq 0 \tag{51}$$

$$w(x) \geq 0 \tag{52}$$

$$q(x) \geq 0 \tag{53}$$

$$a \leq x \leq b \tag{54}$$

Here, the λ_n are the eigenvalues associated with $\phi_n(x)$ and $w(x)$ is the weighting function that is used to define the inner product. The eigenfunctions are complete and orthonormal with respect to these weighting functions

$$\int_a^b w(x)\phi_n(x)\phi_m(x)\, dx = \delta_{nm} \tag{55}$$

We can express $f(x)$ as an infinite series in $\phi_n(x)$ and approximate $f(x)$ with the partial sum $f_N(x)$. The mean square error of the approximation with respect to the weighting function $w(x)$ is $L_2(N)$.

$$f(x) = \sum_{n=0}^{\infty} a_n \phi(x) \tag{56}$$

$$f_N(x) = \sum_{n=0}^{N} a_n \phi(x) \tag{57}$$

$$L_2(N) = \left[\sum_{N+1}^{\infty} |a_n|^2 \right]^{1/2} \tag{58}$$

where

$$a_n = \int_a^b f(x)\phi_n(x)w(x)dx \tag{59}$$

The error is the square of the truncated spectral coefficients. To show how $L_2(N)$ and a_N decrease with N we follow the procedure of Lanczos (1956), and integrate equation (59) twice by parts to obtain

$$a_n = \frac{1}{\lambda_n} p(x) \left[(\phi_n \frac{df}{dx} - f \frac{d\phi_n}{dx}) \right]_a^b - \frac{1}{\lambda_n} \int_a^b \phi_n(x)h(x)w(x)dx \tag{60}$$

where $h(x)$ is defined

$$h(x) \equiv \frac{1}{w(x)} \frac{d}{dx} (p \frac{df}{dx}) - \frac{q(x)f(x)}{w(x)} \tag{61}$$

The surface term in equation (60) vanishes regardless of the values of $f(x)$ and its derivatives at the boundaries (as long as they remain nonsingular) if $p(a) = p(b) = 0$. When $p(a) = p(b) = 0$ we call the Sturm-Liouville operator singular. For singular operators, each time equation (60) is integrated by parts the surface terms vanish as long as the higher derivatives of $f(x)$ remain bounded. Integrating equation (60) by parts 2p times, shows that the spectral coefficient a_n is of order $(1/\lambda_n)^p$.

As $p \to \infty$, the spectral series, $f_N(x)$ converges exponentially.

As an example of a singular operator we consider the Chebyshev polynomials, $T_n(x)$. The Sturm-Liouville equation that generates the Chebyshev polynomials has

$$p(x) = \sqrt{1-x^2} \tag{62}$$

$$w(x) = 1/\sqrt{1-x^2} \tag{63}$$

$$q(x) = 0 \tag{64}$$

$$\lambda_n = n^2 \tag{65}$$

$$-1 \leq x \leq 1 \tag{66}$$

One way to convince ourselves that the Chebyshev polynomials really do have exponential convergence is to approximate a sine function with a Chebyshev series (cf. Gottlieb and Orszag, 1977). Computing the spectral coefficients from equation (59) we find

$$\sin(m\pi x) = 2 \sum_{\substack{n=1 \\ n=\text{odd}}}^{\infty} \left[J_n(m\pi)(-1)^{(n-1)/2}\right] T_n(x) \tag{67}$$

Only the odd Chebyshev polynomials enter the sum in equation (67) because $\sin(m\pi x)$ is an odd function of x. The coefficients of the Chebyshev series are proportional to Bessel functions, $J_n(m\pi)$. Bessel functions of low order behave like sines, but when their order becomes greater than their argument they exponentially decay. Therefore, for $n > m\pi$ the Chebyshev spectral coefficients decay exponentially, and the convergence of the partial sums is exponential. Between -1 and 1, $\sin(m\pi x)$ has m wavelengths. Since $m\pi$ Chebyshev polynomials are required for exponential convergence, we conclude that approximately π Chebyshev polynomials are needed per wavelength of the function being approximated. This factor of π should be compared to the requirements of second-order finite difference methods where we found that approximately 20 grid points per wavelength are needed to obtain 1% accuracy.

As a second example of polynomial expansions, we use the Legendre polynomials as a set of basis functions. The Legendre polynomials are eigenfunctions of a Sturm-Liouville equation with:

$$p(x) = (1-x^2) \tag{68}$$

$$w(x) = 1 \tag{69}$$

$$q(x) = 0 \tag{70}$$

$$\lambda_n = n(n+1) \tag{71}$$

$$-1 \leq x \leq 1 \tag{72}$$

Again, expressing $\sin(m\pi x)$ as a spectral sum we obtain

$$\sin(m\pi x) = \sum_{\substack{n=1 \\ n=\text{odd}}}^{\infty} \left[\frac{(2n+1)(-1)^{(n-1)/2}}{(2m)^{1/2}} J_{n+1/2}(m\pi) \right] P_n(x) \tag{73}$$

The coefficients multiplying the Legendre polynomials are Bessel functions of half integral order; they decrease exponentially when their order is greater than their argument. Again, we need on the order of π polynomials per wavelength to obtain exponential convergence.

One way to heuristically see the resolution properties of a spectral sum is to examine the spacings between zero-crossings of the basis functions. Fourier series have equally spaced zeroes which make them well-suited for approximating periodic functions. A truncated Fourier series resolves boundary layers poorly. Legendre and Chebyshev polynomials have more zero crossings near the boundaries at -1 and 1 than they do at the interior of their domain near 0. Near the interior $P_N(x)$ and $T_N(x)$ both have an average separation between zero crossings of π/N. Near the boundaries, the average spacing reduces to $\pi^2/2N^2$. Partial sums of $P_N(x)$ or $T_n(x)$ are well suited for approximation function with boundary-layers. If a boundary layer has thickness of order δ, then approximately $(1/\delta)^{1/2}$ terms are needed in the spectral sum to resolve the boundary layer.

As our last example of choosing basis functions, we consider the one-dimensional axisymmetric heat equation in cylindrical coordinates with an inhomogeneous forcing term:

$$\frac{\partial T}{\partial t} = \kappa \left(\frac{\partial^2}{\partial r^2} + \frac{1}{r}\frac{\partial}{\partial r} \right) T + 1 \tag{74}$$

with boundary conditions $T(t,r=1) = 0$, $T(t,r=0)$ finite, and initial conditions $T(t=0,r) = f(r)$. An apparently obvious choice of basis functions are the Bessel functions of index zero

$$\phi_n(r) \equiv \sqrt{2}\ \frac{J_0(j_{0n}r)}{J_1(j_{0n})} \tag{75}$$

where j_{0n} is the nth root of J_0. The eigenfunctions are normalized to obey equation (55) with weighting function $w(r) = r$. The $J_0(j_{0n}r)$ Bessel function is an eigenfunction of $\left(\frac{\partial^2}{\partial r^2} + \frac{1}{r}\frac{\partial}{\partial r}\right)$ which is a Sturm-Liouville operator with $p(r) = r$ and eigenvalue j_{0n}. Because the $\phi_n(r)$ are complete, obey the same homogeneous boundary conditions as $T(t,x)$, and are eigenfunctions of the differential operator on the right-hand-side of equation (74), they are the natural basis functions for this problem,

$$T(t,x) = \sum_{n=1}^{\infty} a_n(t)\phi_n(x) \tag{76}$$

The exact solution to equation (74) is

$$T(t,x) = \sum_{n=1}^{\infty} \left[f_n\, e^{-j_{0n}\kappa t} - \frac{\sqrt{2}}{\kappa (j_{0n})^2} \right] \phi_n(r) \tag{77}$$

where

$$f_n = \int_0^1 \phi_n(r) f(r) r\, dr \tag{78}$$

The solution expanded in terms of the Bessel functions does not converge exponentially. From the asymptotic behavior of j_{0n},

$$j_{0n} \sim \pi\,(n - 1/4), \tag{79}$$

we see that the exact solution converges as $1/n^2$ far from the boundary and as $1/n$ near the boundary. The slow convergence is due to the fact that $p(r) \neq 0$ at the boundary and the Sturm-Liouville operator is not singular. For this problem, second-order accurate finite differences are more efficient than a Bessel expansion. Spectral methods can still be used, but Chebyshev or Legendre polynomials should be employed. Although Chebyshev or Legendre functions do not appear to be "natural" choices for cylindrical geometry, they converge rapidly and work well (Marcus, 1983).

IMPLEMENTATION OF SPECTRAL METHODS

Galerkin's Method

Once the spectral numericist has decided on a set of basis functions, his second task is to compute efficiently the spectral coefficients. Generally, this task requires solving a large set of coupled ordinary differential equations. The most straight-forward of all spectral techniques is a modal or Galerkin method. In Galerkin's method each basis function obeys the same boundary conditions as the function that is being approximated. Consider the combined initial value, boundary-value problem

$$\frac{\partial u}{\partial t} = \mathcal{L}(u) \tag{80}$$

where $\mathcal{L}$ is some arbitrary spatial operator. We impose homogeneous boundary conditions on both u and the basis functions, $\phi_i(x)$

$$u(t, x = -1) = u(t, x=1) = 0 \tag{81}$$

$$\phi_n(-1) = \phi_n(1) = 0 \quad \text{for all} \quad n. \tag{82}$$

As usual, the partial sum u_N is defined

$$u_N(t,x) = \sum_{n=1}^{N} a_n(t)\phi_n(x). \tag{83}$$

The easiest case that can be solved with Galerkin's method occurs when $\mathcal{L}$ is a linear differential operator with constant coefficients and where the $\phi_n(x)$ are eigenfunctions of $\mathcal{L}$. In this case, substitution of equation (83) into equation (80) results in a set of N, non-coupled ordinary differential equations which can be solved easily for the $a_n(t)$.

In a slightly more complicated case, we still restrict $\mathcal{L}$ to be a linear differential operator with constant coefficients, but we no longer require that the $\phi_n(x)$ be eigenfunctions of $\mathcal{L}$. For example, we consider the wave equation with $\mathcal{L} = \frac{d}{dx}$

$$\frac{\partial u}{\partial t} = \frac{\partial u}{\partial x}, \tag{84}$$

with $-1 \leq x \leq 1$, and with boundary condition,

$$u(1) = 0. \tag{85}$$

Using the basis functions,

$$\phi_n(x) \equiv [P_n(x) - 1] \tag{86}$$

to expand $u(t,x)$, substituting the series in equation (84), and taking the inner product of both sides of the resulting equation with respect to $\phi_n(x)$, we obtain a set of coupled ordinary differential equations for the spectral coefficients $a_n(t)$. Let $\underline{a}$ be the column vector whose n^{th} element is $a_n(t)$. The equations for the spectral coefficients can be written in matrix form:

$$\underline{\underline{P}}\,\dot{\underline{a}} = \underline{\underline{M}}\,\underline{a} \tag{87}$$

The dot above the vector $\underline{a}$ imples that each element is differentiated with respect to time . The elements of $\underline{\underline{M}}$ and $\underline{\underline{P}}$ are easily written in terms of inner products.

$$P_{mn} = (\phi_m, \phi_n) \tag{88}$$

$$M_{mn} = (\phi_m, \mathcal{L}(\phi_n)) \tag{89}$$

In practice, we would never directly use the matrix equation (87) to solve for $a_n(t)$. Matrix arithmetic (multiplication or inversion) requires of order N^2 or N^3 operations (where N is the number of terms in the spectral sum). Usually $\underline{\underline{P}}$ and $\underline{\underline{M}}$ are sparse or have a special form due to the fact that $\mathcal{L}$ implies a recursion relationship among the spectral coefficients. By exploiting the sparsity or symmetry of $\underline{\underline{M}}$,equation (87) can often be solved in only N operations per timestep.

If $\mathcal{L}$ is a nonlinear differential operator or linear differential operator with non-constant coefficients, then Galerkin's method becomes unwieldy. As an example, consider the nonlinear wave equation

$$\frac{\partial u}{\partial t} = -u\frac{\partial}{\partial x}u \tag{90}$$

over the domain $0 \leq x \leq \pi$ and with boundary condition

$$u(t,x=0) = 0 \tag{91}$$

Writing $u(t,x)$ as a Fourier sine series (which is complete over $0 \leq x \leq \pi$) equation (90) becomes

$$\sum_{n=1}^{N} \dot{a}_n(t)\sin(n\pi x) = -\left[\sum_{k=1}^{N} a_k(t)\sin(k\pi x)\right] \frac{d}{dx}\left[\sum_{m=1}^{N} a_m(t)\sin(m\pi x)\right] \tag{92}$$

$$= -\frac{\pi}{2}\sum_{n=1}^{N-1}\sum_{m=1}^{N-n} m a_m a_{n+m}\sin(n\pi x) - \frac{\pi}{2}\sum_{n=1}^{2N}\sum_{\substack{m=n-N\\ m>0}}^{N} \operatorname{sgn}(n-m)\, m a_m a_{|m-n|}\sin(n\pi x) \tag{93}$$

where $\operatorname{sgn}(n-m)$ is the sign of $(n-m)$ and is equal to zero if $(n-m)=0$. The right-hand side of equation (93) is a convolution product. The convolution leads to three problems: first, it couples all of the ordinary differential equations for $a_n(t)$; second, the convolution product is expensive to compute - it requires (on the order of) N^2 multiplications; thirdly, the convolution product of two partial sums of spectral modes generates spectral modes that are not included in or spanned by the original finite set of basis functions. The convolution of a sine series with N modes and its derivative produces a product containing $2N$ sine modes. In previous cases we found that when a linear differential operator with constant coefficients operates on a finite sum of basis functions, the output is itself spanned by this same finite set of basis functions; in these cases we say that the finite set of basis functions is closed with respect to that operator. In general, a finite set of basis functions is not closed with respect to a nonlinear operator. The lack of closure requires us to project the output of $\mathcal{L}$ back onto the initial set of basis functions. In the example in equation (93), projection is done by setting all of the spectral coefficients a_n with $n > N$ equal to zero. Galerkin's method uses truncation to project the output of $\mathcal{L}$. Other spectral methods use other types of projection. From equation (93), we see that the set of all sine functions is closed under $\mathcal{L}$; no cosines are produced. On the other hand, the set of cosines are not closed under $\mathcal{L}$; $\mathcal{L}$ operating on a cosine series produces a sine series. The closure or lack of closure of an infinite set of sines or cosines is due to the spatial symmetry properties of $\mathcal{L}$. $\mathcal{L}$ is anti-reflection symmetric about $x = 0$ and so are the sines; therefore, the set of sines is closed. The cosines do not have the anti-reflection symmetry of $\mathcal{L}$ and therefore they are not closed. In general, for every incomplete or sparse set of basis functions that is closed with respect to an operator, there is a physical symmetry common to the operator and the sparse set. It is to the numericist's advantage to exploit the fact that some equations together with their boundary and initial conditions have a symmetry. The clever numericist will choose a sparse set of basis functions that shares the symmetry. For example, if we use Galerkin's method to compute thermal convection in a star and if we are interested only in convective cells with duodecahedral symmetry (i.e., the convective cells tesselate the stellar surface in a soccer ball pattern), then using a set of basis functions with duodecahedral symmetry is more efficient than using the spherical harmonics as basis functions. A calculation of duodecahedral convection that uses a spectral sum with all spherical harmonics, $Y_{\ell,m}$ with $|m|<\ell$ and $0 \leq \ell \leq 32$ (1089 modes) can also be done with a much smaller series containing only 19 duodecahedral harmonics to obtain the identical spatial resolution. Each convolution with the smaller set of duodecahedral harmonics is approximately 10,000 times faster than the convolution with the larger set of spherical harmonics.

Tau Method

Galerkin's method cannot be used to approximate a function f(x) if the basis functions do not obey the same homogeneous boundary conditions as f(x) or if f(x) has inhomogeneous boundary conditions. For these boundary-value problems we use the tau method, developed by Lanczos (1956). Again, consider the heat equation (equation 1) over the domain $-1 \leq x \leq 1$ with boundary conditions

$$T(t,-1) = \alpha \tag{94}$$

$$T(t,L) = \beta \tag{95}$$

This time, we write T(t,x) as a spectral sum of Legendre polynomials

$$T(t,x) = \sum_{\ell=0}^{N} a_\ell(t)\, P_\ell(x) \tag{96}$$

The Legendre polynomials do not obey the inhomogeneous boundary conditions of T(t,x); instead, they obey

$$P_\ell(-1) = (-1)^\ell \tag{97}$$

$$P_\ell(1) = 1 \tag{98}$$

Substituting equation (96) into equation (1), multiplying both sides of equation (1) by $P_n(x)$, and integrating the resulting equation from -1 to 1, we obtain (N+1) equations for the (N+1) unknown spectral coefficients, a_n, n=0,1,...N. However, to satisfy the two boundary equations, (94) - (95), we also require that

$$\sum_{\ell=0}^{N} a_\ell(t)P_\ell(-1) = \sum_{\ell=0}^{N} a_\ell(t)(-1)^\ell = \alpha \tag{99}$$

and

$$\sum_{\ell=0}^{N} a_\ell(t)P_\ell(1) = \sum_{\ell=0}^{N} a_\ell(t) = \beta \tag{100}$$

This presents the dilemma that we have (N+3) equations and only (N+1) unknowns. To solve this problem, we must first examine the second derivative operator in more detail. Writing the second derivative of T in terms of the Legendre sum in equation (96) we obtain

$$\frac{d^2}{dx^2} T(x,t) \equiv \sum_{n=0}^{N} b_n(t)P_n(x) = \sum_{n=0}^{N} a_n \frac{d^2}{dx^2} P_n(x) \tag{101}$$

The second derivative of $P_n(x)$ is a linear combination of Legendre polynomials whose order is less than or equal to (n-2). The second derivative operator is a lowering operator; it maps the set of basis functions $\{P_n(x) : n=0,\ldots,N\}$ into the set $\{P_n(x) : n=0,\ldots, N-2\}$. Furthermore, since the operator $\frac{d^2}{dx^2}$ preserves parity, the second derivative of a Legendre polynomial of even (or odd) order is a sum of Legendre polynomials of even (or odd) order. We can write down a relationship between the b_n and a_n of equation (101) as the vector equation

$$\begin{pmatrix} b_0 \\ b_1 \\ b_2 \\ b_3 \\ \cdot \\ \cdot \\ \cdot \\ b_{N-3} \\ b_{N-2} \\ b_{N-1} \\ b_N \end{pmatrix} = \begin{pmatrix} 0 & 0 & x & 0 & x & 0 & . & . & . & x & 0 & x \\ 0 & 0 & 0 & x & 0 & x & . & . & . & 0 & x & 0 \\ 0 & 0 & 0 & 0 & x & 0 & . & . & . & x & 0 & x \\ 0 & 0 & 0 & 0 & 0 & x & . & . & . & 0 & x & 0 \\ \cdot & \cdot & \cdot & \cdot & \cdot & \cdot & \cdot & \cdot & \cdot & \cdot & \cdot & \cdot \\ \cdot & \cdot & \cdot & \cdot & \cdot & \cdot & \cdot & \cdot & \cdot & \cdot & \cdot & \cdot \\ \cdot & \cdot & \cdot & \cdot & \cdot & \cdot & \cdot & \cdot & \cdot & \cdot & \cdot & \cdot \\ 0 & 0 & 0 & 0 & 0 & 0 & . & . & . & 0 & x & 0 \\ 0 & 0 & 0 & 0 & 0 & 0 & . & . & . & 0 & 0 & x \\ 0 & 0 & 0 & 0 & 0 & 0 & . & . & . & 0 & 0 & 0 \\ 0 & 0 & 0 & 0 & 0 & 0 & . & . & . & 0 & 0 & 0 \end{pmatrix} \begin{pmatrix} a_0 \\ a_1 \\ a_2 \\ a_3 \\ \cdot \\ \cdot \\ \cdot \\ a_{N-3} \\ a_{N-2} \\ a_{N-1} \\ a_N \end{pmatrix} \tag{102}$$

or equivalently

$$\underline{b} = \underline{\underline{D}}^2 \; \underline{a} \tag{103}$$

where x stands for any non-zero matrix element. Note that $\underline{\underline{D}}^2$ is singular because the last two rows are zero. The fact that $\underline{\underline{D}}^2$ is non-invertible is due to the fact that $\frac{d^2}{dx^2}$ cannot be inverted unless two boundary conditions are supplied.

We can try to solve the spectral heat equation $\dot{\underline{a}} = \kappa \underline{\underline{D}}^2 \underline{a}$ with a forward Euler integration in time, or

$$a(t+\Delta t) = \underline{a}(t) + \kappa(\Delta t) \, \underline{\underline{D}}^2 \, \underline{a}(t) \tag{104}$$

In solving the heat equation explicitly, we avoid having to invert the singular matrix $\underline{\underline{D}}^2$. Unfortunately, in the explicit equation (104), the last two rows of $\underline{\underline{D}}^2$ make $a_N(t) = a_N(0)$ and $a_{N-1}(t) = a_{N-1}(0)$ for all time. This is undesirable. Furthermore, after one timestep, the $a_n(t)$ no longer satisfy the boundary conditions. The remedy to all of these problems is to modify the matrix $\underline{\underline{D}}^2$ in the heat equation so that the bottom two rows are replaced by the boundary conditions of equations (99) - (100).

$$\begin{vmatrix} \dot{a}_0 \\ \dot{a}_1 \\ \dot{a}_2 \\ \dot{a}_3 \\ \cdot \\ \cdot \\ \cdot \\ \dot{a}_{N-3} \\ \dot{a}_{N-2} \\ \beta \\ \alpha \end{vmatrix} = \begin{vmatrix} 0 & 0 & x & 0 & 0 & 0 & . & . & . & x & 0 & x \\ 0 & 0 & 0 & 0 & 0 & x & . & . & . & 0 & x & 0 \\ 0 & 0 & 0 & 0 & x & 0 & . & . & . & x & 0 & x \\ 0 & 0 & 0 & 0 & 0 & x & . & . & . & 0 & x & 0 \\ \cdot & \cdot & \cdot & \cdot & \cdot & \cdot & \cdot & \cdot & \cdot & \cdot & \cdot & \cdot \\ \cdot & \cdot & \cdot & \cdot & \cdot & \cdot & \cdot & \cdot & \cdot & \cdot & \cdot & \cdot \\ \cdot & \cdot & \cdot & \cdot & \cdot & \cdot & \cdot & \cdot & \cdot & \cdot & \cdot & \cdot \\ 0 & 0 & 0 & 0 & 0 & 0 & . & . & . & 0 & x & 0 \\ 0 & 0 & 0 & 0 & 0 & 0 & . & . & . & 0 & 0 & x \\ 1 & 1 & 1 & 1 & 1 & 1 & . & . & . & 1 & 1 & 1 \\ 1 & -1 & 1 & -1 & 1 & -1 & . & . & . & 1 & -1 & 1 \end{vmatrix} \begin{vmatrix} a_0 \\ a_1 \\ a_2 \\ a_3 \\ \cdot \\ \cdot \\ \cdot \\ a_{N-3} \\ a_{N-2} \\ a_{N-1} \\ a_N \end{vmatrix} \qquad (105)$$

Notice that the matrix on the right-hand side of equation (105) is invertible; if we integrate $\dot{a}_i(t)$ forward in time with a stable implicit method, then the matrix can be inverted at each timestep. Using a backwards Euler method to solve equation (105), we obtain

$$(\hat{\underline{\underline{I}}} - \kappa \Delta t \hat{\underline{\underline{D}}})\ \underline{a}(t+\Delta t) = \hat{\underline{\underline{I}}}\ \underline{a}(t) + \underline{x}(\beta,\alpha) \qquad (106)$$

where $\hat{\underline{\underline{D}}}$ is the matrix on the right-hand side of equation (105), where $\hat{\underline{\underline{I}}}$ is the identity with the last two rows replaced by zeroes and where $\underline{x}(\alpha,\beta)$ is a column vector whose elements are all zero except for the last two which are β and α respectively.

Using equation (106), the $a_n(t)$ exactly satisfy the boundary conditions for all time, but do they still satisfy the heat equation? We evaluate the mean-square error that arises from approximating a function $f(x)$ with boundary conditions $f(1) = \beta$ and $f(-1) = \alpha$ by using the tau method. In a Legendre-tau method we approximate $f(x)$ with $f_N(x)$,

$$f_N(x) \cong f(x) \tag{107}$$

where

$$f_N(x) = \sum_{n=0}^{N} a_n P_n(x). \tag{108}$$

Writing $f(x)$ exactly as an infinite series

$$f(x) = \sum_{n=0}^{\infty} c_n P_n(x) \tag{109}$$

The mean-square error is

$$L_2 = \left[\int_{-1}^{1} |f(x)-f_N(x)|^2 \, dx \right]^{1/2} = \left[\sum_{n=0}^{N} |c_n - a_n|^2 + \sum_{n=N+1}^{\infty} |c_n|^2 \right]^{1/2} \tag{110}$$

With the tau method we evaluate the first (N-2) spectral coefficients of a_n by taking inner products of $f(x)$ with $P_n(x)$ or

$$a_n = \int_{-1}^{1} f(x) P_n(x) dx = c_n \quad \text{for} \quad 0 \le n \le N-2 \tag{111}$$

The last two spectral coefficients are not determined from inner products; instead, they come from the boundary conditions (99)-(100). The mean-square error becomes

$$L_2 = \left[|a_{N-1} - c_{N-1}|^2 + |a_N - c_N|^2 + \sum_{n=N+1}^{\infty} |c_n|^2 \right]^{1/2} \tag{112}$$

The last term on the right-hand side of equation (112) is the same error that arises with Galerkin's method and is exponentially small if $f(x)$ is sufficiently smooth. We must show that the additional two terms $|a_N - c_N|^2$ and $|a_{N-1} - c_{N-1}|^2$ are also exponentially small. The exact function $f(x)$ obeys the exact boundary conditions

$$f(-1) = \sum_{n=0}^{\infty} c_n (-1)^N = \alpha \tag{113}$$

$$f(1) = \sum_{n=0}^{\infty} c_n = \beta \tag{114}$$

Comparing equation (99) with equation (113) and equation (100) with equation (114), we see that

$$(-1)^{N-1} a_{N-1} + (-1)^N a_N = \sum_{n=N-1}^{\infty} (-1)^n c_n \tag{115}$$

and

$$a_{N-1} + a_N = \sum_{n=N-1}^{\infty} c_n. \tag{116}$$

The solution to equations (115) - (116) is

$$a_N = \sum_{k=0}^{\infty} c_{N+2k} \tag{117}$$

$$a_{N-1} = \sum_{k=0}^{\infty} c_{N-1+2k}. \tag{118}$$

Equations (117)-(118) show that the second contribution to L_2 in equation (112) is exponentially small:

$$|a_{N-1} - c_{N-1}|^2 + |a_N - c_N|^2 \leq$$

$$2 \left|\sum_{k=0}^{\infty} c_{N-1+2k}\right|^2 + 2 \left|\sum_{k=0}^{\infty} c_{N+2k}\right|^2 + 2|c_{N-1}|^2 + 2|c_N|^2 \tag{119}$$

The right-hand side of equation (119) decreases exponentially with N.

Pseudo-Spectral Methods

Our last example of a technique for computing the spectral coefficients is known as the pseudo-spectral method, the collocation method or the method of selected points. It is used in nonlinear equations or linear equations with non-constant coefficients. Its purpose is to avoid a convolution product. The method exploits the fact that spectral differentiation is more accurate than finite-differences, but that multiplication of two functions which are tabulated at a set of selected points (or collocation points) is faster than spectral convolution. In the pseudo-spectral methods, all differentiation and quadrature is done with spectral approximations; all multiplication and division are done on a grid of points. The representation of the function goes back and forth between spectral and physical space by use of discrete (and fast) transforms. When transforming a function with N spectral coefficients from spectral space to physical space, the user should sample the function at $M = N$ grid points. Over-sampling with $M > N$ is wasteful; under-sampling with $M < N$ loses information and prohibits reconstruction of the spectral coefficients from the M sampled points. Problems arise in the pseudo-spectral method due to accidental under-sampling. The user can inadvertently under-sample a function if there are nonlinear or non-constant coefficient terms.

To see how under-sampling might arise, consider again the nonlinear wave equation

$$\frac{\partial u}{\partial t} = - u \frac{\partial u}{\partial x} \tag{120}$$

with periodic boundary conditions

$$u(t,x=0) = u(t,x=1) \tag{121}$$

Initially, u and $\frac{\partial u}{\partial x}$ are represented in spectral space as sine series (sines are complete over the interval):

$$u_N(t,x) = \sum_{n=0}^{N} a_n(t) \sin \pi nx \tag{122}$$

$$\frac{\partial u_N}{\partial x}(t,x) = \pi \sum_{n=0}^{N} a_n(t) \cos \pi nx \tag{123}$$

Fourier transforming u_N and $\frac{\partial u_N}{\partial x}$ into physical space, we obtain tabulations of the two functions $u_N(t,x_i)$ and $\frac{\partial u_N}{\partial x}(t,x_i)$ at the collocation points $x_i \equiv \frac{(i-1)}{N}$, $i=1,N$. At this step in the pseudo-spectral method we have enough information to reconstruct the spectral coefficients a_n and $(n\pi a_n)$ from the tabulated functions. The product $-u\frac{\partial u}{\partial x}$ is tabulated at the collocation point in the obvious way:

$$-u_N \frac{\partial u_N}{\partial x}(t,x_i) \equiv -u_N(t,x_i)\frac{\partial u_N}{\partial x}(t,x_i) \tag{124}$$

Now the product $\frac{\partial u_N}{\partial x}$ is tabulated at N points, but if we were to represent it spectrally (by a convolution product) it would require $2N$ terms (see equation 93):

$$-u_N \frac{\partial u_N}{\partial x} = \sum_{n=1}^{2N} b_n \sin(n\pi x) \tag{125}$$

We have accidentally under-sampled because the function $-u_N\frac{\partial u_N}{\partial x}$ has $2N$ spectral coefficients but we have only tabulated $-u_N\frac{\partial u_N}{\partial x}$ at N points. If we had used Galerkin's method to compute the convolution product, we would project the $2N$ spectral coefficients of the product back onto an N-dimensional spectral space by explicitly setting the $b_n = 0$ for all $n > N$ and keeping the coefficients b_n unchanged for all $0 \leq n \leq N$. With the pseudo-spectral method the under-sampled product, $-u_N\frac{\partial u_N}{\partial x}$, is also projected back onto an N-dimensional spectral space when we naively inverse transform the tabulated product back into Fourier space. The discrete inverse transform not only sets the coefficients b_n equal to zero for all $n > N$, but also mixes the spectral coefficient b_n with $0 < n \leq N$ with the spectral coefficients with $n > N$. The mixing is called aliasing. It is due to the fact that the discrete Fourier transform with N points is unable to tell the difference between Fourier modes of wavenumber n and wavenumber $2N-n$. For example, $\sin(n\pi x)$ evaluated at the grid points $x_i = \frac{(i-1)}{N}$ $i=1,N$ is indistinguishable from $-\sin[(2N-n)\pi x]$ evaluated at the same grid points. It should not be surprising that the inverse transform contaminates the

n^{th} spectral coefficient with the $(2N-n)^{th}$ coefficient. There is never contamination or aliasing error if the number of sampling points of a function is greater than or equal to the number of spectral components of the function.

The aliasing error can be avoided or reduced by several methods. One way of removing the alias in equation (121) is to evaluate u_n, $\frac{\partial u_N}{\partial x}$, and the product at $\frac{3}{2}$ N collocation points. The discrete inverse transform of the product will have no aliasing errors in the first N spectral coefficients. Often, aliasing errors can be ignored with no harmful effects. The reason is that aliasing contaminates the exact solution with the spectral coefficients b_n with $n > N$. If N is chosen sufficiently large, then the spectral coefficients with $n > N$ are exponentially small and the aliasing error is exponentially small. For example, aliasing errors in the Navier-Stokes equation can usually be ignored if the equation is solved in a way such that the aliasing error does not violate energy or momentum conservation.

DISCUSSION

There are more types of spectral methods than those discussed in this paper. Most of the other techniques are hybrids of those outlined here. All good spectral methods share the property that their convergence is exponential and are therefore more economical than finite differences. Spectral methods require approximately 7 times fewer degrees of freedom (grid points or modes) per spatial dimension than do second-order finite-differences. Therefore, in 2 or 3 dimensional calculations, spectral methods are often the only practical way of obtaining adequate spatial resolution. In the future, as astrophysicists increasingly want to extend their numerical calculations to 2 and 3 dimensions, spectral methods will increasingly become part of the astrophysicist's standard numerical tools.

REFERENCES

Dahlquist, G. and Bjorke, A.: 1974, Numerical Methods, Prentice-Hall.
Gottlieb, D. and Orszag, S.A.: 1977, Numerical Analysis of Spectral Methods, SIAM.
Lanczos, C. 1956, Applied Analysis, Prentice-Hall.
Marcus, P.S. 1983, submitted to the Journal of Fluid Mechanics.

NUMERICAL MODELING OF SUBGRID-SCALE FLOW IN TURBULENCE, ROTATION, AND CONVECTION

Philip S. Marcus

Massachusetts Institute of Technology

We show that it is impossible to simulate numerically all of the length-scales in astrophysical turbulence. We look at the effects of ignoring the unresolvable, small-scale flow, and show how numerical simulations that neglect the subgrid-scale motions produce erroneous solutions. Then we discuss the "quick fix" remedy of introducing a numerical or eddy-viscosity. We sketch how the analytic theories of turbulence attempt to model the large-scale effects of the small-scale motions. In the last section of this paper we examine four anisotropic, inhomogeneous flows of astrophysical interest for which numerical and eddy-viscosities produce incorrect solutions. Improved models of the subgrid-scale flow are examined. We show how numerical simulation of large Reynolds number (but non-turbulent) flows can guide us in modeling subgrid-scale flows in astrophysical settings.

Turbulence, rotation, and convection are generally treated by astrophysicists in the same manner; they are avoided whenever possible. Jets do not become turbulent, stars do not rotate, and planetary nebulae do not convect - except when the absence of these motions contradicts common sense or when the mixing properties of these motions must be invoked to solve some astrophysical paradox. The reason we avoid calculating these flows is that the numerical computation of turbulence, in even the simplest laboratory settings, is usually impossible because the velocity spans a much larger range of length-scales and has many more degrees of freedom than can be accommodated in present computers. The

K.-H. A. Winkler and M. L. Norman (eds.), Astrophysical Radiation Hydrodynamics, 387–414.

calculation of turbulent flows therefore requires us to model or neglect a large portion of the flow. In this paper, we shall not be concerned with the technical details of computing derivatives or matching boundary conditions but with the more fundamental problem of how the equations of motion should be changed so that the flow at the largest scale can be accurately simulated while the large-scale effects of the numerically unresolvable or subgrid-scale motions are modeled. Our task is analogous to the one of finding a simple spatial boundary condition: it is impractical to compute unbounded flows (for example, a stellar wind) so at some finite distance far from the physics of interest, the calculation is cut off and an artificial but physically appropriate boundary condition is imposed (for example, all characteristics must point outward at the boundary). With turbulent flows it is impractical to compute the flow at all spatial scales or equivalently at every point in Fourier space, so at some large, but finite, wavenumber far away from the physics of interest the calculation is cut off and an artificial boundary condition is imposed that represents the effects of the subgrid-scales. As with all artificially imposed boundary conditions, the less information that propagates from the neglected region of the flow into the computational domain, the better the approximation. Boundary conditions in wave number space are more complicated than those in physical space because the equations of motion in wave space are non-local, allowing the subgrid-scale motions to affect directly the flow everywhere, not just near the boundary.

The philosophy of this paper is that at large scales, astrophysical flows do not exhibit much universality. Large-scale coherent features, such as the solar granulation or the Red Spot of Jupiter, cannot be predicted by an analytic theory of turbulence. They must be simulated numerically for every particular flow, while the small-scale structures in astrophysical flows may be universal. It is a fundamental assumption of this paper that, as energy cascades down from the large energy-producing scales to the small scales, the flow loses information about the large-scales such that the effects of the small scales can be modeled by a universal theory of turbulence.

FORCING DUE TO SUBGRID-SCALE MOTIONS

In most astrophysical flows, we are interested only in the large-scale velocity which is not affected directly by viscosity, so we use the inviscid Euler equation rather than the Navier-Stokes equation. The danger of ignoring viscosity is that it strongly influences the subgrid-scale flow which in turn acts upon the large-scale flow. To see the consequences of the numerically unresolved motions, it is necessary to realize that a numerical simulation implicitly filters the velocity, $\underline{v}$, into a large-scale numerically resolvable component, $\bar{\underline{v}}$, and a subgrid-scale component, $\underline{v}'$. Mathematically, the numerical representation of the velocity convolves $\underline{v}$ with a filtering function $F(x,x')$:

$$\bar{\underline{v}}(x) = \int F(x,x')\underline{v}(x')d^3x' \quad (1)$$

$$\underline{v}'(x) = \underline{v}(x) - \bar{\underline{v}}(x) \quad (2)$$

The filtering function contains a characteristic length-scale, Δ, (e.g., the grid spacing in finite-difference methods) so that $\bar{v}$ represents motions on length-scales larger than Δ and motions smaller than Δ are removed (and don't seriously contaminate $\bar{v}$). $F(x,x')$ does not have to be homogeneous or isotropic, and although it is usually an implicit property of the finite-difference, spectral, or finite element method, it can also be explicitly selected by the numericist (cf. Moin, Reynolds, and Ferziger, 1978). In finite-difference calculations, it is often difficult to express $F(x,x')$ in closed form, but for a pseudo-spectral representation of a 1-dimensional periodic, real velocity field with wavelength 2π and with collocation points $x_p \equiv 2\pi\, p/N$, $p = 0,1,2,\ldots,N-1$, the convolution function $F(x,x')$ is easily determined:

$$\overline{v(x)} = \sum_{k=-N/2}^{N/2} e^{ikx}\, \bar{v}(k) \quad (3)$$

where $\bar{v}(k)$ is the discrete Fourier transform

$$\bar{v}(k) = \frac{1}{N}\sum_{p=0}^{N-1} e^{ikx_p}\, v(x_p) \quad (4)$$

so that

$$\overline{v(x)} = \sum_{p=0}^{N-1} v(x_p)\, \frac{\cos[\frac{N}{2}(x-x_p)]-\cos[(\frac{N}{2}+1)(x-x_p)]}{N[1-\cos(x-x_p)]} \quad (5)$$

Therefore,

$$F(x,x') = \sum_{p=0}^{N-1} \delta(x'-x_p)\, \frac{\cos[\frac{N}{2}(x-x')]-\cos[(\frac{N}{2}+1)(x-x')]}{N[1-\cos(x-x')]} \quad (6)$$

In equation (6), the characteristic length of the filter is $\Delta=2\pi/N$.

The pressure, temperature, density, and all other quantities of the flow are similarly filtered. Nonlinear terms in the equations of motion couple the subgrid flow to the large-scale flow. The numerically filtered Navier-Stokes looks like the Euler equation written in terms of the large-scale pressure, P, density, ρ, and velocity:

$$\partial\underline{\bar{v}}/\partial t = \overline{-(\underline{\bar{v}}\cdot\nabla)\underline{\bar{v}}} - \overline{\nabla\bar{P}/\bar{\rho}} + \overline{1/\bar{\rho}\ \nabla\cdot\underline{\underline{\Pi}}} + \underline{Q} \tag{7}$$

where $\underline{\underline{\bar{\Pi}}}$ is the large-scale component of the viscous stress tensor, and $\underline{Q}$ contains all terms through which the subgrid-scales affect the large-scale flow:

$$\underline{Q} = \overline{-(\underline{v}'\cdot\nabla)\underline{\bar{v}}}\ \ \overline{-(\underline{\bar{v}}\cdot\nabla)\underline{v}'}\ \overline{-(\underline{v}'\cdot\nabla)\underline{v}'} + \overline{(\rho'/\bar{\rho})[1/(\bar{\rho}+\rho')]\nabla(\bar{P}+P')}\ \overline{-\nabla P'/\bar{\rho}} - \overline{(\rho'/\bar{\rho})[1/(\bar{\rho}+\rho')]\ \nabla\cdot(\underline{\underline{\bar{\Pi}}}+\underline{\underline{\Pi}}')} - \overline{\nabla\cdot\underline{\underline{\Pi}}'/\bar{\rho}} \tag{8}$$

For large Reynolds number flows, the viscous contributions to equations (7) and (8) are negligible and will henceforth be ignored, but the subgrid-scale term, $\underline{Q}$, in equation (7) is large and important. Viscosity indirectly influences the value of $\underline{Q}$ by changing $\underline{v}'$. The main danger of neglecting $\underline{Q}$ in numerical simulations is that the computed flow may be numerically stable and exhibit many interesting astrophysical properties, but not look at all like the true stable equilibrium.

IMPOSSIBILITY OF RESOLVING THE SMALL SCALES

To understand our present inability to simulate motion at all scales and the need of modeling the subgrid-forcing, $\underline{Q}$, we need to review some aspects of turbulence theory. A turbulent fluid is made up of motion spanning a wide range of length-scales, ℓ, each scale with its own characteristic velocity $v(\ell)$. In all theories of turbulence (with a small but non-zero viscosity), $v(\ell)$ decreases with decreasing ℓ so there is a characteristic length ℓ_M such that the velocity of scales with $\ell < \ell_M$ have sufficiently small Mach number so, although the large-scale flow is compressible, the small-scale flow is incompressible. If we assume that we have sufficient numerical resolution so that all supersonic and transonic scales are resolved, i.e., $\ell_M > \Delta$, then the subgrid-scale flow is incompressible and we can confine our discussion of turbulence to incompressible flow.

Consider a high Reynolds number flow driven by some external force (e.g., buoyancy) at some length-scale ℓ_b such that the rate of kinetic energy per unit mass per unit time entering the flow is ε. According to the classic Kolmogorov theory of turbulence (1941), there is a net cascade of energy downwards from ℓ_b to a small dissipative length-scale, ℓ_d. In the Kolmogorov picture, the cascade is non-dissipative so that the rate at which energy crosses from motions with scales greater than ℓ into motions less than ℓ is ε, where the rate is approximated as

$$\varepsilon \simeq v(\ell)^3/\ell \quad \text{for} \quad \ell_b \geq \ell \geq \ell_d \tag{9}$$

Equation (9) is the source of Kolmogorov's famous scaling law, $v(\ell) \propto \ell^{1/3}$. For an incompressible fluid, the rate at which viscosity dissipates energy at length-scale ℓ is $\nu\, v(\ell^2)/\ell^2$ where ν is the kinematic viscosity; therefore, since $v(\ell) \propto \ell^{1/3}$, viscous dissipation is most effective at small length-scales. By assuming that all of the kinetic energy is dissipated at the smallest length-scale ℓ_d, we obtain an estimate for ℓ_d.

$$v(\ell_b)^3/\ell_b = \varepsilon = \nu v(\ell_d)^2/\ell_d^2 \tag{10}$$

$$= \nu[v(\ell_b)(\ell_d/\ell_b)^{1/3}]^2/\ell_d^2 \tag{11}$$

or

$$\ell_b/\ell_d = R^{3/4} \tag{12}$$

where R is the Reynolds number based on the largest length-scale

$$R = v(\ell_b)\ell_b/\nu \tag{13}$$

[Equation (12) for determining ℓ_b/ℓ_d can be derived in an alternate manner by assuming $v(\ell) \propto \ell^{1/3}$ and finding the length-scale ℓ_d where the characteristic inertial term of the Navier-Stokes equation, $v(\ell_d)^2/\ell_d$, is equal to the viscous term, $\nu v(\ell_d)/\ell_d^2$.] Equation (12) contains all the information a numericist needs to compute his computer budget for a simulation. It says that to include all of the physically important length-scales from the largest, ℓ_b, to the smallest, ℓ_d, he needs at least $R^{3/4}$ finite-difference points (or spectral modes or finite elements) per spatial dimension. Typical astrophysical flows have large Reynolds numbers. Solar convection has a Reynolds number of 10^{14} which means that a full 3-dimensional simulation requires 10^{31} grid points! With today's supercomputers, calculations with greater than 10^6 grid points are not practical; therefore, in solar convection all scales less than $\Delta = \ell_b/100$ must be neglected or modeled. Although modern experiments and theories of turbulence (see below) show that the Kolmogorov picture is too simple, the estimate in equation (12) for the dissipative length-scale and hence the estimate for the computer budget is still valid.

NEGLECT OF THE SUBGRID-SCALES

Neglecting the subgrid-scales deprives the flow of its natural outlet for dissipating kinetic energy. For simulations of flows with Reynolds numbers of a few hundred in which $\Delta > \ell_d$ the kinetic energy cascades down from ℓ_b to Δ, where it has no place to go and begins to pile up. It continues to accumulate at Δ until the velocity $v(\Delta)$ is sufficiently enhanced so that the rate of dissipation at Δ, $\nu v(\Delta)^2/\Delta^2$, balances ε. Figure 1 shows the kinetic energy spectrum of a pseudo-spectral simulation of Taylor-Couette flow between two cylinders at a Reynolds number of 460. The abscissa in figure 1 is the axial wavenumber, $k \simeq \ell_b/\ell$, and k=15 is the limit of the numerical resolution. In a well-resolved simulation (Marcus, 1983) the energy spectrum for k>3 is a straight line and the upward curl in the energy spectrum at k=15 in figure 1 is due to the fact that $\Delta > \ell_d$ in this calculation. Repeating the calculation with twice as much resolution makes $\Delta < \ell_d$ and the spectrum becomes a straight line from k=4 to k=31.

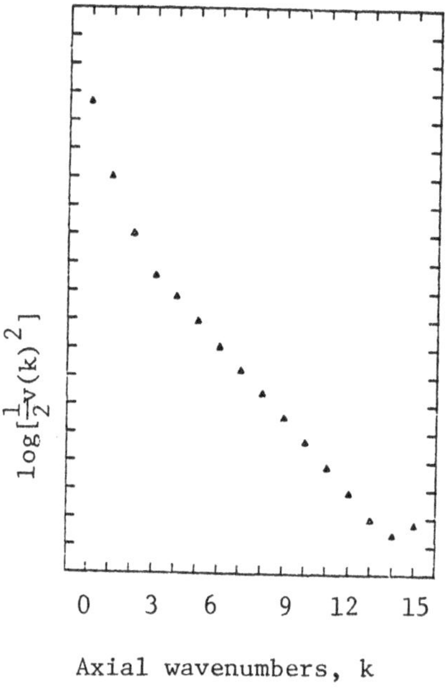

Figure 1 - Under-resolved energy spectrum

In most calculations at Reynolds numbers of a few hundred, insufficient numerical resolution manifests itself by producing artificial small structures with size approximately Δ. In some simulations, the lack of resolution produces other types of spurious results. For example, single-mode theory (Toomre, Gough, and Spiegel, 1977) is a numerical method in which all information in two of the three spatial dimensions is represented spectrally with a basis function (usually a linear combination of Fourier modes or some other eigenfunction of the 2-dimensional Laplacian) and information in the third spatial dimension is computed with a fine mesh of finite-difference grid points. Most astrophysical applications of single-mode theory have been with stellar convection (Toomre, Zahn, Latour, and Spiegel, 1976) in which the vertical direction is treated with finite-differences and the horizontal direction is represented with rolls or hexagonal planforms. The horizontal resolution is $\Delta_h = \ell_b$ and the vertical resolution is $\Delta_v << \ell_b$ with $\ell_v >> \ell_d$. Since the kinetic energy is prohibited by the numerics from cascading into small 3-dimensional motions, the flow forms artificially thin 1-dimensional boundary layers. The vertical thickness of the boundary layers, δ, is estimated by equating the rate of viscous dissipation integrated over the boundary layer to the energy input integrated over the entire volume of the fluid:

$$(\rho \nu v^2/\delta^2)\ell_b^2\delta = \rho\varepsilon\ell_b^3 \qquad (14)$$

or

$$\delta/\ell_b = \nu v^2/\varepsilon\ell_b^2 \simeq \nu/v\ell_b = R^{-1} \qquad (15)$$

The relationship in equation (15) has been confirmed numerically by Marcus (1981).

Under-resolved numerical simulations can have other signatures. A statistically steady state requires that the viscous dissipation balance energy production. If the flow, due to limited resolution, is incapable of dissipating energy efficiently, the numerically computed flow can respond by artificially lowering ε. However, in thermal convection there is a strong constraint preventing ε in the computed flow from deviating significantly from its correct value. The constraint is due to the fact that in thermal convection, the energy input rate per unit mass is proportional to the convective flux, F_c.

$$\varepsilon = F_c/\rho\Lambda \qquad (16)$$

where Λ is the density scale-height. Equivalently,

$$\varepsilon = (F + kc_p\rho \cdot \frac{\partial\langle T\rangle}{\partial z})/\rho\Lambda \qquad (17)$$

where $\partial\langle T\rangle/\partial z$ is the horizontally-averaged vertical temperature gradient, F is the total stellar flux, and k is the thermal diffusivity, ($k = 4acT^3/3\chi\rho$). In stellar convection, $\partial\langle T\rangle/\partial z$ is negative and less than the adiabatic gradient, $\partial\langle T\rangle/\partial z|_{ad}$ (since we assume that the star is not radiatively stable). If the convection is efficient (i.e., ν and k are small), then $\partial\langle T\rangle/\partial z$ cannot be much greater than $\partial\langle T\rangle/\partial z|_{ad}$ (since even a small super-adiabaticity makes the flow advectively unstable if ν and k are small). Since F is fixed and $\partial\langle T\rangle/\partial z$ must be nearly equal to its adiabatic value, the kinetic energy input rate is

$$\varepsilon \simeq (F + kc_p\rho \frac{\partial\langle T\rangle}{\partial z}|_{ad})/\rho\Lambda \tag{18}$$

and is computed correctly by nearly all codes; ε is insensitive to the resolution of the calculation.

In thermal convection, the entropy variance, $s(\ell)^2$, is analogous to the kinetic energy. The variance created at the large scale ℓ_b cascades via the nonlinear inertial terms in the equations of motion to a small thermal dissipation scale, ℓ_T. If $\ell_T < \Delta$, the entropy variance piles up at Δ or the numerically calculated flow must adjust itself so that the rate of entropy variance production, ε_s, is artificially decreased. Unlike ε, the numerical value of ε_s is not strongly constrained. The variance production rate at large scales is proportional to ($\partial\langle T\rangle/\partial z|_{ad}$ - $\partial\langle T\rangle/\partial z$)

$$\varepsilon_s \simeq \frac{Fc_v}{\rho T^2} (\frac{\partial T}{\partial z}\Big|_{ad} - \frac{\partial T}{\partial z}) \tag{19}$$

Although efficient convection makes ($\partial\langle T\rangle/\partial z|_{ad}$ - $\partial\langle T\rangle/\partial z$) small, a change in its value from 0.001 to 0.1 changes the value of ε_s by 100; therefore, ε_s is not strongly constrained. A numerical simulation with $\Delta > \ell_T$ produces not only a pile-up of entropy variance at Δ, but also an erroneously small value of ($\partial\langle T\rangle/\partial z|_{ad}$ - $\partial\langle T\rangle/\partial z$) and thereby reduces the value of ε_s.

Another problem of under-resolved calculations is the computation of the correlation between the temperature and the vertical velocity:

$$C(z) \equiv \frac{\langle T(\underline{x})\ v_z(\underline{x})\rangle}{[\langle T(\underline{x})^2\rangle\langle v_z(\underline{x})^2\rangle]^{1/2}} \tag{20}$$

where angle brackets denote horizontal averaging. In calculations with insufficient horizontal resolution an artificially high value of correlation is produced; in single-mode calculations the correlation is identically equal to unity. Convection experiments in air with Reynolds numbers of a few hundred have a correlation of about 0.6 (Deardorff and Willis, 1967). We have already seen that single-mode calculations are constrained to produce a nearly correct value of ε. The temperature flux, $\langle T(\underline{x})\, v_z(\underline{x})\rangle$ is also constrained to be nearly its correct value since $\bar{\varepsilon}$ is proportional to it. Therefore, since $C(z)$ is over-estimated and $\langle T(\underline{x})\, v_z(\underline{x})\rangle$ is nearly correct, single-mode calculations must necessarily under estimate $\langle T(\underline{x})^2\rangle$ or $\langle v_z(\underline{x})^2\rangle$. It has been shown that single-mode calculations tend to under estimate $\langle T(\underline{x})^2\rangle^{1/2}$ by a factor of about 2 (Marcus, 1981).

In an under-resolved simulation, as the Reynolds number is increased, $v(\Delta)$ increases until it becomes the same order as $v(\ell_b)$. When this happens the energy jumps abruptly back from the small scales to the large scales and then slowly cascades back to the small scales in a periodic oscillation. This periodicity is shown in figure 2, which was calculated with an under-resolved

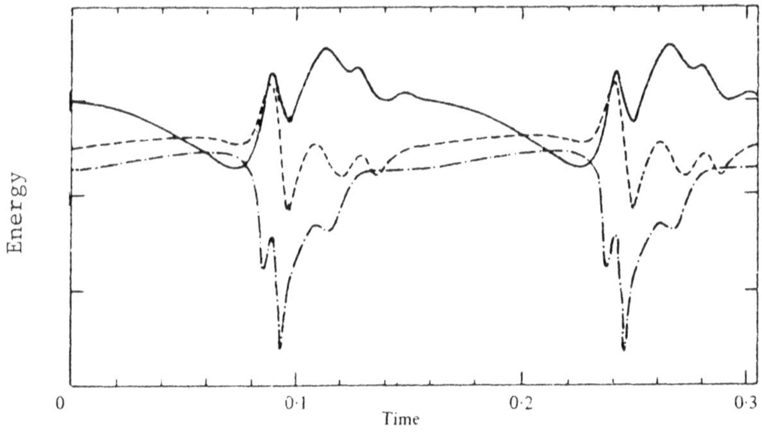

Figure 2

simulation of convection in a sphere. The energy as a function of time at length-scale ℓ_b is shown by the solid curve, at the smallest scale, Δ, by the dash-dot curve, and at an intermediate scale by the dashed curve. The time is in units of the thermal dissipation time. At t=0 the kinetic energy spectrum looks normal with energy decreasing with decreasing ℓ. As time advances, the energy piles up at Δ until the spectrum becomes inverted, with the largest scale containing the least amount of energy. At t=0.08 the energy rushes back into the large scale and the oscillation repeats itself. The period of the oscillation is a function of ℓ_b/Δ. When the numerical resolution is doubled so that $\Delta < \ell_d$ and $\Delta < \ell_T$, the flow settles down to a steady state. Although ℓ_b/Δ controls the period of the oscillation, it is Δ/ℓ_d that determines whether there is a transition from a steady-state to a time-dependent one. Holding all physical quantities fixed, and increasing Δ/ℓ_d causes the flow to become periodic; if Δ/ℓ_d is increased further, the numerically computed flow becomes chaotic in time. All interesting temporal behavior in calculations with $\Delta/\ell_d \gg 1$ or $\Delta/\ell_T \gg 1$ is, of course, a numerical artifact.

SIMPLE EDDY-VISCOSITIES

There are several ways to prevent the pile-up of energy at small scales other than by increasing the resolution so $\Delta < \ell_d$. The easiest method is the introduction (accidental or planned) of numerical viscosity. The subgrid-scales affect the large scales by acting as conduits of kinetic energy from the large motions to the dissipative flow with an approximate rate of $v(\Delta)^3/\Delta$. In Kolmogorov's picture of turbulence, the cascade is local so the subgrid-scales drain energy mostly from the resolvable scales with approximate size Δ. Numerical viscosity can mimic this type of energy drain. Consider solving numerically Euler's equation with constant density by using upwind differencing. In one dimension this difference scheme is

$$\frac{\partial v_i}{\partial t} = -v_i \frac{v_i - v_{i-1}}{\Delta} - \frac{1}{2\rho}(P_{i+1} - P_{i-1}) \tag{21}$$

where $v_i = v(x_i)$, $x_i = i\Delta$, and we have assumed that $v_i > 0$. Equation (21) introduces a numerical viscosity which can be calculated by expanding v_{i-1}, P_{i+1} and P_{i-1} in a Taylor series about x_i. To second order in Δ, equation (21) is equivalent to

$$\begin{aligned} \left.\frac{\partial v}{\partial t}\right|_{x=x_i} &= - v(x_i) \left.\frac{\partial v}{\partial x}\right|_{x=x_i} - \frac{1}{\rho} \nabla P\Big|_{x=x_i} \\ &+ [\tfrac{1}{2} v(x_i)\Delta] \left.\frac{\partial^2 v}{\partial x^2}\right|_{x-x_i} \\ &+ \mathcal{O}(\Delta^2) \end{aligned} \tag{22}$$

The numerical viscosity is $(\frac{1}{2}v(x)\,\Delta)$ and has several desirable properties: it is a positive-definite sink of kinetic energy, it draws its energy primarily from resolvable scales of size Δ, and the rate of dissipation is approximately $v(\Delta)^3/\Delta$ (the last two properties are true if the flow has an energy spectrum such that $v(\ell)^3/\ell$ increases with decreasing ℓ and $v(\ell_b)$ and $v(\Delta)$ are poorly correlated - see below).

Although this crude numerical viscosity drains energy out of the small-scale motions, it usually dissipates a significant amount from the large-scale modes as well. To determine which scales the numerical viscosity dissipates, observe that the rate of energy dissipation at scale ℓ is proportional to

$$\frac{1}{2}\Delta \sum_{\ell',\ell''} C(\ell,\ell',\ell'') \; v(\ell)v(\ell')v(\ell'')/\ell'^2 \tag{23}$$

where $C(\ell\ \ell'\ \ell'')$ is a measure of the correlation among motions of scale ℓ, ℓ', and ℓ''. The velocity, $v(\ell)$, is made up of a band of Fourier modes with wavevectors $\underline{k}$ where $|\underline{k}| \simeq \ell_b/\ell$. The correlation is proportional to the $\underline{\text{volume}}$ integral of the triple product of the Fourier functions in the bands comprising ℓ, ℓ', and ℓ'':

$$C(\ell,\ell',\ell'') \simeq \int dx \int_{k\simeq\ell/\ell_b} d^3k \int_{k'\simeq\ell'/\ell_b} d^3k' \int_{k''\simeq\ell''/\ell_b} d^3k'' \cdot \{ \tag{24}$$
$$e^{(i(\underline{k}+\underline{k}'+\underline{k}'')\underline{x})} |\underline{v}(\underline{k},t)|\,|\underline{v}(\underline{k}',t)|\,|\underline{v}(\underline{k}'',t)|\}$$

where $\underline{v}(\underline{k},t)$ is the Fourier transform of $\underline{v}(\underline{x},t)$. A triad of Fourier modes with wavenumbers $\underline{k}$, $\underline{k}'$, and $\underline{k}''$ contributes to the integral if and only if $(k + k' + k'') = 0$. Certain correlations are kinematically required to be zero. For example, if $\ell >> \ell''$ and $\ell' >> \ell''$ then $C(\ell,\ell',\ell'') = 0$. $C(\ell,\ell',\ell'')$ also reflects the temporal correlation among $v(\ell)$, $v(\ell')$, and $v(\ell'')$. If motions among disparate scales have no phase coherence, then $C(\ell,\ell',\ell'')$ is small. If $C(\ell,\ell',\ell'')$ is insignificant except when $\ell \simeq \ell' \simeq \ell''$, then the rate of numerical dissipation from $v(\ell)$ is approximately

$$\frac{1}{2}\left(\frac{\Delta}{\ell}\right)\frac{v(\ell)^3}{\ell} \tag{25}$$

Equation (25) shows that if $v(\ell)^3/\ell$ is independent of ℓ (Kolmogorov theory) or if $v(\ell)^3/\ell^2$ increases with decreasing ℓ, then the small scales are preferentially dissipated. On the other hand, if $v(\ell)$ decreases exponentially with ℓ (cf. the dissipative spectrum in figure 1), then equation (25) shows that the numerical viscosity is most effective at ℓ_b not Δ. If $C(\ell,\Delta,\Delta)$ is large and if $v(\ell)/\ell$ increases with decreasing ℓ, then the rate of numerical dissipation is proportional to

$$\frac{1}{2}\, v(\ell)v(\Delta)^2/\Delta \tag{26}$$

If $C(\ell,\ell,\ell_b)$ is large and if $v(\ell)/\ell^2$ decreases with decreasing ℓ, then the rate is proportional to

$$\frac{1}{2}\left(\frac{\Delta}{\ell_b}\right) v(\ell)^2 v(\ell_b)/\ell_b \qquad (27)$$

The rates in equations (26) and (27) are most effective at ℓ_b, not Δ, and flows with correlations and spectra that produce these rates are not well simulated with a numerical viscosity of this type. Even in the best case where equation (25) is applicable, there is significant dissipation at ℓ_b. In particular, the Reynolds number based on $v(\ell_b)$, ℓ_b, and the numerical viscosity at ℓ_b implied by equation (25), $1/2\,\Delta\, v(\ell_b)$, is

$$R = 2\,\ell_b/\Delta \qquad (28)$$

For a 3-dimensional calculation with 10^6 grid points, the flow has an effective Reynolds number of only 200, which corresponds to a viscous laminar flow. Therefore, this numerical viscosity is not useful for simulating turbulence.

More sophisticated techniques are needed to dissipate selectively the kinetic energy from the small motions without disturbing the rest of the large-scale flow. For example, Siggia and Patterson (1978) in a pseudo-spectral simulation of turbulence enhanced the viscosity in the modal equations of motion that govern the smallest 15% of the flow and left the equations for the large modes unchanged. The magnitude of their enhancement was adjusted until the energy spectrum obeyed Kolmogorov's $v(\ell) \propto \ell^{1/3}$ scaling law. This technique is useful in spectral calculation but cannot be implemented in an easy way with finite-differences. Furthermore, in complicated flows one does not know _a priori_ the shape of the energy spectrum, so some other method must be found for determining the correct amount of viscous enhancement.

Smagorinsky (1963) has modeled the subgrid-scale forcing, $\underline{Q}$, as an eddy-viscosity in a way that can be easily implemented in finite-difference calculations. The eddy-viscosity is more selective than the numerical viscosity in removing the energy from the small scales while leaving the large scales untouched. Of course, it is not as selective as the modal method of Siggia. To see how an eddy-viscosity works, consider a flow with constant density. The subgrid forcing, $\underline{Q}$, can be written as the divergence, $\nabla\cdot \underline{\underline{R}}$, where $\underline{\underline{R}}$ is the symmetric tensor

$$R_{jk} = \overline{v'_j v'_k} \cdot \overline{\bar{v}_j v'_k} \quad \overline{v'_j \bar{v}_k} \qquad (29)$$

where we have assumed that the viscous terms are negligible and that the derivative operator commutes with the large-scale filter operator. It will be useful to break $\underline{\underline{R}}$ into a trace and a traceless part

$$\tau_{jk} \equiv R_{jk} - R_{mm}\,\delta_{jk}/3 \tag{30}$$

$$\tilde{P}/\rho \equiv R_{mm}/3 \tag{31}$$

$$\underline{Q} = \nabla \cdot \underline{\underline{\tau}} + \nabla\tilde{P}/\rho \tag{32}$$

The filtered Navier-Stokes equation is

$$\frac{\partial\overline{\underline{v}}}{\partial t} = -\overline{(\overline{\underline{v}}\cdot\nabla)\overline{\underline{v}}} - \frac{1}{\rho}\nabla(\overline{P}+\tilde{P}) - \nabla\cdot\underline{\underline{\tau}} + \nu\nabla^2\overline{\underline{v}} \tag{33}$$

Smagorinsky proposed treating $\nabla\cdot\underline{\underline{\tau}}$ like a viscous term by letting $\underline{\underline{\tau}}$ be proportional to the large scale strain, $\overline{\underline{\underline{S}}}$,

$$\underline{\underline{\tau}} = -2\nu_e\,\underline{\underline{S}} \tag{34}$$

where

$$\overline{S}_{jk} = \frac{1}{2}\left(\frac{\partial\overline{v}_j}{\partial x_k} + \frac{\partial\overline{v}_k}{\partial x_j}\right) \tag{35}$$

and ν_e is the eddy-viscosity. Note that we could not let $\underline{\underline{R}}$ itself be proportional to $\underline{\underline{S}}$ because $\underline{\underline{S}}$ is traceless (as is $\underline{\underline{\tau}}$) and $\underline{\underline{R}}$ is not. The turbulent pressure-head, $\tilde{P}$, need not be determined; the quantity $(\overline{P}+\tilde{P})$ is found in the usual manner by requiring that $\underline{v}$ be divergence-free. Taking the divergence of equation (33) we obtain

$$\frac{1}{\rho}\nabla^2(\overline{P}+P) = -\frac{\partial}{\partial x_j}\frac{\partial}{\partial x_k}(\overline{v}_j\overline{v}_k + \tau_{jk}) \tag{36}$$

Smagorinsky proposed an eddy-viscosity of

$$\nu_e = c^2\Delta^2(2\overline{S}_{jk}\overline{S}_{jk})^{1/2} \tag{37}$$

where c is a constant of order unity. His choice of ν_e was motivated by the following argument. We have already shown that ν_e should be proportional to $v(\Delta)\Delta$

$$\nu_e = c'\,v(\Delta)\Delta \tag{38}$$

where c' is a constant of order unity. Although the value of Δ is known, it is not easy to determine $v(\Delta)$ in a finite-difference calculation. The value of $v(\Delta)$ can be estimated from an energy balance. The rate at which energy is drained from the large-scale flow in equation (33) is

$$\nu_e \bar{S}_{jk}\bar{S}_{jk} \tag{39}$$

From the Kolmogorov theory of homogeneous, isotropic turbulence, the rate at which energy enters the subgrid-scales is

$$v(\Delta)^3/\Delta \tag{40}$$

Equating expressions (39) and (40), we find

$$v(\Delta) = (c')^{1/2}\, \Delta\, (\bar{S}_{jk}\bar{S}_{jk})^{1/2} \tag{41}$$

which combined with equation (38) gives the Smagorinsky eddy-viscosity in equation (37). The weak point of Smagorinsky's argument is that if the flow does not behave like homogeneous, isotropic turbulence, then equation (40) is incorrect. We show in a later section how equation (40) can be modified for inhomogeneous flow.

Smagorinsky's isotropic eddy-viscosity is superior to the previously discussed numerical viscosity. If the flow at disparate length-scales is poorly correlated , then the dissipation rate at ℓ is

$$c^2 \left(\frac{\Delta}{\ell}\right)^2 \frac{v(\ell)^3}{\ell} \tag{42}$$

If $v(\ell)^3/\ell$ is independent of ℓ (Kolmogorov theory) or if $v(\ell)^3/\ell^2$ increases with decreasing ℓ, then most of the energy is drained from the smallest resolvable scales. The effective Reynolds number at ℓ_b made from $v(\ell_b), \ell_b$, and the viscosity at ℓ_b implied by equation (42), $c^2\Delta^2 v(\ell_b)/\ell_b$ is

$$R = \frac{1}{c^2} \left(\frac{\ell_b}{\Delta}\right)^2 \tag{43}$$

This Reynolds number is greater than the one computed with the numerical viscosity (equation 28) by a factor of (ℓ_b/Δ). For a 3-dimensional calculation with 10^6 grid points and with $c=0.1$ (see below) the effective large-scale Reynolds number is 10^6 . This Reynolds number is characteristic of a nearly inviscid, turbulent flow, so the Smagorinsky eddy-viscosity may be useful in computing turbulent astrophysical flows.

We note that both Smagorinsky's eddy-viscosity and the numerical viscosity are positive-definite sinks of energy , but for any particular length-scale, ℓ, the viscous term can be a source of energy. However, if $(\bar{S}_{jk}\bar{S}_{jk})$ and Δ are both independent of position in the fluid, then ν_e is a sink of energy at all length-scales. We also note that if the flows at all resolvable length-scales are well-correlated, then it is possible for the Smagorinskii eddy-viscosity to be most dissipative at ℓ_b not Δ .

Eddy-viscosities have been used extensively in meteorological calculations (cf. Smagorinsky , Manake, and Holloway, 1965). Lilly (1967) used Smagorinsky's eddy-viscosity to compute homogeneous, isotropic turbulence and found that c=0.17 in equation (37) gave the best scaling law for the velocity. Deardorff (1970) used an eddy-viscosity to compute an inhomogeneous, anisotropic plane channel flow (plane Poiseuille flow) and found good agreement with the laboratory values of the statistical measures of turbulence by setting c=0.1. With c=0.17 he found the flow too dissipative. Not surprisingly, he also discovered that in regions where the flow is strongly anisotropic or inhomogeneous, the agreement between his calculations and experiments was poor.

ANALYTIC CALCULATION OF SUBGRID-SCALE FLOW

Treating the large-scale effects of the subgrid scales as an eddy-viscosity is clearly inadequate at any place in the fluid or at any scale in the spectrum where the net downward cascade of energy is not approximately equal to $v(\ell)^3/\ell$, or where the cascade is not stationary in time or is not isotropic. But even in stationary, homogeneous, isotropic turbulence the eddy-viscosity may be inaccurate. We need a mathematical framework in which to develop a better model of $\underset{\sim}{Q}$. To understand how the subgrid-scales affect the large scales, it is first necessary to understand how the large-scale motions create the subgrid-scale flow. Kolmogorov's picture of turbulence is based on two major assumptions. One is that small-scale turbulent flow is universal (i.e., the small-scale flow is independent of the detailed structure of the large-scale flow and is therefore the same in every turbulent flow). Strongly related to the idea of universality is the concept of locality, whereby the only scales, ℓ' , that directly affect $v(\ell)$ are those with $\ell' \simeq \ell$. Non-locality is kinematically allowed whenever there is a nonlinear interaction among three or more Fourier modes whose wavevectors add to zero. However, non-locality also requires that there be a strong correlation among disparate length-scales. Universality and locality have never been proven analytically or experimentally, but if it can be shown that universality exists among turbulent flows with very different large-scale coherent structures, then locality is proven.

The second assumption by Kolmogorov is that the energy rate, ε , is the only information that passes from the large-scale to the small-scale flow. Each length-scale has energy cascading downward through it at the same rate, so the subgrid flow has scale-similarity (i.e., apart from a scaling factor, the turbulent velocity at ℓ is indistinguishable from the turbulent velocity at any other small length). Scale-similarity has been shown experimentally to be incorrect (Van Atta and Park, 1972); the intermittent behavior of turbulence at length-scale, ℓ , was measured to be a function of both ε and ℓ/ℓ_b . Intermittency is a familiar sight to anyone who has watched a ripple, or cat's-paw, sweep across the surface of a lake on a gusty day. In laboratory flows, turbulence is accompanied by spatial and temporal bursts that make the flow appear spotty or streaky. For example, in channel flow with Reynolds numbers of several hundred, turbulent spots appear near the walls, and most of the transfer of energy from the large-scale flow to the small-scales is confined to these irregular regions. In turbulent flows, the velocity is a function of time but by ensemble or time-averaging (denoted by double angle brackets) the distribution function of $v(\ell)$ about its mean value can be measured. The velocity and its derivatives have large departures from their mean values for longer amounts of time in an intermittent flow than they do in a non-intermittent flow. The flatness factor of the velocity distribution is defined to be $\langle\langle v(\ell)^4\rangle\rangle/(\langle\langle v(\ell)^2\rangle\rangle)^2$ and is one measure of the likelihood of large departures from the mean; hence, the flatness measures the intermittency of the flow. Experiments show that the flatness factor increases with $(\ell_b-\ell)/\ell_b$, which means that the velocity at ℓ is not only a function of ε but also has a memory of the length of the cascade from ℓ_b to ℓ. Heuristically, it appears that the longer the energy cascades, the more likely it is to produce a large deviation from the mean.

Besides ε and ℓ/ℓ_b , it is not known what other quantities associated with the large-scale flow are needed to determine the small-scale flow. It has been conjectured (Lorenz, 1969) that all of the details of the large-scale flow are needed to compute the small-scale flow, and vice versa; if any minor change occurs in the small-scales the large-scale flow changes dramatically (e.g., a butterfly flapping its wings determines tomorrow's weather). If the "butterfly effect" is important, then numerical simulation and modeling of turbulent astrophysical flows is impossible. Even if the mean values of a large-scale turbulent flow are amenable to subgrid-scale modeling, if the large-scale intermittency depends upon detailed knowledge of the small scales, many astrophysical calculations are impossible. For example, consider the calculation of stellar convection. If we are interested in computing the extent of mixing due to the convective overshoot, then knowledge of the mean convective flow may not be useful. If intermittent bursts carry the convective penetration hundreds of times farther than the mean overshoot and if the intermittency occurs on a time-scale much longer than an eddy-turnover time (but much shorter than a stellar lifetime) then it is the intermittency, not the mean convective overshoot, that is relevant to stellar mixing. Throughout the remainder of this paper, by necessity, we shall assume that the large-scale quantities of interest can be computed with subgrid-scale models.

There are several analytic theories of turbulence based on the ideas of universality, localness, and scale-similarity that attempt to calculate the subgrid-scale forcing. To provide a theoretical basis for the previous phenomenological discussion of $\underline{Q}$, we now sketch briefly the analytic determination of the subgrid-scale forcing. Almost all theories of incompressible turbulence begin with the Fourier transform of the Navier-Stokes equation.

$$\left(\frac{\partial}{\partial t} + \nu k^2\right) v_\alpha(\underline{k},t) = -\frac{i}{2} P_{\alpha B\gamma}(\underline{k}) \sum_{\underline{P}} v_\beta(\underline{P},t) V(\underline{k}-\underline{P},t) \tag{44}$$

where the tensor $P_{\alpha\beta\gamma}(\underline{k})$ is defined

$$P_{\alpha\beta\gamma}(\underline{k}) \equiv k_\beta P_{\alpha\beta}(\underline{k}) + k_\gamma P_{\alpha\beta}(\underline{k}) \tag{45}$$

with

$$P_{\alpha\beta}(\underline{k}) \equiv \delta_{\alpha.\beta} - k_\alpha k_\beta / |k|^2 \tag{46}$$

The nonlinear term in equation (44) is the Fourier transform of the divergence-free component of $-(\underline{v}\cdot\nabla)\underline{v}$ (thereby including the effects of the pressure term). Exact evaluation of the nonlinear term is impossible. The type of approximation used is what distinguishes each theory of turbulence. Multiplying equation (44) by $v(k,t)$ one obtains an equation for the energy $E(k,t)$. In principle, all of the well-known theories are capable of casting the energy equation into the general form

$$\left\{\frac{\partial}{\partial t} + k^2[\nu + \nu_e(k,t)]\right\} E(k,t) = k^4 A(k,t) \tag{47}$$

where the nonlinearities are absorbed into an eddy-viscosity, $\nu_e(k,t)$ and a forcing term, $A(k,t)$, where ν_e and A depend only weakly on k. The ν_e term in equation (47) is readily understood, at least on a phenomenological level, since we have already shown that the effect of the small-scales on the large-scales (via the nonlinear terms) is to act like a viscosity. The forcing term that is proportional to k^4 is heuristically understood by observing that in the limit of stationary, inviscid turbulence there will be an equi-partition of energy among all the modes in Fourier space (Rose and Sulem, 1978). Since the number of Fourier modes with wavenumber k is proportional to $4\pi k^2$, the energy spectrum of a stationary, inviscid flow is

$$E(k) \propto k^2 \qquad (48)$$

The k^4 forcing term is needed in equation (47) to be consistent with equation (48). A(k,t) is created by the non-local nonlinear interaction of Fourier modes. It is the ability of A(k,t) and $\nu_e(k,t)$ to selectively drain and force energy out of or into selective wavevectors that permits the small-scale flow to control the large-scale motions in a way that can be qualitatively different from the way ordinary viscosity controls laminar flow. The Red Spot of Jupiter, stellar dynamos, and super-granulation are probably all due to the non-viscous-like properties of the small-scale forcing.

One method of approximating the nonlinear terms in equation (44) is by using the renormalization group (Forster, Nelson, and Stephen, 1977), which uses the assumption of scale-similarity with propagator techniques from field theory. Intermittency cannot be treated correctly because of the self-similar assumption. The more widely used method of solving equation (44) is Kraichnan's direct interaction theory (1959) and its offspring: Lagrangian history direct interaction, (Kraichnan, 1965), the test field model (Kraichnan, 1971, Newman and Herring, 1979), and eddy-damped quasi-normal Markovian theory (Orszag, 1970, 1974). One proceeds by approximating cumulants or moments of the velocity. Equation (44) expresses the time rate of change of the velocity as a function of second-order moments of the velocity. Multiplying equation (44) by $\underline{v}(\underline{k}')$ yields an equation for the second-order moments in terms of the third-order moments. By continuing to multiply equation (44) by the velocity, a hierarchy of moment equations is built. Approximations are made when the hierarchy is truncated by either discarding the highest moments or approximating them as functions of lower order moments.

With respect to numerical modeling of $\underline{Q}$, the most usable results of analytic turbulence theory are the evaluation of A(k,t), $\nu_e(k,t)$, and (when the heat equation is also used) $k_e(t)$, the eddy-thermal-diffusivity. Closed form expressions for these quantities as integrals of the energy and thermal variance spectra weighted by functions of the turbulent time-scales have been obtained by Herring(1973), Basdevant, et al. (1978), Leslie and Quarini (1978), Chollet and Lesieur(1981), and others. The predictions are in reasonable agreement with measurements of decay times in wind tunnel grid-turbulence and with experimental measurements of ν_e/k_e, the eddy Prandtl number, in a turbulent boundary-layer (Fulachier and Dumas, 1976). Unfortunately, most of the analytic theories of turbulence include arbitrary constants whose values must be determined by comparison with experiment. Furthermore, none of the theories are easily generalizable to flows with inhomogeneous or anisotropic motions of the large-scale. No one has found a method of numerically simulating the large-scale flow with a subgrid-scale forcing based on an analytic theory of turbulence that is more useful than one of the phenomenological treatments of $\underline{Q}$. We conclude this paper with four examples of calculations where an eddy-viscosity model of $\underline{Q}$ is inadequate. We improve the model of $\underline{Q}$ in a heuristic manner while bearing in mind the theoretical implications of this section.

IMPROVED MODELS OF THE SUBGRID FLOW

Computation of the constant in the eddy-viscosity

In a simulation of laboratory flow, the value of the constant, c', which appears in equation (38) for the eddy-viscosity, is determined by adjusting its value until the numerical calculation agrees with the laboratory data. In simulations of astrophysical flow, we do not have this luxury. We require that the numerical method itself determine the value of c'. Such a method is of special importance in stellar convection where the subgrid flow diffuses entropy as well as momentum. The values of the constants that appear in the eddy-viscosity and in the eddy entropy diffusivity will not be the same, and their ratio determines the eddy-Prandtl number. One method for determining the constants uses scale-similarity (Marcus, 1980). This technique, like the others discussed previously, divides the velocity field into a resolvable component $\overline{\underline{v}}$ with length-scales greater than Δ and a subgrid component $\underline{v}'$. However, it also introduces an intermediate length-scale, Δ', with $\ell_b > \Delta' > \Delta$. The length, Δ', is small enough so that indigent numericists who can only afford to use the coarser resolution, Δ', in their simulation will still produce accurate results if they use an eddy-viscocity. For simplicity, we consider a flow with periodicity ℓ_b and use a spectral method where $\overline{\underline{v}}$ is defined to contain all Fourier modes $\underline{k}$ with $2\pi/\Delta > |\underline{k}| \geq 2\pi/\ell_b$ and where $\underline{v}'$ contains modes with $2\pi/\ell_d \geq |\underline{k}| \geq 2\pi/\Delta$. Define $\underline{v}''$ to be the part of the velocity field that is resolvable with the fine resolution, Δ, but unresolvable with the coarse resolution, i.e., $\underline{v}''$ contains modes $2\pi/\Delta > |\underline{k}| \geq 2\pi/\Delta'$. The large-scale filtering function used in the fine resolution calculation, indicated by an overbar, discards all modes with $|\underline{k}| \geq 2\pi/\Delta$. The filtering function in the coarse calculation, indicated by a double overbar, discards modes with $|\underline{k}| \geq 2\pi/\Delta'$. In a constant density fluid, the coarse resolution calculation solves the equation

$$\partial\bar{\bar{\underline{v}}}/\partial t = \overline{\overline{-(\bar{\bar{\underline{v}}}\cdot\nabla)\bar{\bar{\underline{v}}}}} - \nabla(\bar{\bar{P}}+\tilde{\tilde{P}})/\rho + \nu\nabla^2\bar{\bar{\underline{v}}} \\ \overline{\overline{+\nabla\cdot[c'\Delta' v(\Delta')\nabla\bar{\bar{\underline{v}}}]}} \tag{49}$$

and the fine resolution solves

$$\partial\bar{\underline{v}}/\partial t = \overline{-[(\bar{\bar{\underline{v}}}+\underline{v}'')\cdot\nabla](\bar{\bar{\underline{v}}}+\underline{v}'')} - \nabla(\bar{P}+\tilde{P})/\rho \\ +\nu\nabla^2(\bar{\bar{\underline{v}}}+\underline{v}'') \\ \overline{+\nabla\cdot[c'\Delta v(\Delta)\nabla(\bar{\bar{\underline{v}}}+\underline{v}'')]} \tag{50}$$

where $v(\Delta)$ and $v(\Delta')$ are the characteristic velocities at Δ and Δ', respectively, and where $\tilde{P}$ and $\tilde{\tilde{P}}$ are determined so that the right sides of equations (49) and (50) are divergence-free. Note that

$$\bar{\underline{v}} = \bar{\bar{\underline{v}}} + \underline{v}'' \tag{51}$$

Since both calculations are assumed accurate, the computed value of $\bar{\bar{\underline{v}}}$ is the same in both calculations, and since scale-similarity is assumed, the values of c' in equations (49) and (50) are the same. By multiplying equation (49) by $\bar{\underline{v}}$ and integrating over the volume, we obtain the rate of change of energy in $\bar{\bar{\underline{v}}}$

$$\partial\bar{\bar{E}}/\partial t = -\int d^3x \; 2\nu\bar{\bar{S}}_{jk}\bar{\bar{S}}_{jk} + 2c'\Delta' v(\Delta')\bar{\bar{S}}_{jk}\bar{\bar{S}}_{jk} \tag{52}$$

while equation (50) predicts

$$\partial\bar{\bar{E}}/\partial t = -\int d^3x \; 2\nu\bar{\bar{S}}_{jk}\bar{\bar{S}}_{jk} + 2c'\Delta v(\Delta)\bar{S}_{jk}\bar{\bar{S}}_{jk} \\ -\int d^3x \; \bar{\bar{\underline{v}}}\cdot\left[\overline{(\underline{v}\cdot\nabla)\underline{v}''}\right] \tag{53}$$

The correct value of c' is determined at each time step by requiring that the two expressions for $\bar{\bar{E}}(t)$ are equal

$$c'(t) = \frac{\int d^3x \; \bar{\bar{\underline{v}}}\cdot\left[\overline{(\bar{\underline{v}}\cdot\nabla)\underline{v}''}\right]}{2\int d^3x[\Delta' v(\Delta')\bar{\bar{S}}_{jk} - \Delta v(\Delta)\bar{S}_{jk}]\bar{\bar{S}}_{jk}} \tag{54}$$

The numerator in equation (54) is the rate at which energy flows from $\bar{\bar{\underline{v}}}$ into $\underline{v}''$, i.e., from motion with length-scales greater than Δ' into scales ℓ'' where $\Delta' \geq \ell'' > \Delta$. In almost all physically realistic calculations, this rate is positive. In a spectral simulation of Boussinesq convection, Marcus demonstrated that this method of determining c' is stable and rapidly convergent. The reason for rapid convergence is shown by the following argument. If $\Delta' v(\Delta')$ and $\Delta v(\Delta)$ are independent of position, then

$$\int d^3x\, \Delta v(\Delta)\, \overline{S}_{jk}\, \overline{\overline{S}}_{jk} = \Delta v(\Delta) \int d^3x\, \overline{\overline{S}}_{jk}\, \overline{\overline{S}}_{jk} \tag{55}$$

and

$$c'(t) = \frac{\frac{1}{2}\int d^3x\; \overline{\overline{\underline{v}}}\cdot\left[(\overline{\underline{v}}\cdot\nabla)\underline{v}''\right] \Big/ \int d^3x\, \overline{\overline{S}}_{jk}\, \overline{\overline{S}}_{jk}}{\Delta' v(\Delta') - \Delta v(\Delta)} \tag{56}$$

The quantity, $\overline{\overline{S}}_{jk}\overline{\overline{S}}_{jk}$, is positive, and $[\Delta' v(\Delta') - \Delta v(\Delta)]$ can be assumed positive, since in any realistic flow $v(\ell)$ decreases with decreasing ℓ. If the value of c' is too small, the flow responds by piling up energy in the smallest scale Δ, making $v(\Delta)$ large and $[\Delta' v(\Delta') - \Delta v(\Delta)]$ small. The small denominator in equation (56) then makes c' increase at the next time step. Similarly, if the value of c' is too large, the energy is preferentially dissipated from the smallest mode, the denominator in equation (56) becomes large, and the value of c' is reduced. We have never found the pile up of energy at Δ so severe that the denominator changes sign and produces a negative eddy-viscosity. By requiring that the rate of change of the entropy variance be the same in calculations with fine and coarse resolution, an equation analogous to equation (54) can be derived for the constant in the eddy entropy diffusivity. Marcus used this technique to determine numerically the eddy-Prandtl number of convection in water.

Inhomogeneous flow

The above method can be used in inhomogeneous flows provided that Δ, Δ' $v(\Delta)$, $v(\Delta')$, and c' are treated as functions of position. For example, $v(\Delta, \underline{x})$ is the characteristic velocity at $\underline{x}$, with length-scale Δ, averaged over a volume of fluid of size λ where $\ell_b \gg \lambda \gg \Delta$. $c'(\underline{x})$ is determined by requiring that the rates of change of energy in the coarse and fine resolution calculations are the same at each position in the fluid. This method fails in inhomogeneous flows where the physical mechanism by which the subgrid-scale flow removes energy from the large-scale flow is a function of position. For example, consider a numerical simulation of large-scale turbulent solar convection. Equation (38) for the eddy-viscosity and equation (39) for the energy loss rate from the large-scale flow are valid. However, equation (40) for the dissipation rate is not correct because the subgrid-scale effects at the center and at the base of the convective zone are very different. Near the center, the numerically resolvable components of the velocity that are smaller than a density scale-height behave like incompressible flow when they lose their kinetic energy to the subgrid-scales. In this region, equation (40) is adequate. At the base of the convective zone, the downward velocity runs into a stably stratified layer and causes kinetic energy to change into potential and thermal energy via small-scale density and pressure fluctuations, respectively. In this region, the subgrid-scales do not act like an incompressible flow and, the dissipation in equation (40) must be replaced with one that models the effects of a subgrid-scale compressible flow.

We illustrate the use of a spatially dependent form of the eddy-viscosity with a simpler inhomogeneous flow for which there is an abundance of laboratory data. In channel flow, the velocity breaks up into small, highly dissipative vortices near the boundaries. In these regions, the subgrid-scale dissipation rate depends directly on the viscosity. Following the work of Moin, et al. (1978), in this boundary region we replace equation (40) for the dissipation rate with

$$\nu v(\Delta)^2/\Delta^2 \tag{57}$$

Equation (57) combined with equations (38) and (39) yields

$$\nu_e = 2c'' \Delta^4 \bar{S}_{jk}\bar{S}_{jk}/\nu$$

where c" is a constant of order unity. Equation (58) is valid only in the boundary regions where it replaces Smagorinsky's eddy-viscosity (equation 37). Moin et al. define y_c as the distance from the wall where the average value of ν_e from equation (58) is equal to the average value of ν_e from equation (37). In the region between the wall and y_c , they use equation (58) and exterior to this region, they use equation (37). This improved eddy-viscosity reproduces the statistics of the turbulent flow near the wall much more faithfully than the numerical simulations that use Smagorinsky's eddy-viscosity everywhere (Deardorff, 1970).

Intermittency

None of the previously discussed subgrid-scale models allow intermittency. As we have shown, it may be necessary to model the intermittency in $\underset{\sim}{Q}$ to simulate accurately the mixing properties of convection. We propose a model of the subgrid-scale flow based on work by Bell and Nelkin (1977). Bell and Nelkin calculated intermittency in homogeneous, isotropic turbulence by solving a set of heuristic equations that was designed to mimic the nonlinear cascade of energy. Although they did not develop their model with the intention that it be used with a direct numerical simulation of the large-scale velocity, the incorporation can be done easily. In Bell and Nelkin's model, the velocity is represented by N scalars v(i,t), where v(i,t) is the characteristic velocity of a band of modes with wavelengths between $\ell_b/2^{i-1}$ and $\ell_b/2^i$. The characteristic wavelength of the ith band is $\ell(i)$ and $\ell(i+1)=\ell(i)/2$. N is chosen large enough so that $\ell(N)$ is less than ℓ_d. Solar convection, with a Reynolds number of 10^{14}, can be modeled with this method because the number of variables, N, need only be greater than 35. The v(i,t) satisfy the following evolution equations

$$\frac{dv(i,t)}{dt} = a_1\{v(i-1),t)^2 - 2v(i,t)v(i+1,t) + a_2[v(i-1,t)v(i,t) - 2v(i+1,t)^2]\}/\ell(i) - \nu v(i,t)/\ell(i)^2 \quad i = 1,2,\ldots,N \tag{59}$$

where a_1 and a_2 are constants of order unity and where $v(0,t) = v(N+1,t) = 0$. These equations force the cascade to be local since v(i,t) is influenced only by itself and by v(i-1,t) and v(i+1,t). The nonlinear terms in equation (59) represent the nonlinear terms in the Navier-Stokes equation that allow the energy to cascade through the different length-scales. The constant, a_2 determines the relative importance of the downward cascade (large to small scales) to the upward cascade (small to large scales). Like the integral of the nonlinear terms in the Navier-Stokes equation, the nonlinear terms in equation (59), when summed over all i, conserve energy. The last term in equation (59) is the viscous dissipation. Bell and Nelkin found that this simple set of equations reproduce many of the statistical properties of intermittent turbulence. Marcus, et al. (1983) generalized the cascade model to inhomogeneous, compressible convection for use in astrophysical calculations.

Incorporating the Bell-Nelkin cascade model in a numerical simulation is easiest in a spectral or pseudo-spectral computation. Motions with length-scales greater than Δ are governed by the Navier-Stokes equation and directly simulated. Motions smaller than 2Δ are treated with the cascade model (equation 59), where the ith band is now defined to contain length-scales between $\Delta/2^{i-2}$ and $\Delta/2^{i-1}$. The velocity with scales between Δ and 2Δ is represented both in the direct large-scale simulation and in the cascade model. Equation (59) for v(i,t) with i=1 is discarded and replaced with

$$v(1,t) = [\int d^3x\, \underline{v}_\Delta(\underline{x},t)\cdot\underline{v}_\Delta(\underline{x},t)/\int d^3x]^{1/2} \tag{60}$$

where

$$\underline{v}_\Delta(\underline{x},t) \equiv \int_{\pi/\Delta \leq |\underline{k}| < 2\pi/\Delta} d^3k \; \bar{\underline{v}}(\underline{k},t) \qquad (61)$$

and where $\bar{\underline{v}}(\underline{k},t)$ is the Fourier transform of the large-scale velocity. The Navier-Stokes equation for the large-scale velocity is solved as usual except that an eddy-viscosity term, $\nabla \cdot \nu_e \nabla \underline{v}$ is used with the modes with wavelengths between Δ and 2Δ. The eddy-viscosity is determined by conservation of energy. The rate at which kinetic energy enters the subgrid-scales from the large scales is determined by multiplying equation (59) by v(i,t) and summing the product from i=2 to i=N.

$$a_1 v(1,t)v(2,t)[v(1,t)+a_2 v(2,t)]/\ell(2) \qquad (62)$$

The rate at which the eddy-viscosity removes energy from the large-scale flow is

$$\nu_e \int d^3x \; (\nabla \underline{v}_\Delta)\cdot(\nabla \underline{v}_\Delta) \qquad (63)$$

Equating these two rates we obtain

$$\nu_e = \frac{a_1 v(1,t)v(2,t)[v(1,t)+a_2 v(2,t)]}{\ell(2) \int d^3x \; (\nabla \underline{v}_\Delta)\cdot(\nabla \underline{v}_\Delta)} \qquad (64)$$

This method of computing ν_e, allows intermittency in the subgrid-scales and has been implemented by the author. However, at the present time, no comparisons have been made between experimental values of the intermittency (flatness factor, etc.) and the numerical simulations.

Eddy-viscosity in rotating flows

Our last example of the need for a more sophisticated eddy-viscosity is rotating flow. Not only the magnitude, but also the functional form of the model of the subgrid-scale forcing must be correct. In particular, it is the detailed form of $\underline{Q}$ that determines the way in which angular momentum is expelled in a contracting proto-star, permits a dynamo to form in the sun, and is responsible for the longevity of the Red Spot of Jupiter. $\underline{Q}$ is usually chosen to be the sum of a gradient and the divergence of a traceless tensor, $\underline{\underline{\tau}}$, (see equations 30 - 32) so that mass, momentum, and angular momentum are conserved. A bad choice of $\underline{\underline{\tau}}$ will ruin a calculation. For example, consider a computation of a collapsing proto-star that uses Smagorinsky's $\underline{\underline{\tau}}$ which is defined in equations (34) and (37). Changing the value of c changes the relative rates of mass infall and angular momentum redistribution, but the stable equilibrium rotation curve of a newborn star will remain the same. The form, not the magnitude, of $\underline{\underline{T}}$ determines whether a star rotates as a solid body, has constant angular momentum per unit mass ($v_\phi \propto r^{-1}$), or has a more exotic rotation law. Without knowing the correct form of $\underline{\underline{T}}$, it is pointless to undertake numerical calculations where the results depend strongly on the angular momentum distribution.

Modeling $\underline{Q}$ as the divergence of a tensor implies that the subgrid-scales affect the large-scale flow by diffusing some physical quantity. Therefore, the correct choice of $\underline{\underline{T}}$ depends on the transport properties of the small scale flow. Smagorinsky's eddy-viscosity diffuses momentum, but momentum does not always appear to be the correct quantity to diffuse. To see what the small-scale flow transports, let us for the moment consider a simpler eddy diffusivity. In compressible flows, the entropy is an adiabatic invariant - i.e., in the limit of no dissipation, the covariant derivative of the entropy is zero.

$$\left(\frac{\partial}{\partial t} + \underline{v}\cdot\nabla\right) s = 0 \qquad \text{when} \quad \nu = k = 0 \tag{65}$$

The small-scale turbulent flow carries entropy fluctuations with it. If the time-averaged turbulence is homogeneous, the flow becomes isentropic. In compressible fluids, the temperature is not an adiabatic invariant, so the flow does not become isothermal. Since the large-scale effect of the subgrid-flow is the diffusion of entropy, not temperature, an eddy entropy diffusivity must be included in the equations of motion, and it would be incorrect to include an eddy thermal diffusivity.

Even in incompressible flow, momentum is not an exact adiabatic invariant because of the pressure. However, in rotating cylindrical flows, there is experimental evidence that the angular momentum per unit mass (with respect to the axis of rotation) acts like an adiabatic invariant (cf. DiPrima and Swinney, 1981). In rotating spherical flows, experiments indicate that this is not true (Wimmer, 1976), and it has been suggested that in rapidly rotating spherical flows, the effective adiabatic invariant is the enstrophy (i.e., the square of the vorticity). It can be easily shown that for axisymmetric flows, the angular momentum per unit mass is an exact adiabatic invariant, and for 2-dimensional flows (no velocity component parallel to the axis of rotation) the enstrophy is an exact adiabatic invariant. At the present time, there is no way of analytically or observationally determining what the correct invariant (if one exists!) is in a rotating star. However, numerical simulations (with no eddy-viscosity) of cylindrical Couette flow show clearly that the small, but numerically resolvable, scales advect angular momentum as an adiabatic invariant (Marcus, 1983); a high resolution numerical simulation might reveal what quantity behaves like an adiabatic invariant in a rotating compressible sphere of fluid and thereby allow us to choose $\underline{\underline{I}}$ properly in a stellar calculation.

DISCUSSION

We have shown that in many astrophysical flows it is impossible to simulate all of the scales of interest. Ignoring the small scales sometimes causes a numerical instability. Their neglect does not always produce an unstable code, which is unfortunate because we are often lulled into thinking that if a calculation does not blow up, it produces an accurate solution. We have given some examples of spurious results produced in under-resolved computations, and we warn the reader to be wary of them. In homogeneous, isotropic flows without intermittent behavior, a simple eddy-viscosity is probably sufficient to model the large-scale effects of the subgrid-scale flow. However, in most flows of astrophysical interest, the subgrid-scale effects are due to complicated physical processes that are a function of position and time. Analytic theories of turbulence can provide us with some guidance in modeling the small scales, but, at the present time, phenomenological models based on the physics of the subgrid-scale flow are necessary.

This work was supported in part by NSF grants MEA-8215695 and AST-8210933. Numerical computations by the author were done on the CRAY-1 at the National Center for Atmospheric Research, operated by the National Science Foundation.

REFERENCES

Basdevant, C., Lesieur, M. & Sadourny, R. 1978 J. Atmos. Sci. 35, 1028.

Bell, T. L. & Nelkin, M. 1977 Phys. Fluids 20, 345.

Chollet, J. P. & Lesieur, M. 1981 J. Atmos. Sci. 38, 2747.

Deardorff, J. W. 1970 J. Fluid Mech. 41, 453.

Deardorff, J. W. & Willis, G. E. 1967 J. Fluid Mech. 28, 657.

DiPrima, R. C. & Swinney, H. L. 1981 in Hydro. Instabilities and the Transition to Turbulence, Topics in Applied Physics, Vol. 45 (ed. H. L. Swinney and J.P. Gollub) p. 139 Springer.

Forster, D., Nelson, D. R., & Stephen, M. J. 1977 Phys. Rev. A 16, 732.

Fulachier, L. & Dumas, R. 1976 J. Fluid Mech. 77, 257.

Herring, J.R. 1973 in Proc. Langly Working Conf. on Free Turbulent Shear Flows. NASA SP321 (1973) Langley Research Center, Langley, VA.(1973).

Kraichnan, R. H. 1959 J. Fluid Mech. 5, 497.

Kraichnan, R. H. 1965 Phys. Fluids 8, 575.

Kraichnan, R. H. 1971 J. Fluid Mech. 47, 513.

Kolmogorov, A. N. 1941 C. R. Acad. Sci. U. R. S. S. 30, 299.

Leslie, D. C. & Quarini, G.L. 1978 J. Fluid Mech. 91 65.

Lilly, D. K. 1967 in Proc. of the IBM Scient. Comp. Symp. on Env. Science. IBM Form No. 320-1951.

Lorenz, E. N. 1969 Tellus 21, 289.

Marcus, P. S. 1980 Ap. J. 240, 203.

Marcus, P. S. 1981 J. Fluid Mech. 103, 241.

Marcus, P. S. 1983 submitted to J. Fluid. Mech.

Marcus, P. S., Press, W. H. & Teukolsky, S. A. 1983 Ap. J. 267, 795.

Moin, P., Reynolds, W. C., & Ferziger, J. R. 1978 NASA Technical Report Number TF-12 NASA Ames.

Newman, G. R. & Herring, J. R. 1979 J. Fluid Mech. 94, 163.

Orszag, S. A. 1970 J. Fluid Mech. 41 363

Orszag, S. A. 1974 in Proc. 1973 Les Houches Summer School (ed. R. Balian & J.-L. Peabe), pp. 237-374 Gordon & Breach.

Rose H. & Sulem, P. L. 1978 J. Phys. (Paris) 39, 441.

Siggia, E. D. & Patterson, G. S. 1978 J. Fluid. Mech. 86, 567.

Smagorinsky, J. 1963 Monthly Weather Rev. 91, 99.

Smagorinsky, J., Manake, S. & Holloway, J. 1965 Monthly Weather Rev. 93, 727.

Toomre, J., Gough, D.O., & Spiegel, E.A. 1977 J. Fluid Mech. 79, 1

Toomre, J., Zahn, J.-P., Latour, J. & Spiegel, E. A. 1976 Ap. J. 207, 545.

Van Atta, C. W. & Park, J. 1972 in Statistical Models and Turbulence (ed. M. Rosenblatt and C. Van Atta) p. 402 Springer.

Wimmer, M. 1976 J. Fluid Mech. 78, 317.

PARTICLE METHODS

J.W.Eastwood

Culham Laboratory,Abingdon,Oxon.OX14 3DB,England.
(Euratom/UKAEA Fusion Association).

ABSTRACT

There are many methods of transforming the differential equations to discrete forms suitable for numerical computations. In this chapter we show how particle methods relate to other approaches (finite difference, finite element and spectral methods) and survey particle methods. Sections 2, 3 and 4 focus respectively on particle-mesh (PM) methods for collisionless systems, particle-particle/particle-mesh (P^3m) methods for correlated systems and fluid particle methods for fluids and magnetofluids.

1. INTRODUCTION

Computational astrophysics is concerned primarily with the study of time evolution by means of computer simulation. The physical phenomena under investigation are described by mathematical models : These models are in turn represented by computer programs, which together with the computer provide the means for the computational astrophysicist to perform his *computer experiments*. The step from mathematical model to computer program almost always requires the approximation of continuous variables and differential equations by sets of values and algebraic equations. There are basically three methods of making the step from the differential continuum to the numerically computable discrete system, the *finite difference approximation*(FDA), the *finite element method* (FEM) and *particle methods*.

In the finite difference approximation, [1], functions of a continuous variable are replaced by a mesh of points at which values are approximated by differences of those mesh values. The space-

K.-H. A. Winkler and M. L. Norman (eds.), Astrophysical Radiation Hydrodynamics, 415–447.

time continuum of the initial value/boundary value problem becomes a lattice of points in space and time. Derivatives appearing in the differential equations are replaced by differences, and the resulting difference equations only apply at mesh points. The attraction of the FDA is its simplicity: Difference equations are easily constructed, and coding is easy and fast in regular geometries. Set against these are difficulties arising with irregular boundaries and anisotropic media. Also, the large number of mesh points per wavelength required for accurate representation of collective phenomena may make FDA less cost effective than more sophisticated alternatives.

The finite element method [2], provides greater accuracy and flexibility than the FDA. The FEM replaces continuous variables by restricted classes of trial functions. These functions are usually either the union of simple piecewise polynomials or are orthogonal global polynomials. In the latter case, the FEM is also known as the *spectral method* [3]. Instead of representing continuous functions by sets of mesh values as in the FDA, a set of amplitude factors for the trial functions are used. These amplitudes may be, for example, nodal values and derivative or fourier harmonic amplitudes. The benefits of FEM are that it gives optimal schemes, and generalises readily to high accuracies and irregularly shaped regions.

Finite difference (or finite element) schemes are further classified into *Eulerian*, *Lagrangian* and *mixed* models depending on the motion of the mesh (or elements). Eulerian schemes use meshes fixed to some observer frame, Lagrangian schemes use meshes which move with the fluid velocity and mixed schemes use grid velocities which are neither Eulerian or Lagrangian. Eulerian models are simplest, but have difficulty with advection. Lagrangian schemes accurately treat advection, but in more than one dimension rapidly run into difficulties with mesh shearing. Mixed schemes represent a compromise aimed at getting the accuracy of the Lagrangian method without its pathological meshes.

Particle methods [4] provide the most effective means to date of getting the advantages of both Lagrangian and Eulerian methods. To achieve this end they exploit the real characteristics of hyperbolic systems. Hyperbolic like terms in the differential equations are treated in a Lagrangian fashion, and Eulerian schemes are used for parabolic and elliptic like terms. In the Lagrangian part of the calculation, the continuum is replaced by a set of sample points (particles) each carrying attributes such as mass, position,entropy, charge, etc. Ideally, these attributes are conserved quantities of the equations retaining only hyperbolic terms, so their time advancement consists simply of moving the particles. Parabolic and elliptic equations of the Eulerian part of the calculation are handled using either FDA or FEM. The two parts of the calculation

(Lagrangian and Eulerian) are linked together at each timestep by interpolation (from mesh to particles) and inverse interpolation (from particle to mesh).

The feature which distinguishes particle methods from FDA and FEM is the use of a set of particles moving according to some equations of motion. The most obvious application of particle methods is to the study of correlations in N-body classical systems, where there is a one to one correspondence between physical particles and computer simulation particle. Examples of such systems include dense plasmas, ionic liquids, stellar clusters and galaxy correlation studies. Particle methods for such correlation studies are discussed in Section 3. The particle-mesh (PM) method for collisionless phase fluids described in section 2 is similarly motivated by the corpuscular nature of the material being simulated. However, in this instance, each computer particle corresponds to tens or hundred of million of physical particles. The third class of particle models, those for simulating fluids and magnetofluids are further divorced from the particulate nature of the material, introducing the particles as a computationally convenient means of describing the motion of continuum fluids: Models of this type will be described in Section 4.

Particle methods, like FDA and FEM, prove effective in some applications and not in others. For all three approaches, we may lay down criteria to help judge particular schemes in the context of particular problems. Essential properties of any numerical scheme are *consistency* with the differential equations and numerical *stability*. Relative merits of schemes may be assessed using criteria of *accuracy*, *efficiency* and *conservation*. Accuracy is measured by order, by convergence, by linearised dispersion analysis and by test problems. Efficiency is concerned with optimising the cost/quality compromise. Conservation criteria pertain to positivity (of density) and integrals of the motion (mass, momentum, etc). Only in special circumstances can the non-linear properties of the numerical schemes be fully tested. Throughout this chapter, the above criteria will be repeatedly used in discussing alternative schemes and identifying the 'best' choices.

In addition to criteria, we may call upon past computations to indicate where particle methods are appropriate. Particles are the obvious choice for naturally corpuscular systems. Their simplicity and robustness make them attractive for multidimensional calculations: They provide the only successful method to have simulated six dimensional phase fluid (collisionless) plasmas. In fluid calculations, their main application has been to supersonic flow: Particle codes can readily handle fragmentation, voids, free surfaces and multi material calculations which are beyond the capabilities of almost all other codes.

2. PARTICLE-MESH METHODS.

Particle-mesh (PM) methods use a combination of particle and mesh representations. The material - a plasma or gravating mass - is represented by an ensemble of particles. The evolution of the system is described by the motions of these particles. The purpose of the mesh is to provide a discrete representation of the charge (or mass) density, the potential and the force fields. As will be shown below (Sec.2.3), the use of fields on a mesh gives enormous computational speed gains over direct particle-particle force summation.

The original and simplest PM method, the nearest-grid-point (NGP) scheme [5] is used in Sec.2.1 to illustrate the basic features of the PM calculational cycle. The relationship of the PM model to the collisionless Boltzmann equation is outlined in Sec.2.2. Sections 2.4 to 2.7 look in some detail at particular aspects of the PM calculational cycle. Section 2.8 contains an outline of the physical properties of PM schemes; there it is shown how the PM calculation gives a better approximation to a collisionless system than meshless point particle calculations ! Combining the results of the physical analysis with costings gives the algorithm optimisation described in Sec.2.9.

2.1. The NGP Scheme.

The NGP model we shall consider is that of a one dimensional electron gas embedded in a fixed neutralising background. The reason for this choice is that it allows us to see all the essential features without complications of isolated boundaries and initial equilibria. PM schemes generalise trivially to two and three dimensions [4,5,6] so one dimension is used to avoid confusion of extra indices. PM algorithms for gravitating systems differ only in the sign appearing in Poisson's equation, so one may obtain gravitational PM models by reading 'mass' for 'charge' and $-1/4\pi G$ for ε.

The state of the 1-D NGP model at any timelevel is described by the set of particle positions and velocities $\{x_i, v_i, i \in [1, N_p]\}$. Initial conditions are prescribed by distributing the particles of mass M and charge Q throughout the computational box, $x \in [0,L]$, according to the initial distribution, $f(x,v,0)$, of density and velocity. Boundary conditions are prescribed for both particles and fields at the boundaries $x = 0$ and $x = L$ of the computational box. For simplicity we shall assume periodic conditions, so particles moving through $x = 0$ re-enter the computational box at $x = L$ and vice-versa.

The evolution of the PM model is computed by integrating the equations of motion of the particles:-

$$\frac{dx_i}{dt} = v_i \quad ; \quad M\frac{dv_i}{dt} = F(x_i) = Q\,E(x_i) \qquad (1)$$

The electric field E is obtained from the charge density distribution ρ of the particles plus background charge by solving Poisson's equation

$$\frac{d^2\phi}{dx^2} = -\frac{\rho}{\varepsilon_o} \quad ; \quad E = \frac{-d\phi}{dx} \qquad (2)$$

Equations (2) are represented by FDA's: The computational box is divided into a set of cells of width H, where $H = L/N$. At the centre of each cell is a mesh point at which values of charge density, potential and electric field are stored. Derivatives are approximated by differences of mesh values.

The timestep cycle of the NGP model is as follows:-

a) Start : Given $\{x_i^n,\ v_i^{n-\frac{1}{2}};\ \ i\ \varepsilon[1,N_p]\}$, where n denotes timelevel, time $t = n\Delta t$.

b) Assign charge to mesh: Charge density is charge per unit volume. In the NGP scheme, charge density at mesh point p at position $x_p = pH$ is computed by summing all the charge in cell $p(x_p-H/2 < x < x_p + H/2)$ and dividing by the cell volume

$$\rho_p = \frac{1}{H} \sum_{\substack{\text{particles in}\\ \text{cell } p}} Q + \rho_o \qquad (3)$$

In practice, $\{\rho_p,\ p\ \varepsilon\ [0,N_g-1]\}$ are computed by sweeping through the particles, locating the cell, p, containing each particle and incrementing ρ_p by Q/H.

c) Solve for the potential: The simplest FDA to Poisson's equation gives

$$\phi_{p-1} - 2\phi_p + \phi_{p+1} = -\rho_p H^2/\varepsilon_o \quad ; \ p\varepsilon[0,N_{g-1}] \qquad (4)$$

Equation (4) with periodic boundary conditions may be inverted to obtain potentials $\{\phi_p;\ p\varepsilon[0,N_{g-1}]\}$

d) Accelerate particles: NGP force interpolation takes the force field at any point in a cell to be equal to the value at the mesh point at the centre of that cell. Thus, for particle i in cell p the force is given by

$$F(x_i) = QE(x_p) = Q\,\frac{(\phi_{p-1}-\phi_{p+1})}{2H} \qquad (5)$$

and velocities are updated using the difference approximation

$$v_i^{n+\frac{1}{2}} - v_i^{n-\frac{1}{2}} = F^n\Delta t/m \qquad (6)$$

The acceleration phase, Eqs.(5) and (6), are performed by sweeping through particles rather than cell-wise.

e) Move particles : The new velocities $\{v_i^{n+\frac{1}{2}}\}$ are used to obtain new positions

$$x_i^{n+1} = x_i^n + v_i^{n+\frac{1}{2}} \Delta t \tag{7}$$

and it is at this stage that particle boundary conditions are applied.

2.2. The Mathematical model.

A common misconception concerning PM simulations is that of the nature of the physical process being simulated. NGP (and more sophisticated variants) provide means to simulate *collisionless* systems, and for a given number of simulation particles give a more accurate representation than would a set of point particles in a meshless system. This conclusion is true both for plasmas where there is Debye screening and for simulation of galaxies where no screening occurs.

The mathematical model for a 1-D Vlasov-Poisson system plasma is [7]

$$\frac{\partial f}{\partial t} + v\frac{\partial f}{\partial x} + \frac{F}{m}\frac{\partial f}{\partial v} = 0 \tag{8}$$

$$F = qE = -q\frac{d\phi}{dx} \tag{9}$$

$$\frac{d^2\phi}{dx^2} = -\frac{\rho}{\varepsilon_o} \tag{10}$$

$$\rho = \rho_o + q\int f dv \tag{11}$$

where $f(x,v,t)$ is the one particle electron distribution function. In the NGP scheme, equations (9) and (10) are discretised by FDA, Eq.(8) is represented by particles and the integral in Eq.(11) is evaluated by 'Monte-Carlo' methods [8].

One interpretation of the particles in the PM models is simply as a mathematical artefact for transforming the awkward advection equation (8) into a set of ordinary differential equations (1). Equations (1) are the characteristics of Eq.(8), along which f is constant: Given f at some time t the equations of motion (1) may be used to map f to some later time t′. The discrete approximation to f is a set of samples:

$$\tilde{f} = \sum_{i=1}^{N_p} \delta(x-x_i)\ \delta(v-v_i) \tag{12}$$

where $\{x_i, v_i\ ;\ i \in [1,N_p]\}$ evolve according to Eq.(1). $\tilde{f}$ comprises samples of a *smooth* distribution function f, so estimates of moments of f are obtained using weighted means of samples, i.e.

$$\int dv\, v^n f \simeq \int dv\, v^n \left\{ \frac{N_s}{\lambda} \int_{x-\lambda/2}^{x+\lambda/2} W(x-x')\tilde{f}(x',v,t)dx' \right\}$$

$$= \frac{N_s}{\lambda} \sum_i W(x-x_i)\, v_i^n \qquad (13)$$

W is a weighting factor, λ is the range of the averaging and N_s is a scaling factor determined by mass (or charge) conservation. Setting $\lambda = H$ and the weighting factor

$$W(x) = \begin{cases} 1 & -H/2 < x \leq H/2 \\ 0 & \text{otherwise} \end{cases} \qquad (14)$$

reduces Eq.(13) to the NGP assignment scheme.

A physical interpretation of collisionless PM models is to view each particle as a finite sized cloud of electrons where internal degrees of freedom are suppressed. The complete description of the cloud (slab in one dimension) is given by its velocity, position, density profile and charge/mass ratio. The cloud must have a finite width to soften short range interactions to the extent that they freely pass through each other, otherwise binary collisional effects will dominate collective effects. Each cloud corresponds to N_s electrons, where N_s is typically 10^5, so without softening short range forces qualitatively incorrect behaviour would result. Generally, a cloud width ($\simeq$H) of order the Debye length is used: This minimises collisional effects without seriously affecting collective properties. The value of charge density at a given point is given by the sum of contributions of clouds overlapping that point, so the range of neighbouring particles for purposes of finding weighted mean values may be interpreted as the cloud width (c.d.Eq.(13)). To avoid statistical fluctuations in such mean values, we need a large number (say $\sim$ 10) of clouds to overlap : This factor leads to the requirement that the number of particles per Debye slab (square or cube) is large.

2.3. PM vs. direct summation.

The finite sized particle interpretation of the NGP scheme begs the question, why use a mesh at all ? The beneficial reduction in collisional effects arising from smoothing of forces by the mesh could equally well be achieved by summing interparticle forces (Coulomb's Law) between finite sized clouds. In this manner, all assignment, interpolation and mesh errors would be avoided. The answer is that the direct summation method is an untenable cost/quality compromise.

The cost savings that can be achieved by employing particle-mesh schemes is readily seen by considering a simple example. The direct particle-particle (PP) scheme computes forces by pairwise sums : $\underset{\sim}{F}_i = \Sigma\, \underset{\sim}{f}_{ij}$. Take, for example, a system of point charges where $f_{ij} = x_i - x_j / |x_i - x_j|^3$. Assume + − ÷ and x each count one operation and $(....)^{1.5}$ counts as three, then the operations count for evaluating F_i for N_p particles and advancing positions and velocities by one timestep $i \sim 10N_p^2$. To be compared with this is an operations count of $\alpha N_p + \beta(N)$ for PM schemes, where the term αN_p arises from charge assignment, force interpolation and integration of equations of motion. The term $\beta(N)$ is the operations count for solving Poissons equation. ($\sim 5N^3\log_2 N^3$ in 3-D). Using typical values $\alpha = 20$, $N = 32$, $N_p = 10^5$ and time, τ per floating point operation of $1\mu s$ we obtain cpu time estimates of

$$\text{PM:} \quad (\alpha N_p + 5N^3\log_2 N^3)\tau \simeq 4\tfrac{1}{2} \text{ sec/timestep}$$

$$\text{PP} \quad 10N_p^2 \quad \tau = 10^5 \text{sec}/\Delta t \simeq 1 \text{ day/timestep}$$

The enormous speed advantage of PM over PP is obtained at the cost of increased algorithm complexity, but with negligible loss of accuracy in the case of more sophisticated PM schemes.

2.4. Assignment and Interpolation.

The NGP scheme outlined in Sec.2.1 is the lowest ('zeroth') order member of the hierarchy of PM schemes. It gives a force between particles which varies stepwise, which is zero for particles within the same cell (in accord with the finite sized particle interpretation), which is not displacement invariant and which depends on direction with respect to the mesh('angular anisotropy').

Generalisation of NGP assignment and interpolation may be obtained as follows. Using the 'top-hat' function, Eq.(14), and the density of particle centres $n(x) = \int \tilde{f}\, dv$ we may rewrite NGP assignment (Eq.(3)) as

$$\rho_p = \rho_o + \frac{Q}{H}\sum_i W(x_p - x_i) = \rho_o + q\frac{N_s}{H}\int W(x_p - x')n(x')dx' \quad (15)$$

and interpolation as

$$F(x_i) = N_s\, q \sum_{p=0}^{n-1} W(x_i - x_p) \quad E_p \quad (16)$$

Higher order schemes result from replacing the NGP form of W in Eqs.(15) and (16) by other appropriate functions.

Criteria used in selecting functional forms for W are i) at large separations compared to H, fluctuations and angular dependence of interparticle forces become small, ii) charge assigned and force interpolated should vary smoothly as particles are displaced w.r.t. the mesh iii) momentum should be conserved and iv) computational cost. Criterion (i) leads to a multipole expansion hierarchy of schemes, [9,10] with successively higher order schemes having dipolar, quadrupole, etc, leading error and involving increasing numbers of mesh points. Criterion ii) selects from those schemes satisfying (i) the subset which lead to smoothly varying forces(and thence smaller collisional effects and better energy conservation). Criterion (iii) demands that the force interpolation function be identical to the charge assignment function to avoid self forces and possible related numerical instabilities. The cost criterion favours compact assignment schemes of product form.

Applying all four criteria one is left with the cost/quality hierarchy of product form charge assignment/force interpolation functions summarised in table 1. In table 1 d denote dimensionality, Π is the 'top-hat' function, * denotes convolutions and the abbreviations CIC, TSC, PQS refer to cloud-in-cell (or area weighting), triangular shaped cloud and piecewise quadratic shaped. The names associated with the schemes arises from a graphical description of the assignment schemes. Figure 1 illustrates this for the case of TSC in one dimension. If one imagines that each particle carries a triangle with it, the fraction of its charge

TABLE 1. Hierarchy of assignment/interpolation schemes.

SCHEME	ORDER	NUMBER OF POINTS	ASSIGNMENT FUNCTION	SMOOTHNESS OF FORCE
NGP	0	1^d	Π	Stepwise
CIC	1	2^d	$\Pi*\Pi$	Continuous piecewise linear
TSC	2	3^d	$\Pi*\Pi*\Pi$	Continuous value and 1st derivative
PQS	3	4^d	$\Pi*\Pi*\Pi*\Pi$	Continuous to 2nd derivative

assigned to a particular mesh point is equal to the overlap of the triangle with the cell containing the mesh point. Inspection of figure 1 then reveals

$$\begin{aligned} W_o(x) &= \tfrac{3}{4} - x^2 \\ W_1(x) &= \tfrac{1}{2}(x + \tfrac{1}{2})^2 \\ W_{-1}(x) &= \tfrac{1}{2}(x - \tfrac{1}{2})^2 \end{aligned} \tag{17}$$

which by displacement invariance, $W_p(x) = W(x-p)$ yields $W(x) = \Pi*\Pi*\Pi$ as given in table 1.

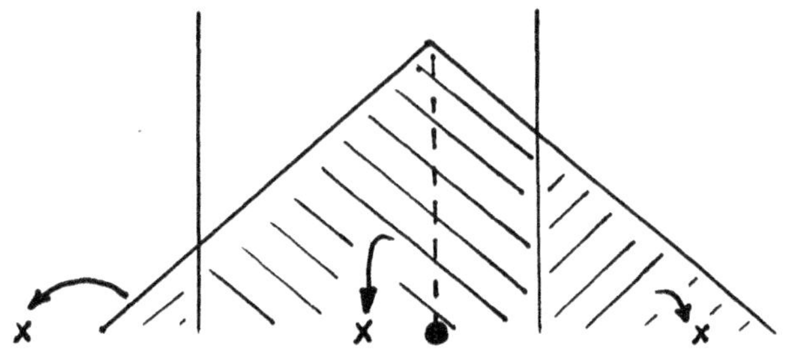

FIGURE 1. In TSC assignment and interpolation, the fraction assigned/interpolated to or from a mesh point is equal to the overlap area of the triangle with the mesh point's cell.

2.5. Time integration.

Constraints on the choice of ordinary differential equation solvers for use with the large number of particles involve in PM simulations favour the use of simple low order schemes. Stability requires $\Omega\Delta t \sim 1$ for most usable explicit schemes independent of their order. Limitations of storage make second order schemes involving only two timelevels the most attractive, where greater accuracy is obtained by reducing the timestep. Indeed, it is questionable as to whether higher order schemes do introduce any gains where field and their derivatives are defined numerically. A further factor favouring the lower order schemes is the absence of the potentially troublesome nonphysical (parasitic) roots of higher order schemes.

The most widely used integration scheme is the time symmetrical second order accurate leapfrog scheme (Eqs.(6) and (7)). Leapfrog is stable for timesteps where $\Omega\Delta t < 2$, where Ω is the maximum characteristic frequency of the system. It requires simultaneous storage of particle positions and velocity at only one timelevel ($\{\underset{\sim}{x}_i^n, \underset{\sim}{v}_i^{n-\frac{1}{2}}\ ;\ i\ \varepsilon[1,N_p]\}$). Numerical errors arising from finite Δt appear as phase errors (oscillation frequency too large)without dissipation.

An example of the pitfalls that may arise in choosing time integration schemes without due attention to accuracy,stability and efficiency is the 'Beeman' algorithm [11]

$$\begin{aligned} x^{n+1} &= x^n + v^n\Delta t + \frac{\Delta t^2}{6m}\,[4F^n - F^{n-1}] \\ v^{n+1} &= v^n + \frac{\Delta t}{6}\,[2F^{n+1} + 5F^n - F^{n-1}] \end{aligned} \qquad (18)$$

This scheme has been favoured in certain circles as a more accurate (!?) alternative to leapfrog for particle calculations. However, by eliminating v from Eqs.(18) we find

$$x^{n+1} - 2x^{n} + x^{n-1} = F^{n}\Delta t^{2}/m \qquad (19)$$

which is identical to leapfrog ! Thus, the 'Beeman' algorithm will give the same results as leapfrog (apart from roundoff), but require twice the storage and four times the amount of computer time.

Lorentz force terms are usually treated implicitly:-

$$\frac{\underset{\sim}{v}^{n+\frac{1}{2}}-\underset{\sim}{v}^{n-\frac{1}{2}}}{\Delta t} = \frac{q\underset{\sim}{E}^{n}}{m} + \frac{\underset{\sim}{v}^{n+\frac{1}{2}}+\underset{\sim}{v}^{n-\frac{1}{2}}}{2} \times \underset{\sim}{\Omega}^{n} \qquad (20)$$

An interesting property of Eq.(20) is that for large Δt, the component perpendicular to the magnetic field reduces to the adiabatic drift approximation ($\underset{\sim}{v} = \underset{\sim}{E} \times \underset{\sim}{B}/B^2$). A further sophistication of Eq.(20) that is often used is a frequency correction factor to compensate for the finite Δt phase errors in the cyclotron motion: This amounts to replacing the cyclotron frequency, $\underset{\sim}{\Omega}$, in equation (20) by $\alpha\underset{\sim}{\Omega}$, where $\alpha = \tan(|\Omega|\Delta t/2)/(|\Omega|\Delta t/2)$.[12,13].

The extension of the equations of motion to describe viscous force terms is described in Sec.3.5. Another variant of the equations of motion are met in the vortex fluid model (Sec.4.1), where the rotational forcing term causes simple explicit schemes to be unstable. A further class of schemes we shall not cover here are integration schemes for implicit PM model time integration: Such PM models are now under active investigation [14,15] as attempts are made to push particle schemes into longer timescale calculations, where details of the faster frequencies are unimportant and the timestep restriction $\Omega\Delta t<2$ of explicit integration schemes severely limits the feasibility of many calculations.

2.6. Energy Conserving Schemes.

'Energy conserving' schemes [16,17,18] are a class of collisionless PM schemes derived using a variational finite element method. The advantages of this formulation are that it leads to PM models which are optimal for the basis functions chosen, it gives Hamiltonian schemes in the limit $\Delta t \to 0$, and it can be applied to fully electromagnetic systems, awkward geometries and arbitrarily spaced meshes.

To illustrate the method of obtaining the discretised finite element equations, we again use the one-dimensional plasma model as our example. The action integral is a functional of potential

Φ and particle co-ordinate, x :-

$$I[\Phi,\underset{\sim}{x}] = \int_{t_o}^{t_1} L\,dt \tag{21}$$

where the combined field and particle Lagrangian is [19]

$$L = \int d\mu\, f[\tfrac{1}{2}mv^2 - q\Phi] + \int dx'\left(\frac{\varepsilon_o|\nabla\Phi|^2}{2} - \rho_o\Phi\right) \tag{22}$$

The first integral is over phase space and the second over physical (x) space.

The finite element approximation to L is obtained by replacing f by a set of samples, $\tilde{f}$ (c.f.Eq.(12)) and replacing the potential Φ by trial functions

$$\Phi \simeq \phi = \sum_p \phi_p(t)\, W_p(x) \tag{23}$$

Amplitudes $\{\phi_p\}$ correspond to mesh potential values and trial (basis) functions W_p correspond to charge assignment/force interpolation functions. The discretised approximation, $\tilde{L}$ to Eq.(22) is

$$\tilde{L} = \sum_{i=1}^{N_p}\left[\tfrac{1}{2}\, m\, v_i^2 - q \sum_p \phi_p\, W_p(x_i)\right]$$
$$+ \int dx'\left[\frac{\varepsilon_o}{2}\left|\sum_p \phi_p \frac{dW_p}{dx}\right|^2 - \rho_o \sum_p \phi_p W_p\right] \tag{24}$$

The approximate action integral, $\tilde{I}$, arising from $\tilde{L}$ is a functional of the nodal amplitudes $\{\phi_p\}$ and the set of particle coordinates $\{x_i\}$. These lead to the Euler-Lagrange equations

$$\frac{\partial\tilde{L}}{\partial x_i} - \frac{d}{dt}\frac{\partial\tilde{L}}{\partial v_i} = 0\ ; \quad \frac{\partial\tilde{L}}{\partial\phi_p} = 0 \tag{25}$$

Equations (25) specify the optimal (in the sense that it minimises $\tilde{I} \gtrless I$) energy conserving scheme once W is chosen. The lowest order conforming choice for W is the CIC scheme [4,20] ($W = \Pi * \Pi$), which give

$$m\dot{v}_i = -q\,\frac{(\phi_p - \phi_{p-1})}{H}\ ; \quad x_{p-1} < x_i \leqslant x_p \tag{26}$$

$$\frac{\phi_{p-1} - 2\phi_p + \phi_{p+1}}{H^2} = \frac{1}{\varepsilon_o}\left[\rho_o + \frac{q}{H}\sum_{i=1}^{N_p} W_p(x_i)\right] \tag{27}$$

The right hand side of Eq.(27) describes the usual CIC assignment scheme and left hand side shows the three point fda to the Laplacian. The difference from the momentum conserving PM scheme comes in force interpolation (lhs of Eq.(26)), which is NGP centred at cell boundary points rather than mesh points. A consequence of the mixed CIC/NGP scheme is that momentum conservation is lost: An isolated particle placed arbitrarily on a mesh will oscillate in its own force field. This factor has generally little effect on the collective behaviour of the simulation model, and usually may be ignored [18].

2.7. The field Equation.

The success of particle-mesh models relies on techniques to rapidly solve elliptic equations, and in particular to solve Poisson's equation. In one dimension, fast direct methods such as the tridiagonal solver (a variant of Gauss elimination for tri-diagonal matrices) and cyclic reduction give operations counts proportional to the number of mesh points, N. Slightly slower (operations count $\sim 5N\log_2 N$) but capable of handling arbitrary cyclic convolutions is the fast fourier transform (FFT) method.

Fourier analysis and cyclic reduction methods readily generalise to two (and three) dimensions: DCR uses double cyclic reduction, FACR uses a combination of fourier analysis in one direction and cyclic reduction in the second and FFT fourier transforms in both directions. FACR is fastest, but the flexibility of double FFT outweighs the speed advantage of FACR in many applications.The 2-D analogue of the tridiagonal scheme is Nested dissection : This scheme gives minimal sparse matrix infill for matrices arising from simple FDA on regular rectangular meshes.

For an extensive survey of elliptic solver techniques we refer readers to Chapter 6 of [4]. Here we restrict our discussion to treatment of isolated boundary condition. The FFT method of treating isolated b.c. transforms Poisson's equation to integral form and uses zero padding to bring the integral to a periodic convolution form [21,22]. An alternative scheme, James' algorithm, [23] uses the capacity matrix method [24].

In one dimension, if we have a distribution of charge (or mass) which is nonzero only on the interval $x\varepsilon[0,L]$, then we may represent it by the set of values $\{q_p;p\varepsilon[0,N-1]\}$. A mesh approximation to the potential is given by

$$\phi_p = \sum_{p'=0}^{N-1} G_{p-p'}\, q_{p'} \qquad (28)$$

where $\{G_p\}$ are values obtained from the Greens function solution to Poisson's equation with appropriate (isolated) boundary conditions. Equation (28) is true for all p. A PM calculation requires values of ϕ for $p\ \varepsilon[0,N-1]$, so we are free to modify Eq.(28) to bring it to a more convenient form, provided that $\{\phi_p;p\varepsilon[0,N-1]\}$ are unchanged. The most convenient modification to periodic convolution form is as follows: Let

$$\rho_p = \begin{cases} q_p & p\ \varepsilon\ [0,N-1] \\ 0 & p\ \varepsilon\ [N,2N-1] \end{cases}$$

$$g_p = \begin{cases} G_p & p \in [0,N-1] \\ G_{2N-p} & p \in [N,2N-1] \end{cases}$$

then

$$\psi_p = \sum_{p'=0}^{2N-1} g_{p-p'}\, \rho_{p'} \; ; \; p \in [0,2N-1]$$

$$\equiv \phi_p \quad \text{if} \quad p \in [0,N-1]$$

The FFT algorithm for computing ϕ_p, $p \in [0,N-1]$ is thus

(i) Transform $q_p \supset \hat{q}_k$ (2.5NlogN)

(ii) Multiply $\hat{\psi}_k = \hat{g}_k \, \hat{q}_k$ (N)

(iii) Transform $\hat{\psi}_k \supset \psi_p$ (2.5NlogN)

Operations count $\approx$ $5N\log_2 N$

Computational savings are made by exploiting the property q_p, ψ_p real $\supset \hat{q}_k, \hat{\psi}_k$ hermitian. The method immediately generalised to 2-D and 3-D, where suitable ordering of operations leads to storage requirements of $2(N^d)$ for an N^d 'active' mesh, where d is dimensionality [22].

James' [23] variant of the capacity matrix method is derived from the procedure:-
i) Solve for potential ψ, where $\nabla^2\psi = \rho$ and $\psi = 0$ on the square (in 2-d) boundary ∂. The potential ψ is the superposition of the desired solution ϕ with the potential due to surface charge σ on ∂.i.e.

$$\phi = \psi - \int G(\underset{\sim}{x};\underset{\sim}{x}')\sigma(\underset{\sim}{x}')d\underset{\sim}{x}' \qquad (29)$$

where G is the Green's function solution of Poisson's equation with $G \to 0$ as $|x| \to \infty$ ($G \sim \ln r$).
ii) Given ψ on ∂, compute the induced charge σ.
iii) Compute ϕ on ∂ (using Eq.(29)).
iv) Given ϕ on ∂, solve the Dirichlet b.c. problem in the interior.
The implementation of these steps by James assumes a 5 point FDA Laplacian, uses sine fourier transform methods and only partially solves step (i) to avoid unnecessary computations.

2.8. Physical properties.

Particle mesh models are amenable to the same analysis as the mathematical models of the physical system from which they derive. Consequently, the models may be subjected to physical analysis in addition to numerical analysis, with the result that much better representations of the physical system are achieved. Transform space analysis gives an exact description of the errors incurred in the force calculation cycle. Linearised analysis shows how the numerical effects change wave dispersion and damping. Kinetic theory for finite sized particle ensembles describe effects of the 'graininess' introduced by discretisation of the phase fluid, and a simple stochastic theory accurately describes the loss of energy conservation due to non-conservative numerical errors in the force. Space precludes all but a brief outline here: Details may be found in references [4,25,26,27].

For a uniform periodic (or infinite) spatial mesh, the charge assignment step may be recognised as a convolution followed by a sampling and the steps to solve for potential, difference to find fields and interpolate forces may be seen as convolutions. Application of the convolution theorem then reveals that for a density of particles centres, $n(x)$, whose transform is $\hat{n}(k)$; the force field (for $q = 1$, $H = 1$) is

$$\hat{\underset{\sim}{F}}(\underset{\sim}{k}) = \hat{W}\,\hat{\underset{\sim}{D}}\,\hat{G}\left(\sum_{\underset{\sim}{n}}\hat{W}(-\underset{\sim}{k}_{\underset{\sim}{n}})\;\hat{n}(\underset{\sim}{k}_{\underset{\sim}{n}})\right) \tag{30}$$

Successive terms in the r.h.s. of Eq.(30) are the transforms of the operations force interpolation, potential differencing, solving Poisson's equation and charge assignment, respectively. Comparing Eq.(30) with the expression for the correct force harmonic amplitudes

$$\hat{F}_{exact} = -i\underset{\sim}{k}\;\hat{g}\;\hat{n} \tag{31}$$

enables the combined effect of all the errors on the forces to be evaluated. All terms contribute to errors in the magnitude of the force. Errors in the force direction arise from potential differencing ($\hat{\underset{\sim}{D}}$) and the (non-conservative) force fluctuations are due solely from charge assignment. The sum over $\underset{\sim}{n}$ in Eq.(30) is known as the alias sum. Aliasing arises because the mesh can only resolve wavelength lying in the Principal (first Brillouin) zone. Shorter wavelengths in the density distribution masquerade as wavelengths in the Principal zone. Optimally matching elements of the force calculation using Eqs.(30) & (31) can give orders of magnitude improvements in force accuracy without any increase in computational cost.

Using Eq.(30) in linearised dispersion analysis reveals that the non-physical mode coupling (aliasing) can cause numerical instability under certain conditions: In plasmas these conditions arise when the Debye length $\lambda_D \ll H$. The consequence of this instability is a rapid increase in total energy, until the plasma temperature is sufficient for $\lambda_D \sim H$.i.e. the mathematically convenient 'cold' plasma approximation is inaccessible to the PM simulation experiment (as is its laboratory counterpart !).

Collisions are analysed by interpreting the PM model as a

finite sized particle ensemble. Collisionless behaviour is obtained if the number of particles per particle is large. For plasmas, the collision time, τ_c, is related to the plasma period, τ_p, number of particles per Debye square (in 2D), N_D, and particle width w by

$$\frac{\tau_c}{\tau_p} = N_D \left[1 + \left(\frac{w}{\lambda_D}\right)^2\right] \tag{32}$$

In computations a balance has to be struck between collisional and dispersive properties. Computational costs demands as few particles as possible. Long collision times are obtained with wider particles, whereas dispersive properties are better treated by smallest practical particle width $\sim H$ ($\simeq \lambda_D$ for plasmas).

Numerically stable PM calculations generally exhibit slow secular increases of total energy [28]. The dominant contribution to this stochastic heating is the fluctuating part of the force - this part corresponds to the $n \neq 0$ terms in the alias sum in Eq.(30). The size of the fluctuating force is determined by the rate of decay of $\hat{W}$, or, in x-space, the smoothness of W: This is one reason why the schemes in the NGP-CIC-TSC hierarchy are generally superior to multipole schemes. Typically, energy conservation is $\sim$ 25 times better for CIC than for NGP [29].

Angular momentum and flux of angular momentum are important quantities in the numerical simulation of galaxies. The total angular momentum, $\underset{\sim}{L}$, of an ensemble of N particles satisfies

$$\frac{d\underset{\sim}{L}}{dt} = \sum_{i=1}^{N-1} \sum_{j=i+1}^{N} (\underset{\sim}{x}_i - \underset{\sim}{x}_j) \times \underset{\sim}{f}_{ij} \tag{33}$$

where $\underset{\sim}{x}_i$ and $\underset{\sim}{x}_j$ are particle coordinates and $\underset{\sim}{f}_{ij}$ are interparticle forces. For central conservative forces, Eq.(33) equals zero and $\underset{\sim}{L}$ is conserved. Leapfrog time integration still permits conservation of $\underset{\sim}{L}$ if forces are central and conservative, for if we define torque and angular momentum, respectively,

$$\sum_i x_i^n \times \underset{\sim}{F}_i^n, \quad m \sum_i \underset{\sim}{x}_i^n \times \underset{\sim}{v}_i^{n+\frac{1}{2}}$$

then leapfrog gives

$$\frac{\underset{\sim}{L}^{n+\frac{1}{2}} - \underset{\sim}{L}^{n-\frac{1}{2}}}{\Delta t} = \sum_{\substack{i>j \\ i,j}} (\underset{\sim}{x}_i^n - \underset{\sim}{x}_j^n) \times \underset{\sim}{f}_{ij}^n$$
$$= 0 \text{ if } \underset{\sim}{f}_{ij} = |\underset{\sim}{f}| \frac{(\underset{\sim}{x}_i - \underset{\sim}{x}_j)}{|x_i - x_j|} \tag{34}$$

Interparticle forces $\underset{\sim}{f}_{ij}$ are not generally parallel to $\underset{\sim}{x}_i - \underset{\sim}{x}_j$ in PM models because of directional errors in the gradient and aliasing. Again, increasing the order of the scheme and optimising the force calculation can bring dramatic improvements over the commonly used algorithms for simulating galaxies [6,30,31].

2.9. Final Remarks.

The process of setting up an optimal PM simulation model comprises the following steps:-

a) choose Δt for stability and accuracy
b) choose computational box to cover longest wavelengths
c) choose W and mesh size to get best quality/cost compromise, where quality is measured by dispersive/conservation compromise.
d) choose N_p to get acceptable collisional effects
e) Given W for assignment/interpolation, choose a differencing operator $\underset{\sim}{D}$ to get commensurate accuracy in the force direction, and choose an optimal $\hat{G}$ (cf.sec.3.4) to offset reshaping effects of W and $\underset{\sim}{D}$ for wavelengths of importance, and set $\hat{G}$ to zero for large wavenumbers to suppress the strong short wavelength alias fluctuations.

The outcome is a numerical scheme which identically conserves mass, charge and momentum, and guarantees density positivity. Stability is assured if $\omega\Delta t < 2$ ($\sim .5 - 1$). Long collision times, accurate wave dispersion and good energy and angular momentum conservation are readily obtained using the Q-minimising optimisation procedure [32]. Numerical instabilities in PM models are relatively benign: They announce their presence by poor energy conservation and saturate when physical parameters have moved to values for which numerical effects are small.

3. PARTICLE-PARTICLE/PARTICLE-MESH (P^3M) METHODS.

If the discrete structure of a medium is of importance then the (collisionless) PM method becomes uneconomical: To handle rapidly varying short range interparticle forces would demand a prohibitively fine mesh. Costs also militate against direct particle-particle summation methods unless only a small number of particles ($\lesssim 500$) are involved or forces are only short range. In ionic liquids, dense plasmas, stellar clusters, galaxy clustering, molecular structure, etc., both rapidly varying short range and slowly varying ($\sim 1/r^2$) long range forces are important, and the assumptions of small ensembles of particles often do not hold: It is in such circumstances that P^3M methods [4,33,34,35] are appropriate, for they enable the N_p^2 operations count of the PP method (cf.(Sec.4.3 & 4.4) to be reduced to an operations count which is proportional to the number of particles, N_p!

The trick used in P^3M is similar in concept to the Ewald summation method. The interparticle force, $\underset{\sim}{f}$, expressed as the sum of a short range part, $\underset{\sim}{f}_{sr}$ and a smoothly varying part, $\underset{\sim}{R}$

$$\underset{\sim}{f} = \underset{\sim}{f}_{sr} + R \qquad (35)$$

$\underset{\sim}{f}_{sr}$, which is non zero only for small particle separations($r<r_e$),

is computed by direct pairwise summation. If N_n is the number of neighbouring particles in the range ($= r_e$) of each particle, the operations count for this step is $\alpha N_p N_n$. The smoothly varying force, $\underset{\sim}{R}$, is computed using the PM method, with operation count $\sim \alpha N_p + \beta(N)$. Generally, a pseudo-spectral treatment of the PM part of the calculation is employed to simplify the minimization of numerical errors. Problems in setting up computationally economical P^3M calculations are i) to make the cutoff radius r_e of $\underset{\sim}{f}_{s.r.}$ small and ii) to make anisotropies and fluctuations in $\underset{\sim}{R}$ negligible. These have been solved using the Q-minimising optimisation method [4,25,32,34,35].

3.1. The timestep loop.

The timestep loop of P^3M differs from that of PM schemes only in the addition of the short range particle-particle (PP) force calculation. Again, leapfrog provides the most convenient time integration scheme, giving the timestep loop:-

i) Start : $\left\{\underset{\sim}{x}_i^n,\ \underset{\sim}{p}_i^{n-\frac{1}{2}}\ ;\quad i\ \varepsilon[1,N_p]\right\}$

ii) PM force calculation:

$$p_i^* = \underset{\sim}{p}_i^{n-\frac{1}{2}} + \underset{\sim}{F}_i^{mesh}\ ;\ i\ \varepsilon[1,N_p] \tag{36}$$

iii) PP force calculation:

$$\underset{\sim}{p}_i^{n+\frac{1}{2}} = \underset{\sim}{p}_i^* + \sum_j \underset{\sim}{f}_{ij}^{sr} \tag{37}$$

iv) Move:

$$\underset{\sim}{x}_i^{n+1} = \underset{\sim}{x}_i^n + \underset{\sim}{p}_i^{n+\frac{1}{2}}/m_i \tag{38}$$

v) Go to (i)

The momentum, $\underset{\sim}{p}$, rather than velocity is used in P^3M to reduce the computational costs of step (iii).

Step ii) follows the PM cycle: assign charge to mesh, solve for potential, interpolate forces and update momenta. Cost/quality compromise favours TSC for almost all P^3M calculations, whereas for PM calculations, CIC is usually the best choice.

Step iii) requires a fast algorithm for locating the neighbours of each particle to accumulate the short range force sums: Described in the next section is a linked list technique which reduces the search to a negligible portion of the timestep cycle. In practical implementations of the algorithm, it is advantageous to also include diagnostic calculations (radial distribution function, interaction pressure, etc) in the PP calculation cycle [35].

3.2. The short range force.

The total short range part of the force on a particle i is given by the sum

$$F_i^{sr} = \sum_{j=1}^{N_p} f_{ij}^{sr} \qquad (39)$$

The elementary approach to evaluating F_i^{sr} is to sweep through all particles, test for separations, r, less than r_e and if $r < r_e$, compute f_{ij}^{sr} and add it to F_i^{sr}: Such an approach is impractical, since operations counts scale as N_p^2.

The computational cost of locating particles j contributing to F_i^{sr} is greatly reduced if coordinates are ordered such that tests for neighbours need only be performed over a small subset, N_n, of the total number of particles. It is for this reason that the *chaining mesh* is introduced. The chaining mesh is a regular lattice with cells whose sides have length $\geq r_e$, so searches for neighbours of a particle in some cell q are restricted to particles in cell q, and neighbouring cells. Associated with each chaining cell, q is an entry in the head-of-chain table (HOC) giving the address of the first particle in cell q. Addresses of subsequent particles are given by the value of the link list (LL) coordinates.

The first stage of the short range force calculation is fillin the HOC and LL tables. In 3-D, this takes 3 arithmetic operations per particle [4,35]. Once the tables are filled a zero entry in HOC(q) indicates cell q is empty, and a non-zero entry give the co-ordinate of the first particle, j. LL(j) gives the address of the next particle, or is zero to indicate the end of list. Given HOC and LL, the PP momentum update step comprises the nested loops, sweeping over particles i in chaining cell q and neighbours j in cell q and cells adjacent to q.

3.3. Error analysis and optimisation.

The errors introduced into the force calculation in the PP ste is generally negligible. The residual error is that due to the failure of the mesh calculated approximation $F(x = x_2 - x_1; x_1)$ to exactly model the desired smooth part of the force $R(x)$ between two particles at positions x_1 and x_2. A measure of the error is given by

$$Q = \frac{1}{V_c} \int_{V_c} dx_1 \int_{V_b} dx \, |F(x;x_1) - R(x)|^2 \qquad (40)$$

Applying the Power theorem [36] enables Eq.(40) to be written in terms of fourier transforms $\hat{W}$, $\hat{D}$, and $\hat{G}$ (c.f. Eq.(30)). For a given choice of charge assignment function, $\hat{W}$, this reduces Eq.(40)

to $Q = Q(\hat{G}, \hat{\underset{\sim}{D}})$. Minimising Q w.r.t.harmonics $\hat{G}$ of the Greens function and over a restricted class of difference operator $\underset{\sim}{D}$ gives optimal combinations of elements of the PM part of the force calculation as a function of r_e.

Subsidiary error measures can also be evaluated to identify which component of the calculation cycle is the dominant error source, thus enabling elements of the calculation to be optimally matched. Indeed, the optimisation may be completely automated, reducing the choice of scheme to a choice of cycle time/particle number/accuracy compromise : This one choice alone is sufficient to determine W, r_e, $\hat{G}$ and mesh size. A full account of this procedure, and a 3-D implementation of the P^3M algorithm may be found in references [4] and [35].

The operations count, T, for P^3M is given by

$$T = \alpha N_p + \beta N^3 \log N^3 + \gamma N_n N_p \qquad (41)$$

The first term, αN_p, is the count for charge assignment, force interpolation, filling HOC and LL tables and integrating the equations of motion, the second terms is for solving the field equations (using FFT) and the third term is the count for the short range PP force summation. Constants α, β, and γ are determined by the particular choice of W, $\underset{\sim}{D}$, etc. The number of neighbours, N_n is given by the product of particle density, $n = N_p/V_b$, where $V_b = (NH)^3$. Substituting for N_n in Eq.(41) gives the cost per particle

$$\Gamma = \frac{T}{N_p} = \alpha + \beta \frac{N^3 \log N^3}{N_p} + \gamma \left(\frac{r_e}{N}\right)^3 N_p \qquad (42)$$

Minimising Γ w.r.t. N_p gives the optimal operating point, which by differentiating Eq.(42) occurs where

$$\beta N^3 \log N^3 = \gamma \left(\frac{r_e}{N}\right)^3 N_p^2 \qquad (43)$$

i.e. at the optimal operating point, the potential solver time equals the short range force calculation time, In addition, Eqs.(42) and (43) give the optimal number of particles per cell

$$\left(\frac{N_p}{N^3}\right)_{opt} = \left[\frac{\beta \log N^3}{\gamma r_e^3}\right]^{\frac{1}{2}} \qquad (44)$$

and the optimal operations per particle per timestep

$$\Gamma_{opt} = \alpha + 2[\gamma \beta r_e^3 \log N^3]^{\frac{1}{2}} \qquad (45)$$

Breakeven for P^3M with PP schemes occurs at $\sim$ 300 particles. For 1000 particles, P^3M shows a speed gain $\sim$ 10, rising to $\sim$ 450 for 30,000 particles !

3.4. Clustering of galaxies.

The application of the published triply periodic P^3M program P3M3DP[35] to the study of clustering of galaxies in a Friedman cosmology [37] is facilitated by transforming the equations to the expanding frame. Assuming uniformity on a large scale, and a uniform expansion a(t) = r(t)/r(0), force balance gives

$$\dot{a}^2 - \frac{8\pi}{3} G \frac{\hat{\rho}(o)}{a(t)} = -k \qquad (46)$$

where k < 0 and k > 0 for open and closed universes, respectively.

Transformation to the expanding frame is achieved by setting $r' = r/a$, potential $\phi' = a\phi + a\ddot{a}_{\frac{1}{2}}$, etc. The result is a Poisson's equation

$$\nabla^2\phi' = 4\pi G(\rho' - \bar{\rho}(0)) \qquad (47)$$

and equations of motion

$$\dot{\underset{\sim}{v}}' + 2H\underset{\sim}{v}' = -\frac{1}{a^3}\underset{\sim}{\nabla}\phi' \qquad (48)$$

$$\dot{\underset{\sim}{x}}' = \underset{\sim}{v}' \qquad (49)$$

The effect of the transformation is to insert a 'negative mass' background into the field equation ($\bar{\rho}(0)$) and a viscous drag term ($\simeq 2H\underset{\sim}{v}'$, where H = Hubble 'constant' = $\dot{a}/a$) into the equations of motion. In consequence, correlations in an infinite system up to a maximum scalelength of order of half the computation box side may be studied using a 'mass neutral' triply periodic model.

4. PARTICLE FLUID METHODS.

Particle methods for simulating fluids are in many respects similar to the particle methods described above for collisionless systems and for correlated systems. The vortex model (Sec.4.1) differs from the PM model in that velocities, rather than forces are the relevant vector field. The PIC fluid model extends the use of particles to non-conservative systems, using a mesh to handle diffusion like terms. SPH and Larson's schemes (Secs.4.3 & 4.4) avoid meshes, instead describing diffusion by using a physical model of kinetic processes in a PP model. A major difference between particle fluid algorithms and the schemes of Secs.3 and 4 is the extent of theoretic understanding and sophistication of algorithms. The basis of much of the fluid work is heuristic, with dissipative low order schemes being favoured: In Sec.4.5 we show that, if needed, high order accurate (finite element) PIC models may be obtained.

4.1. The vortex model.

The 'vortex' fluid model represents an incompressible inviscid fluid by an ensemble of vortices [38,39]. The assumptions of incompressibility and zero viscosity reduce the fluid equations in the absence of body forces to

$$\nabla \cdot \mathbf{u} = 0 \quad ; \quad \rho \frac{d\mathbf{u}}{dt} = - \nabla p \tag{50}$$

For two dimensional flows (in the x-y plane, say) we introduce stream function $\boldsymbol{\psi}$ and vorticity $\boldsymbol{\omega}$, both of which have nonzero components, ψ and ω, only in the z direction.

$$\mathbf{u} = \nabla \times \boldsymbol{\psi}, \quad \boldsymbol{\omega} = \nabla \times \mathbf{u} \tag{51}$$

Taking the curl of the equation of motion and using Eqs.(51) shows vorticity is conserved along fluid particle trajectories, $d\omega/dt=0$, and combining Eqs.(51) gives a Poisson equation for the stream function:-

$$\nabla^2 \psi = - \omega \tag{52}$$

A mathematically identical system arises in the guiding centre model of a plasma [40].

Discretisation follows in exactly the same manner as for the collisionless PM models. The fluid is described by a set of particles whose attributes are position and vorticity. The timestep cycle comprises the steps i) assign vorticity to the mesh ii) solve for the stream function, ψ iii) compute mesh velocity field ($\mathbf{u} = \nabla \times \psi \hat{z}$) iv) interpolate velocity field and move particle. Magnitude errors, directional errors and aliasing now appear in the velocity field: Their consequence is compressibility and degradation of energy conservation ($E = \frac{1}{2} \int \omega\psi d\tau$). Analysis and optimisation techniques used in PM models carries over with little change to the vortex model, although a new difficulty does arise with the equations of motion : Simple explicit schemes are unstable for the equation $\dot{\mathbf{x}} = \nabla \wedge \psi\hat{\mathbf{z}}$, so more complicated implicit predictor-corrector schemes generally used.

4.2. Particle-in-Cell (PIC).

The fluid (and magnetohydrodynamic - MHD) equations may be formally written

$$A \frac{\partial u}{\partial t} = L_1(u) + L_2(u) + s \tag{53}$$

where the r.h.s. is the sum of advective (hyperbolic), diffusive (parabolic) and source terms. In PIC, [41,42,43,44], Eqs.(53) are replaced by a fractional timestep scheme ; the diffusion stage of the timestep

$$A \frac{\partial v}{\partial t} = L_2(v) + s \tag{54}$$

is solved on an Eulerian spatial mesh and its result is used as initial conditions for the advective stage

$$A \frac{\partial w}{\partial t} = L_1 (w) \tag{55}$$

Equation (55) is solved using particle methods, and the two parts of the timestep are coupled using interpolation and inverse interpolation. A complete timestep cycle using Eqs.(54) and (55) is first order accurate, although by symmetrising in time, second order accuracy can be obtained.

To illustrate the PIC method, let us consider the representation of smooth flows in an ideal isentropic fluid. The equations for mass momentum and entropy function ($e = p\rho^{-\gamma}$) conservation are:-

$$\frac{\partial \rho}{\partial t} + \nabla . \rho \underset{\sim}{u} = 0 \tag{56}$$

$$\frac{\partial}{\partial t} \rho \underset{\sim}{u} + \underset{\sim}{\nabla} . \rho \underset{\sim}{u}\underset{\sim}{u} = -\underset{\sim}{\nabla} p \tag{57}$$

$$\frac{\partial}{\partial t} \rho e + \underset{\sim}{\nabla} . \rho e \underset{\sim}{u} = 0 \tag{58}$$

Applying timesplitting of Eqs.(54) and (55) to Eq.(56) gives the two parts of the timestep for mass transport

$$\frac{\partial \rho}{\partial t} = 0 \quad : \quad \text{diffusion stage} \tag{59}$$

$$\frac{d}{dt} \int \rho d\tau = 0 \quad : \text{advection stage} \tag{60}$$

The diffusion stage says do nothing, and the advection stage states that mass is conserved along the particle trajectory. The same argument applies to the entropy function. Momentum transport, Eq.(57) differs in that the diffusion stage has the source term, $-\nabla p$:-

$$\frac{\partial}{\partial t} \rho \underset{\sim}{u} = - \underset{\sim}{\nabla} p \quad : \quad \text{diffusion stage} \tag{61}$$

Spatial discretisation follows that used for PM models. The fluid is represented by a set of sample points. Associated with each sample point (or particle) are a set of attributes, which for the isentropic fluid will be mass, m_j, position $\underset{\sim}{x}_j$ momentum $\underset{\sim}{p}_j$ and entropy function e'_j ($= m_j e_j$). The computational box (see fig. 2) is divided into a set of cells, each cell containing a mesh point. Field quantities (pressure p, density ρ and velocity $\underset{\sim}{v}$) are stored on the mesh.

The timestep cycle follows established patterns of assignment/ interpolation linking the mesh and particle parts of the calculation:-

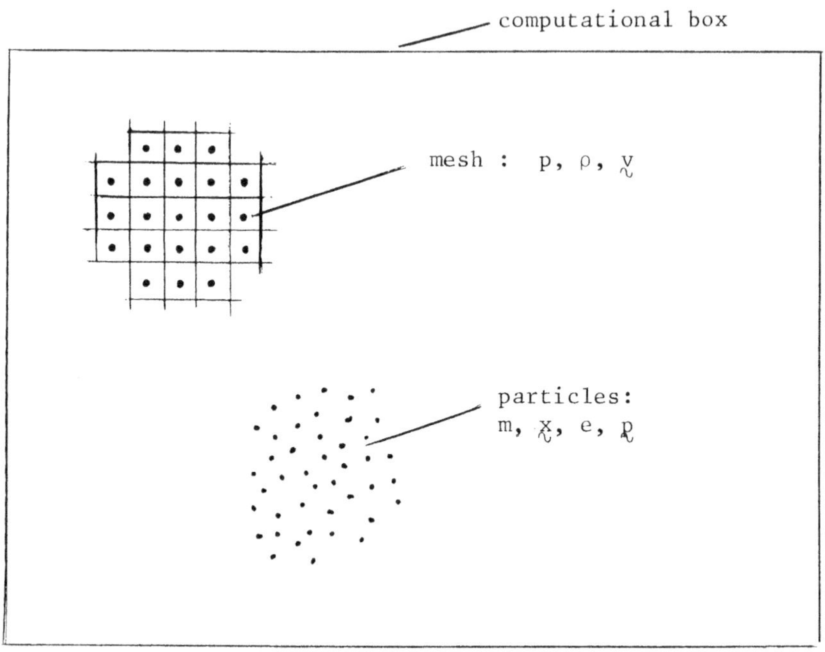

Figure 2: The computational box in PIC calculations is covered by a mesh on which fields are computed. Also filling the box are a set of particles which are used to advect mass, momentum, etc.

i) start : given $\{p^n_{\underset{\sim}{p}}, \rho^n_p, \underset{\sim}{v}^n_p\}$ on mesh

$\{m_j, \underset{\sim}{x}_j, e'_j\}$ on particles.

ii) Diffusion stage : update mesh velocity field

$$\underset{\sim}{v}^{\sim n+1}_{\underset{\sim}{p}} = \underset{\sim}{v}^n_p - \frac{\Delta t}{\rho^n_{\underset{\sim}{p}}} \underset{\sim}{\nabla}_d \, p^n_{\underset{\sim}{p}} \tag{62}$$

Equation (62) is simply a FDA to Eq.(61): $\underset{\sim}{\nabla}_d$ is used to denote a difference gradient, indices $\underset{\sim}{p}$ and n denote mesh point and time-level.

iii) Assign momentum to particles

$$\underset{\sim}{p}_j = m_j \underset{\sim}{v}_j = m_j \sum_{\underset{\sim}{p}} W(\underset{\sim}{x}^n_j - \underset{\sim}{x}_{\underset{\sim}{p}}) \; \underset{\sim}{v}^{\sim n+1}_p \tag{63}$$

iv) Advection stage : interpolate velocities and move particles

$$\underset{\sim}{u}_j = \sum_p \tilde{\underset{\sim}{v}}_p^{n+1} W^* (\underset{\sim}{x}_j^n - \underset{\sim}{x}_p) \quad (64)$$

$$\underset{\sim}{x}_j^{n+1} = \underset{\sim}{x}_j^n + \underset{\sim}{u}_j \Delta t \quad (65)$$

v) Assign to mesh

density : $$\rho_p^{n+1} = \frac{1}{v_c} \sum_j m_j W(\underset{\sim}{x}_j^{n+1} - \underset{\sim}{x}_p) \quad (66)$$

momentum : $$(\rho \underset{\sim}{v})_p^{n+1} = \frac{1}{v_c} \sum_j \underset{\sim}{p}_j W(\underset{\sim}{x}_j^{n+1} - \underset{\sim}{x}_p) \quad (67)$$

entropy function : $$(\rho e)_p^{n+1} = \frac{1}{v_c} \sum_j e_j' W(\underset{\sim}{x}_j^{n+1} - \underset{\sim}{x}_p) \quad (68)$$

vi) Compute new mesh fields

pressure $$p_p^{n+1} = e_p^{n+1} (\rho_p^{n+1})^\gamma \quad (69)$$

velocity $$\underset{\sim}{v}_p^{n+1} = (\rho \underset{\sim}{v})_p^{n+1} / \rho_p^{n+1} \quad (70)$$

The functions W and W* are the NGP and CIC assignment/interpolation functions, respectively. Equations (62)-(70) have been formulated this way to show that the PIC schemes for fluids can be geleralised to higher order in exactly the same manner as the collisionless PM models, although as yet no practical implementation of higher order schemes has been published.

It is relatively straightforward to show, by summing Eqs.(62) - (70) over particles and/or mesh points as appropriate, that the PIC scheme described has the properties that mass, momentum and entropy function are identically conserved and density positivity is assured. Linearised analysis gives the sound wave dispersion relation for a stationary PIC fluid

$$\frac{\sin^2(\omega\Delta t/2)}{(\Delta t/2)^2} = c_s^2 k^2 \frac{\sin^2(kH/2)}{(kH/2)^2} \cos(\frac{kH}{2}) \quad (71)$$

which tends to the differential model dispersion $\omega^2 = c_s^2 k^2$ as Δt and mesh size H become small. It follows from Eq.(71) that the timestep, Δt, must satisfy the stability criterion $c_s \Delta t/H < 1$ for stability.

If flows are not smooth, then the entropy function e is not conserved and the formulation described above becomes inappropriate.

To handle shocked flows, the entropy conservation equation (58) must be replaced by the energy equation

$$\frac{\partial}{\partial t} \rho E + \underset{\sim}{\nabla} . ((\rho E + p)\underset{\sim}{u}) = 0 \qquad (72)$$

where $E = \frac{1}{2} v^2 + I$ is the specific energy density and I is the specific internal energy density. Equation (72) is time split into a diffusion stage part where the work terms $p \underset{\sim}{\nabla} . \underset{\sim}{u}$ (plus any source terms and heat flows, if present) are evaluated on a mesh, and an advection term, where energy is transported by particles [42].

The sequency of interpolating from the mesh to particles, moving particles and assigning from particles to mesh at each time-step introduces numerical diffusion. In the momentum equation these errors appear as numerical viscosity and in the energy equation as heat flow. Using the momentum and energy equation formulation ensures that the numerical viscous heating is properly accounted for in the energy equation. If the usual NGP scheme is employed for assignment and interpolation, the resulting diffusion is similar to that obtained using upwind differencing [45] for advection : For this case, additional numerical viscosity is usually unnecessary in PIC shock calculations. However, higher order schemes allow the numerical diffusion of PIC advection to be substantially reduced, giving much better representation of smooth flows but requiring additional viscosity for shocks.

The original PIC scheme devised by Harlow [41] has remained in use for over twenty years with only minor changes. Its most extensive application is to supersonic flows, where the strong diffusion is of little consequence. It can handle voids and free surfaces with ease, and the absence of diffusion in the mass transport equations enables multifluids with contact surfaces to be treated. Heat flow, viscosity sources and sinks may be added to the diffusion stage of the calculation without affecting the overall scheme.

4.3. Smooth Particle Hydrodynamics (SPH).

The SPH scheme [46,47] uses the sample point representation of the fluid (cf.PM and PIC schemes), but uses no spatial mesh; A parallel would be to use the particle-particle summation method for an ensemble of finite sized particles rather than employing a mesh. In gravitating fluids, where forces are long ranged, SPH can rapidly become prohibitively expensive because of the N^2 operation count scaling, although using the methods of section 3 the N^2 scaling problem can be overcome.

The basis of the method is to represent the fluid by a set of sample points (or finite sized fluid elements) which move according to

$$\frac{d\underset{\sim}{v}}{dt} = \frac{-\underset{\sim}{\nabla}p}{\rho} + \underset{\sim}{F} \; ; \quad \frac{d\underset{\sim}{x}}{dt} = \underset{\sim}{v} \tag{73}$$

Associated with a particle at $\underset{\sim}{x}'$ is a density profile $(M/N)w(\underset{\sim}{x}-\underset{\sim}{x}')$, where (M/N) is the mass per particle. If we let $n(\underset{\sim}{x})$ be the density of particle centres, then the computed density for the N particle system

$$\rho_N(\underset{\sim}{x}) = \frac{M}{N}\sum_j w(\underset{\sim}{x}-\underset{\sim}{x}_j) = \frac{M}{N}\int w(\underset{\sim}{x}-\underset{\sim}{x}')n(\underset{\sim}{x}')d\underset{\sim}{x}' \tag{74}$$

approximates the 'smoothed' density

$$\rho_s(\underset{\sim}{x}) = \int w(\underset{\sim}{x}-\underset{\sim}{x}')\rho(\underset{\sim}{x}')d\underset{\sim}{x}' \tag{75}$$

where $\rho(\underset{\sim}{x})$ is the true density

$$\rho(\underset{\sim}{x}) = \lim_{n\to\infty} \frac{M}{N} n(\underset{\sim}{x}) \tag{76}$$

The condition for $\rho_N \to \rho_s$ is that the number of particles per particle is large and the condition required for $\rho_s \to \rho$ are that $\Delta\rho/h$ is small, where h is the 'width' of the profile w and $\Delta\rho$ is the change of ρ over scale length h.

Gravitational forces are computed by means of pairwise force sums. Pressure forces are obtained by (analytic) differentiation of the density $\rho_N(x)$ (Eq.(74)). Given a polytrope of index n, $p = \kappa\rho^{1+1/n}$, then

$$\frac{\underset{\sim}{\nabla}p}{\rho} = \kappa \; \rho^{1/n-1}\left(\frac{n+1}{n}\right)\underset{\sim}{\nabla}\rho \tag{77}$$

where ρ is approximated by ρ_N. The method straightforwardly generalises to include magnetic field computations and Lorentz forces, centripetal coriolis and viscous forces [46].

At its present stage of development, SPH is of limited application. For the small number of particles (<100) suggested, [46], only smooth, almost structureless profiles can satisfy conditions where $\rho_N \simeq \rho_s \simeq \rho$. For large numbers of particles, the N^2 scaling of the operations count makes the computations prohibitively expensive. Timings quoted [46] for N = 40 are 0.25 seconds per timestep on an IBM 360/165. N^2 scaling take this figure to 2.6 minutes/step for N = 1000 and 2.8 hours/step for N = 8000. The pressure computation using Eq.(77) does not conserve momentum, although it is possible to overcome this bv symmetrising the calculation [47]. Unlike PIC, velocity is treated as a particle, rather than a field quantity, so individual particle velocities may deviate significantly from the local smoothed velocity field unless relatively large viscosities are assumed.

4.4. Larson's method.

Larson's method [48] is in practice almost the same as SPH, but is obtained by physical argument rather than by sampling the mathematical description of a continuum fluid. The model used is to regard the fluid as an ensemble of gravitating colliding 'gas bag' particles. Interparticle forces comprise a softened gravitational attraction plus a short ranged repulsion. The repulsion term contains viscous damping term which appears only when neighbouring gasbags are moving towards each other. The radial acceleration, a_r, between two neighbouring particles is thus

$$a_r = \frac{Gm}{r^2+\varepsilon^2} + \begin{cases} (c^2+Qu_r^2)/r & ; \quad u_r < 0 \\ c^2 / r & ; \quad u_r > 0 \end{cases} \tag{78}$$

The repulsion term is only applied for nearest and second nearest neighbours.

The pairwise accumulation of forces in Larson's scheme ensures momentum conservation. If only a small number ($\sim$ 100) particles are used, discrete particle effects will dominate and only smooth viscous motions can be adequately treated. The N^2 scaling of operation counts confronts Larson's scheme with the same problem as is met with SPH, although variable timesteps used by Larson [48] can give some savings. If one adopts the P^3M solution [35] to the N body force summation problem, then one is faced with the question as to whether one should use PIC, where velocity is treated as a field, rather than particle attribute.

4.5. EPIC

The particle-in-cell(PIC) fluid simulation method is most suited to supersonic flows, where numerical diffusion inherent in the 'zeroth' order assignment/interpolation scheme is of little import. It can be, however, extended to accurately model less extreme cases. In this section, a brief outline of some such generalisations of the PIC scheme [49] within a finite element framework will be given. For simplicity, we shall restrict the discussion to ideal fluids, although addition of sources and transport terms present no serious obstacles (cf.Sec.4.2).

The kinematic equations for the transport of density, ρ,pressure p, and magnetic field, $\underset{\sim}{B}$, in an adiabatic perfectly conducting fluid may be written as

$$\rho(\underset{\sim}{x}) = \rho_o(\underset{\sim}{x}_o)\ |D|^{-1} \tag{79}$$

$$p(\underset{\sim}{x}) = p_o(\underset{\sim}{x}_o)\ |D|^{-\gamma} \tag{80}$$

$$\underset{\sim}{B} = |D|^{-1}\ \underset{\approx}{D}\cdot \underset{\sim}{B}_o(\underset{\sim}{x}_o) \tag{81}$$

where $\underset{\sim}{x} = \underset{\sim}{x}(\underset{\sim}{x}_o, t)$ is the Lagrangian coordinate of a fluid element and $\underset{\approx}{D} = \partial\underset{\sim}{x}/\partial\underset{\sim}{x}_o$ is the deformation gradient tensor. Subscripts 'o' indicate values before the deformations described by $\underset{\approx}{D}$. Given $\underset{\approx}{D}$, values of ρ_o, p_o, $\underset{\sim}{B}_o$ may be mapped forward to a later time using Eqs.(79) - (81).

Let us now focus our attention on the mass equation. Multiplying Eq.(79) by a basis function $w_k(\underset{\sim}{x})$, integrating over volume, and noting $|D|$ is the Jacobian for volume transformation we find

$$\int w_k(\underset{\sim}{x})\rho(\underset{\sim}{x})\, d\underset{\sim}{x} = \int w_k(\underset{\sim}{x})\rho_o(\underset{\sim}{x}_o)\, d\underset{\sim}{x}_o \tag{82}$$

If we assume a Lagrangian basis function, $w_k(x,t) = w_k(x_o,t_o)$ (i.e. the 'mesh' is carried along with the fluid) and approximate ρ by the trial functions $\sum_\ell \rho_\ell\phi_\ell$ then Eq.(82) becomes

$$\left(\int w_k(\underset{\sim}{x})\phi_\ell(\underset{\sim}{x})\, d\underset{\sim}{x}\right)\rho_\ell = \int w_k(x_o)\rho_o(x_o)\, d\underset{\sim}{x}_o \tag{83}$$

i.e. $A_{k\ell}\,\rho_\ell = M_k$ (84)

For the *Lagrangian* choice of basis function, the algorithm for advancing ρ is
1) compute node masses $\{M_k\}$
2) move nodes to new positions $\underset{\sim}{x}_o \rightarrow \underset{\sim}{x}$
3) compute 'mass' matrix, A.
4) solve Eq.(84) for new $\{\rho\}$ on displaced nodes.

The *eulerian* basis function $w_k(x)$ = constant gives

$$\left(\int w_k\, \phi_\ell\, d\underset{\sim}{x}\right)\rho_\ell = \int w_k(x)\, \rho_o(\underset{\sim}{x}_o)\, d\underset{\sim}{x}_o \tag{85}$$

The r.h.s. of Eq.(85) may be evaluated by quadrature if we let $m_i = \rho_o \Delta\underset{\sim}{x}_o$ be the mass in element i, then Eq.(85) becomes

$$A_{k\ell}\rho_\ell = \sum_i m_i\, w_k\, (x_i) \tag{86}$$

The sum in Eq.(86) may immediately be recognized as mass assignment where w_k is the assignment/interpolation function, as A may be interpreted as a 'sharpening' operator which offsets the spreading effects of assignment.

The central points to be extracted from Eqs.(82)-(86) are that the finite element formalism (EPIC schemes) gives optimal prescription for given choice of functions w and generates algorithms from purely Lagrangian to Eulerian mesh PIC according to the choice of time dependence of w. Using a non-Lagrangian time dependent w and quadrature gives moving mesh PIC schemes. Using different basis functions enables local mesh refinement to be included.

The discretised kinematic equations obtained using the above procedure are unconditionally stable. Timestep criteria will be introduced if explicit approximations to the equations of motion are used, in exactly the same manner as for standard PIC schemes (Sec.4.2). However, when disparate timescales are encountered, as is often the case for magnetofluids, it may be advantageous to make the approximation to the equation of motion implicit to avoid timestep stability limits. The equations of motion

$$\rho \frac{d\underset{\sim}{v}}{dt} = -\underset{\sim}{\nabla} p + \underset{\sim}{j} \times \underset{\sim}{B} \tag{87}$$

are the Euler-Lagrange equations arising from the variational integral

$$I = \int d\underset{\sim}{x} \left[\tfrac{1}{2} \rho |\underset{\sim}{v}|^2 - \left(\frac{p}{\gamma - 1} + \frac{B^2}{2\mu_o} \right) \right] \tag{88}$$

$$= \int d\underset{\sim}{x}_o \, \mathcal{L} \, (\xi_i, \xi_{i,j}) \tag{89}$$

where $\underset{\sim}{\xi} = \underset{\sim}{x} - \underset{\sim}{x}_o$ is the displacement field. Replacing $\underset{\sim}{\xi}$ in Eq.(89) by trial functions, forming approximate Euler-Lagrange equations (cf.Sec.2.6) and using implicit FDA in time give non-linear equations for nodal displacements which may be solved by, for example, conjugate gradient search methods [50].

5. FINAL REMARKS.

Particle methods will undoubtedly remain an integral part of the arsenal of the computational physicist. The particle-mesh method for collisionless system has now a well developed 'finite sized particle' theory. Wave dispersion and growth, drag and diffusion coefficients and conservation properties show agreement between theory and experiment. Well defined methods for analysis and optimisation of PM algorithms now exist : It is no longer justifiable to use NGP and simple finite differences for Poisson's equation except for gross velocity space instabilities.

The P^3M algorithms are a further result of the analysis techniques developed for PM models. P^3M opens a whole new area of previously inaccessible study: Reducing the N body force sum from an $\sim N^2$ to an $\sim N$ operation count process has made possible the numerical study of large scale correlation in systems with long range forces.

The fluid PIC scheme, despite being only 'zeroth' order and lacking the theoretical backing that PM models have, has remained in use for over two decades: Its greatest strength is in handling particularly awkward supersonic flows. The relative ease with which PIC handles multimaterial boundaries, voids and open surfaces far outweigh the difficulties of numerical diffusion and relatively high computational costs. The meshless fluid particle models (secs.4.3 & 4.4) seem less likely to overcome computational cost barriers in the case where gravitational forces are present. Generalisations of PIC discussed in Sec.4.5 indicate the scope for extensions to higher accuracy, to moving meshes, to implicit time integration and to magnetohydrodynamic flows.

A great attraction of particle models is the natural way in which they incorporate the conservation laws: The usually difficul advection terms appear as movement of discrete 'parcels' of conserved quantities around the computational box. The close parallels between particles and the underlying corpuscular nature of the continuum system being modelled mean that numerical errors appear as physics. If poor choices of parameters are made, particle models give the right answer to the wrong problem, rather than the wrong answers to the posed problem ! An example of this robustness of particle methods is the PM simulation of a cold plasma: Numerical errors 'warm' up the plasma until the physical scale lengths are resolved by the spatial mesh.

6. REFERENCES.

[1] Richtmyer, R.D., and Morton, K.W.: 1967, *Difference Methods for Initial-Value Problems*, Wiley, N.Y.
[2] Strang, G., and Fix, G.: 1973. *An Analysis of the Finite Element Method*, Prentice-Hall, N.J.
[3] Gottleib, D., and Orszag, S.A.: 1977, *Numerical Analysis of Spectral Methods*, SIAM, Pa.
[4] Hockney, R.W., and Eastwood, J.W.: 1981, *Computer Simulation using Particles*, McGraw-Hill, N.W.
[5] Hockney, R.W.: 1970, Methods Comput. Phy., 9, pp.135-211.
[6] Hockney, R.W., and Brownrigg, D.R.K.: 1974, Mon.Not.R. Astron. Soc., 167, pp.351-357.
[7] Clemmow, P.C., and Dougherty, J.P.: 1969, *Electrodynamics of Particles and Plasmas*, Addison-Wesley, Mass.
[8] Hammersley, J.M., and Handscombe, D.C.: 1964, *Monte Carlo methods*, Methuen, London.
[9] Kruer, W.L., Dawson, J.M., and Rosen, B.: 1973, J.Comput. Phys., 13, pp. 114-129.
[10] Eastwood, J.W., and Hockney, R.W.: 1974, J.Comput.Phys., 16, pp.342-359.
[11] Beeman, D.: 1976, J.Comput.Phys, 20, pp.130-139.
[12] Boris, J.P.: 1970, Proc. 4th Conf. on Numer.Simulation of Plasma, pp.3167, NRL.
[13] Birdsall, C.K., and Langdon, A.B.: to be published (1982), *Plasma Physics via Computer Simulation*, McGraw-Hill, N.Y.
[14] Cohen, B.I., Freis, R.P., and Thomas, V.: 1982, J.Comput. Phys, 45, pp.345-366.
[15] Cohen, B.I., Langdon, A.B., and Friedman, A.: 1982, J.Comput, Phys, 46, pp.15-38.

[16] Lewis, H.R.: 1970, J.Comput.Phys, 6, pp.136-141.
[17] Lewis, H.R.: 1970, Methods Comput.Phys, 9, pp.309-338.
[18] Langdon, A.B.: 1973, J.Comput.Phys, 12, pp.247-268.
[19] Goldstein, H.: 1959, *Classical Mechanics*, Addison-Wesley, Mass.
[20] Birdsall, C.K., and Fuss, D.: 1969, J.Comput.Phys, 3, pp.494-511.
[21] Eastwood, J.W., and Jesshope, C.R.: 1977, Comput.Phys, Common, 13, pp.233-239.
[22] Eastwood, J.W., and Brownrigg, D.R.K.: 1979, J.Comput. Phys, 32, pp.24-38.
[23] James, R.A.: 1977, J.Comput.Phys, 15, pp.71-93
[24] Hockney, R.W.: 1978, *Computers, Fast Elliptic Solvers and Applications* (ed.U.Schumann), Advance Publications, London, pp.141-169.
[25] Eastwood, J.W.: 1975, J.Comput.Phys, 18, pp.1-70.
[26] Langdon, A.B.: 1970, J.Comput.Phys, 6, pp.247-267.
[27] Birdsall, C.K.,Langdon,A.B., and Okuda, H.: 1970. Meth. Comput. Phys., 9, pp.241-258.
[28] Hockney, R.W.: 1971, J.Comput.Phys., 8, pp.19-44.
[29] Pieravi, A., and Birdsall, C.K.: 1978, Report UCB/ERL M78/32, College of Engineering, Univ. of Calif.,Berkeley.
[30] James, R.A., and Sellwood, J.A.: 1978, Mon.Not.R.Astron Soc., 182, pp 331-344.
[31] Miller, R.H.: 1971, Astrophys. and Space Sci., 14, pp.73-90.
[32] Eastwood,J.W.: 1976, Comput.Meth. in Classical and Quantum Phys, pp.196-205, Advance Publications,London.
[33] Hockney, R.W., Goel, S.P., and Eastwood,J.W.:1973, Chem Phys.Lett., 21, pp.589-591.
[34] Eastwood,J.W.: 1976, Comput.Meth. in Classical and Quantum Phys, pp.206-228, Advance Publications,London.
[35] Eastwood,J.W., Hockney, R.W., and Lawrence, D.N.: 1980 Comput.Phys, Common, 19, pp.215-261.
[36] Bracewell, R.: 1965, *The Fourier Transform and its Application*, McGraw-Hill, N.Y.
[37] Efstathiou, G., and Eastwood, J.W.: 1981, Mon.Not.R.Astron Soc., 194, pp.503-526.
[38] Christiansen, J.P.: 1973, J.Comput.Phys, 13, pp.363-379.
[39] Christiansen, J.P., and Zabusky, N.B.: 1973, J.Fluid.Mech, 61, pp.219-243.
[40] Lee, W.W., and Okuda, H.: 1978, J.Comput.Phys, 26, pp.139-152.
[41] Harlow, F.H.:1964, Meth.Comput.Phys,3, pp.319-343.
[42] Amsden, A.A.: 1966, LASL Report LA - 3466.
[43] Harlow, F.H., Amsden, A.A., and Nix, J.R.: 1976, J.Comput. Phys. 20, pp.119-129.
[44] Cook, T.L., Demuth, R.B., and Harlow, F.A.: 1981, J.Comput Phys, 41, pp.51-67.
[45] Eastwood, J.W.: 1982, *Computational Plasma Physics*,lecture notes, 19th Culham Plasma Phys, Summer School, Culham Laboratory.

[46] Gingold, R.A., and Monaghan, J.J.: 1977, Mon.Not.R.Ast. Soc, 181, pp.375-389.
[47] Monaghan, J.J.: 1981, J.Comput.Phys, 44, pp.397-399.
[48] Larson, R.B.: 1978, J.Comput.Phys, 27, pp.397-409.
[49] Eastwood, J.W.: (1982) in preparation.
[50] Eastwood,J.W.: 1977, Computer Science Report RCS67, Reading Univ., Reading, U.K.

WHY ULTRARELATIVISTIC NUMERICAL HYDRODYNAMICS IS DIFFICULT

Michael L. Norman and Karl-Heinz A. Winkler

Los Alamos National Laboratory
and
Max-Planck-Institut fuer Physik und Astrophysik

ABSTRACT

An implicit, adaptive-mesh numerical technique is described for modeling ultrarelativistic ($\gamma >> 1$) gas flows in flat spacetime in one space dimension. The numerical code is an adaptation of the WH80s Newtonian radiation hydrodynamics code described by Winkler and Norman in these proceedings. Shock fronts are treated by the method of artificial viscosity. We derive the equations of motion for an ideal gas with artificial viscosity in Eulerian and arbitrarily-moving coordinates. We show through numerical examples the systematic errors that result from inconsistently omitting or including the artificial viscous "pressure" in the Eulerian equations of motion. These errors are eliminated in the present code, which gives good results on three test problems involving strong relativistic shock waves.

1. INTRODUCTION

The term ultrarelativistic hydrodynamics refers to flows in which the bulk Lorentz factor $\gamma \equiv (1-v^2/c^2)^{-1/2}$ greatly exceeds unity; here v is the flow speed and c is the speed of light. In such flows, v/c can no longer be taken as a small parameter, and therefore the fully relativistic equations of fluid flow must be solved. In the present paper we address the numerical difficulties associated with solving the equations of ultrarelativistic hydrodynamics in a flat background spacetime, i.e. special relativistic hydrodynamics, although our findings

K.-H. A. Winkler and M. L. Norman (eds.), Astrophysical Radiation Hydrodynamics, 449–475.

have relevance to numerical techniques for studying general relativistic fluid flows as well.

Relativistic fluid flows arise in a variety of astrophysical settings: in the blast waves of supernova explosions (e.g., Colgate and Johnson 1960), during gravitational collapse and accretion (e.g. May and White 1967; Wilson 1972), and in active galactic nuclei (e.g., Blandford and McKee 1976). Perhaps the most compelling observational evidence we have for bulk relativistic flows occurring in nature are the superluminal radio sources (see, e.g., Kellerman and Pauliny-Toth 1981), which are interpreted as relativistic jets pointing nearly along our line of sight (Blandford and Königl, 1979). Typical values for γ inferred from VLBI measurements of radio component proper motion lie in the range $2 \leq \gamma_{jet} \leq 45$ (Kellerman and Pauliny-Toth, 1981).

Numerical models of relativistic flows have generally proceeded along two lines: one-dimensional Lagrangean calculations (May and White 1967; McKee and Colgate 1973; Anile et al. 1983), and two-dimensional Eulerian calculations (Wilson 1972, 1979; Smarr et al. 1980; Dykema 1980; Hawley et al. 1984a,b; Evans et al., this volume). As emphasized by Shapiro (1979), Lagrangean techniques are generally inapplicable for modeling multidimensional relativistic flows, such as nonspherical black hole accretion and jets, in which large fluid shear would lead to unacceptable distortions of the numerical mesh. The Eulerian techniques of the Wilson school avoid these difficulties by solving the relativistic fluid equations on an orthogonal mesh which is either stationary or, at best, moves so as to follow the fluid in an average sense (i.e., follow overall contraction or expansion). However, when specialized to one-dimensional flat spacetime, Wilson's Eulerian techniques are found to be less accurate than Lagrangean techniques on ultrarelativistic flows involving strong shock waves (Wilson 1972; Centrella and Wilson 1984).

This paper is the result of an investigation to determine why this is the case. The nature of the inaccuracies suggested to us an error in formulation of the Eulerian equations used by Wilson and his collaborators. This we verified by solving the equations of special relativistic hydrodynamics in an arbitrary coordinate system which we could specify to be either Lagrangean, Eulerian or fully adaptive (cf. Winkler and Norman, this volume). By insisting on coordinate-independent solutions to standard test problems, we were able to discover the inconsistencies in Wilson's Eulerian approach which never become an issue in Lagrangean coordinates.

In section 2 we present the Eulerian equations of special relativistic hydrodynamics, paying particular attention to how the artificial viscous stress terms used to treat shock fronts appear in the equations. A test problem is defined against which numerical solutions are compared. In section 3, the adaptive mesh equations of special relativistic hydrodynamics are derived, and our numerical technique is described. Results of our calculations are presented and analysed in section 4. The insight thus gained is used in section 5 to critique existing techniques for multidimensional relativistic flow and discuss their limits of applicability. Suggestions for further research along these lines are also given.

2. EULERIAN SPECIAL RELATIVISTIC HYDRODYNAMICS WITH ARTIFICIAL VISCOSITY

Shock fronts which invariably arise in compressible fluid flows are generally treated numerically by the method of artificial viscosity (see, e.g., Richtmyer and Morton, 1967), although alternate forms of numerical dissipation are now being used with great success in Newtonian calculations (e.g., Woodward, these proceedings). We adopt here the former approach for stabilizing relativistic shocks because of its robustness. In so doing, we must consider the relativistic dynamics of a nonideal fluid in order to formulate our numerical equations. Although artificial viscosity introduces nonideal effects only in the narrow shock transition zones embedded in an otherwise ideal gas flow, the additional complications involved in implementing artificial viscosity in Eulerian calculations carry over to physical viscosities as well. Thus, we are making a small incursion into the numerics of nonideal relativistic hydrodynamics.

Our starting point is the stress-energy tensor $T^{\mu\nu}$ of a nonideal fluid, from which the conservation laws of momentum and energy are derived via

$$T^{\mu\nu}{}_{;\nu} = 0 \quad . \tag{1}$$

Here the greek indices vary over the four spacetime coordinates $x^{\mu} \equiv (x^{i}, ct)$, i varying over the three spatial coordinates, and the semicolon denotes covariant differentiation over the repeated index, which is summed (Einstein summation convention). The stress-energy tensor for a viscous, heat conducting fluid can be written (Misner, Thorne and Wheeler 1973) as

$$T^{\mu\nu} = \rho(1+e/c^{2})U^{\mu}U^{\mu} + (P-\zeta\theta)h^{\mu\nu} - \eta\sigma^{\mu\nu} + q^{\mu}U^{\nu} + q^{\nu}U^{\mu}, \tag{2}$$

where ρ, e and P are the proper mass density, specific internal energy and scalar pressure of the fluid, respectively. U^μ is the fluid four-velocity given by

$$U^\mu = \gamma(v^i, c) \quad , \tag{3}$$

where v^i is the fluid three-velocity, and γ is determined by the four-normalization condition $U^\mu U_\mu = -c^2$, whence

$$\gamma = (1 - v^i v_i / c^2)^{-1/2} \quad . \tag{4}$$

The spatial projection tensor $h^{\mu\nu}$ is given by

$$h^{\mu\nu} = c^{-2} U^\mu U^\nu + g^{\mu\nu} \quad , \tag{5}$$

where $g^{\mu\nu}$ is the metric tensor.

The nonideal processes of viscosity and heat conduction contribute the last four terms in Eq. (2); $\zeta\theta$ is the bulk viscous "pressure," $\eta\sigma^{\mu\nu}$ is the shear stress tensor, and q^μ is the heat flux four-vector. The rate of compression θ is given by

$$\theta = U^\mu{}_{;\mu} \quad . \tag{6}$$

The stress-energy tensor for an ideal fluid with artificial viscosity is obtained by ignoring shear viscosity and heat conduction in Eq. (2), yielding

$$T^{\mu\nu} = [\rho + c^{-2}(e + P + Q)]\, U^\mu U^\nu + (P + Q) g^{\mu\nu} \quad , \tag{7}$$

where we have used Eq. (5) and made the substitution

$$Q = -\zeta^a \theta \quad . \tag{8}$$

ζ^a is the coefficient of artificial viscosity, to be specified below. We note that Eq. (7) is the usual expression for the stress-energy of an ideal fluid with the substitution $P \rightarrow P + Q$, a fact that will aid us in deriving subsequent equations.

Henceforth we shall consider only flat spacetime and derive the specific equations of special relativistic hydrodynamics that we solve numerically. Adopting the Lorentz metric, i.e. $g^{\mu\nu} \equiv \eta^{\mu\nu}$ where

$$\eta^{\mu\nu} = \begin{pmatrix} 1 & 0 & 0 & 0 \\ 0 & 1 & 0 & 0 \\ 0 & 0 & 1 & 0 \\ 0 & 0 & 0 & -1 \end{pmatrix} \quad , \tag{9}$$

then letting $\mu = j$ in $T^{\mu\nu}{}_{;\nu} = 0$ yields

$$\partial_t[(\rho h + Q/c^2)\gamma^2 v_j] + \partial_i[(\rho h + Q/c^2)\gamma^2 v_j v^i] + \partial_j(P + Q) = 0 \quad , \tag{10}$$

expressing conservation of momentum; letting $\mu = 4$ in $T^{\mu\nu}{}_{;\nu} = 0$ gives

$$\partial_t[(\rho h + Q/c^2)\gamma^2] + \partial_i[(\rho h + Q/c^2)\gamma^2 v^i] - c^{-2}\partial_t(P + Q) = 0 \quad , \tag{11}$$

expressing conservation of total energy. In Eqs. (10) and (11), $h = 1 + c^{-2}(e + p/\rho)$, the normalized specific total enthalpy.

Conservation of fluid rest mass follows from

$$(\rho U^\mu)_{;\mu} = 0 \quad , \tag{12}$$

which yields

$$\partial_t(\rho\gamma) + \partial_i(\rho\gamma v^i) = 0 \quad . \tag{13}$$

An alternative to Eq. (11) is an evolution equation for the internal energy of the gas, which can be derived from Eqs. (10), (11) and (13), giving

$$\partial_t(\rho e\gamma) + \partial_i(\rho e\gamma v^i) + (P + Q)(\partial_t\gamma + \partial_i\gamma v^i) = 0 \quad . \tag{14}$$

The system of dynamical equations (10), (11) or (14), and (13), in conjunction with definition equations (4) and (8), is complete provided ζ^a appearing in Eq. (8) is specified and the equation of state $P = P(\rho, e)$ is given. In the following we will adopt an ideal gas equation of state $P = (\Gamma - 1)\rho e$, where Γ is the ratio of specific heats.

A reasonable choice for ζ^a is (Dimitri Mihalas, private communication)

$$\zeta^a = -\rho\gamma\ell_E^2 \min(\theta, 0) \quad , \tag{15}$$

where ℓ_E is the desired shock thickness in the Eulerian frame, and θ is given by Eq. (6). The minimum operation insures that the artificial viscosity switches on only during compression, in analogy to the von Neumann-Richtmyer (1950) viscosity.

The factor of γ in Eq. (15) can be understood as follows. The quantity $\gamma\ell_E$ is the shock thickness in the proper frame, which we shall call ℓ_p. Equation (15) can then be rewritten

$$\zeta^a = -\rho\ell_p^2 \min(\theta, 0)/\gamma \quad , \tag{16}$$

which agrees with May and White's definition [see their eq. (98)] provided ℓ_p is of order the width of a Lagrangean mass zone. However, we want to specify a fixed shock thickness in the inertial frame, hence Eq. (15).

We describe in the next section our implicit numerical technique for solving Eqs. (10), (13) and (14) in one spatial dimension, and specify our chosen form for Q which differs somewhat from the above. Here we merely wish to comment on some aspects of the dynamical equations in their present form which render a solution by explicit numerical techniques exceedingly difficult.

First, recognize that Q is generally a nonlinear function of derivatives of v, meaning that the quantity $Q\gamma^2 v$ appearing in the relativistic momentum density in Eq. (10) is a very complicated nonlinear function of v and its derivatives. Imagine now attempting to solve Eq. (10) for the velocity distribution within a numerically resolved strong relativistic shock wave. Within the shock transition $Q\gamma^2 v$ is the dominant contribution to the relativistic momentum density of the gas, and Q, γ and v each are varying strongly. Conventional explicit techniques have great difficulties in determining Q consistently with the other variables and therefore typically keep Q fixed, using instead a value from the previous time level, and then solve Eq. (10) explicitly for the quantity $\gamma^2 v$. Since $\gamma^2 v$ is only an algebraic nonlinear function of v, it can be inverted easily to give v.

Second, the appearance of γ in Eqs. (13) and (14) and of P and e in Eq. (10) means that the relativistic equations are fundamentally more tightly coupled than their Newtonian counterparts. Conventional explicit techniques typically break this coupling, which is especially close within a relativistic shock transition, by solving each equation sequentially holding the other dependent variables fixed.

It has been our experience, and independently that of John Hawley's (private communication), that solution of Eqs. (10, 13 and 14) by such conventional inconsistent explicit techniques leads to spurious numerical results and instabilities for ultrarelativistic flows with strong shocks. Stable but physically inaccurate solutions can nevertheless be achieved by dropping the numerically difficult viscous pressure term under the time derivative and/or the advection term of the momentum equation (10). That is, the incomplete form of the relativistic momentum density

$$S_j^{incomplete} = (\rho h \qquad)\gamma^2 v_j \tag{17}$$

is used in Eq. (10) in place of the complete momentum density

$$S_j^{complete} = (\rho h + Q/c^2)\gamma^2 v_j \quad . \tag{18}$$

This is the approach taken by Wilson and his collaborators. However, omitting Q in the first two terms of Eq. (10) leads to considerable errors in shock compression and heating if γ is much larger than 2. Figure 1 illustrates this systematic error using data compiled by Centrella and Wilson (1984). Fortunately, the numerical simulations of general relativistic collapse investigated by Wilson, Smarr, Dykema, Evans and Hawley rarely develop flow speeds much in excess of 0.5 c, which means that these systematic errors remain small.

The above-mentioned difficulties are absent in 1-D Lagrangean calculations since, in this coordinate system, Q enters only as forcing terms in the equations of motion, and not as an integral part of one of the dependent variables as in Eq. (18). May and White (1967) report that their calculations of strong relativistic shock fronts show perceptible

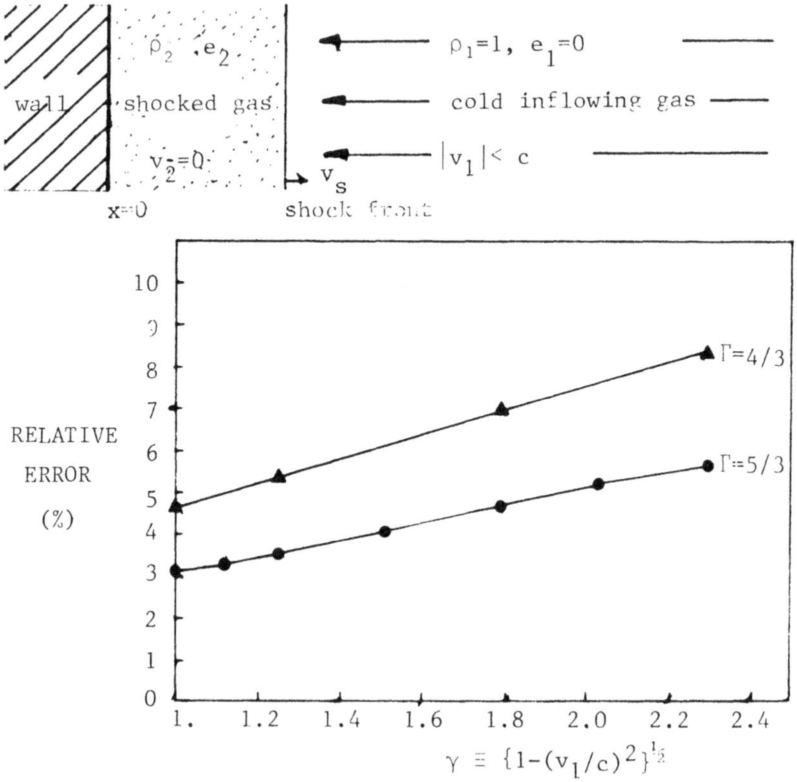

Fig. 1 Centrella's test problem and numerical results showing the relative error in shock heating of cold, relativistically inflowing gas against a wall using the explicit Eulerian techniques of Centrella and Wilson (1984). a) schematic of test problem and definition of terms. Analytic solution for postshock state: $\rho_2^*=(\Gamma+1)/(\Gamma-1)+(\Gamma+1)/\Gamma e_2^*$, $e_2^*=\gamma-1$ where $\gamma=(1-(v_1/c)^2)^{-\frac{1}{2}}$ and $v_s=\gamma|v_1|/(\rho_2^*-\gamma)$. b) dependence of relative error $=|\rho_2-\rho_2^*|/\rho_2^*$ versus Lorentz foctor γ for $\Gamma=4/3$ and $\Gamma=5/3$ gases. Data from Centrella and Wilson (1984).

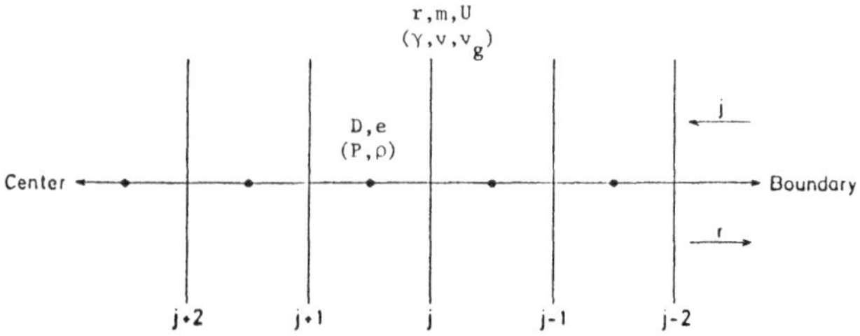

Fig. 2 Centering of the physical variables on the adaptive grid. Derived quantities are indicated in parenthesis.

errors in shock compression only beyond $\gamma \approx 10$ for a $\Gamma = 4/3$ gas, reaching a 30% relative error at $\gamma = 100$. These errors are much smaller than an extrapolation of the upper line in Fig. 1 would predict, leading us to surmise that May and White's errors were the result of discretization, and not of formulation.

3. ADAPTIVE MESH EQUATIONS

We solve Eqs. (10), (13) and (14), supplemented with the definition for γ and the ideal gas equation of state, in one space dimension on an adaptive mesh using a numerical method of solution fully analogous to that given in Winkler, Norman and Mihalas (1985), and Winkler and Norman (this volume) for the solution of the equations of Newtonian radiation hydrodynamics. Before presenting our finite difference equations, we first write down the equations of special relativistic hydrodynamics in an arbitrary coordinate system in integral form from which they are derived.

Expressing the dynamical equations in vector notation, integrating them over a moving volume, and applying the adaptive mesh transport theorem (Winkler, Norman and Mihalas 1984)

$$\frac{d}{dt}\left(\int_V f dV\right) \equiv \int_V \left[\frac{\partial f}{\partial t} + \underline{\nabla} \cdot (\underline{v}_g f) dV\right]$$
$$\equiv \int_V \frac{\partial f}{\partial t}\, dV + \oint_{\partial V} f \underline{v}_g \cdot d\underline{s} \tag{19}$$

then letting $f = \underline{S}^{\text{complete}}$, $\rho\gamma$ and $\rho e\gamma$, respectively, we arrive at the adaptive mesh equations of special relativistic hydrodynamics:

momentum equation

$$\frac{d}{dt}\left[\int_V (\rho h + Q/c^2)\gamma^2 \underline{v} dV\right] + \oint_{\partial V} (\rho h + Q/c^2)\gamma^2 \underline{v}\underline{u}_{rel} \cdot d\underline{s}$$
$$+ \oint (P + Q) d\underline{s} = 0 \tag{20}$$

continuity equation

$$\frac{d}{dt}\left(\int_V \rho\gamma dV\right) + \oint_{\partial V} \rho\gamma \underline{u}_{rel} \cdot d\underline{s} = 0 \tag{21}$$

internal energy equation

$$\frac{d}{dt}\left(\int_V \rho e \gamma dV\right) + \oint_{\partial V} \rho e \gamma \underline{u}_{rel} \cdot d\underline{s} \tag{22}$$

$$+ \int_V (P + Q)\left(\frac{d\gamma}{dt} - \underline{v}_g \cdot \underline{\nabla}\gamma + \underline{\nabla} \cdot \gamma \underline{v}\right) dV = 0 \quad .$$

For completeness, we also include the total energy equation which we do not solve directly, but of course indirectly through Eqs. (20) and (22).

$$\frac{d}{dt}\left[\int_V (\rho h + Q/c^2)\gamma^2 dV\right] + \oint_{\partial V} (\rho h + Q/c^2)\gamma^2 \underline{u}_{rel} \cdot d\underline{s}$$

$$- c^{-2}\left\{\frac{d}{dt}\left[\int_V (P + Q) dV\right] - \oint_{\partial V} (P + Q)\underline{v}_g \cdot d\underline{s}\right\} = 0 \tag{23}$$

Here $\frac{d}{dt}$ represents a time derivative taken with respect to fixed values of the adaptive mesh coordinates, whereas $\frac{\partial}{\partial t}$ is the Eulerian time derivative taken with respect to fixed coordinates in the laboratory frame. $\underline{v}_g$ is the grid velocity measured in the laboratory frame, $\underline{u}_{rel} \equiv \underline{v} - \underline{v}_g$, the relative velocity of the matter with respect to the moving mesh, and V and $d\underline{s}$ are the volume and surface element of a moving zone.

The one-dimensional difference equation representations of Eqs. (20-22) and (23) in the form in which we solve them are

$$\delta[(h + Q/\rho c^2)\, U\Delta\xi] + \delta t\Delta[(-\frac{\delta m}{\delta t})(h + Q/\rho c^2)U]$$

$$+ \delta t r^{\mu}\Delta(P + Q) = 0 \tag{24}$$

$$\delta[\Delta\xi] + \delta t\Delta[r^{\mu}u_{rel}D] = 0 \quad , \tag{25}$$

$$\delta[e\Delta\xi] + \delta t\Delta[(-\frac{\delta m}{\delta t})e] \tag{26}$$

$$+ \delta t\ (P + Q)[\Delta V(\frac{\delta\gamma}{\delta t} - v_g\frac{\Delta\gamma}{\Delta r}) + \Delta(r^{\mu}U)] = 0 \quad ,$$

which are solved together with

$$\Delta m - D\Delta V = 0 \tag{27}$$

and the equation for the adaptive mesh described in detail in Winkler and Norman, this volume. The independent variables in these equations are the time index n and the grid index j; δ denotes a time difference at fixed j, and Δ denotes a spatial difference with respect to j at fixed n. The primary dependent variables are $U \equiv \gamma v$, $D \equiv \gamma\rho$, e, m and r; secondary variables derivable from the primary set are $v_g \equiv \delta r/\delta t$, $u_{rel} \equiv v - v_g$, $\gamma \equiv (1 + U^2/c^2)^{1/2}$, $v \equiv U/\gamma$, and $\rho \equiv D/\gamma$. In Eqs. (24-27), $\Delta\xi$ is shorthand notation for $D\Delta V$, where ΔV is the zone volume given by

$$\Delta V \equiv \frac{1}{\mu+1}\,\Delta\left(r^{\mu+1}\right) , \tag{28}$$

and μ = 0, 1, 2 for cartesian, cylindrical or spherical coordinates. The spatial centering of the primary and secondary (in parenthesis) variables is sketched in Fig. 2.

With the exceptions of Eq. (27) and the adaptive mesh equation, which are solved at the new time level, all physical variables are first centered in time as

$$\langle x\rangle_t \equiv z x^{n+1} + (1 - z)x^n , \tag{29}$$

z lies in the range $0.5 < z \leq 1.0$, with a typical value of 0.55. Thus the code uses differential operators on time-averages rather than time-averages of differential operators. With the exception of advected quantities, simple spatial averages are used where necessary for proper centering; non-linear terms are represented as products of averages rather than averages of products.

The difference equations (24-27) and the adaptive mesh equation constitute a coupled set of nonlinear algebraic equations in the unknown primary variables, which is solved implicitly using block Gaussian elimination within a Newton-Raphson iteration procedure. The reader is referred to section V of Winkler and Norman, these proceedings, for details of the advection scheme, the numerical reasons for selectively using $\Delta\xi$, Δm and $\delta m/\delta t$, and a description of the control procedure.

We encountered numerical difficulties when using the artificial viscosity formulation given in Eqs. (8) and (16) on test problems involving ultrarelativistic shock fronts. In particular, we found that this prescription fails to place enough grid points in the high γ side of the shock transition and consequently leads to spurious oscillations. An alternate prescription for Q which is motivated by Jim Wilson's work and used in subsequent numerical examples, and which we find gives satisfactory numerical results, is

$$Q = -\zeta^a \, \partial_x v \quad , \tag{30}$$

$$\zeta^a = c_2 \rho h \ell^2 \gamma \min(\partial_x v, 0) \quad .$$

The unsatisfactory aspect about Eq. (30) is that Q thus defined is not Lorentz invariant. At the time this work was done, we did not appreciate the difference between ℓ_E and ℓ_p. We now believe that the Lorentz invariant form Eq. (7) will have the desired shock-resolving properties if Eq. (15) is used for the coefficient of artificial viscosity. The additional factor of γ^2 between Eq. (15) and Eq. (16) will have the effect of putting more grid points in the high γ side of the shock transition, as desired.

4. NUMERICAL EXAMPLES

As a first test of our numerical method, we computed the test problem given in Fig. 1 for the shock thermalization of cold, relativistically moving gas hitting a wall. In order to assess the importance of including the contribution of artificial viscosity to the relativistic momentum density, we performed three separate calculations wherein the following three momentum equations, given in differential form for simplicity, were solved:

Test 1: $\partial_t S^{\text{incomplete}} + \partial_x(S^{\text{incomplete}} v) + \partial_x(P + Q) = 0 \quad ,$

Test 2: $\partial_t S^{\text{incomplete}} + \partial_x(S^{\text{complete}} v) + \partial_x(P + Q) = 0 \quad ,$

Test 3: $\partial_t S^{\text{complete}} + \partial_x(S^{\text{complete}} v) + \partial_x(P + Q) = 0 \quad ,$

where the definitions for $S^{\text{incomplete}}$ and S^{complete} are given in Eqs. (17) and (18). An Eulerian (i.e. fixed) mesh of 200 grid points was used in each case. The resulting distributions of specific internal energy are shown in Fig. 3.

In Fig. 3a, we see that the omission of Q in both the time derivative and advection terms of the equation of motion results in a substantial overheating (~13%) of the shock processed gas compared to the theoretical value of $e_2^* = \sqrt{2} - 1$. The overshoot next to the wall is the familiar wall-heating effect discussed in Winkler and Norman (this volume). The two isolated regions of overshoot were produced by increasing the amount of artificial viscosity as the calculation proceeded. This was done to see whether reducing discretization errors in the shock transition would bring us closer to the analytical value. The fact that the computed solution diverges from rather than converges to the theoretical solution demonstrates that the 13% discrepancy derives from a fundamental error in formulation of the basic equations.

Although the discrepancy is reduced to ~4% when Q is retained in the advection term of the equation of motion used in Test 2, the solution also degrades as the amount of artificial viscosity is increased (Fig. 3b), again revealing a fundamental error in formulation.

When Q is retained in all three terms of the equation of motion, excellent results are obtained. In Fig. 3c, we display the results of a $\gamma = 10$ cold inflow. The computed post-shock energy differs from the theoretical result $e_2^* = 9$ by 0.2%.

This discrepancy is the result of discretization errors, which can be reduced by increasing the number of grid points in the shock transition. This we accomplish while maintaining a physically thin shock transition through the use of an adaptive mesh. This is illustrated in Fig. 4. In this example the viscous length ℓ in Eq. (30) is set to approximately 10^{-2} of an equidistant zone size. The adaptive mesh is constructed such that artificial viscous energy generation is automatically detected and resolved, which insures a numerically well-resolved shock front. The relative error in the post shock energy is found to be < 0.01%.

As a second test of our code, we computed the breakup of an initial pressure discontinuity in a shock tube into its three constituent nonlinear waves, i.e., a Riemann problem. The initial conditions are those used by Sod (1978); referring

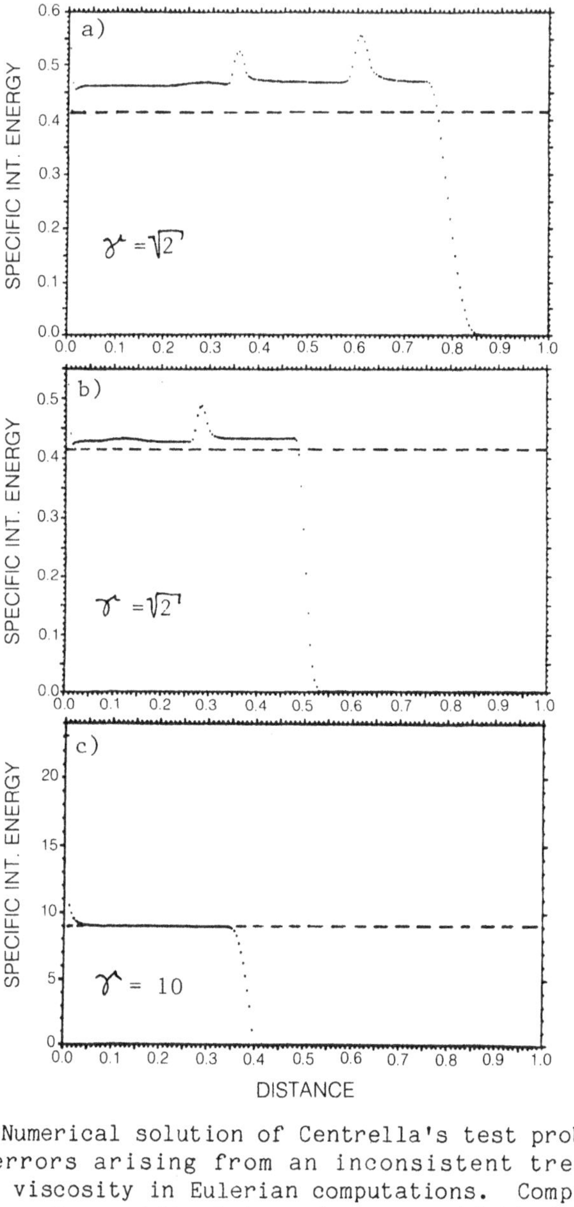

Fig. 3 Numerical solution of Centrella's test problem showing the errors arising from an inconsistent treatment of artificial viscosity in Eulerian computations. Computed distributions of specific internal energy (dotted line) and analytic solution (dashed line) are displayed for three cases described in the text. a) Case 1: $\gamma = \sqrt{2}$, relative error = 13%, b) Case 2: $\gamma = \sqrt{2}$, relative error = 4%. c) Case 3: (consistent implementation) $\gamma = 10$, relative error = 0.2%. All solutions computed implicitly on an Eulerian mesh of 200 zones.

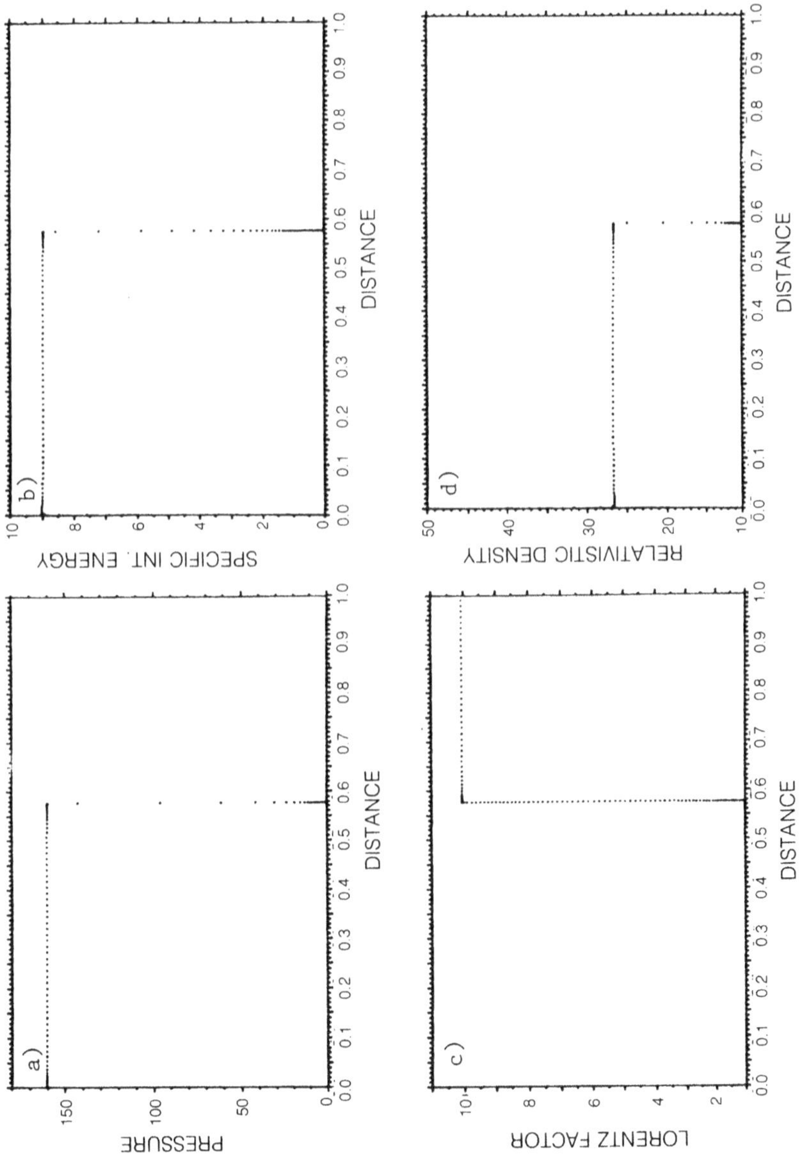

Fig. 4 Implicit adaptive-mesh solution to Centrella's test problem for γ = 10 inflow. Plotted are the distributions of a) pressure, b) specific internal energy, c) Lorentz factor, and d) relativistic mass density $D \equiv \rho\gamma$. Relative error < 0.01%.

to the states on the left and right of the initial discontinuity as L and R, respectively, we assume

$$P_L = 1.0, \ \rho_L = 1.0, \ v_L = 0 \ \text{for} \ 0. \leq x \leq 0.5$$

$$P_R = 0.1, \ \rho_R = 0.125, \ v_R = 0 \ \text{for} \ 0.5 < x \leq 1.0$$

$$\Gamma_L = \Gamma_R = 7/5 \qquad (31)$$

By taking $c = 1$, the resultant flow is mildly relativistic, whereas if we set $c >> (P/\rho)^{1/2}$ we recover the Newtonian results we have reported elsewhere (Winkler, Norman and Mihalas 1985).

Figure 5 shows the state of the gas at $t = 0.24$ for the relativistic problem using an adaptive mesh of 200 zones. The resulting profiles are qualitatively similar to the Newtonian results, and differ quantitatively in the following respects. The post shock velocity is 0.43 instead of 0.9 in the Newtonian case, yielding $\gamma \approx 1.1$. The positions of the right-facing shock wave, the left-facing rarefaction wave, and the contact discontinuity are retarded compared to the Newtonian solution at the same evolutionary time as measured by an Eulerian observer. The post-shock internal energy is 2.83 versus 2.85 in the Newtonian case, and agrees with the theoretical value to < 0.1%.

In order to convince ourselves that we indeed have a coordinate independent representation of the dynamical equations, we ran the shock tube problem in both Eulerian and Lagrangean coordinates. This exercise proved instructive, for we had initially omitted the $v_g \Delta\gamma/\Delta r$ term in the internal energy equation (26) which converts between the Eulerian and adaptive mesh time derivatives. Our solutions revealed O(10%) errors in the Lagrangean and adaptive calculations of the final distribution of e (see Figs. 6d-f), clearly too large to be ascribable to discretization errors. When the correction term was included, we produced the coordinate independent results shown in Figs. 6a-c. Although the Eulerian and Lagrangean calculations fail to capture the discontinuity and rarefaction wave profiles as crisply as the adaptive mesh calculation, their positions agree as do the post-shock values of e and ρ.

As a third and final test of our numerical methodology,

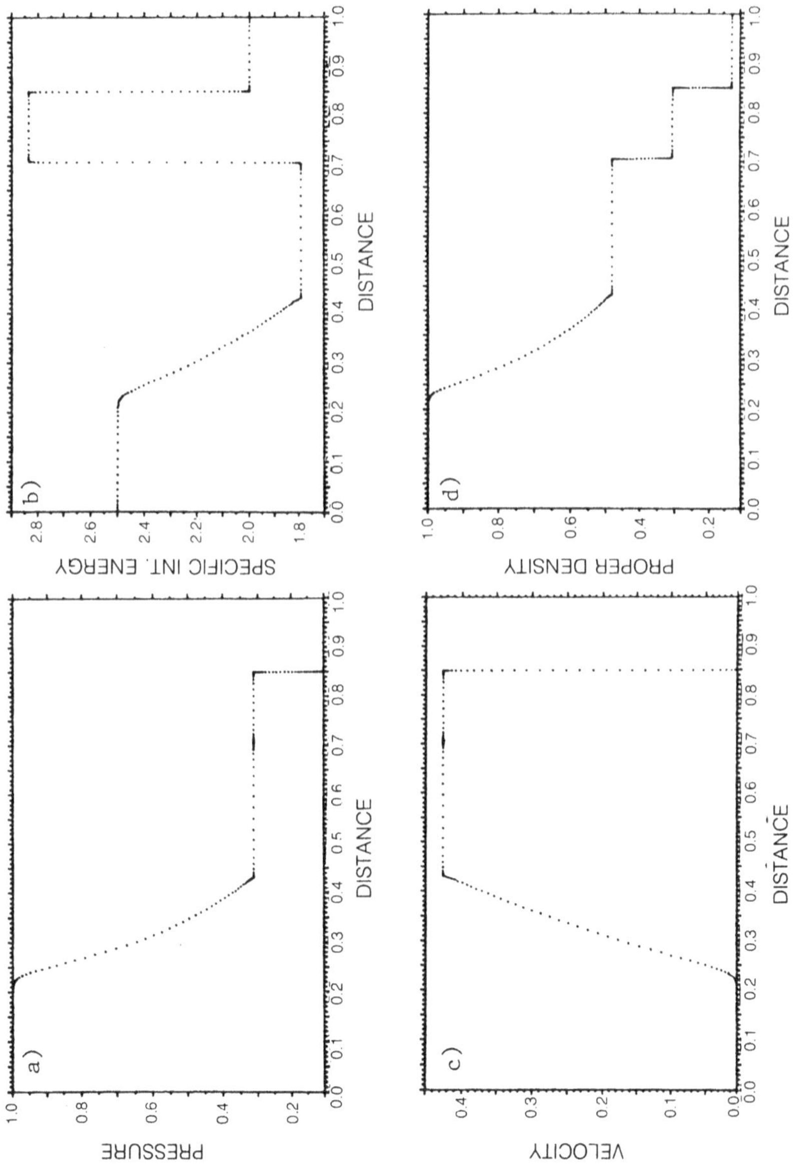

Fig. 5 Adaptive-mesh numerical solution to the relativistic shock tube test problem at t = 0.24.

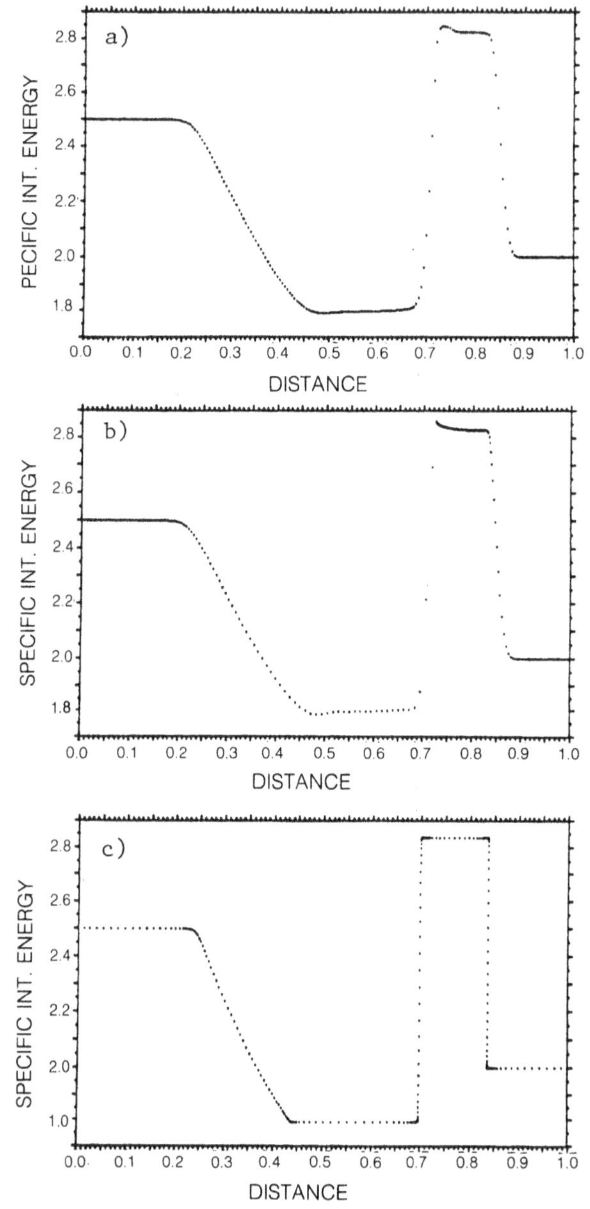

Fig. 6 Comparison of a) Eulerian, b) Lagrangean and c) adaptive-mesh numerical solutions to the relativistic shock tube test problem described in the text. Specific internal energy at t = 0.24 is plotted at every grid point in a mesh of 200 zones. d-f) same as a-c) but without $v_g \Delta Y/\Delta r$ term in Eq. (26).

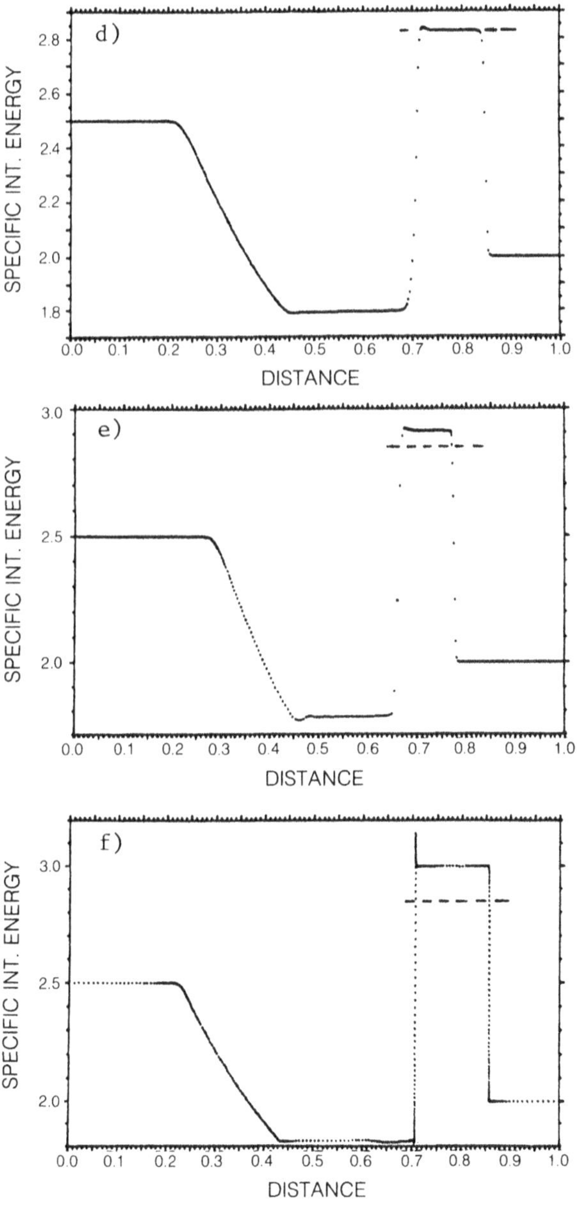

Fig. 6 continued

we computed the formation of a relativistic blast wave by increasing the initial pressure discontinuity of the previous example to 10^5. We used as initial conditions

$$P_L = 10^3, \ \rho_L = 1, \ v_L = 0 \ \text{for} \ 0 \le x \le 0.5 \ ,$$

$$P_R = 10^{-2}, \ \rho_R = 1, \ v_R = 0 \ \text{for} \ 0.5 < x \le 1.0 \ ,$$

$$\Gamma_L = \Gamma_R = 4/3.$$

The solution at an advanced time computed with an adaptive mesh of 400 zones is shown in Fig. 7. We notice several distinctively relativistic features. In Fig. 7c, we see that the velocity profile in a strong relativistic rarefaction wave is not linear, as it is in Newtonian planar flow, but rather curls over as v approaches c. As $c = 1$ in this example, the post-shock velocity of $v = 0.96$ corresponds to a Lorentz factor of $\gamma = 3.5$ for the shocked gas. The bulk of this gas resides in a thin shell bounded by the leading shockfront and trailing contact discontinuity as can be seen in Fig. 7d. The proximity of these two discontinuities is partially due to Lorentz contraction and partially due to the limiting velocity of light; the shock speed is 0.986c, corresponding to a Lorentz factor of $\gamma = 6$.

Although the dense shell is physically thin, it is numerically well-resolved on the adaptive mesh. Fig. 8 shows the four state variables of Fig. 7 plotted against zone index j rather than the spatial coordinate r. As can be seen in Fig. 8d, approximately 140 zones have been concentrated in the shock and contact discontinuities and the intervening constant state, representing nearly 30% of the available grid points. However, from Fig. 9a, we see that this region occupies a mere 1% of the physical domain. The distribution of zone sizes as a function of zone number is displayed in Fig 9b and confirms what was stated earlier: although 50 zones resolve the shock transition, their combined thickness of 50×10^{-6} is only 1/50 of an equidistant zone width $1/400 = 2.5 \times 10^{-3}$.

5. DISCUSSION AND CONCLUSIONS

The blast wave example just described captures the essential difficulties faced in numerically modeling ultrarelativistic gas flows in any number of dimensions: strong relativistic shocks and narrow physical structures. We

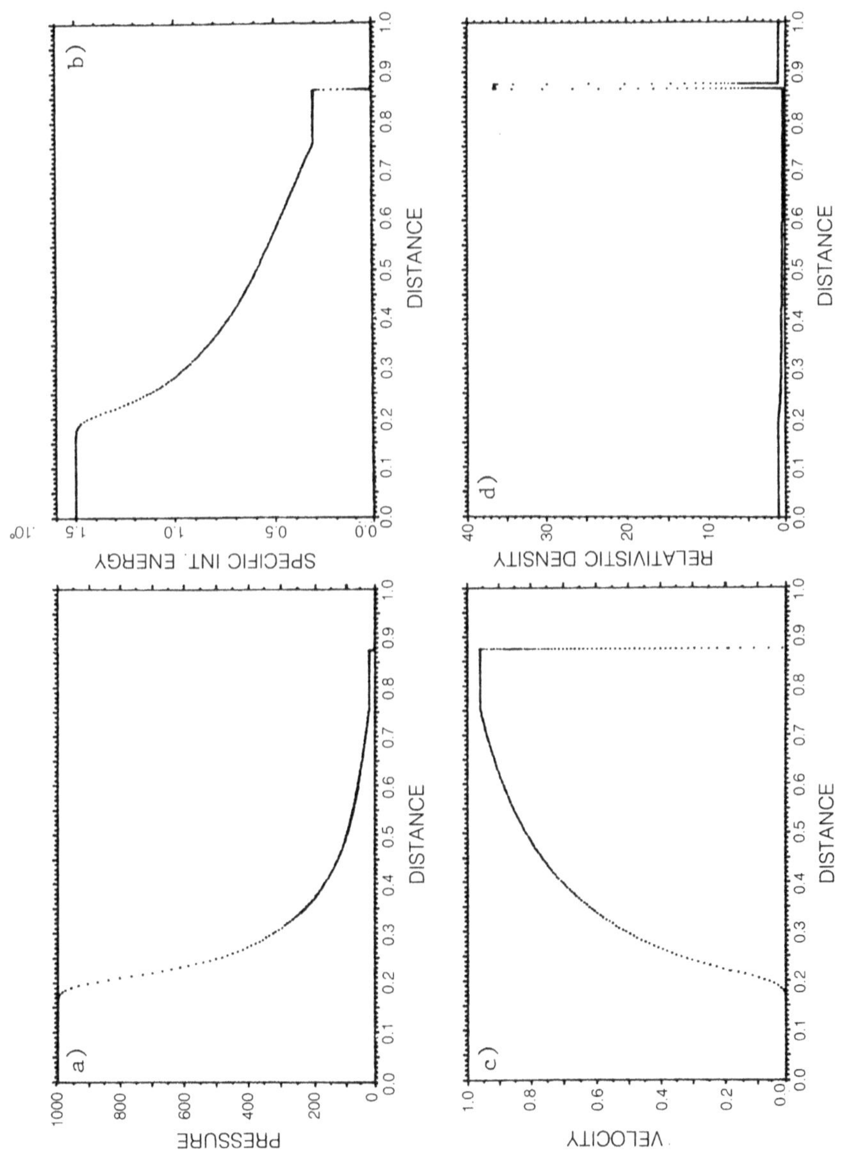

Fig. 7 Adaptive-mesh numerical solution to the relativistic blast wave test problem at t = 0.385. Note the formation of a thin dense shell of gas just behind the blast wave in d). V_{shock} = 0.986c, V_{shell} = 0.96c. A mesh of 400 zones was used in this computation.

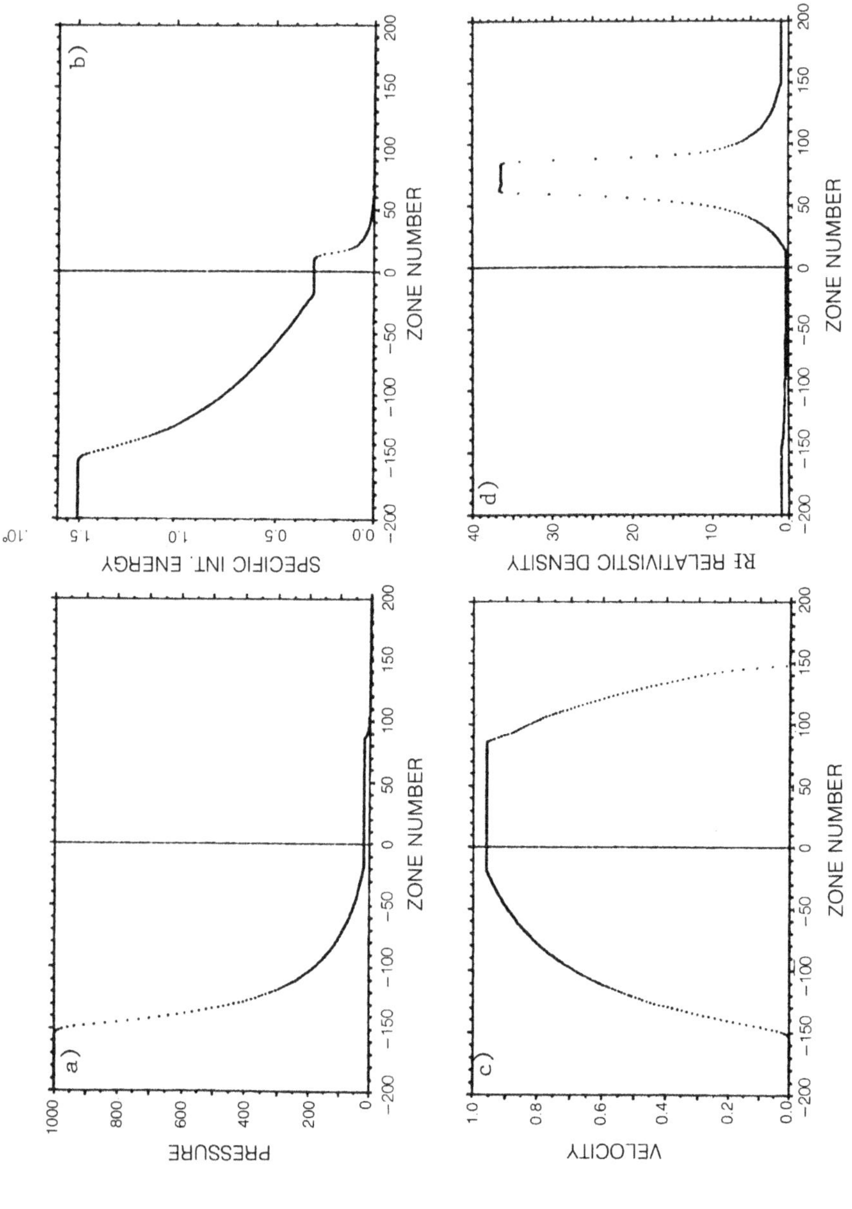

Fig. 8 Relativistic blast wave solution of Fig. 7 plotted against zone index j of the adaptive mesh.

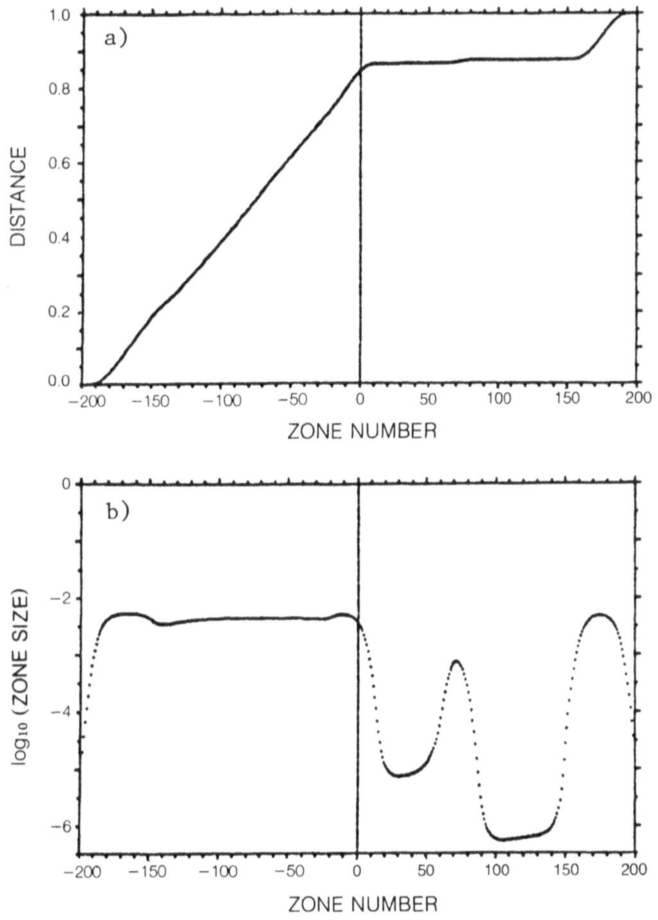

Fig. 9 a) Distribution of the spatial coordinate X as a function of zone index j for the relativistic blast wave solution of Fig. 7. b) Distribution of zone size as a function of zone index. The local minima around j = 30 and j = 100 indicate the refined zoning in the contact discontinuity and shock front, respectively.

have shown that if artificial viscosity is used to mediate the shock transition, self-consistency demands that the artificial viscous "pressure" Q be included everywhere in the dynamical equations where the thermal pressure P appears making the formal substitution $P \to P + Q$. In an Eulerian or adaptive coordinate system this introduces numerical complications, for now Q enters into the definition of the relativistic momentum and total energy densities [cf. Eqs. (11, 12)], making these quantities complicated nonlinear functions of the fluid velocity and its first derivative. Only with implicit techniques is one able to solve the set of dynamical equations in finite-difference form to physical and numerical self-consistency, which, as we have seen, is required for accurate shock jumps.

It is nevertheless possible to obtain correct shock jumps using explicit techniques if one adopts the following non-self-consistent physical formulation: a) omit Q from the first two terms of the momentum equation (11) (i.e., use $S^{incomplete}$ rather than $S^{complete}$), and b) solve the total energy equation (12) instead of the internal energy equation (14), omitting Q entirely. The resulting system of equations conserves mass, momentum and total energy thereby insuring correct shock jumps. The penalty one pays for the lack of physical self-consistency is a nonphysical distribution of internal energy within the shock transition. Negative values of internal energy are not uncommon within strong shock transitions and necessitate such fix-ups as imposing a positive lower bound (floor) on the range of values the internal energy may assume. Such an approach would be appropriate only for studying ideal relativistic gas flows where one is not interested in the physical processes internal to the shock transition.

If one <u>is</u> interested in the physics of relativistic shock waves, then one must numerically track and resolve the shock transition and self-consistently model the dissipative processes within it. At present, implicit adaptive mesh techniques are the best we have for doing this correctly. Lagrangean techniques, although in principle applicable in one dimension, have the disadvantage that inherent in the choice of coordinate system, one has direct control only over how well the mass distribution is resolved. With an adaptive mesh, one has control over how well any and all physical quantities are resolved, be they primary dependent variables or derived quantities (for an example and discussion, see Winkler et al. 1985).

In multidimensions, a number of avenues need to be explored. As pointed out in the introduction, we consider Lagrangean techniques unsuitable for modeling relativistic flows of astrophysical interest because of the unavoidable shear-induced mesh tangling difficulties.

Under the category of explicit Eulerian techniques, we need to investigate the application to relativistic hydrodynamics of new algorithms which handle shock fronts without artificial viscosity, such as Woodward's Piecewise Parabolic Method (Woodward, this volume). Gone are the difficulties of self-consistently incorporating Q into the difference equations and solving them. But what new difficulties may arise as we generalize to multidimensional relativistic flows?

We have implemented the non-self-consistent approach outlined above in a 2-dimensional explicit Eulerian hydrodynamics code similar to that described in Norman and Winkler, this volume, and encounter numerical difficulties in modeling flows with $\gamma \geq 2$. The difficulties manifest in flow velocities exceeding light speed, and seem to result from directional splitting the momentum equation. It is entirely plausible that directional splitting breaks down as $v \to c$ because the tight physical coupling between the momentum components contained in the coupling constant γ is not reflected in the solution procedure. Further work is needed to determine whether these difficulties can be avoided by a simple reformulation of the difference equations, or whether they are inherent to explicit techniques.

Even if these difficulties can be overcome, one still faces the problem of numerically resolving the highly localized structures produced in ultrarelativistic flows, such as the thin shell of matter in the blast wave problem shown in Fig. 7d. In several space dimensions, this problem may be so severe as to negate the usefulness of the computation unless some form of grid adaptation is used. Given the high degree of physical coupling between the flow variables as $v \to c$, we anticipate that rezoning "between the timesteps" will not work. If this is the case, then we must contemplate using an implicitly-coupled adaptive mesh in two or three dimensions such as we have developed in one dimension. We are currently extending our technique to multidimensions. Solving the associated sparse matrix problem will require the use of fast iterative techniques to be economically feasible, and will demand computers with larger memories in order to store the matrices.

However, the greatest advantage of implicit techniques for relativistic hydrodynamics, regardless of the coordinate system, is the ability to solve the dynamical equations selfconsistently. Although this paper illustrates the importance of doing so only for the special case of ideal relativistic flows with artificial viscosity, the conclusion carries over to nonideal relativistic flow with physical viscosities and heat conduction. Such effects will be important in modeling disk accretion around black holes (Hawley and Smarr 1985). In addition, magnetic and electrodynamic effects may be involved in the acceleration of relativistic jets (Blandford 1985) observed on parsec scales with very long baseline interferometry. The mixing of the space and time components of the stress tensors associated with these physical effects when the equations are expressed in non-comoving coordinates make their solution by explicit techniques difficult and their regime of applicability limited to mildly relativistic velocities. Using implicit techniques, it should be possible to expand the range of flow velocities accessible to numerical study, from the slow viscosity-induced inward drift of matter in an accretion disk to the near light-speed expansion of a compact radio source.

REFERENCES

1. Anile, A. M., Miller, J. C. & Motta, S. 1983, Phys. Fluids 26, p. 1450.
2. Blandford, R. D. 1985, in "Numerical Astrophysics," eds. J. Centrella, J. LeBlanc & R. Bowers, (Jones and Bartlett: Portola Valley), p. 6.
3. Blandford, R. D. & Königl, A. 1979, Astrophys. J. 232, p. 34.
4. Blandford, R. D. & McKee, C. F. 1976, Phys. Fluids 19, p. 1130.
5. Centrella, J. & Wilson, J. R. 1984, Astrophys. J. Supp. Ser. 54, p. 229.
6. Colgate, S. A. & Johnson, M. H. 1960, Phys. Rev. Letters 5, p. 235.
7. Dykema, P. G. 1980, Ph.D. Dissertation, University of Texas at Austin.
8. Hawley, J. F., Smarr, L. L. & Wilson, J. R. 1984, Astrophys. J. Supp. Ser. 55, p. 211.
9. Hawley, J. F. & Smarr, L. L. 1985, in "Numerical Astrophysics," eds. J. Centrella, J. LeBlanc & R. Bowers, (Jones and Bartlett: Portola Valley), p. 30.
10. Kellerman, K. I. & Pauliny-Toth, I. 1981, Ann. Rev. Astron. Astrophys. 19, p. 373.
11. May, M. M. & White, R. H. 1967, Meth. Comp. Phys. 7, p. 219.

12. McKee, C. R. & Colgate, S. A. 1973, Astrophys. J. 181, p. 903.
13. Misner, C. W., Thorne, K. S. & Wheeler, J. A. 1973, "Gravitation," (Freeman: San Francisco).
14. Richtmyer, R. D. & Morton, K. W. 1967, "Difference Methods for Initial Value Problems," (Interscience: New York).
15. Shapiro, P. R. 1979, in "Particle Acceleration Mechanisms in Astrophysics," eds. J. Arons, C. Max, & C. McKee, (American Institute of Physics: New York), p. 295.
16. Smarr, L. L., Taubes, C. & Wilson, J. R. 1980, in "Essays in General Relativity," ed. F. Tipler, (Academic Press: New York).
17. Sod, G. A. 1978, J. Comp. Phys., 27 pl.
18. Von Neumann, J., & Richtmyer, R. D. 1950, J. Appl. Phys., 21, p. 232.
19. Wilson, J. R. 1972, Ap. J., 173, p. 431.
20. Wilson, J. R. 1979, in "Sources of Gravitational Radiation," ed. L. Smarr, (Cambridge Univ. Press: Cambridge), p. 423.
21. Winkler, K.-H. A., Norman, M. L. & Mihalas, D. 1984, J. Quant. Spectrosc. Radiat. Transfer, 31, p. 473.
22. Winkler, K.-H. A., Norman, M. L. & Mihalas, D. 1985, to appear in "Computational Techniques: Multiple Timescale Problems," eds. J. I. Brackbill & B. I. Cohen, (Academic Press: New York).

NEUTRINO TRANSPORT IN RELATIVITY

James R. Wilson

Lawrence Livermore National Laboratory

ABSTRACT: Two methods of solving the neutrino transport equation in general relativity are presented. The first method is for the spherical collapse of stars to black holes. It uses a Lagrangian hydrodynamic formulation as a framework. The method is of mediocre accuracy, but it can handle all cases efficiently. The second method of solving the transport equation is for use in cosmology. In this case all velocities are of the same size and an explicit monotonic second order in space differencing is used. This latter method includes the effects of neutrino rest mass.

1. SPHERICAL COLLAPSE

I will present two methods that have been used for the solution of the neutrino transport equation in one space dimension for use with general relativistic hydrodynamic computer codes. First I will consider the case of spherical symmetry appropriate for the study of stellar collapse. The metric in this case is taken as in May and White (1966) as

$$ds^2 = a^2dt^2 - R^2(d\theta + \sin^2\theta d\phi^2) - b^2dm^2 \tag{1}$$

where m is the radial mass coordinate. The energy momentum tensor, $T^{\nu}{}_{\mu}$, is taken as

$$\begin{vmatrix} -P & 0 & 0 & -Gb/a \\ 0 & -P-H & 0 & 0 \\ 0 & 0 & -P-H & 0 \\ Ga/b & 0 & 0 & \rho(1+E) \end{vmatrix} \tag{2}$$

K.-H. A. Winkler and M. L. Norman (eds.), Astrophysical Radiation Hydrodynamics, 477–490.

The pressure, P, and thermal energy, E, are the sum of the matter values and the neutrino contribution, P_ν, and E_ν, and G is the flux of neutrino energy, and $H = 1/2(E_\nu - 3P_\nu)$. The neutrino distribution function is represented by $F(\mu,\nu,m,t)$ where μ is the cosine of the angle with respect to the radius vector, ν is the energy a neutrino would have at spacial infinity if it left the star without further collisions. The local neutrino energy is ν/a, and t is the time. Usually a is less than one. This energy definition allows a limited range of neutrino groups to span the neutrino energy range because at high densities (deep gravitational potentials) $a \ll 1$ and thus the energy groups represent high energy neutrinos then present. Also it takes care of the gravitational red shift automatically. The moments of the distribution function needed for the energy momentum tensor are

$$
\begin{aligned}
E_\nu &= \int F d\mu d\nu / a \\
P_\nu &= \int F \mu^2 d\mu d\nu / a \\
G &= \int F \mu d\mu d\nu / a
\end{aligned}
\tag{3}
$$

Following May and White (1966) we arrive at the equations of motion

$$
\begin{aligned}
M &\equiv 4\pi \int_0^R \left(\rho + \rho\varepsilon + \frac{GU}{\Gamma}\right) R^2 dR \\
\Gamma &\equiv \frac{\partial R}{b\partial m} = \left(1 + U^2 - \frac{2M}{R}\right)^{1/2} \\
U &\equiv \frac{1}{a}\frac{\partial R}{\partial t} \\
\frac{1}{a}\frac{\partial U}{\partial t} &= -\frac{4\pi R^2 \Gamma}{(1+\varepsilon+P/\rho)}\left[\frac{\partial P}{\partial m} + \frac{2H}{R}\frac{\partial R}{\partial m} + \frac{\partial}{\partial t}\left(\frac{Gb}{a}\right)\right] - \frac{M}{R^2}\left(1 + 4\pi P R^3/M\right) \\
a &= \exp \int_R^\infty \left[\frac{\partial P}{\partial m} + \frac{2H}{R}\frac{\partial R}{\partial m} + \frac{\partial}{\partial t}\left(\frac{Gb}{a}\right)\right] \frac{dm}{\rho(1+e+P/\rho)} \\
\frac{1}{a}\frac{\partial}{\partial t}\log(\rho R^2) &= \frac{\partial U}{\partial m} \Big/ \frac{\partial R}{\partial m} \\
\frac{\partial \varepsilon m}{\partial t} + P^m \frac{\partial}{\partial t}\left(\frac{1}{\rho}\right) &= \text{neutrino collision terms} = T\frac{\partial S}{\partial t}
\end{aligned}
\tag{4}
$$

The velocity of light and the gravitational constant are each taken to be one.

The neutrino transport equation is taken from Lindquist (1966) and it becomes in our variables:

$$\frac{\partial F}{a\partial t} = -\frac{\mu\Gamma}{\alpha R^2}\frac{\partial}{\partial R}(aR^2F) - \Gamma\left(\frac{1}{R} - \frac{\partial}{\partial R}\log a\right)\frac{\partial}{\partial\mu}\left[F(1-\mu^2)\right]$$

$$+ \frac{F}{a\rho}\frac{\partial\rho}{\partial t} + R\frac{\partial}{\partial R}(U/R)\frac{\partial}{\partial\mu}\left\{\left[\mu(1-\mu^2)F\right] + \mu^2\nu\frac{\partial F}{\partial\nu}\right\}$$

$$+ \frac{\nu\partial F}{\partial\nu}\frac{1}{a}\frac{\partial}{\partial t}\log(R/a) + K\rho(B-F)\left[1 + \exp(-\nu/\alpha T)\right] \qquad (5)$$

$$\frac{\partial S}{a\partial t} = \frac{1}{T}\int(1 + \exp(-\nu/at))K(B-F)d\mu d\nu/a + \text{hydrodynamic terms}$$

$$B = (\nu/a)^3/\left[1 + \exp(\nu/aT)\right]$$

S is the matter entropy, $K(\nu,T,P)$ is the opacity of matter, and B is the black body distribution function for neutrinos. I will only consider one type of neutrinos in this paper and only pure absorption interactions.

The terms on the right hand side of equation (5) can be interpreted as follows. The first term is the radial flux term. The second is the acceleration term. The first part of which is the usual centrifugal acceleration and the second part is the gravitational acceleration. The third term is the volume expansion. The fourth term is a kind of aberation effect. In the fifth term the $\partial a/\partial t$ part is the change in F due to change in meaning of ν. The velocity part of the fifth term is of the same type as the fourth term.

The method of differencing the hydrodynamic part of the equations will not be discussed since that is available in May and White (1966). Equation (5) will be dealt with by operator splitting the equation into the matter speed dependent terms and the light speed terms.

The light speed part is

$$\frac{1}{a}\frac{\partial F}{\partial t} = -\frac{\mu\Gamma}{aR^2}\frac{\partial}{\partial R}(aR^2F) - \Gamma\left(\frac{1}{R} - \frac{\partial a}{a\partial R}\right)\frac{\partial}{\partial\mu}\left[(1-\mu^2)F\right] + K'\rho(B-F) \qquad (6)$$

$$K' = K\left[1 + \exp(-\nu/aT)\right]$$

This part of the equation must be solved implicitly since we use the hydrodynamic time step determined by the Courant condition. All the other terms are solved explicitly in time since they are all proportional to the matter velocity which is usually of the order of sound speed. In conjunction with (6) we simultaneously solve the heat exchange term with the matter

$$\rho c v \frac{\partial T}{\partial t} = - \int K\rho (B-F) d\mu d\nu \tag{7}$$

The method of partial temperatures is used which consists of changing T each time and F is advanced for any particular μ, ν zone. First consider the equations

$$\frac{\partial F}{\partial t} = aK'\rho(B-F) \tag{8}$$

$$C_v \frac{\partial T}{\partial t} = - K'(B-F)\Delta\mu\Delta\nu$$

where $\Delta\mu$ and $\Delta\nu$ are the zone sizes of μ and ν. For time step, Δt, the equations are evaluated at the new time level and with $B^{new} = B^{old} + (T^n - T^o)\ \partial B^{old}/\partial t$ the time differenced version of (8) is

$$F^n - F = \frac{aK'\rho\Delta t(B^o - F^n)}{1 + \frac{K'\Delta t}{C_v} \frac{\partial B^o}{\partial T} \Delta\nu\Delta\mu} \tag{9}$$

$$T^n - T = - \frac{K'\rho\Delta t(B^o - F^n)\Delta t\Delta\nu\Delta\mu}{C_v + K'\Delta t \frac{\partial B^o}{\partial T} \Delta\nu\Delta\mu}$$

The collision term $K'\rho a\Phi t$ will henceforth be replaced by

$$k = \frac{K'\rho\alpha\Delta t}{1 + \frac{K'\Delta t}{C_v} \frac{\partial B^o}{\partial T} \Delta\nu\Delta\mu}$$

As each μ, ν zonal value of F is advanced T, B, and $\partial B/\partial T$ are advanced. The F and T equations can now be advanced sequentially.

Now the time differenced equation for F is

$$F^n - F^o = - \frac{\partial\Gamma\Delta t}{R^2} \frac{\partial}{\partial R} (aR^2F^n) - \Gamma\Delta t \left(\frac{1}{R} - \frac{\partial a}{\partial R}\right) \frac{\partial}{\partial\mu} \left[F^n(1-\mu^2)\right] + k(B-F^n) \tag{10}$$

In order to assure that the spacially differenced equation yield the diffusion like equation for the case of $\Delta R >> \lambda$ (λ = mean free path), a factor $f = 1/(1 + |\mu|\lambda/\Delta R)$ is placed in front of the spacial derivative terms. Also the spacial terms are upwind differenced. First we inward directed beam equation

$$F_k^n(m = -1) = k_k \ (B_k - F_k^n)$$

$$+ \frac{\Delta t}{Vol_k} (f_{k+1/2} \ a_{k+1/2} \ R_{k+1/2}^2 \ F_{k+1}^n - f_{k-1/2} \ a_{k-1/2} \ R_{k-1/2}^2 \ F_k^n) \tag{11}$$

is solved from the outside in where the subscript k is the radial zone number. The zonal volume is

$$Vol_k = (R_{k+1/2}^3 - R_{k-1/2}^3)/(3\Gamma)$$

Then for the successive angular groups the acceleration term is evaluated by use of the upwind in angle differencing of F. We let the angular derivative be differenced as,

$$\frac{\partial}{\partial \mu} \left[(1-\mu^2)F\right] \rightarrow \frac{1}{\Delta\mu_i} \left[(1-\mu_{i-1}^2)F_{i-1}^n - (1-\mu_i^2)F_i^n\right]$$

where i is the angular zone index. i = 1 designates the inward directed flux. In the equation for outward directed fluxes the spacial derivative term is taken to be

$$\frac{\Delta t}{Vol_k} (f_{k+1/2} \ a_{k+1/2} \ R_{k+1/2}^2 \ F_k^n - f_{k-1/2} \ a_{k-1/2} \ R_{k-1/2}^2 \ F_{k-1}^n) \tag{12}$$

For energy exchange with matter the $\Delta\mu$ associated with F(i=1) is half the $\Delta\mu$ to the next zone. For subsequent F's, the $\Delta\mu$ for energy deposition is the average of the $\Delta\mu$'s above and below in zone number. The $\Delta\mu$ and ΔR are chosen so that the equations are conservative for the $\int F dVol$. As mentioned above for each zone in μ and ν the temperature is changed when an F is changed. It is necessary to update the black body distribution after each update of F to preserve numerical stability. The opacity is calculated only once for each time step. In the problems of interest for neutrino flow, the opacity is a weak function of temperature. For accuracy the maximum and the minimum of the matter temperature are checked and if the total temperature swing is larger than some input number, typically 5 to 10 %, then the time step is reduced. All the light speed terms are now completed. After each advance of the neutrino flux, F, the momentum absorbed by the matter $p = k\mu F \Delta\mu\Delta\nu/a$ is added to the matter momentum.

The compression term in equation (5) is simply differenced by preserving the product ρF at the time in the code where ρ is changed. The remaining velocity terms in equation (5) are differenced by a neutrino number conserving upwind in μ or ν manner using the old time values for the F.

This system is rather diffusive due to the upwind differencing of all terms. In practice, the sensitivity to number of angles was checked by varying the number of angular groups. In the calculations made thus far, the number of space and energy groups was much larger than the number of angular zones and sensitivity of zoning in these variables was not checked.

When an apparent horizon forms the acceleration terms reverse sign and the angular differencing in principle is unstable. However, in practice for collapsing stars the acceleration terms reverse sign at such high density that the neutrino mean free path is very small compared to the inverse of the acceleration terms, thus, the acceleration term is too small to produce instability.

For examples of calculations made with this program see Wilson (1971).

2. COSMOLOGY

The second example of solution of the relativistic neutrino transport I will consider is the case of a toroidal space-time (spacially periodic) time with spatial variations in one direction only, and no shear. I take the metric as given by

$$ds^2 = -dt^2\left(\alpha^2 - \frac{\beta_x^2}{A^2}\right) + A^2(dX^2 + dy^2 + dz^2) + 2\beta_X dX dt$$

For a discussion of the hydrodynamics without radiation flow in this type of space time, see the section by the author on numerical hydrodynamics. The metric functions α, β_X, A are functions of X and t only. We consider only space-times periodic in a fixed length in the X direction. We wish to consider massive neutrinos. We take the convention that $G = 1/(8\pi)$, $C = 1$, and $m = 1$ where m is the rest mass of the neutrino. To find the appropriate transport equation, we start from Lindquist (1966),

$$\frac{DF}{Ds} = U^i\left(\frac{\partial F}{\partial X^i} - \Gamma^\alpha_{i\beta}\, U^\beta \frac{\partial F}{\partial U^\alpha}\right) \qquad (13)$$

where U^i is the four velocity of the neutrino. We choose momentum W and angle cosine μ such that

$$U^X = \left(\mu W \frac{\beta}{\alpha} \frac{\sqrt{A^2 + W^2}}{A}\right) / A^2$$

$$U^t = \sqrt{A^2 + W^2} / (\alpha A) \tag{14}$$

$$U^{\perp} = \sqrt{(1 - \mu^2)} W/A^2 = U^y = U^z$$

The transport equation can be written in conservative form as

$$\frac{\partial G}{\partial t} + \frac{\partial}{\partial X} G \left(\frac{\mu v}{A} - \beta\right)$$

$$+ \frac{\partial}{\partial \mu} \left(G(1 - \mu^2) \left[\frac{v}{A^2} \frac{\partial A}{\partial X} + \mu \frac{\partial \beta}{\partial X} - \frac{\alpha}{Av} \frac{\partial \alpha}{\partial X}\right]\right) \tag{15}$$

$$+ \frac{\partial}{\partial W} \left[G\mu W \left(\frac{v}{A^2} \frac{\partial A}{\partial X} + \mu \frac{\partial \beta}{\partial X} - \frac{\alpha}{Av} \frac{\partial \alpha}{\partial X}\right)\right] = \text{collision terms}$$

where the velocity $v = \alpha W/\sqrt{A^2 + W^2}$, $\beta = \beta_X/A^2$ and $G(X,t,\mu,W) = W^2 F$. The coordinate number density is $N = \int G dW d\mu$. The choice of variable has led to a very simple equation. For a Friedman universe $\beta = 0$, $\alpha = 1$, A is the expansion scale factor. The cooling of the neutrinos for fixed W comes about through the A terms in U^t [see Eq. (14)].

To find the metric functions A, α, β, I must solve the gravity equations (see Centrella and Wilson, 1983). The gravity equations are:

$$\frac{\partial}{\partial X} \left(\frac{A^2 \partial \alpha}{\partial X}\right) - \frac{\alpha A^3}{2} \left(\rho_\alpha + 3K_x^{x\,2} - 2kK_x^x + k^2\right) + A^3 \dot{k} = 0 \tag{16}$$

$$\frac{\partial \beta}{\partial X} = \frac{\alpha}{2} (3K_x^x - k) \tag{17}$$

$$2A \frac{\partial^2 A}{\partial X^2} - \left(\frac{\partial A}{\partial X}\right)^2 + \frac{A^4}{2} \left(\frac{3}{2} K_x^{x2} - kK_x^x - \frac{k^2}{2} + 2\rho_H\right) = 0 \tag{18}$$

$$\frac{\partial}{\partial X} \left[(K_x^x - k) A^3\right] = A^3 S_X \tag{19}$$

$$\frac{\partial A}{\partial t} = \frac{\partial A}{2} (K_x^x - k) + \beta \frac{\partial A}{\partial X} \tag{20}$$

where K_x^x is the X, X component of the extrinsic curvature tensor and k is the trace of the extrinsic curvature, which is taken to be a function of time only.

Equation 16 is from the coordinate condition that the trace of the extrinsic curvature tensor be constant in space. Equation

17 is the coordinate condition that the space part of the metric be isotropic. Equation 18 is the Hamiltonian or energy constraint and equation 19 is the momentum constraint. Equation 20 is an evolution equation arising from the defintions of extrinsic curvature. There is no gravitational dynamics since gravitational shear has been rejected. Equation 16 is solved for both α and k. The requirement of periodicity of all functions leads to equation 16 determining both α and $\dot{k}$.

From the neutrino equations we need the Hamiltonian density, ρ_H, the lapse density, ρ_α, and the momentum density S_x. In terms of the distribution function, G, the neutrino contributions to these densities are:

$$\rho_H^\nu = \int \frac{G\alpha^2 W d\mu dW}{vA(\alpha^2 - \beta\mu vA)} \qquad (21)$$

$$\rho_\alpha^\nu = \int \frac{Gv(A^2+2W^2)d\mu dW}{AW(\alpha^2 - \beta\mu vA)} \qquad (22)$$

$$S^\nu = \int \frac{G\mu W\alpha d\mu dW}{(\alpha^2 - \beta\mu vA)} \qquad (23)$$

The collision terms are comprised of an opacity, K, a velocity of approach, δv, an energy in the matter frame, ε, and a matter temperature, T.

$$\delta v = \frac{\alpha A}{\sqrt{A^2+W^2}} \sqrt{(U_{rad} \cdot U_{mat})^2 - 1} \qquad (24)$$

$$\varepsilon = - U_r \cdot U_m \qquad (\text{. is four vector product})$$

where $U_r \cdot U_m = \alpha U^t_{mat} \left(- \frac{\sqrt{A^2+W^2}}{A} + \frac{\mu\beta W}{\alpha A^2} \right) + \mu W U^z_{mat}$ and U_{mat} is the four velocity of matter which has opacity K and density ρ.

$$\left. \frac{\partial G}{\partial t} \right|_{col} = K\rho\delta v \; (B(\varepsilon/T) - G) \qquad (25)$$

B is a black body distribution appropriate to the particle under consideration.

For ordinary matter the time component of the energy momentum tensor is

$$T_{tt} = (\rho + P + \rho\varepsilon) \, U_t^2 + g_{tt} P \qquad (26)$$

for a perfect gas $P = (\gamma - 1)\rho\varepsilon$ and thus, the rate of change of T_{tt} due to emission and absorption is,

$$\dot{T}_{tt} = \rho\dot{\varepsilon}_m(\gamma U_t^2 - (\gamma - 1)\ (\alpha^2 - \beta^2/A^2)) \tag{27}$$

while the time component of the neutrino energy momentum tensor is

$$\dot{T}^{\nu}_{tt} = \dot{G}\left(\alpha \frac{\sqrt{A^2 + W^2}}{A} - \beta\mu W\right) dWd\mu \tag{28}$$

Thus, for the coupling of each neutrino group we have the energy exchange

$$\rho\dot{e}_m(\gamma U_t^2 - (\gamma-1)(\alpha^2 - \beta^2/A^2)) = \dot{G}_{col}\left(a \frac{\sqrt{A^2+W^2}}{A} - \beta\mu W\right)\Delta W\Delta\mu \tag{29}$$

$$\varepsilon_m = C_v T$$

As in the previous section on neutrinos in spherical geometry we first solve for a reduced coupling by an implicit forward in time differencing of the equations for energy exchange between neutrinos and matter. And as before we also use a partial temperature approach to solve the coupling in a simple but stable manner.

In applications in which we are interested it is presumed that the derivative term of equation (15) is the dominant transport term. We operator split by first solving the equation

$$\frac{\partial G}{\partial t} + \frac{\partial}{\partial X}\ (G\ (\frac{\mu v}{A} - \beta)) = \text{collisions} \tag{30}$$

In difference form this is

$$G_k^n - G_k^o + \frac{1}{\Delta X}\ (\text{Flux}_{k-\frac{1}{2}} - \text{Flux}_{k+\frac{1}{2}}) = k(B_k^o - G_k^n) \tag{31}$$

k is now the reduced coupling coefficient as discussed in the spherical section.
(We will only indicate zonal indices for the independent variable being treated at the moment.)

To evaluate the flux terms we use a monotonic advection scheme developed by R. Barton. Consider the case where the net velocity $\mu v/A - \beta$ is positive, the value of G at the interface between the K-1 and K zones is evaluated as follows.
Let $G_1 = 3/2\ G_{k-1} - 1/2\ G_{k-2}$, $G_2 = 1/2\ (G_k + G_{k-1})$.
If G_{k-1} is greater than G_k choose G_3 as the greater of G_1 and G_2 and $G_{k-\frac{1}{2}}$ as the lesser of G_3 and G_{k-1}. A similar algorithm is used for the case that $\mu v/A-\beta$ less than zero. The flux at the k-1/2 boundary is

$$F_{k-\frac{1}{2}} = \frac{\bar{G}_{k-\frac{1}{2}} \left(\frac{\mu v}{A} - \beta\right)}{1 + \frac{(\lambda_k + \lambda_{k-1})\mu}{2\Delta X}} \tag{32}$$

In the present form of the code the old G's are used to form the F's. In the calculations expected to arise in cosmology the speed of sound and the velocity of neutrinos are comparable so if the hydrodynamics is done explicitly, explicit spacial transport is adequate. The system of equations (31) is now easily solved for the new G. After each advance of G the proper energy is put into or removed from the matter so as to conserve energy.

In order to conserve momentum the material momentum density is changed in accordance with

$$\dot{S}_X^{mat} = \frac{\mu W \alpha \Delta\mu \Delta W}{(\alpha^2 - \beta\mu v A)} \dot{G}_{col} \tag{33}$$

where G_{col} is the part of the change of G associated with emission and absorption.

After equation (30) has been used to advance G for all μ and W then the equation

$$\frac{\partial G}{\partial t} + \frac{\partial}{\partial \mu} \left[(G(1-\mu^2) \left(\frac{V}{A^2} \frac{\partial A}{\partial X} + \frac{\mu \partial\beta}{\partial X} - \frac{\alpha}{Av} \frac{\partial \alpha}{\partial X}\right) \right] = 0 \tag{34}$$

is used to evaluate the G advection in angle. The angle cosine associated with the I th value of G is

$$\mu_I = \frac{-1 + 2 (I - 1)}{(IMAX - 1)} \tag{35}$$

For G_1 and G_{IMAX} the $\Delta\mu$ for phase space integrals is taken as 1/(IMAX-1). For all other zones this $\Delta\mu$ is taken as 2/(IMAX-1). As in the X direction advection we define an advective velocity

$$V_{I-\frac{1}{2}} = (1 - \mu^2_{I-\frac{1}{2}}) \left[\frac{v}{A2} \frac{\partial A}{\partial X} + \frac{\mu \partial\beta}{\partial X} - \frac{\alpha}{Av} \frac{\partial \alpha}{\partial X} \right]. \tag{36}$$

The end fluxes $F_{\frac{1}{2}}$ and $F_{IMAX+\frac{1}{2}}$ are zero. If the velocity V is positive $G_{1+1/2}$ for the calculation of $Flux_{1+1/2}$ is taken as the lesser of G_1 and 1/2 $(G + G_2)$. For I greater than 1 the same monotonic method is used as was used in the X advection to form $G_{I+1/2}$.

$$F_{I+\frac{1}{2}} = \bar{G}_{I+\frac{1}{2}} V_{I+\frac{1}{2}} \tag{37}$$

G is advanced explicitly by the angle and energy advection terms by forming the fluxes with old Gs. If V is negative the above procedure is reversed in I.

Finally the advection in specific momentum, W, is calculated in a manner completely similar to the method used for the angular advection. The velocity for momentum advection is

$$V = \mu W \left(\frac{v}{A^2} \frac{\partial A}{\partial X} + \frac{\mu \partial \beta}{\partial X} - \frac{\alpha}{Av} \frac{\partial a}{\partial X} \right). \tag{38}$$

The fluxes at the maximum and minimum momentum zones are treated similarly to the way the fluxes in angles were handled for the extremes of angle. The splitting of the transport equation of the angle and momentum terms should be good in practice because the angle and momentum transport velocities (see equation 38) are given in terms of derivatives of metric coefficients which in turn obey second order elliptic equations. The transport velocities should be smooth functions.

Thus far only a few calculations have been made with this program. A calculation was made with a very large amplitude perturbation in matter density where the length of the calculational space was about ten times the horizon size at the initial time and the opacity and density were such that the coupling was strong per cycle, and the matter temperature was about ten times the neutrino rest mass. The calculation was run until the matter and neutrinos were decoupled, the calculational space was much less than horizon size, and the neutrinos were non relativistic. While nothing unseemly appeared in this calculation it was such a complicated test that the code accuracy is not tested. Only code rugosity was tested. We will now discuss a much simpler problem that can be compared to analytic results. We started with a non-relativistic neutrino distribution with no matter, a very small sinusoidal amplitude in the density of neutrinos, a problem length, L, small compared to the horizon and the thermal velocity low compared to the expansion velocity,

$$T << L \frac{\partial \log A}{\partial t}^{2}$$

Since the matter is relatively cold the perturbation should grow in the same manner as a dust perturbation. That is, after a small initial time for the distribution to adjust into the fundamental growth mode, the amplitude should grow as $t^{2/3}$. From figure 1 we see this relation holds quite well over a large range of time. We have only plotted amplitude for the time interval over which nonlinear gravitational effects should be small. While figure 1 appears very simple, when one looks at the details of the distribution function it goes through large changes in structure.

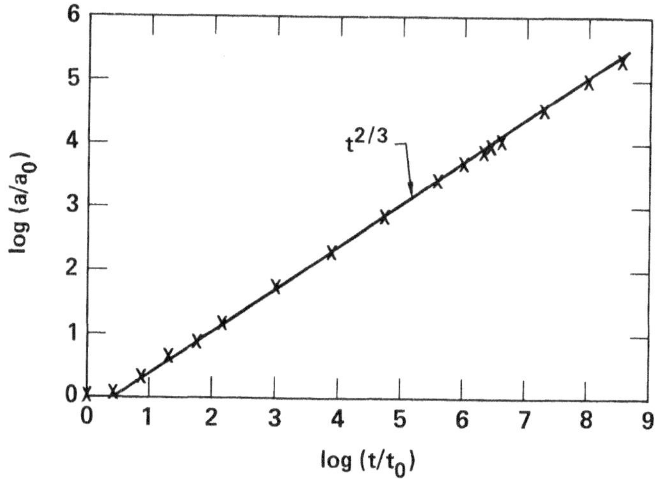

Figure 1. In this graph the amplitude of a density perturbation is plotted versus time. The solid line is a fit of $t^{2/3}$ to the amplitude as calculated by the computer program. The computer calculated amplitudes are indicated by the crosses.

In the panels of figure 2 we show how several averages over the distribution function change with time.

The first graph of A versus Z at time 1.4458×10^5 is the first graph to show any non-uniformity among the three types of graphs. The graph of B versus Z at time 1.0474×10^6 is the first time non-uniformity shows up on the B type graph. At time 7.5887×10^6 the first visable spreading of the curves in C type plot can be seen. At time 2.0517×10^8 all three types of graphs are shown. The distributions have a complicated structure at this time, but the density amplitude of the perturbations is still very close to the $t^{2/3}$ curve of figure 1.

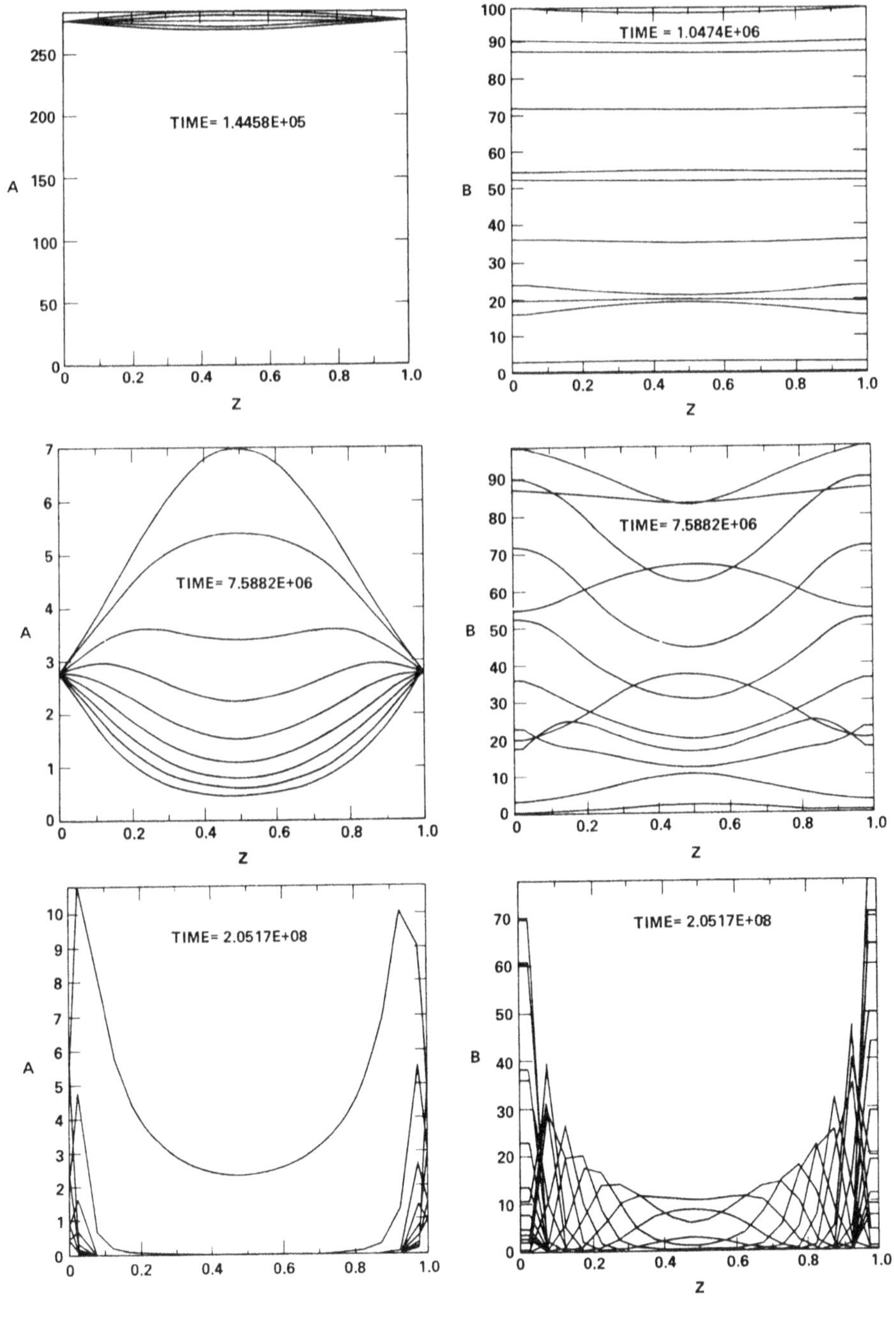

Figure 2 (caption next page)

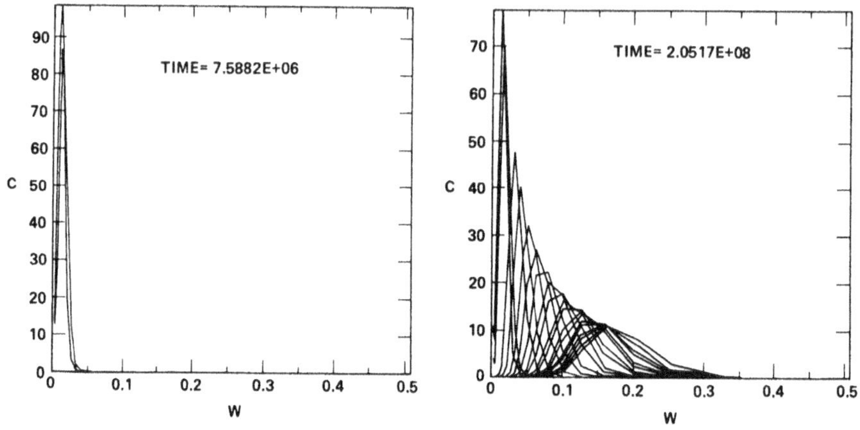

Figure 2 cont'd. This set of graphs show how several averages over the distribution function evolve with time. The quantity Z is always the spacial coordinate and W is the specific momentum of the particles. The quantities plotted on the ordinate scale are defined as follows:

$A(\mu,Z) = \int GdW$ plotted for

$B(W,Z) = \int Gd\mu$ plotted for

$C(Z,W) = \int Gd\mu$ plotted for the several Z zones

The times on the graphs correspond to the ratio t/t_o of Figure 1.

References

Lindquist, R.W.: 1966, Ann. Phys., 37, 487
May, M.M., and White, R.H.: 1966, Phys. Rev., 141, 1232
Wilson, J.R.: 1971, Astrophys. J., 163, 209

NUMERICAL RELATIVISTIC GRAVITATIONAL COLLAPSE WITH SPATIAL TIME SLICES

Charles R. Evans
Department of Astronomy
University of Illinois
Urbana, IL 61801
and
Lawrence Livermore National Laboratory
P.O. Box 808 L-35
Livermore, CA 94550

Larry L. Smarr*
Departments of Astronomy and Physics
University of Illinois
Urbana, IL 61801

James R. Wilson
Lawrence Livermore National Laboratory
P.O. Box 808 L-35
Livermore, CA 94550

1. INTRODUCTION

In the usual flatspace hydrodynamics, there is a unique split between Eulerian and Lagrangian observers. An Eulerian observer is one at rest in space, while the Lagrangian observer is one at rest in the fluid. We can define an observer at an event in spacetime by giving his 4-velocity at that event. The path through spacetime taken by the observer is called the timeline of that observer. In flat Minkowski spacetime (the spacetime of special relativity), the Eulerian observer's 4-velocity is exceedingly simple, because the time slices (3-spaces representing a given instant of time) are parallel flat 3-planes. Since the Eulerian

*Alfred P. Sloan Fellow

K.-H. A. Winkler and M. L. Norman (eds.), Astrophysical Radiation Hydrodynamics, 491–529.

observer is at rest, his 4-velocity is normal to the time slice. He is not shearing, converging, accelerating, or rotating relative to nearby Eulerian observers. In contrast, the Lagrangian observer at the same event is doing all these things, relative to nearby Lagrangian observers. While the timelines of the Lagrangian observers, following the fluid particles, are complicated curved trajectories through spacetime, the Eulerian observer's timelines are straight and parallel to each other.

This situation changes drastically in curved spacetime, where Einstein's equations of general relativity must be solved in tandem with the hydrodynamic equations. The preferred flat time slices of Minkowski spacetime no longer exist, therefore the preferred Eulerian observers no longer exist. One must choose some smooth time slicing of spacetime, which will then define a set of Eulerian observers at rest in those time slices (Smarr, 1977; Smarr and York, 1978b). Thus, the first thing we lose is the uniqueness of the Eulerian observers. Furthermore, the resulting Eulerian observers will be shearing, converging, and accelerating in general. They will not be rotating since they are normal to the time slices. The 3-spaces that the Eulerian observers are at rest in will no longer be flat, but instead will have curvature (measured by the 3-Ricci tensor R_{ij}). The spatial partial derivatives of flatspace must be replaced by covariant derivatives with Christoffel symbols. Both the timelines of the Eulerian observers and of the Lagrangian observers are curved trajectories in the curved spacetime.

However, having accepted this conceptual generalization of the notion of Eulerian observers, Einstein's equations can be written in a form not very different from those of ordinary self-gravitating hydrodynamics. That is, there are evolution equations containing the first time derivatives of some basic variables coupled to elliptic (or parabolic) equations for some auxiliary variables. Once one has this form of Einstein's equations, one can apply standard finite-difference techniques on a spatial grid to reduce the coupled set of partial differential equations to a large set of algebraic equations.

Numerical relativity thus discretizes the spacetime continuum manifold into a spacetime lattice, the events of which are organized into a spacetime coordinate system. The unknown variables are the components of the 4-metric tensor at the lattice points. As the lattice is refined to smaller scales, the solutions converge to the continuum solution of Einstein's differential equations.

In practice, this lattice of events is built up from an initial data hypersurface by constructing a family of hypersurfaces to the future of that data. There are two fundamentally different approaches to this construction: the spacelike and the null. In the first, the spacetime is built up from spacelike hypersurfaces starting on a Cauchy initial value surface (for a detailed definition see Eardley and Smarr, 1979). In the second, characteristic initial data is evolved using null hypersurfaces. Our paper deals with the former case using spacelike hypersurfaces. For a method which illustrates the latter case see the companion article in this volume by Stewart.

Before going into the rather complicated details of numerical relativity, we will physically motivate why such an all-out approach is necessary. Our primary objective for this code is to calculate the dynamics and emitted gravitational radiation from violent, strongfield, asymmetric, and high (internal) velocity gravitational collapse. The calculation of such energetic sources is certainly out of reach of analytic techniques. Yet it is precisely these violent events which are likely to be the most efficient sources of gravitational radiation and therefore (provided they occur frequently enough) may be the first sources detected experimentally.

Some numerical calculations, e.g. Müller (1982) and Müller and Hillebrandt (1981), have used a Newtonian hydrodynamics code with Newtonian gravity to calculate very nonspherical gravitational collapse. Then they use the standard quadrupole formula to calculate the gravitational wave form and energy loss by gravitational radiation. Their approach is to carefully treat the microphysics, but to use approximations for the gravity. Such an approach loses two major pieces of gravitational physics.

First, as the gravitational field becomes relativistic, i.e. when $\phi/c^2 \ll 1$ no longer holds, then the scalar potential ϕ representation of the gravitational field becomes invalid and the tensor representation of Einstein must be used. Another way to say this is that one must replace the gravitational acceleration $\nabla\phi$ in the hydro equations by a complicated family of potential gradients. These new terms are significant in determining the flow of the fluid in relativistic regimes. If these terms are not included, the flow calculated by a Newtonian code can begin to deviate significantly from the true flow. We call these parts of the gravitational field "longitudinal." A code which corrects this difficulty by solving the nonradiative Einstein equations is discussed by Wilson (1975).

Second, there is no useful method for calculating the gravitational radiation emitted from a relativistic problem except by calculating the full Einstein equations as we do. The use of the approximate method called the "quadrupole formula" can only be justified in nonrelativistic regimes. This formula states that the angle averaged gravitational radiation flux is given by

$$dE/dt = 0.3\ (\dddot{I}_{zz})^2 \quad (1)$$

and the gravitational wave amplitude $h_+^{(Q)}$ is given by

$$h_+^{(Q)} = 1.5r^{-1}\ \ddot{I}_{zz}\ \sin^2\theta \quad (2)$$

where a raised dot denotes time derivative and units are G = c = 1. The quadrupole moment itself is given by

$$I_{zz} = 2\ \pi \int dr\ d\theta\ r^4\ \sin\theta\ (\cos^2\theta - 1/3)\ \rho \quad (3)$$

where the volume integral is taken over the matter density $\rho(r,\theta)$ expressed in spherical coordinates.

This quadrupole formula was derived by Landau and Lifshitz (1941) as an approximation to the Einstein equations. As Thorne (1980a, p. 289) carefully points out, the derivation breaks down if either the "slow motion" $v/c \ll 1$ or "weak field" $\phi/c^2 \ll 1$ restrictions are violated. For self-gravitating bodies where $\phi \sim v^2$, these are seen to be physically self consistent restrictions. As Thorne (1980a) points out neutron stars are therefore outside the regime of validity of the quadrupole formula.

In gravitational collapse to nuclear densities or beyond, one can get velocities of order $v/c \sim 0.2$, which violates the "slow motion" assumption. Such high velocities can cause phase shifts in the emitted radiation. This can suppress the actual amount of radiation generated, over that expected by the quadrupole moment, by destructive interference of waves. For calculations of this effect see Nakamura and Sasaki (1981) and Haugan et al. (1982).

Although the surface gravitational potential of stellar interiors at nuclear density may be "only" $\phi/c^2 \sim 0.15$, the central value can reach $\phi/c^2 \sim 0.5$, clearly violating the "weak field" condition. For such high gravitational fields, the very definition of the quadrupole moment itself, equation (3), becomes ill-defined, since the integral is not covariantly defined. The corrections which curved space makes to the lengths and time derivatives used in equation (3) are of order ϕ/c^2; their appearance as high powers in equation (1) renders the definition meaningless. For detailed examples see Evans (1984b). More intricate and sophisticated definitions of the quadrupole moment exist, which do not depend on a volume integral over strong field regions, although these still require slow motion. For a comprehensive analysis see Thorne (1980b).

In conclusion, there seems to be no "short cut" to studying the physics of relativistic core collapse. Nature is well described here only by the full theory of general relativity. The only way known to solve the theory in the dynamic strong field regime is numerical relativity. Therefore, our approach is to calculate the gravitational physics correctly first, then later add more detailed microphysics.

The outline of our paper is as follows. In section 2 we write out the Einstein equations as they appear on a series of spacelike hypersurfaces (the "3+1" approach). We describe how the coordinate freedom (general covariance) shows up in this form of Einstein's equations. The initial value problem and the closely related problem of the constraint equations in general relativity are discussed.

Our code is discussed in section 3. It will handle any axisymmetric collapse problem as long as there is no rotation. For two different examples of how to handle rotation, see Nakamura, et al. 1980; Nakamura, 1981; and Bardeen and Piran, 1983. The particular coordinate system we use yields a set of coupled differential equations (elliptic plus hyperbolic) which we must solve for the gravitational field. Some aspects of the finite differencing of this system of partial differential equations is considered.

Finally a brief conclusion and prospectus for the future is contained in section 4.

2. DYNAMICAL APPROACH TO GENERAL RELATIVITY: GENERAL TECHNIQUES

a) Einstein equations in (3+1) formalism

The approach we will take for the Einstein equations is to evolve them as a Cauchy problem, just as one normally does for hydrodynamics. That is, one starts at t=0, in an initial data 3-space, on which one gives the starting values of the physical variables. Then one integrates forward a small amount of time using the evolution equations. This gives a sequence of initial data 3-spaces, the union of which gives the spacetime. There are only two novelties in this procedure when one is solving Einstein's equations, although both novelties are also present in Maxwell's equations of electrodynamics. The first is that the initial data for Einstein's equations are constrained, i.e. there are several elliptic type differential equations that the variables must satisfy on each time slice. The second is that one must maintain a coordinate gauge as one advances the evolution equation. In practice, this means solving some more elliptic equations on each time step.

Consider an initial data 3-space for Einstein's equations. The fundamental variable representing the gravitational field on that time slice is the 3-metric γ_{ij}. This is a symmetric 3 by 3 matrix of functions, with six independent components. Some of these functions may be identically zero in special cases. The Einstein equations yield evolution equations for those metric functions which are nonzero. Since Einstein's equations are a second order hyperbolic system, we must also have a "velocity-like" variable to be conjugate to the 3-metric which is the "position-like" variable. This turns out (Arnowitt, Deser, and Misner, 1962) to be the extrinsic curvature tensor K_{ij} which tells how the 3-space is embedded in the spacetime (Wheeler, 1963).

However, there is another interpretation (Smarr, 1975; Smarr, 1977; Smarr and York, 1978b; Smarr, Taubes, and Wilson, 1980) of K_{ij} which may appeal more to researchers in hydrodynamics. The extrinsic curvature tensor turns out just to be the negative of the expansion tensor formed from the Eulerian observer's 4-velocity. That is, if we think of the timelines of the Eulerian observers as the timelines of some imaginary fluid in the spacetime, then the trace of K_{ij} is the rate of expansion of the fluid locally, while the tracefree part of K_{ij} is the negative of the local shear tensor of the fluid. We will set our time slicing gauge by demanding that the Eulerian observer fluid be incompressible (Lichnerowicz, 1944), which will force the trace of K_{ij} to zero. The Einstein equations will then yield evolution equations for the tracefree part of K_{ij}.

We now write out formally these two evolution equations. The time derivative of the 3-metric is given by:

$$\partial_t \gamma_{ij} = \underbrace{-2\alpha K_{ij}}_{\text{shear}} + \underbrace{D_i\beta_j + D_j\beta_i}_{\text{convective derivatives}} \tag{4}$$

while the time derivative of K_{ij} is given by:

$$\begin{aligned}\partial_t K_{ij} = \alpha\, [&\underbrace{R_{ij}}_{\text{tides}} \underbrace{- 2 K_{im} K^m_j + K^m_m K_{ij}}_{(\text{shear})^2}] \\ &\underbrace{- D_i D_j \alpha}_{\text{differential acceleration}} + \underbrace{\beta^m D_m K_{ij} + K_{im} D_j \beta^m + K_{mj} D_i \beta^m}_{\text{convective derivatives}} \\ &\underbrace{- 8\pi\, \alpha\, [S_{ij} + 1/2\, \gamma_{ij} (\rho_H - S^m_m)]}_{\text{matter stress}}\end{aligned} \tag{5}$$

where $c = 1$, $G = 1$. From the first equation, one can see that K_{ij} acts as a "velocity" for γ_{ij}. We will discuss the terms one by one.

The derivative operator D_i is the spatial covariant derivative. This is the same as the partial derivative on a scalar quantity, but has additional Christoffel symbols when acting on a vector:

$$D_i\beta_j = \partial_i\beta_j - \Gamma_{ij}^{k}\beta_k. \tag{6}$$

The Christoffel symbol Γ^i_{jk} is built out of the 3-metric by:

$$\Gamma^k_{ij} = 1/2\,\gamma^{km}\,(\partial_i\gamma_{mj} + \partial_j\gamma_{im} - \partial_m\gamma_{ij}), \tag{7}$$

where γ^{ij} is the matrix inverse of γ_{ij}. Thus, the Christoffel symbol is itself nonlinear in the 3-metric. Even more nonlinear is the 3-Ricci tensor in the second equation:

$$R_{ij} = \partial_m\Gamma^m_{ij} - \partial_j\Gamma^m_{im} + \Gamma^m_{nm}\,\Gamma^n_{ij} - \Gamma^m_{nj}\,\Gamma^n_{im}. \tag{8}$$

The first two terms of the Ricci tensor contain the second spatial derivatives of γ_{ij}. One can see this more clearly by rewriting (see, e.g. Smarr, 1979) the Ricci tensor as:

$$\begin{aligned} R_{ij} = -1/2\,\gamma^{km}\,(\partial_i\partial_j\gamma_{km} + \partial_k\partial_m\gamma_{ij} - \partial_i\partial_m\gamma_{kj} \\ - \partial_k\partial_j\gamma_{im}) - \gamma^{km}\gamma_{pq}\,(\Gamma^p_{km}\Gamma^q_{ij} - \Gamma^p_{kj}\Gamma^q_{im}). \end{aligned} \tag{9}$$

Therefore, if one substitutes $\partial_t\gamma_{ij}$ for K_{ij} (i.e. using eqn. (4)) in equation (5), one sees roughly that the resulting equation contains a wave operator on γ_{ij}. One of the great difficulties of numerical relativity is that in the presence on all the other terms in equations (4) and (5), the hyperbolic character of the evolution equations is much more difficult to exhibit.

In fact, the equations (4) and (5) can be rewritten in either completely second order form or in completely first order form. All three of these forms have been finite differenced and used for problems in numerical relativity (for references see Smarr, 1979). In our application using these techniques in section 3, we directly utilize the form of the equations given above.

Physically, the Ricci tensor represents the gravitational tidal stresses which cause a time rate of shear ($\partial_t K_{ij}$) in the Eulerian observers. Note that in addition to the second spatial derivative terms in Ricci, there are terms quadratic in the Christoffel symbols, which mean terms quadratic in $\partial_i\gamma_{ij}$. These have a precise analog in the next term in equation (5), which is quadratic in K_{ij}, or in $\partial_t\gamma_{ij}$. It is already clear then, that we are going to have trouble with a time explicit finite-differencing code in getting the spatial and time derivative terms in $\partial_t K_{ij}$ centered correctly so that the balancing between the two principal nonlinear terms occurs correctly in strong field regions.

Note that the nonlinear terms in the evolution equation are of two kinds (Smarr, 1979). First, the velocity matrix in the second order wave operator on γ_{ij} depends on the 3-metric ($\sim \alpha^2 \gamma^{ij}$). Second, there are the nonlinear terms quadratic in the first space and time derivatives of the 3-metric. There is no detailed analysis of how these terms cause steepening or dispersion of wave forms in strong field regions. Furthermore, because these are tensor equations, unlike the scalar Korteweg-de Vries equation, the nonlinear terms can be reshuffled by index manipulation. For instance, it is often useful to rewrite the evolution equation (5) for $\partial_t K^i_j$ which eliminates the quadratic term $-2 K_{im} K^m_j$ in equation (5). Then on a tr(K) = 0 slice, there is no term quadratic in the time derivatives.

The matter source term in equation (5) is the material stress as seen in the rest frame of the Eulerian observers. Again, this stress causes a time rate of shear ($\partial_t K_{ij}$) in the Eulerian observers. The rest mass density ρ, the specific internal energy ε, and the pressure p of a perfect fluid are all defined in the Lagrangian rest frame of the fluid. These quantities must be Lorentz boosted to the Eulerian frame (by matter velocity V_i) to get the values of the density, momentum, and stress that are seen by the Eulerian observer. The results are (Smarr, Taubes, and Wilson, 1980):

$$\rho_H = \rho h \Gamma_2^{\ 2} - p, \qquad \Gamma_2 \equiv (1 - V_i V^i)^{-1/2} \tag{10a}$$

$$S_i = -\rho h \Gamma_2^{\ 2} V_i, \qquad h = 1 + \varepsilon + p/\rho \tag{10b}$$

$$S_{ij} = p\gamma_{ij} + \rho h \Gamma_2^{\ 2} V_i V_j . \tag{10c}$$

The matter terms are evolved by the general relativistic equations of hydrodynamics. We will not treat this part of the problem in this paper since it is exhaustively covered elsewhere (Wilson, 1979; Dykema, 1980; Smarr, Taubes, and Wilson, 1980; Nakamura, et al 1980; Evans, 1984a,b). The equations can be written in a form which resembles very closely the Newtonian equations of Eulerian hydrodynamics. The manner in which the transport terms [e.g. $\partial_i(AV^i)$] are differenced varies from group to group. For our approach to differencing the hydrodynamic equations see Dykema (1980). This technique is also used for treating the transport terms that appear in the reduction of equation (5) in our model (see section 3).

b) Gauge conditions

Finally, we come to the gauge terms in equations (4) and (5). The time gauge is set by the scalar lapse function α. Although two successive time slices are separated by a constant coordinate time interval Δt, the proper time separating the two slices varies over the 3-space. At any point in the 3-space the proper time interval to the next slice is $\alpha\Delta t$. Thus, the lapse function acts as a spatially variable time step, which can slow down the proper time evolution in strong field regions, thus avoiding running the time slices into the singularities inside black holes. Because the proper time step varies in space, the Eulerian observers find themselves accelerated by an amount $\partial_i \ln\alpha$, even though they remain at rest. This acceleration is required to counteract the gravitational acceleration which would otherwise make them move inward. The lapse function appears in two manners in the evolution equations (4) and (5). First, it multiplies certain terms to correct the coordinate time derivative to proper time. Second, the differential spatial acceleration $(D_i D_j \alpha)$ causes a time rate of change of the shear of the Eulerian observers.

It is often convenient to have a moving spatial coordinate system, e.g. a corotating system. This means that the timelines along which the spatial coordinates are constant do not coincide with the timelines of the Eulerian observers. At t=0 one can use any convenient spatial coordinate grid (cartesian, cylindrical, spherical, etc.) to label all points in the initial 3-surface. The coordinate timeline through a given point in the initial surface will intersect all future time slices and label one point in each time slice with the spatial coordinate that it had at t=0. In this way, the coordinate time lines provide a map which lays down a copy of the original coordinate grid on successive time slices. Just as in the case of matter discussed above, there is a local boost velocity which transforms one from the Eulerian rest frame to the coordinate rest frame. This is called the shift vector β_i. It acts in the evolution equations as a convective derivative, just as the matter velocity appears in the Euler equations of hydrodynamics.

Thus we see that the functions (α, β_i) which specify, respectively, the time and spatial coordinates are integrated directly into the partial differential evolution equations (4) and (5). However, the Einstein equations leaves one the freedom to decide which time and space coordinates are used on any given problem. That is, we must specify α and β_i on each time slice, in order to be able to evolve (γ_{ij}, K_{ij}) into the correct coordinate system.

The equation we use to specify α is the trace of equation (5):

$$\partial_t[\mathrm{tr}(K)] = \beta^i \partial_i[\mathrm{tr}(K)] - \Delta\alpha + \alpha[K_{ij}K^{ij} + 4\pi(\rho_H + S^i_i)]. \qquad (11)$$

This equation gives the time rate of change of the expansion [tr(K)] of the Eulerian observers. For noncosmological spacetimes, there is no overall expansion of the physical space, so it seems natural to set the expansion equal to zero. For the cosmological case see Centrella and Wilson (1983, 1984) and Piran (1980). As remarked above, Lichnerowicz (1944) points out that setting tr(K) = 0 will make the imaginary fluid of Eulerian observers incompressible. In order to keep the trace of K_{ij} always zero, equation (11) tells us that we must solve a linear elliptic equation for α on each time slice:

$$\Delta\alpha - \alpha[K_{ij}K^{ij} + 4\pi(\rho_H + S^i{}_i)] = 0. \qquad (12)$$

This is analogous to how imposing the radiation gauge in electromagnetism on each time slice leads to an elliptic equation for the scalar potential (see Smarr and York, 1978a).

The time slicing condition in equation (12) is termed "maximal slicing" in the literature, because it maximizes the volume of the 3-space relative to other slicings. Its major practical advantage is that it allows the evolution to penetrate inside of black holes, while slowing down the evolution in those regions enough to almost always avoid hitting singularities [see, e.g. Estabrook, et al. 1973; Smarr and York, 1978b; Eardley and Smarr, 1979; Shapiro and Teukolsky, 1980; Evans, 1984a,b]. Other possible slicings are discussed in Smarr and York, 1978b; Smarr, 1979; Eardley, 1979; Smarr, Taubes, and Wilson, 1980; Nakamura, 1981; and Bardeen and Piran, 1983.

With the time slicing set by the choice of the lapse function, one has to decide how to move the spatial coordinates from one slice to the next by choosing the shift vector. This choice is secondary to the choice of the time slicing, because if a bad time slice hits a singularity, then no further evolution is possible. However, given a good time slicing, one can lay spatial coordinates down on it using a variety of shift vectors. Equation (4) can be used to determine the conditions on β_i which must hold on each time level to enforce the desired coordinate condition. A rather complete discussion of the alternatives exists in the literature (Smarr and York, 1978b; Smarr, 1979; Wilson, 1979; Smarr, Taubes, and Wilson, 1980; Bardeen and Piran 1983; and Evans 1984b), as well as comparisons of different shift vectors used on the same time slicing (Eppley, 1979; Piran, 1980). One particular choice will be illustrated in section 3.

In summary, the Einstein equations yield a quasilinear system of evolution equations for (γ_{ij}, K_{ij}), which are intimately coupled to a set of (typically) elliptic equations for (α, β_i). The solutions to one set of equations are coefficients in the other set of equations, and vice versa. Therefore, they must be solved together, step by step into the future. This requirement of solving for the spacetime coordinates, simultaneously with the physical variables being evolved in those coordinates, is the major conceptual and numerical difference from ordinary flatspace problems.

c) Initial value equations and constraints

As mentioned above, the evolution equations (4) and (5) are only part of Einstein's equations. They are the analogs to the Maxwell equations for the time derivatives of the electric and magnetic fields (see e.g. Smarr and York, 1978a). However, there are also analogs of the constraint equations $D^i E_i = 4\pi\rho$ and $D_i B^i = 0$, which contain no time derivatives. The direct analog is the linear momentum constraint equation:

$$D^i[K_{ij} - \gamma_{ij}\mathrm{tr}(K)] = 8\pi S_j. \tag{13}$$

Because gravity is spin-2 instead of spin-1 like electromagnetism, there is an extra index on the constraint equation. In addition, there is a nonlinear constraint equation in general relativity which has no direct analog in electromagnetism, termed the Hamiltonian constraint:

$$R^i{}_i + [\mathrm{tr}(K)]^2 - K_{ij}K^{ij} = 16\pi\rho_H. \tag{14}$$

The existence of the constraint equations means that not all components of (γ_{ij}, K_{ij}) are true dynamical degrees of freedom (just as for A_i in electromagnetism). There are twelve functions in the pair (γ_{ij}, K_{ij}). Four of these can be fixed by the coordinate conditions imposed on (α, β_i). Four more are fixed by the four constraint equations. This leaves two free functions in each of γ_{ij} and K_{ij}. These represent the two degrees of freedom possessed by a propagating transverse gravitational wave field.

There are two fundamentally different ways to evolve numerically the Einstein equations (see Piran, 1980, for a complete discussion). The unconstrained evolution solves equations (4) and (5) for (γ_{ij}, K_{ij}) and inserts them into equations (13) and (14) as a check on the accuracy of the evolution. The constrained evolution uses equations (13) and (14) to solve for certain constrained pieces of (γ_{ij}, K_{ij}), while using a subset of equations (4) and (5) to evolve "dynamical" components of (γ_{ij}, K_{ij}). It is an unsolved problem as to which method is superior. Bardeen and Piran (1983) describe a partially constrained evolution scheme.

Our method, given below, uses the fully constrained approach (see Evans 1984a,b).

When the constraint equations are written in the form of equations (13) and (14), they are not very useful for numerical work, although some attempts have been made to use them (see Piran, 1980). An elegant method (see, e.g. York, 1979) from a theoretical as well as a numerical perspective, is to recast equations (13) and (14) as elliptic equations on "longitudinal" pieces of (γ_{ij}, K_{ij}). The 3-metric is split into a conformal factor ϕ and a conformally related 3-metric:

$$\gamma_{ij} = \phi^4 \hat{\gamma}_{ij}. \tag{15}$$

The conformal factor is the "Newtonian" part of the gravitational field, which determines the mass of the gravitating system, while the conformally related metric contains the gravitational radiation information (York 1971, 1972). The extrinsic curvature is scaled by the conformal transformation as

$$K^{ij} = \phi^{-10} \hat{K}^{ij} \tag{16}$$

and split into transverse and longitudinal parts:

$$\hat{K}^{ij} = \hat{K}_T^{ij} + \hat{K}_L^{ij}. \tag{17}$$

In what follows we assumed maximal slicing is applied. The longitudinal part can be derived from a vector potential W^i:

$$\hat{K}_L^{ij} = (\hat{L}W)^{ij} = \hat{D}^i W^j + \hat{D}^j W^i - 2/3\, \hat{\gamma}^{ij} \hat{D}_k W^k. \tag{18}$$

Here $\hat{D}_i$ is the covariant derivative built from $\hat{\gamma}_{ij}$.

The procedure to solve for a constrained initial data 3-space is as follows (York, 1979; Evans, 1984a,b). First, pick the unconstrained data $(\hat{\gamma}_{ij}, \hat{K}_T^{ij})$ and the "bare" matter variables $(\rho_H, \hat{S}^i)$. By substitution of equations (15), (16), (17), and (18) into equation (13) one transforms the momentum constraint into a linear vector elliptic equation:

$$(\hat{\Delta}_L W)^i = \hat{D}_i (\hat{L}W)^{ij} = \hat{D}_k \hat{D}^k W^i + 1/3\, \hat{D}^i(\hat{D}_k W^k) + \hat{R}_k{}^i W^k = 8\pi\, \hat{S}^i, \tag{19}$$

which one solves for the potential W_i. Here, $\hat{R}_{ij}$ is the Ricci tensor built out of $\hat{\gamma}_{ij}$. With $\hat{K}_T^{ij}$ and W_i, one can construct $\hat{K}^{ij}$, using equations (17) and (18). Substituting equations (15) and (16) into equation (14) one transforms the Hamiltonian constraint into a quasilinear elliptic equation:

$$\hat{\Delta}\,\phi = (\phi/8)\,[\hat{R}_i^{\ i} - 16\pi\,\hat{\rho}_H\,\phi^{-2} - (\hat{K}_{ij}\hat{K}^{ij})\,\phi^{-8}], \tag{20}$$

which one solves for the conformal factor. Here, $\hat{\Delta}$ is the covariant Laplacian built out of $\hat{\gamma}_{ij}$. The conformal factor ϕ can now be used to "dress" all the "bare" free data by equation (15), (16) and:

$$\rho_H = \phi^{-6}\,\hat{\rho}_H\,,\quad S^i = \phi^{-10}\,\hat{S}^{\,i}. \tag{21}$$

By construction, the data (γ_{ij}, K_{ij}, ρ_H, S^i), now satisfies the constraint equations. How this scheme can be used to provide a constrained evolution, will be the subject of section 3.

Our emphasis will not be on the astrophysical significance of the calculations, but rather on the form of the Einstein equations and how one can finite difference them. We will only discuss our dynamical axisymmetric code. For descriptions of the three possible one-spatial dimensional codes see: spherical (Misner and Sharp, 1964; May and White, 1967; Estabrook, et al. 1973; Wilson, 1979; Shapiro and Teukolsky, 1979, 1980), planar (Centrella and Wilson, 1983, 1984), and cylindrical (Piran, 1980). For discussions of other two-spatial dimension relativity codes see e.g. Smarr, 1975; Smarr, 1979; Smarr, 1984; Eppley, 1979; Nakamura, et al. 1980; Nakamura, 1981; and Bardeen and Piran, 1983.

3. AXISYMMETRIC MODEL

a) Symmetries, gauge, and differential equations

In this section we describe the equations and numerical method used in a computer code we have developed to study, in a fully self-consistent manner, general relativistic hydrodynamic systems. Our calculation is restricted to configurations exhibiting axisymmetry, no axial rotation, and equatorial plane symmetry. This discussion is intended to provide an update on some of the progress we have made since Wilson (1979) and Dykema (1980). The emphasis here is on presenting for the numerical astrophysicist the form of the differential equations to be solved and a survey of the numerical techniques employed, particularly several new aspects, which have allowed us to obtain stable and regular solutions. A more complete description of the mathematical algorithm can be found in Evans (1984a,b) which includes the hydrodynamic equations not given here.

A numerical method for calculating axisymmetric models has been presented previously (Wilson, 1979) using cylindrical spatial coordinates (ρ, z, ϕ). Shortly thereafter we switched to using spherical coordinates (r, θ, ϕ) which are better suited to the study of black hole formation (Dykema, 1980). While cylindrical coordinates would be better for calculating a two neutron star collision, the new code can be used to simulate problems such as the bounce of aspherical neutron stars, the collapse of stars to form black holes, and the formation of black holes by pure gravity waves (Brill, 1959). The reason for this change has been the need to well cover the "throat" which forms in the interior of the black hole once an horizon appears. The proper length of the throat increases monotonically with external time, forcing us to put more of our grid points into the black hole (Estabrook, et al. 1973; Shapiro and Teukolsky, 1980).

These collapse calculations are very "stiff"; there are disparate length scales in the problem which require many zones to accurately simulate. For example in the exterior of a newly formed black hole, the gravitational radiation emitted during the ring down has a far field wavelength of approximately 20M where M is the mass of the hole. It is necessary therefore to have the grid extend out to a distance on the order of 100M in order to be able to analyze the radiation in the wave zone. When the black hole forms, the coordinate radius of the surface of the collapsing star goes to zero exponentially with external time. This effect is due to an impending coordinate singularity brought about by our 3-gauge choice combined with maximal slicing. Maximal slicing exponentially slows the advance of proper time which effectively freezes the star at a finite proper radius. Nonetheless the interior of the star must be well zoned which requires many radial zones to span the two regions. This represents an extreme situation which our code must handle. For further discussion of the crucial role of accuracy and grid resolution see Smarr (1984).

The methods used for solving the hydrodynamic equations have been virtually unchanged, despite the new coordinates, since Wilson (1979). However, the techniques to solve for components of the gravitational field have undergone significant revision. We will therefore only discuss these techniques, which are rather different from those described in Wilson (1979), used in the new code for the solution of the gravitational field.

The time coordinate is chosen by employing maximal slicing, equation (12). In addition, we pick the spatial coordinate gauge by choosing conditions on the shift vector which fix the form of the 3-metric in time. There are several choices of this type and the one we have used fixes the form of the line element to be:

$$ds^2 = -\alpha^2 dt^2 + A^2(dr + \beta^r dt)^2 + A^2 r^2 (d\theta + \beta^\theta dt)^2 + B^2 r^2 \sin^2\theta \, d\phi^2, \tag{22}$$

where boundary conditions are employed to make (r, θ, ϕ) spherical coordinates. The advantage of such simplifying gauges is that the shift can be used to minimize the number of independent components in the 3-metric, which vastly simplifies the number of terms involved in calculating 3-covariant derivatives.

The constraint equations (13) and (14) are superfluous during an evolution in the sense that once they have been used to obtain consistent initial data, the evolution equations (4) and (5) guarantee their continued satisfaction at subsequent times. Numerically this will not be true, as the degree to which the constraints are satisfied tends to decrease with time due to numerical drift. Since we do not understand the stability of non-Einstein data (data off the constraint surface), we have chosen to invert all the constraint equations on each successive time level and evolve a reduced dynamical problem (fully constrained approach). The constrained approach, using a simplifying gauge choice has the further advantage that the number of evolved components of the gravitational field will reflect precisely the number of dynamical degrees of freedom. We can therefore choose to evolve components which are closely related to gravitational radiation variables in the weak field, wave zone. With our symmetries only one dynamical degree of freedom exists (one independent polarization state) and the component of the metric we evolve is immediately relatable to the weak field transverse-traceless gravitational radiation amplitude h_+^{TT} $(r \pm t, \theta, \phi)$ (Evans, 1984b).

The technique used for solving the momentum constraints uses the full York (1973) transverse-traceless decomposition of the extrinsic curvature tensor. This includes inverting a transverse part, ordinarily not done on the initial slice, but which must be done on subsequent time slices to provide the fully constrained approach. In addition, the solution for the longitudinal part on the initial slice provides more physically meaningful initial data.

In what follows we give the equations for the gravitational field in the form most appropriate for finite differencing. The presentation of the differential equations will be brief since their derivation can be found elsewhere (Evans, 1984a,b). Also we do not discuss here the hydrodynamic equations of motion, though hydrodynamic quantities will enter as source terms in the other equations. Of the possible metric simplifying gauges, the choice (22) places the metric in diagonal and quasi-isotropic form (reduces for a static spherical configuration to the isotropic coordinates for Schwarzschild geometry). This amounts to demanding that $\partial_t \gamma_{r\theta} = \partial_t [r^2\gamma_{rr} - \gamma_{\theta\theta}] = 0$ and that $\gamma_{r\theta} = r^2\gamma_{rr} - \gamma_{\theta\theta} = 0$ on the initial slice, since $\gamma_{r\phi} = \gamma_{\theta\phi} = 0$ due to the symmetry assumptions. The 3-metric with the definitions $\gamma_{rr} = A^2$ and $\gamma_{\phi\phi} = B^2r^2\sin^2\theta$ then has the form

$$\gamma_{ij} = \mathrm{diag}\ (A^2,\ A^2 r^2,\ B^2 r^2 \sin^2\theta), \tag{23}$$

though we also use the alternative form,

$$\gamma_{ij} = \phi^4\ \mathrm{diag}\ (T^{2/3},\ T^{2/3} r^2,\ T^{-4/3} r^2 \sin^2\theta) = \phi^4\ \hat{\gamma}_{ij}, \tag{24}$$

with the definition

$$\phi^6 = A^2 B\quad ,\quad T = \frac{A}{B}\ . \tag{25}$$

The metric $\hat{\gamma}_{ij}$ is the conformally related metric for which $[\det (\hat{\gamma}_{ij})]^{1/2} = r^2 \sin\theta$, ϕ is the conformal factor used for all conformal scalings, and lnT, which vanishes in spherical symmetry, gives a measure of the anisotropy of the 3-space.

The gauge conditions which allow the metric to be written in the form (23) fix two relationships between the shift components β^r and β^θ ($\beta^\phi = 0$ with no rotation). These are

$$r\ \partial_r \left(\frac{\beta^r}{r}\right) - \partial_\theta \beta^\theta = \alpha\ (2\ K^r_r + K^\phi_\phi)\ , \tag{26}$$

$$r\ \partial_r\ \beta^\theta + \partial_\theta \left(\frac{\beta^r}{r}\right) = 2\ \alpha\ \frac{K^r_\theta}{r}\ . \tag{27}$$

This first order system of partial differential equations (PDE) is elliptic (Evans 1984b). This system is not directly numerically inverted but rather is rewritten as a pair of second order PDEs. Potentials, χ and Φ, are introduced by

$$\frac{\beta^r}{r} = r\ \partial_r\ \chi + \partial_\theta\ \Phi, \tag{28}$$

$$\beta^\theta = r\ \partial_r\ \Phi - \partial_\theta\ \chi, \tag{29}$$

which allow (26) and (27) to be written as separate second order equations

$$\frac{1}{r}\ \partial_r (r\ \partial_r \chi) + \frac{1}{r^2}\ \partial_\theta \partial_\theta \chi = \frac{\alpha}{r^2}\ (2\ K^r_r + K^\phi_\phi), \tag{30}$$

$$\frac{1}{r}\ \partial_r (r\ \partial_r \Phi) + \frac{1}{r^2}\ \partial_\theta \partial_\theta \Phi = \frac{2\alpha}{r^2} \left(\frac{K^r_\theta}{r}\right)\ , \tag{31}$$

where the differential operators are flat space 2-dimensional Laplacians. With our assumed symmetries the numerical calculation is carried out over the single quadrant $0 \leq \theta \leq \pi/2$ in the (r,θ) plane. The angular boundary conditions for χ and Φ are $\partial_\theta \chi = \Phi =$

0 for $\theta = 0, \pi/2$. The radial boundary conditions at $r = 0$ and at large radius will be left until the numerical techniques for treating this system are described below.

We give the evolution equations for the metric components ϕ and $\ln T$, derived from the general evolution equations (4) and (5), though in our approach only the component $\ln T$ is directly evolved, with ϕ found by solving the Hamiltonian constraint on successive time slices. These are,

$$\partial_t(\phi^6) = \frac{1}{r^2}\,\partial_r\,[r^2\phi^6\beta^r] + \frac{1}{\sin\theta}\,\partial_\theta\,[\sin\theta\;\phi^6\beta^\theta], \tag{32}$$

$$\partial_t \ln T = \beta^r\partial_r \ln T + \beta^\theta\partial_\theta \ln T + \partial_\theta\beta^\theta - \cot\theta\;\beta^\theta + \alpha\,\lambda, \tag{33}$$

where $\lambda = K^r_r + 2\,K^\phi_\phi$ is a particular linear combination of the extrinsic curvature tensor which we have found useful.

With our assumed symmetries and maximal slicing only the three components K^ϕ_ϕ, λ, and K^r_θ of the extrinsic curvature are independent. Using equations (5) and (32) we evolve the conformally related quantity $\hat{\lambda} = \phi^6\lambda$ by,

$$\begin{aligned}\partial_t\,\hat{\lambda} &= \frac{1}{r^2}\,\partial_r(r^2\beta^r\hat{\lambda}) + \frac{1}{\sin\theta}\,\partial_\theta(\sin\theta\;\beta^\theta\,\hat{\lambda}) \\ &+ \frac{\hat{K}_\theta{}^r}{r}\,[r\partial_r\beta^\theta - \partial_\theta(\frac{\beta^r}{r})] + \frac{8\,\pi\,\alpha\,B\,S_\theta^2}{r^2(D + E + pU)U} \\ &+ \frac{1}{r^2}\,\partial_r[r^2\alpha\,B\partial_r \ln T] + \frac{AB^2\sin\theta}{r^2}\,\partial_\theta\,[\frac{1}{AB\sin\theta}\,\partial_\theta\alpha] \\ &+ \frac{\alpha B^2\sin\theta}{r^2}\,\partial_\theta\,[\frac{1}{AB\sin\theta}\,\partial_\theta A],\end{aligned} \tag{34}$$

where D, E, S_θ, p, and U are hydrodynamic quantities (Evans, 1984a,b; Dykema, 1980) and alternate metric variables are used to provide the simplest result. Note that λ and $\ln T$ satisfy the boundary conditions $\partial_\theta\lambda = 0$ for $\theta = 0, \pi/2$, $\partial_r\lambda = 0$ at $r = 0$, and an outgoing radiation (Sommerfeld) condition is applied at $r = r_m$.

We also obtain from equation (12) the elliptic lapse equation for α due to the maximal slicing conditions $\partial_t \mathrm{tr}(K) = \mathrm{tr}(K) = 0$:

$$\frac{1}{r^2}\partial_r[r^2 B\,\partial_r\alpha] + \frac{1}{\sin\theta}\partial_\theta[\sin\theta\; B\,\partial_\theta\alpha]$$
$$= \alpha\,\phi^6\{2\lambda^2 - 6\lambda(K^\phi_\phi) + 6(K^\phi_\phi)^2 + 2\,(\frac{K^r_\theta}{r})^2 \tag{35}$$
$$+ 8\pi\,(D + E)\,(U - \frac{1}{2U}) + 8\pi\,pU(U + \frac{1}{2U})\}.$$

The lapse satisfies boundary conditions $\partial_\theta\alpha = 0$ for $\theta = 0, \pi/2$, $\partial_r\alpha = 0$ at $r = 0$. The exterior condition at large radius is discussed in Evans (1984b).

The Hamiltonian constraint (20) would ordinarily be written as an equation for the conformal factor ϕ, but we use it instead to solve for the metric component $\psi \equiv B^{1/2}$, where the two are related by $\phi^6 = T^2\psi^6$. The source of the Hamiltonian constraint is written in terms of conformally related quantities $\hat{K}_j^{\ i}$, $\hat{D}$, $\hat{E}$ and the metric T which are already obtained. The quasilinear elliptic equation becomes,

$$\frac{1}{r^2}\partial_r(r^2\partial_r\psi) + \frac{1}{r^2\sin\theta}\partial_\theta(\sin\theta\;\partial_\theta\psi)$$
$$= -\frac{1}{4}\psi\,\Big[\frac{1}{r}\partial_r(r\partial_r\,\ln T) + \frac{1}{r^2}\partial_\theta\partial_\theta\,\ln T \tag{36}$$
$$+ T^{-2}\psi^{-8}(\hat{\lambda}^2 - 3\hat{\lambda}\hat{K}^\phi_{\ \phi} + 3\hat{K}^{\phi\ 2}_{\ \phi} + (\frac{\hat{K}^r_\theta}{r})^2)$$
$$+ 8\pi\,\psi^{-2}\hat{D}U + 8\pi\,T^{2(1-\Gamma)}\,\psi^{4-6\Gamma}\,\hat{E}\big(\Gamma U + \frac{1-\Gamma}{U}\big)\Big],$$

where Γ is the ratio of specific heats. The equation in this form can be used for solution of the initial value problem as well as on subsequent time slices. This is in fact the reason conformally related quantities are evolved instead of the physical values. The boundary conditions used are: $\partial_\theta\psi = 0$ at $\theta = 0, \pi/2$, $\partial_r\psi = 0$ at $r = 0$, and the Robin condition $\partial_r\psi + \frac{1}{r}(\psi-1)\big|_{r=r_m} = 0$ (York and Piran, 1982). Once the Hamiltonian constraint is solved and ϕ is known, the physical quantities are obtained by their respective conformal scalings: $K^i_j = \phi^{-6}\,\hat{K}_j{}^i$, $S_i = \phi^{-6}\,\hat{S}_i$, $D = \phi^{-6}\,\hat{D}$, and $E = \phi^{-6\Gamma}\,\hat{E}$.

We evolve only one component of the extrinsic curvature; the other two are obtained by inversion of the momentum constraints [see equation (13)]. The conformally related extrinsic curvature $\hat{K}_j{}^i = \phi^6 K^i_j$ is decomposed into transverse and longitudinal parts using (17), with the transverse part satisfying $\hat{D}_i\,\hat{K}^i_{Tj} = 0$. The vector potential W_i for the longitudinal part of $\hat{K}_j{}^i$ is obtained by solution of the second order elliptic "vector Laplacian" (York, 1973, 1979), equation (19). Since $W^\phi = S_\phi = 0$, this system yields two coupled equations:

$$\begin{aligned}
&\frac{4}{3}\frac{1}{r^2}\,\partial_r[r^4\partial_r\,(\frac{W^r}{r})] + \underline{\frac{1}{\sin\theta}\,\partial_\theta[\sin\theta\;\partial_\theta(\frac{W^r}{r})]} + F_1\,(\frac{W^r}{r})\\
&\quad + \frac{r}{3T^2\sin\theta}\,\partial_\theta\,[T^2\,\sin^2\theta\,\cos\theta\;\partial_r(\frac{W^\theta}{\sin\theta\cos\theta})]\\
&\quad - \frac{2}{3}\sin\theta\,\cos\theta\,(3+\frac{r}{T}\,\partial_r T)\,\partial_\theta\,(\frac{W^\theta}{\sin\theta\cos\theta})\\
&\quad + F_2(\frac{W^\theta}{\sin\theta\cos\theta}) = 8\pi\,r\,\hat{S}_r,
\end{aligned} \tag{37}$$

$$\begin{aligned}
&\frac{\sin\theta\cos\theta}{r^2}\,\partial_r[r^4\partial_r(\frac{W^\theta}{\sin\theta\cos\theta})]\\
&\quad + \frac{4}{3}\,\frac{1}{\sin^2\theta\cos\theta}\,\partial_\theta\,[\sin^3\theta\cos^2\theta\;\partial_\theta(\frac{W^\theta}{\sin\theta\cos\theta})] + \underline{F_3(\frac{W^\theta}{\sin\theta\cos\theta})}\\
&\quad + \frac{1}{3}\,r\,T^2\,\partial_\theta\,[\frac{1}{T^2}\,\partial_r(\frac{W^r}{r})] + \underline{(3+\frac{2}{3}\,\frac{r}{T}\,\partial_r T)\,\partial_\theta(\frac{W^r}{r})}\\
&\quad + F_4\,(\frac{W^r}{r}) = 8\pi\,\hat{S}_\theta,
\end{aligned} \tag{38}$$

with

$$F_1 = \frac{2}{3}\,\frac{T^2}{r^2}\,\partial_r\,[\frac{r^4}{T^3}\,\partial_r T],$$

$$F_2 = \underline{-4+6\,\sin^2\theta} + \frac{2}{3}\,\frac{r}{T}\,\partial_r T + \frac{2}{3}\,\frac{T^2\sin\theta\cos\theta}{r^2}\,\partial_r\,[\frac{r^3}{T^3}\,\partial_\theta T], \tag{39}$$

$$F_3 = \underline{-\,6\,\sin\theta\cos\theta} + \frac{2}{3}\frac{T^2\cos\theta}{\sin^4\theta}\,\partial_\theta\,[\frac{\sin^5\theta}{T^3}\,\partial_\theta T],$$

$$F_4 = \frac{2}{3}\frac{T^2 r}{\sin^3\theta}\,\partial_\theta\,[\frac{\sin^3\theta}{T^3}\,\partial_r T].$$

These equations are written in a form different from Evans (1984a) as the form given here is appropriate for finite differencing. Also, the underlined terms are for later reference in considering how to best difference this system.

Once W^r and W^θ are known, the longitudinal components are obtained from (18)

$$\hat{\lambda}_L = -\frac{2r}{T}\,\partial_r T\,(\frac{W^r}{r}) - \frac{2\sin\theta}{T}\,\partial_\theta\,[T\cos\theta\,(\frac{W^\theta}{\sin\theta\cos\theta})], \tag{40}$$

$$\hat{K}_{L\phi}{}^{\phi} = -\frac{2}{3}\frac{r}{T^2}\,\partial_r\,[T^2\,(\frac{W^r}{r})] - \frac{2}{3}\frac{\sin^2\theta}{T^2}\,\partial_\theta\,[T^2\cot\theta\,(\frac{W^\theta}{\sin\theta\cos\theta})], \tag{41}$$

$$\frac{\hat{K}_{L\theta}{}^{r}}{r} = \partial_\theta\,(\frac{W^r}{r}) + r\sin\theta\cos\theta\,\partial_r\,(\frac{W^\theta}{\sin\theta\cos\theta})\,. \tag{42}$$

Since $\hat{\lambda}$ has been evolved, equations (17) and (40) give the transverse part of $\hat{\lambda}$ by $\hat{\lambda}_T = \hat{\lambda} - \hat{\lambda}_L$. The transverse part of $\hat{K}_j{}^i$, which satisfies $\hat{D}_j\hat{K}_T{}^j{}_i = 0$, is also identically the solution of the second order elliptic system, the "tensor Laplacian" (Evans, 1984a,b; York, 1973),

$$(\hat{\Delta}_T\hat{K}_T)^i_j \equiv [\hat{L}(\hat{D}\cdot\hat{K}_T)]^i_j = 0. \tag{43}$$

Taking the (r,r) and (r,θ) components, this reduces to two coupled PDEs for $\hat{K}_{T\phi}{}^{\phi}$ and $\hat{K}_{T\theta}{}^{r}$ using $\hat{\lambda}_T$ as source:

$$4T^{2/3}\,\partial_r\,[\frac{r^2}{T^{2/3}}\,\partial_r\,\hat{K}_{T\phi}{}^{\phi}] + \frac{T^{2/3}}{\sin\theta}\,\partial_\theta\,[\frac{\sin\theta}{T^{2/3}}\,\partial_\theta\,\hat{K}_{T\phi}{}^{\phi}] + \underline{G_1\hat{K}_{T\phi}{}^{\phi}}$$

$$-\frac{r}{T^{5/3}\sin\theta}\,\partial_\theta\,[\sin^2\theta\,\cos\theta\,T^{5/3}\,\partial_r\,(\frac{\hat{K}_{T\theta}{}^{r}}{r\sin\theta\cos\theta})]$$

$$+\,(7 + \frac{r}{3T}\,\partial_r T)\,\sin\theta\,\cos\theta\,\partial_\theta\,(\frac{\hat{K}_{T\theta}{}^{r}}{r\sin\theta\cos\theta}) + \underline{G_2\,(\frac{\hat{K}_{T\theta}{}^{r}}{r\sin\theta\cos\theta})}$$

$$= 2\, T^{1/6} \sin^2\theta\, \partial_r \left[\frac{r^2}{T^{1/6}} \partial_r \left(\frac{\hat{\lambda}_T}{\sin^2\theta}\right)\right] + \sin^2\theta\, \partial_\theta \partial_\theta \left(\frac{\hat{\lambda}_T}{\sin^2\theta}\right)$$

$$+ \underline{G_3 \left(\frac{\hat{\lambda}_T}{\sin^2\theta}\right)}\ , \qquad (44)$$

$$T^{2/3} \sin\theta\cos\theta\, \partial_r \left[\frac{r^2}{T^{2/3}} \partial_r \left(\frac{\hat{K}_{T\theta}^{\ r}}{r\sin\theta\cos\theta}\right)\right]$$

$$+ \frac{T^{2/3}}{\sin^2\theta\cos\theta} \partial_\theta \left[\frac{\sin^3\theta\cos^2\theta}{T^{2/3}} \partial_\theta \left(\frac{\hat{K}_{T\theta}^{\ r}}{r\sin\theta\cos\theta}\right)\right] + \underline{G_4 \left(\frac{\hat{K}_{T\theta}^{\ r}}{r\sin\theta\cos\theta}\right)}$$

$$-rT^{7/3} \partial_\theta [T^{-7/3} \partial_r \hat{K}_{T\phi}^{\ \phi}] + (-8 + \frac{r}{3T} \partial_r T)\, \partial_\theta \hat{K}_{T\phi}^{\ \phi} + G_5\, \hat{K}_{T\phi}^{\ \phi}$$

$$= (\sin\theta\cos\theta + \frac{2}{3} \sin^2\theta\, \frac{1}{T}\, \partial_\theta T)\ r\partial_r \left(\frac{\hat{\lambda}_T}{\sin^2\theta}\right)$$

$$- (5 + \frac{2r}{3T} \partial_r T) \sin^2\theta\, \partial_\theta \left(\frac{\hat{\lambda}_T}{\sin^2\theta}\right) + \underline{G_6 \left(\frac{\hat{\lambda}_T}{\sin^2\theta}\right)}\ , \qquad (45)$$

with

$$G_1 = \underline{12\, r^3\, T^{1/6}\, \partial_r \left[\frac{1}{r^2} \partial_r \left(\frac{r}{T^{1/6}}\right)\right]} + \frac{T^{5/3}}{\sin\theta} \partial_\theta \left[\frac{\sin\theta}{T^{8/3}} \partial_\theta T\right],$$

$$G_2 = \underline{(3 \cos^2\theta - 1)(7 + \frac{r}{3T} \partial_r T)} - 5\sin\theta\cos\theta\, \frac{1}{T}\, \partial_\theta T,$$

$$G_3 = \underline{6\cos^2\theta - 15\sin^2\theta} - 5\sin\theta\cos\theta \frac{1}{T}\partial_\theta T - r\sin^2\theta \frac{1}{T}\partial_r T ,$$

$$G_4 = -\sin\theta\cos\theta \left[\underline{12} + \frac{2r}{T}\partial_r T + \frac{2}{3}(2\cot\theta - \tan\theta)\frac{1}{T}\partial_\theta T\right] ,$$

$$G_5 = r^3 T^{2/3} \partial_r \left[\frac{1}{r^2 T^{5/3}}\partial_\theta T\right] - \frac{3T^{2/3}}{r}\partial_\theta \left[\frac{1}{T^{1/3}}\partial_r \left(\frac{r^2}{T^{1/3}}\right)\right] ,$$

$$G_6 = \underline{-12\sin\theta\cos\theta} + 2\sin^2\theta \frac{1}{T}\partial_\theta T - 2r\sin\theta\cos\theta \frac{1}{T}\partial_r T. \quad (46)$$

The boundary conditions for $\hat{K}_{T\phi}^{\phi}$ and $\hat{K}_{T\theta}^{r}$ are $\hat{K}_{T\theta}^{r} = \partial_\theta \hat{K}_{T\phi}^{\phi} = 0$ for $\theta = 0, \pi/2$, $\partial_r \hat{K}_{T\phi}^{\phi} = \partial_r(\frac{\hat{K}_{T\theta}^{r}}{r}) = 0$ at $r = 0$, and an approximate weak field transverse-traceless radiation boundary condition at $r = r_m$: $\hat{K}_{T\phi}^{\phi}(r_m) = \frac{1}{2}\hat{\lambda}_T(r_m)$, $\hat{K}_{T\theta}^{r}(r_m) \approx 0$ (Evans 1984b). Having evolved only $\hat{\lambda}$, $\hat{S}_r$ and $\hat{S}_\theta$, the entire (conformal) extrinsic curvature tensor has now been reconstructed.

The equations (37), (38) and (44), (45) are rather complicated but their use allows the momentum constraints (13) to be represented by second order elliptic systems. The foremost reasons for their use is that 1) the solution of (37) and (38) for the initial value problem provides physically more interestng initial data and that 2) the use of second order equations allows iterative numerical matrix inversion techniques to be applied and the proper (asymptotic) boundary conditions to be formulated.

b) Finite differencing

The numerical algorithm employed is a time explicit, Eulerian scheme utilizing a partially adaptive mesh. The calculation is approximately second order accurate in the two spatial directions (deviations due to a graded mesh in the radial direction) and generally first order accurate in time (though transport is largely second order accurate).

The single quadrant of the (r,θ) plane over which the calculation is performed is zoned by giving the discrete set of radii r_j and angles θ_k for zone corners with $1 \leq j \leq jm$ and $1 \leq k \leq km$. The angles are fixed by $\theta_{k=2} = 0$, $\theta_{k=km} = \pi/2$ and use of equal angular increments $\Delta\theta$. In addition, zone centered angles are given by $\theta_{k+1/2}$ where $\theta_{k+1/2} = \theta_k + 1/2\ \Delta\theta$ and dummy zone centers at $\theta_{3/2}$ and $\theta_{km+1/2}$ are employed to facilitate application of angular boundary conditions. Radial positions of zone corners are given by a graded mesh $r_{j+1} = r_j + \Delta r_{j+1/2}$ with, typical, $\Delta r_{j+1/2} > \Delta_{j-1/2}$ where $\Delta r_{j+1/2}$ is the zone width between faces at j and j+1. These positions are altered during the calculation to put radial zones where they are needed and leads to an associated grid velocity, $V_g{}^r$. Two radial locations remain fixed: $r_2 = 0$ and $r_{jm} = r_m$. A set of zone centered positions $r_{j+1/2}$ are carried as well with $r_{j+1/2} = 1/2(r_j + r_{j+1})$ and associated widths Δr_j such that $r_{j+1/2} = r_{j-1/2} + \Delta r_j$. Dummy zone centers are located at $r_{3/2}$ and $r_{jm+1/2}$.

In many situations there will be disparate physical length scales that require $r_{jm}/r_3 > jm$ (jm the number of alloted radial zones). This necessitates a graded mesh and we (usually) use a uniform grading $\Delta r_{j+1/2} = \sigma\ \Delta r_{j-1/2}$, σ = constant. The constant is determined initially by specifying r_3 and r_{jm}. To generate the mesh on a new slice, r_3 typically is changed to roughly track the material compression in the inner-most zone:

$$r_3{}^{n+1} = r_3{}^n + \xi\ \overline{V}_3{}^r \Delta t\ , \qquad \xi \cong 1.0, \tag{47}$$

where $\overline{V}_3{}^r$ is the average material velocity in the first zone. Then the new value of r_3 is used with r_{jm} (held fixed) to obtain a new σ, which generates the new mesh $(r_j, r_{j+1/2})$. Finally, a grid velocity is produced by

$$V_g{}^r{}_j = \frac{1}{\Delta t}\,(r_j{}^{n+1} - r_j{}^n), \tag{48}$$

which is included in all evolution equations as giving a hydrodynamic-like convective derivative (Dykema 1980). This method of continuous remapping maintains well resolved, smooth zoning, even in the interior of black holes.

The natural spatial centering from the difference equations considers the following variables as zone centered: the hydrodynamic quantities D, E, p and U; the gravitational field quantities T, ϕ, λ, α, $K^\phi{}_\phi$, and the alternate metric set A and B. The variables centered on radial zone interfaces (r_j) are S_r, V^r, V_g^r, β^r, and W^r. Those centered on angular zone interfaces (θ_k) are S_θ, V^θ, β^θ, and W^θ. Finally, $K^r{}_\theta$ is naturally centered at the zone corners (θ_k, r_j). The time centering of velocity-like variables i.e., λ, $K^{\phi}{}_{\phi}$, $K^{r}{}_{\theta}$, S_r, S_θ, V^r, $V^r{}_g$, β^r, β^θ can be thought of at a time level $\Delta t/2$ offset from all others. Operator splitting is used extensively throughout the code, though we will not explicitly indicate the details here (see Dykema 1980 for further details).

We will treat some of the numerical issues in what follows. Our method for handling the transport terms that appear in the evolution equations for the gravitational field, (33) and (34), is briefly described. We use for these equations a transport scheme very similar to that used for the hydrodynamic equations (Dykema 1980; Wilson 1979) and a more complete discussion can be found in Dykema (1980). We then concentrate on discussing a number of questions surrounding the finite differencing of the various elliptic equations that arise from our analytic method. We do not present the complete difference equations in each case since many aspects of the differencing and averaging are largely arbitrary. Rather, we will devote attention to the general techniques used except for those points which have been developed which have contributed significantly to obtaining stable, regular results.

The first step in updating the dynamical quantities λ and lnT to a succeeding time level is to advance the values at the outer boundary. Here the boundary condition is modeled as an outgoing spherical wave,

$$\partial_t \hat{\lambda} = -\frac{c}{r}\,\partial_r(r\hat{\lambda}) \tag{49}$$

where $c = -\beta^r + \alpha/A$ is the local speed of light. The difference equation becomes (with angular index suppressed)

$$\hat{\lambda}^{n+1}_{jm+1/2} = \hat{\lambda}^{n}_{jm+1/2} - c_{jm}\frac{\Delta t}{\Delta r_{jm}}\left[\hat{\lambda}^{n}_{jm+1/2} - \frac{r_{jm-1/2}}{r_{jm+1/2}}\hat{\lambda}^{n}_{jm-1/2}\right] \tag{50}$$

with c_{jm} computed at the zone interface. The use of (50) has typically resulted in a very low impedance mismatch; our gravity waves propagate off the mesh very cleanly. We handle lnT in a similar way.

The transport terms of equation (33) are differenced as follows. An update of lnT in the radial direction is accomplished by,

$$(\ln T)^{n+1}_{j+1/2} = (\ln T)^{n}_{j+1/2} + (\Delta t/2)\overline{V}^{r}_{j+1/2}[(\partial_r \ln T)^n_j\,(1-\eta_{j+1/2})$$

$$+ (\partial_r \ln T)^n_{j+1}\,(1+\eta_{j+1/2})], \tag{51a}$$

where,

$$\bar{V}^r_{j+1/2} = 1/2[(\beta^r_j + V_g{}^r{}_i) + (\beta^r_{j+1} + V_g{}^r{}_{j+1})], \tag{51b}$$

$$(\partial_r \ln T)^n_j = \frac{1}{\Delta r_j}[\ln T^n_{j+3/2} - \ln T^n_{j+1/2}], \tag{51c}$$

$$\eta_{j+1/2} = (1-\mu_{j+1/2}) \frac{\Delta t \bar{V}^r_{j+1/2}}{\Delta r_{j+1/2}} + \mu_{j+1/2} \frac{\bar{V}^r_{j+1/2}}{|\bar{V}^r_{j+1/2}|}, \tag{51d}$$

$$\mu_{j+1/2} = \frac{|(\partial_r \ln T)^2_{j+1} - (\partial_r \ln T)^2_j|}{(\partial_r \ln T)^2_{j+1} + (\partial_r \ln T)^2_j} \tag{51e}$$

which is mixed first order, second order transport and the grid velocity (for the radial direction only) is evident. The transport in the θ-direction is immediately obtainable, however the combined update is accomplished by three sweeps through the mesh, updating partially $\Delta t/2$ in the radial direction, Δt in the θ-direction, and a final $\Delta t/2$ in the radial direction. This is due to the observation by Leith (1965) that the above method is von Neumann stable and formally more accurate.

A similar weighted first order, second order method is used for the transport terms in equation (34) for $\hat{\lambda}$, except these have the appearance of a divergence of a flux. Conservative differencing is employed so that the sum $\sum_{j,k} \hat{\lambda}_{k+1/2,j+1/2}\,(r^3_{j+1}-r^3_j)(\cos\theta_k-\cos\theta_{k+1})/3$ is maintained during the transport update. To do so, the relevant differential operators in (34) must be replaced according to

$$\frac{1}{r^2}\,\partial r\,\Big|_{j+1/2} \rightarrow \frac{3\Delta_{j+1/2}}{(r^3_{j+1} - r^3_j)}\,, \tag{52a}$$

$$\frac{1}{\sin\theta}\,\partial_\theta\,\Big|_{k+1/2} \rightarrow \frac{\Delta_{k+1/2}}{\cos\theta_k - \cos\theta_{k+1}}\,. \tag{52b}$$

The reader is referred to the complete discussion by Dykema (1980) for the finite difference equation for (34). More up-to-date transport algorithms could be used for both (33) and (34) as well as the relativistic hydrodynamics equations (see Hawley, Smarr, and Wilson (1984) for comparisons of these methods). The use of transport algorithms of this type for the gravitational field quantities stems from the conclusion that the transport will be mostly second order accurate since these fields show no tendancy to steepen, unlike the fluid, into shock-like structures.

The symmetry boundary conditions for $\hat{\lambda}$ are $\hat{\lambda}_{j=3/2} = \hat{\lambda}_{j=5/2}$ at the origin, $\hat{\lambda}_{k=3/2} = \hat{\lambda}_{k=5/2}$ at $\theta=0$ and $\hat{\lambda}_{k=km-1/2} = \hat{\lambda}_{k=km+1/2}$ on the equator. Note that $\hat{\lambda} = 0$ on axis as well as satisfying $\partial_\theta \hat{\lambda} = 0$ there. Use of the Neumann conditions allows $\hat{\lambda}$ to drift near $\theta=0$, but in practice this is always relatively small. The conditions for lnT are the same except $(\ln T)_{k=5/2} = 0$ is maintained.

We turn next to the issue of finite differencing the elliptic systems. The equation (35) for the lapse function, α, is a straightforward scalar elliptic equation. There are no fundamental problems in obtaining solutions and we solve it by using successive over relaxation (SOR). In order to reduce the number of relaxation iterations, an extrapolation acceleration for α is used of the form

$$\alpha(t_n)_{\text{first guess}} = \frac{\alpha^2(t_{n-1})}{\alpha(t_{n-2})} . \tag{53}$$

This of course requires storing an extra time level. Geometrical extrapolation is used for α because once a black hole forms, α decreases exponentially in time in the strong field region (Smarr and York, 1978b; Evans, 1984a,b). The lapse is considered zone-centered and so the Neumann conditions on the axis, equator and at the origin are naturally applied as was done for $\hat{\lambda}$ above.

The Hamiltonian constraint (36), used to obtain the metric variable $w = B^{1/2}$, is nonlinear. To find its solution, we use a best available guess for ψ to evaluate the source (the right-hand-side). Then the linear operator is solved by SOR to obtain a temporary function $\bar{\psi}$. The next approximation to the solution is formed by taking a weighted linear combination of $\bar{\psi}$ and the last guess. This is a modified Picard (outer) iteration. Writing equation (36) as $\Delta\psi = N(\psi)$, where Δ is the 3-dimensional scalar Laplacian, the algorithm becomes,

$$\Delta\bar{\psi}^{n+1} = N(\psi^n), \tag{54a}$$

$$\psi^{n+1} = \psi^n + \omega(\bar{\psi}^{n+1} - \psi^n). \tag{54b}$$

During the early stages of collapse, before a black hole forms, the value of this weight is largely unimportant and can be taken as unity. Once a black hole forms, the solutions for the conformal factor are in a very nonlinear regime and the use of a correct weight is crucial to finding a solution. The optimal weight appears to be $\omega = 0.22$ once a black hole forms and values $\omega > 0.5$ appear to be divergent.

An evolution equation for ψ is obtained by combining equations (32) and (33) and this is used to provide a first approximation for the elliptic solver. Since such an equation exists, we use it rather than extrapolate as was done for the lapse function α above. The boundary conditions for ψ on axis, equator, and origin are similar to those for the lapse and are straightforwardly applied. The Robin condition at the outer boundary (given in section 2) is enforced by

$$\psi_{jm+1/2} = [\psi_{jm-1/2}(1 - \frac{\Delta r_{jm}}{2r_{jm}}) + \frac{\Delta r_{jm}}{r_{jm}}] (1 + \frac{\Delta r_{jm}}{2r_{jm}})^{-1} \qquad (55)$$

It is important to note before going on what behavior the various vector and tensor components are near r=0 and θ=0, the singular points of the coordinate system. This is necessary both to understand the boundary conditions on axis and at the origin and to facilitate "regularization" of the differencing near these points. The latter point is most important in handling the vector and tensor elliptic systems. The local dependence of these variables is determined (Evans 1984b) by enforcing the symmetry assumptions and by assuming the local cartesian components are Taylor expandable about r=0 in cartesian coordinates. These series are then transformed back to spherical polar coordinates and written as an order expansion in powers of r.

Only the lowest order dependence will be given. The vector components β^r, β^θ have the form

$$\frac{\beta^r}{r} = c_0 + c_1 \sin^2\theta + O(r^2), \quad \beta^\theta = c_1 \sin\theta \cos\theta + O(r^2) \qquad (56)$$

where c_0, c_1 are constants (note β^θ angular dependence determined locally by β^r/r). The behavior of W^i and V^i are of the same form. The momentum density is similar except given in covariant form:

$$S_r = r(c_2 + c_3 \sin^2\theta) + O(r^3), \quad S_\theta = c_3 r^2 \sin\theta\cos\theta + O(r^4). \qquad (57)$$

The metric component lnT has the property

$$\ln T = c_4 r^2 \sin^2\theta \; [1 + O(r^2)]. \qquad (58)$$

The extrinsic curvature components are linked together locally by

$$\lambda = 3c_5 \sin^2\theta \; [1 + O(r^2)],$$
$$K^\phi{}_\phi = c_5 \; [1 + O(r^2)], \qquad (59)$$
$$\frac{K^r{}_\theta}{r} = 3c_5 \sin\theta\cos\theta \; [1 + O(r^2)].$$

All other quantities behave as transformed scalars; they are analytic functions of the form $f(r^2\cos^2\theta, r^2\sin^2\theta)$, which satisfy the assumed symmetries.

The second order shift potential equations (30) and (31) are decoupled in the potentials χ and Φ. Furthermore the appearance of 2-dimensional Laplace operators allows these equations to be finite Fourier decomposed in the angular direction and replaced by a set of ordinary differential equations (ODEs). This avoids the use of iterative or relaxation solution methods, as was used for the lapse equation and the Hamiltonian constraint equation above, and significantly speeds the solution of these equations on the computer.

A finite Fourier transform of χ, Φ, and the source terms in (26) and (27):

$$P = \alpha\,(2K^r{}_r + K^\phi{}_\phi), \tag{60}$$

$$Q = 2\alpha\,\frac{K^r{}_\theta}{r}, \tag{61}$$

gives

$$\chi_{k+1/2} = a_0 + 2\sum_{m=1}^{km-3} a_m \cos 2m\theta_{k+1/2},$$

$$\Phi_k = 2\sum_{m=1}^{km-3} b_m \sin 2m\theta_k ,$$

$$P_{k+1/2} = p_0 + 2\sum_{m=1}^{km-3} p_m \cos 2m\theta_{k+1/2},$$

$$Q_k = 2\sum_{m=1}^{km-3} q_m \sin 2m\theta_k . \tag{62}$$

These transforms are consistent with the parities of the functions under axisymmetry and equatorial plane symmetry. The inverse transforms are

$$p_m = \left(\frac{1}{km-2}\right)\sum_{k=2}^{km-1} P_{k+1/2} \cos 2m\theta_{k+1/2},$$

$$q_m = \left(\frac{1}{km-2}\right)\sum_{k=3}^{km-1} Q_k \sin 2m\theta_k , \tag{63}$$

with identical expressions for the amplitudes a_m and b_m. In the above expressions, the radial index is suppressed; all amplitudes are implicitly functions of r.

From the local dependence (56) for the shift vector near $r = 0$, it can be shown (Evans 1984b) that the definitions (28) and (29) for the potentials imply a singular local behavior in χ which shows up in a_0:

$$a_0 = \text{constant} \times \ln r, \tag{64}$$

near $r = 0$. Hence the amplitudes a_0 and p_0 are handled separately. The p_0 amplitude in (63) is just the angle average of P. So by defining

$$\langle \frac{\beta^r}{r} \rangle = \left(\frac{1}{km-2}\right) \sum_{k=2}^{km-1} \left(\frac{\beta^r}{r}\right)_{k+1/2}, \tag{65}$$

this part of the decomposition of the shift vector can be found by directly integrating (26):

$$\langle \frac{\beta^r}{r} \rangle = - \int_r^{r_m} p_0 \frac{dr}{r} + \langle \frac{\beta^r}{r} \rangle \Big|_{r=r_m}. \tag{66}$$

A discussion of the choice of boundary conditions for (66) is found in Evans (1984b). The amplitude a_0, containing the singular behavior of the potential χ, is never explicitly calculated.

With the transforms defined in (62) and (63), the decomposition of (30) and (31) becomes

$$r\partial_r(r\partial_r a_m) - 4m^2 a_m = p_m, \tag{67}$$

$$r\partial_r(r\partial_r b_m) - 4m^2 b_m = q_m. \tag{68}$$

These are a set of uncoupled ODEs for each $m = 1, \ldots, km-3$. With boundary values given at both $r = 0$ and $r = r_m$, they are solved using the tridiagonal algorithm (Roache, 1976). The inner boundary conditions are $\partial_r a_m = \partial_r b_m = 0$ at $r = 0$. Discussion of the physical conditions on the shift at large radius is beyond our purpose here and we refer instead to Evans (1984b).

Finally we close with a discussion of the numerical problems encountered with the accuracy of finite differencing coupled elliptic systems near the singular points of curvilinear coordinate systems and the techniques we have used to circumvent these numerical irregularities. We employ spherical coordinates in our calculations and hence problems in the finite difference approximation appear at the origin, $r = 0$, and to a lesser extent on the symmetry axis, $\theta = 0$. Similar numerical regularity problems also occurred in an earlier version of this code which employed topological cylindrical coordinates (Wilson, 1979). In that case the irregularities showed up along the entire (coordinate singular) symmetry axis. Similar numerical irregularities have long been a source of problems in using 2-dimensional numerical relativity codes. We have developed techniques for eliminating this aberrant behavior in our spherical coordinate numerical relativity code. The general method however appears immediately applicable to other coordinate topologies. We have identified two causes of these numerical irregularities near symmetry points. The first is due to the loss of finite difference accuracy and increase of truncation errors near these regions in curvilinear coordinate systems. The second is related to the appearance of nontrivial, local homogeneous solutions to coupled elliptic systems.

The usual arguments on the formal size of truncation errors in finite difference approximations are based on the use of a Taylor expansion. For a problem in cartesian-like coordinates with zone increments Δx, centered spatial differences will produce truncation errors with respect to the corresponding derivatives they replace of order $(\Delta x/\ell)^2$, where ℓ is the characteristic physical length. These errors can be arbitrarily reduced with increase in the resolution (i.e. $\Delta x \to 0$). The problem with finite differencing in curvilinear coordinates, near the special points or regions of the coordinate system, is that Taylor series do not in general exist there. One encounters in spherical coordinates differential operators of the form $\frac{1}{r}\,\partial_r$, for example (similarly one has $\frac{1}{\rho}\,\partial_\rho$ in cylindrical coordinates). With the explicit appearance of the radial coordinate, attempts to produce centered finite differences of such operators will produce truncation errors that are the maximum of $(\Delta r/\ell)^2$ or $(\Delta r/r)^2$, where ℓ is again the characteristic physical length. Far from the origin of coordinates these errors can be made sufficiently small. However in the vicinity of $r = 0$, the errors associated with most finite difference representations will be of order unity (or worse).

We can illustrate this problem and its solution with a specific example. Consider the radial second order operator of the flat space 3-dimensional Laplacian:

$$\frac{1}{r^2}\,\partial_r(r^2\partial_r\psi). \tag{69}$$

Let us consider that this operator produces zone centered values, i.e. located at positions $r_{j+1/2}$ and that ψ is zone centered as well: $\psi_{j+1/2}$. Using our mesh definitions, one can straightforwardly difference (69):

$$\frac{1}{r_{j+1/2}^2}\frac{1}{\Delta r_{j+1/2}}[r_{j+1}^2\frac{1}{\Delta r_{j+1}}(\psi_{j+3/2}-\psi_{j+1/2})$$

$$-r_j^2\frac{1}{\Delta r_j}(\psi_{j+1/2}-\psi_{j-1/2})]. \tag{70}$$

Let us assume further that we can expect ψ to have the local radial dependence

$$\psi(r) = c_0 + c_1 r^2 + O(r^4), \tag{71}$$

i.e. like the gravitational potential or our conformal factor. Application of (69) on this ψ gives locally the analytic value $6c_1$. If however we take $\psi_{j+1/2} = c_0 + c_1 r_{j+1/2}^2$ and use (70) we obtain

$$c_1\frac{1}{r_{j+1/2}^2}\frac{1}{\Delta r_{j+1/2}}[r_{j+1}^2(r_{j+3/2}+r_{j+1/2})$$

$$-r_j^2(r_{j+1/2}+r_{j-1/2})]. \tag{72}$$

Evaluating this at the first three radial locations, j = 2, 3, 4, yields the values 8.0 c_1, 6.22 c_1, 6.08 c_1, respectively. Hence, it can be seen that the errors near the origin in such a straightforward finite differencing (70) of (69) are of order unity.

The solutions to these problems is to choose the specific finite difference representation for the operator in question which reproduces the expected analytic result under the assumption that the general local dependence of the function is known. This altered finite difference operator for (69) for example will differ from the straightforward one (70) away from the origin only at the level of second order, which is already conceded to the approximation as a whole. For (69), we would use instead

$$\frac{3}{r_{j+1}^3-r_j^3}[\frac{2r_{j+1}^3}{r_{j+3/2}+r_{j+1/2}}\frac{1}{\Delta r_{j+1}}(\psi_{j+3/2}-\psi_{j+1/2})$$

$$-\frac{2r_j^3}{r_{j+1/2}+r_{j-1/2}}\frac{1}{\Delta r_j}(\psi_{j+1/2}-\psi_{j-1/2})], \tag{73}$$

which, assuming (71) as the expected form for ψ, reproduces the analytic value.

A second problem which will be typically encountered in solving coupled elliptic systems is local homogeneous solutions. We can illustrate why this is not normally a problem by considering the scalar Poisson equation

$$\Delta\phi = 4\pi\rho. \tag{74}$$

The homogeneous solution for this equation is just a constant, say c_0. Assuming our symmetry conditions, the local dependence of ϕ near $r = 0$ would be $\phi = c_0 + r^2(c_1 + c_2 \sin^2\theta) + \theta(r^4)$, in spherical coordinates. Regardless of whether this equation (74) is written in cartesian coordinates or spherical coordinates prior to finite differencing, the difference representations of the partial derivatives in the Laplacian will identically cancel off the constant c_0. The finite differencing of (74) can then concentrate on reproducing the expected next order behavior in ϕ. The value c_0 is of course only determined by the global solution of (74) once suitable boundary values have been prescribed.

The situation changes drastically when one considers vector or tensor (or in general multi-component coupled) elliptic systems. Consider a 2-tensor K^i_j which exhibits our assumed symmetries including a tracefree condition (i.e. like the extrinsic curvature tensor). The cartesian components in the limit $r \to 0$ (Evans, 1984b) have the form

$$\lim_{r\to 0} K^i_j = \text{diag}\,[c_5, c_5, -2c_5], \tag{75}$$

where c_5 is a constant. If a tensor-like Laplace operator, like the "tensor Laplacian" (43), in cartesian coordinates operates on (75), this "constant" or homogeneous solution will be annihilated. The same is true of a finite difference representation. However, if the tensor (75) is transformed to spherical coordinates, this local homogeneous solution will be of the form (59) which has angular dependence! If the finite difference operator is to produce accurate results, these terms must still be identically annihilated. Figure 1 shows the irregular behavior near $r = 0$ produced in the component $K^\phi{}_\phi$ when no attention is paid to these details. An anomalous numerical "spike" occurs in the first five to ten radial zones. The solution to the problems arising from local homogeneous solutions is to rewrite the differential equations in a form which allows the difference equations to cancel off these terms. This is the reason for writing equations (37), (38) and (44), (45) in the form given in this paper. The reader can verify that the underlined terms in these equations allow the expected local dependence of W^i, (56), and K^i_j, (59), to drop through. Once both of these numerical techniques are applied, the local behavior of the solutions near $r = 0$ is greatly improved. This can be seen in Figure 2 which shows the effect of these

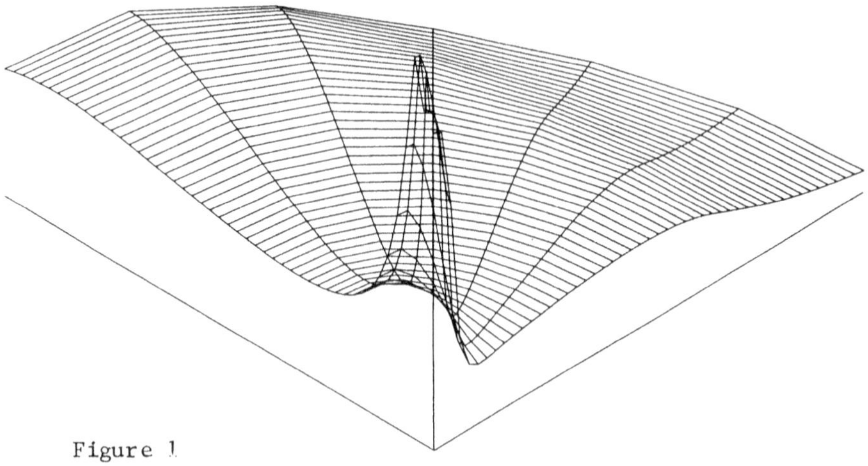

Figure 1

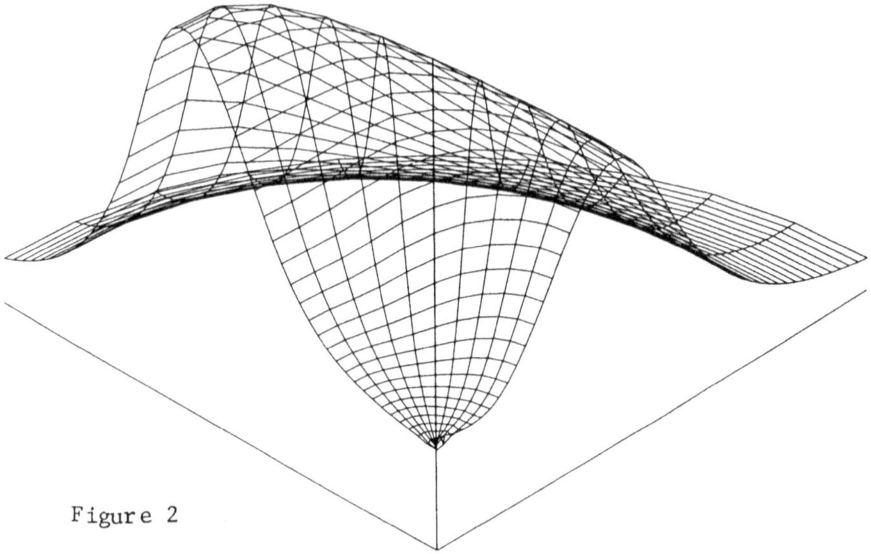

Figure 2

Figure 1. The local behavior near r = 0 in the extrinsic curvature component $K^{\phi}{}_{\phi}$. An anomalous numerical "spike" is evident at the origin affecting the first 5-10 zones. In this figure, the symmetry axis is to the right and the equatorial plane to the left.

Figure 2. The same component plotted after the finite differencing has been "regularized." Note that the vertical scales in the two figures are different, with Figure 1 being dominated by the numerical error. Hence, there is considerable improvement in Figure 2.

changes on the same component plotted in Figure 1. More details of these techniques can be found in Evans (1984b).

4. CONCLUSION

We have presented a general approach for solving the Einstein equations of general relativity coupled to hydrodynamics. This coupled nonlinear system of partial differential equations naturally split into equations of constraint and of evolution. The constraints can be solved for the longitudinal pieces of the gravitational field which generalizes the Newtonian scalar potential. The evolution equations determine the generation and propogation of gravitational radiation, a feature not found in Newtonian gravity. For relativistic systems, such as supernova core collapse, this full approach is the only reliable calculational tool available.

The general framework is then specialized to the axisymmetric case in which no rotation occurs. In addition, the spacetime is assumed to be asymptotically flat. A particular choice of coordinates simplifies the differential equations considerably. A computer code was written using finite difference techniques to solve these equations. Some crucial details are spelled out which are unique to differencing the Einstein equations.

Although space does not allow us to give examples of the use of this code, it is a working production code. Some runs simulating highly oblate collapse to post-nuclear densities are discussed in Evans (1984a,b). These runs are the first to be able to calculate the gravitational radiation emitted in highly nonspherical relativistic collapse. Other studies to elucidate the physics of strong dynamic gravitational fields are currently underway.

The approach we use will be extended to include rotation. This will yield an alternate technique to those used by Nakamura, et al. (1980) or Bardeen and Piran (1983). Going beyond that one has good near-term prospects to use our approach to study full four dimensional spacetimes (three space + time). This can be done in either the asymptotically flat or the cosmological regime. This final step in studying gravitational physics will require the new generation of supercomputers with multiprocessors and large memories. In addition, more efficient numerical algorithms to solve the Einstein equations will be developed. A more detailed and realistic treatment of other microphysics will also be added. We believe this will create a rich new field of physics: computational relativity.

ACKNOWLEDGEMENTS

We have benefited from discussions with Joan Centrella, Pieter Dykema, Richard Isaacson, James LeBlanc, Richard Matzner, Michael Norman, Karl-Heinz Winkler, and James W. York, Jr. We wish to thank the Max-Planck-Institüt für Physik und Astrophysik, Institüt für Astrophysik for their hospitality. Much of this research was conducted at the Lawrence Livermore National Laboratory and the Center for Relativity, University of Texas at Austin. Financial support was partially provided by the Associated Western Universities Graduate Fellowship Program, the National Science Foundation under grant number PHY 83-08826, and the Department of Energy contract number W-7405-ENG-48.

BIBLIOGRAPHY

Arnowitt, R., S. Deser, and C.W. Misner: 1962, "The Dynamics of General Relativity," in Witten (ed.), Gravitation: An Introduction to Current Research, John Wiley, New York, pp. 227-265.

Bardeen, J.M. and T. Piran: 1983, "General Relativistic Axisymmetric Rotating Systems: Coordinates and Equations," Phys. Reports **96**, pp. 206-250.

Brill, D.R.: 1959, "On the Positive Definite Mass of the Bondi-Weber-Wheeler Time-Symmetric Gravitational Waves," Ann. Phys. **7**, pp. 466-483.

Cantor, M.: 1979, "A Necessary and Sufficient Conditon for York Data to Specify and Asymptotically Flat Spacetime," J. Math. Phys. **20**, pp. 1741-1744.

Centrella, J. and J.R. Wilson: 1983, "Planar Numerical Cosmology I. The Differential Equations," Astrophys. J. **273**, pp. 428-435.

Centrella, J. and J.R. Wilson: 1984, "Planar Cosmology II. The Difference Equations and Numerical Tests," Astrophys. J. Suppl. **54**, pp. 229-250.

Dykema, P.G.: 1980, "Numerical Simulation of Axisymmetric Gravitational Collapse," Ph. D. thesis, Univ. of Texas at Austin (unpublished).

Eardley, D. and L. Smarr: 1979, "Time Functions in Numerical Relativity: Marginally Bound Dust Collapse," Phys Rev. **D19**, pp. 2239-2259.

Eardley, D.: 1979, "Global Problems in General Relativity," in L. Smarr (ed.), Sources of Gravitational Radiation, Cambridge University Press, Cambridge, pp. 127-138.

Eppley, K.: 1979, "Pure Gravitational Waves," in L. Smarr (ed.), Sources of Gravitational Radiation, Cambridge University Press, Cambridge, pp. 275-292.

Estabrook, F., H. Wahlquist, S. Christensen, B. DeWitt, L. Smarr, and E. Tsiang: 1973, "Maximally Slicing a Black Hole," Phys. Rev. **D7**, pp. 2814-2817.

Evans, C.R.: 1984a, "A Method for Numerical Simulation of Gravitational Collapse and Gravitational Radiation Generation," in J. Centrella, R. Bowers, J. LeBlanc, and M. Le Blanc (eds.), Numerical Astrophysics: A Meeting in Honor of James Wilson, Jones and Bartlet, Portola Valley CA, (in press).

Evans, C.R.: 1984b, "A Method for Numerical Relativity: Simulation of Axisymmetric Gravitational Collapse and Gravitational Radiation Generation," Ph. D. thesis, University of Texas at Austin (unpublished).

Haugan, M.P., S.L. Shapiro, and I. Wasserman: 1982, "The Suppression of Gravitational Radiation from Finite-Size Stars Falling Into Black Holes," Astrophys. J. **257**, pp. 283-290.

Hawley, J.F., L.L. Smarr, and J.R. Wilson: 1984, "A Numerical Study of Nonspherical Black Hole Accretion. II. Finite Differencing and Code Calibration," Astrophys. J. Suppl. **55**, pp. 211-246.

Landau, L.D. and E.M. Lifshitz: 1941, Teoriya Polya, Nauka, Moscow, (English translation The Classical Theory of Fields), Addison-Wesley, Cambridge.

Leith, C.E.: 1965, "Numerical Simulation of the Earth's Atmosphere," Methods of Computational Physics, **4**, pp. 1.

Lichnerowicz, A.: 1944, "L'integration des Equations de la Gravitation Relativiste et le Probleme des n Corps," J. Math. Pure Appl. **23**, pp. 37-63.

May, M. and R.H. White: 1966, "Stellar Dynamics and Gravitational Collapse," in B. Alder, S. Fernbach, and M. Rotenberg (eds.), Methods in Computational Physics, Vol. 7, Academic Press, New York, pp. 219-258.

Misner, C.W. and D.H. Sharp: "Relativistic Equations for Adiabatic, Spherically Symmetric Gravitational Collapse," Phys. Rev. **136**, pp. B571-B577.

Müller, E. and W. Hillebrandt: 1981, "The Collapse of Rotating Stellar Cores," Astron. Astrophys. **103**, pp. 358-366.

Müller, E.: 1982, "Gravitational Radiation From Collapsing Rotating Stellar Cores," Astron. Astrophys. **114**, pp. 53-59.

Nakamura, T., K. Maeda, S. Miyama and M. Saski: 1980, "General Relativistic Collapse of an Axially Symmetric Star. I. The Formulation and The Initial Data," Prog. Theor. Phys. **63**, pp. 1229-1244.

Nakamura, T. and M. Saski: 1981, "Is Collapse of a Deformed Star Always Effectual for Gravitational Radiation," Phys. Lett. **106**, pp. 69-72.

Nakamura, T.: 1981, "General Relativistic Collapse of Axially Symmetric Stars Leading to the Formation of Rotating Black Holes," Prog. Theor. Phys. **65**, pp. 1876-1890.

O'Murchadha, N. and J.W. York: 1974, "Initial-Value Problem of General Relativity. I. General Formulation and Physical Interpretation," Phys. Rev. **D10**, pp. 428-436.

Piran, T.: 1980, "Numerical Codes for Cylindrical General Relativistic Systems," J. Comp. Phys. **35**, pp. 254-283.

Piran, T.: 1983, "Methods of Numerical Relativity," in N. Deruelle and T. Piran (eds.), Gravitational Radiation, North Holland, Amsterdam, 203.

Roache, R.J.: 1976, Computational Fluid Dynamics, Hermosa Publishers, Albuquerque.

Shapiro, S.L. and S.A. Teukolsky: 1979, "Gravitational Collapse of Supermassive Stars to Black Holes: Numerical Solution of the Einstein Equations," Astrophys. J. Lett. **234**, pp. L177-L181.

Shapiro, S.L. and S.A. Teukolsky: 1980, "Gravitational Collapse to Neutron Stars and Black Holes: Computer Generation of Spherical Spacetimes," Astrophys. J. **235**, pp. 199-215.

Smarr, L.L.: 1975, "The Structure of General Relativity with a Numerical Illustration: The Collision of Two Black Holes," Ph.D. Dissertation, The University of Texas at Austin, (unpublished).

Smarr, L.L.: 1977, "Spacetimes Generated by Computers: Black Holes with Gravitational Radiation," Ann. N.Y. Acad. Sci. **302**, pp.569-604.

Smarr, L.: 1979, "Gauge Conditions, Radiation Formulae and the Two Black Hole Collision," in L. Smarr (ed.) Sources of Gravitational Radiation, Cambridge University Press, Cambridge, pp. 245-274.

Smarr, L.L.: 1984, "Computational Relativity: Numerical and Algebraic Approaches," to appear in the Proceedings of the Tenth International Conference on General Relativity and Gravitation.

Smarr, L.L., C. Taubes, and J.R. Wilson: 1980, "General Relativistic Hydrodynamics: The Comoving, Eulerian, and Velocity Potential Formalisms," in F. Tipler (ed.) Essays in General Relativity: A Festschrift for A. Taub, New York, Academic Press, pp. 157-183.

Smarr, L. and J.W. York: 1978a, "Radiation Gauge in General Relativity," Phys. Rev. **D17**, pp. 1945-1956.

Smarr, L. and J.W. York: 1978b, "Kinematical Conditions in the Construction of Spacetime," Phys. Rev. **D17**, pp. 2529-2551.

Thorne, K.S.: 1980a, "Gravitational-Wave Research: Current Status and Future Prospects," Rev. Mod. Phys. **52**, pp. 285-297.

Thorne, K.S.: 1980b, "Multipole Expansions of Gravitational Radiation," Rev. Mod. Phys. **52**, pp. 299-339.

Wheeler, J.A.: 1964, "Geometrodynamics and the Issue of the Final State," in C. DeWitt and B.S. DeWitt (eds.), Relativity, Groups, and Topology, Gordon & Breach, New York, pp. 317-522.

Wilson, J.R.: 1977, "A Numerical Study of Rotating Relativistic Stars," in R. Giaconni and R. Ruffini (eds.), Proceedings of the International School of Physics Enrico Fermi LXV, Amsterdam, N. Holland, pp.644-675.

Wilson, J.R.: 1979, "A Numerical Method for Relativistic Hydrodynamics," in L. Smarr (ed.), Sources of Gravitational Radiation, Cambridge University Press, Cambridge, pp. 423-446.

York, J.W.: 1971, "Gravitational Degrees of Freedom and the Initial-Value Problem," Phys. Rev. Lett. **26**, pp. 1656-1658.

York, J.W.: 1972, "Role of the Conformal Three-Geometry in the Dynamics of Gravitation," Phys. Rev. Lett. **28**, pp. 1082-1085.

York, J.W.: 1973, "Conformally Invariant Orthogonal Decomposition of Symmetric Tensors on Riemannian Manifolds and the Initial-Value Problem of General Relativity," J. Math. Phys. **14**, pp. 456-464.

York, J.W.: 1979, "Kinematics and Dynamics of General Relativity," in L. Smarr (ed.), Sources in Gravitational Radiation, Cambridge University Press, Cambridge, pp. 83-126.

York, J.W. and T. Piran: 1982, "The Initial Value Problem and Beyond," in R. Matzner and L. Shepley (eds.), Spacetime and Geometry, University of Texas Press, Austin, pp. 147.

THE CHARACTERISTIC INITIAL VALUE PROBLEM IN GENERAL RELATIVITY

John M. Stewart

Dept. Applied Mathematics and Theoretical Physics
Silver Street, Cambridge CB3 9EW, UK

Abstract

A new approach to the numerical solution of Einstein's Equations based on the characteristic initial value problem is outlined. Because this approach requires integration through caustics and up to singularities, this topic is described and some novel finite-differencing techniques are introduced for integrating up to singularities of non-linear differential equations. Finally some preliminary results from this new approach are presented.

0 INTRODUCTION

At present gravitational wave astronomy is in its infancy. Thus the role of numerical calculations in general relativity must be to elucidate that fascinating subject. There are a number of topics for which neither exact solutions nor approximative techniques have yet produced significant insights, and where numerical relativity may be able to help. Th word "yet" is important here, for numerical work may suggest new ideas or approaches to aid those died-in-the-wool theorists who abhor numerical computation. (As concrete examples we may cite the new perturbation techniques being developed for the colliding black hole problem, and the existence of singularities behind impulsive plane-fronted waves mentioned later.)

There are two main approaches to the initial value problem in general relativity. One is to select a spacelike hypersurface and set the initial data on that. This, the Cauchy problem, is

K.-H. A. Winkler and M. L. Norman (eds.), Astrophysical Radiation Hydrodynamics, 531–568.

the approach adopted by Dr. Smarr and others, and so I need not review it here, but refer the reader to Dr. Smarr's lectures. The second main approach is to give data on two intersecting null hypersurfaces, the characteristic initial value problem (CIVP). (Mixed problems, in which data is given on a combination of spacelike, null or timelike surfaces, are not yet in favour in relativity). In the CIVP the initial data is essentially the specification of the radiation and matter crossing the initial hypersurfaces. Thus the CIVP is almost ideally suited to study interacting gravitational waves, (see e.g. fig. 9). There is a second, more subtle, reason for considering the CIVP. The concept of the energy of an isolated system in general relativity is of course notoriously difficult. However there is an unambiguous measure, the Bondi energy-momentum 4-vector. Its timelike component, the Bondi mass, is positive and decreases with time as the system emits gravitational radiation, thus providing a measure of the radiation's energy. (For a recent review see e.g. (1).) The Bondi mass is defined and has to be measured at future null infinity ($\mathscr{I}^+$), roughly speaking, "infinity along outgoing light rays". Numerically we cannot integrate over infinite grids. We can compactify our coordinates; e.g. $x \mapsto y = \tan^{-1} x$ reduces the infinity line $-\infty < x < \infty$ to the finite segment $-\frac{1}{2}\pi < y < \frac{1}{2}\pi$. Two problems however remain:

i) as we go out to $\mathscr{I}^+$ spacetime approaches asymptotically the flat Minkowski spacetime, and most quantities of physical interest approach zero. In a numerical calculation they would be swamped by numerical errors,

ii) in the new coordinates the equations are singular at $\mathscr{I}^+$.

An ingenious idea due to Penrose (see e.g. (2,3,4), can be used to circumvent these problems. Instead of working with the physical line element

$$d\tilde{s}^2 = \tilde{g}_{ab} dx^a dx^b,$$

Penrose introduces an "unphysical" line element

$$ds^2 = \Omega^2 d\tilde{s}^2 = g_{ab} dx^a dx^b.$$

The conformal factor Ω is positive and vanishes only at null infinity. It falls off in such a way that g_{ab} is regular at $\mathscr{I}^+$, and all quantities constructed from it are finite. Automatically a suitable scaling is introduced which solves problem i). An elegant argument due to Friedrich (5) solves provlem ii). We have gauge freedom in the choice of Ω. $\Omega \to \theta\Omega$, where $\theta > 0$, θ finite on $\mathscr{I}^+$ is an equally good choice, and Friedrich showed how to choose θ so that the Einstein equations become regular at null infinity. (There is however a price to be paid for working with

"unphysical" spacetimes. Although the Weyl curvature, "the radiation part of the gravitational field", (i.e. the part not determined by matter) transforms homogeneously, the Ricci curvature (the part of the gravitational curvature determined by matter transforms inhomogeneously, so that although the physical spacetime may be vacuum, we have to include Ricci terms in the unphysical spacetime.) Notice that only null separations are unaltered by conformal transformations. Thus a coordinate system and other concepts adapted to null hypersurfaces are particularly adapted to conformal compactification.

These lectures are designed to be a reasonably self-contained introduction to the numerical solution of the CIVP in relativity. The first task is to establish the equations. This turns out to be a complicated task, and so chapter 1 outlines the problem for Maxwell's equations so as to orient the reader for chapter 2. Chapter 2 and the appendices form an introduction to the paper of Stewart & Friedrich (6), which in turn is closely based on work of Friedrich. Friedrich showed how to reduce the CIVP to the solution of a symmetric hyperbolic quasilinear system of equations followed by the solution of a symmetric hyperbolic linear system. Unfortunately the price to be paid for this is that the systems are rather large, and thus although numerical codes are very fast they do require quite large store regions.

One of the disadvantages of tying problems to null hypersurfaces in general relativity is that these surfaces develop caustics, where the surfaces (and possibly spacetime) develop singularities. An introduction to this topic, being a (possibly) oversimplified account of Friedrich and Stewart (7) is presented in chapter 3. It is clear from that chapter that any robust numerical algorithm for the CIVP must be able to locate and analyse the inevitable singularities in the solution of the differential equations.

Since numerical techniques for the solution of non-linear differential equations near singularities appeared not to exist in the literature they had to be constructed. In chapter 4 we present a variety of new methods for integrating ordinary differential equations near singularities. Chapter 5 extends this work to hyperbolic equations in one space variable. Because this work may have considerable utility in other applications it is written without reference to relativity and should be accessible to any numerical analyst.

One point which was made at the workshop is that it takes about 5 man-years to write a general code and about one man-year for each problem in this subject. This programme has only been running for 4 man-years, and so the final chapter on results is much shorter than we would like. However, our new techniques allow us to locate and integrate right up to singularities and we illustrate this with the Schwarzschild and colliding plane wave spacetimes, and an outline of work in progress on more general radiation-filled spacetimes. Chapters 4-6 are based on work done in collaboration with Corkill. They will be published in expanded form in (8), and in Corkill's doctoral dissertation.

As is clear from this introduction the ideas for this programme were developed in close collaboration with Helmut Friedrich and Richard Corkill whose generous cooperation is gratefully acknowledged. Bernd Schmidt also contributed greatly with copious constructive criticism. Abundant computer facilities were made available by the Max Planck Institute for Astrophysics, Munich.

1 MAXWELL's EQUATIONS

Because of the complexity of the Einstein equations it is a good idea to consider first the source-free Maxwell equations in flat spacetime

$$\nabla_b F^{ab} = 0, \qquad \nabla_{[a} F_{bc]} = 0. \tag{1.1}$$

Here F^{ab} is the antisymmetric Maxwell tensor with 6 independent components. Usually one chooses an orthonormal tetrad of vectors $\{e^{(i)}\}$,

$$\eta^{ab} e_a^{(0)} e_b^{(0)} = 1, \quad \eta^{ab} e_e^{(0)} e_b^{(\alpha)} = 0, \quad \eta^{ab} e_a^{(\alpha)} e_b^{(\beta)} = -\delta_{\alpha\beta},$$

where $a, b, \ldots = 0,1,2,3, \alpha, \beta, \ldots = 1,2,3, \eta^{ab} = \mathrm{diag}(1,-1,-1,-1)$, and then one can label the components of F^{ab} as $E^{(\alpha)}$, $B^{(\alpha)}$, or more simply $\underset{\sim}{E}$, $\underset{\sim}{B}$. Equations (1.1) then become

$$\underset{\sim}{\nabla}.\underset{\sim}{E} = 0, \qquad \underset{\sim}{\nabla}.\underset{\sim}{B} = 0, \tag{1.2a}$$

$$\underset{\sim}{\nabla} \mathrm{x} \underset{\sim}{E} = -\dot{\underset{\sim}{B}}, \qquad \underset{\sim}{\nabla} \mathrm{x} \underset{\sim}{B} = \dot{\underset{\sim}{E}}, \tag{1.2b}$$

in the usual notation. To solve an initial value problem one first specifies part of $\underset{\sim}{E}, \underset{\sim}{B}$ at t=0. One then solves the <u>constraint equations</u> (1.2a) in the surface t=0 to get $\underset{\sim}{E}, \underset{\sim}{B}$ there. Now the <u>evolution equations</u> (1.2b) determine $\underset{\sim}{E}, \underset{\sim}{B}$ at all later times. Given a perfectly accurate integration scheme there is no need to reconsider the constraint equations. For

$$(\underset{\sim}{\nabla}.\underset{\sim}{E})^{\bullet} = \underset{\sim}{\nabla}.(\dot{\underset{\sim}{E}}) = \underset{\sim}{\nabla}.\underset{\sim}{\nabla} \times \underset{\sim}{B} = 0,$$

and so equations (1.2a) are satisfied for all times. In practice however $\underset{\sim}{\nabla}.\underset{\sim}{E}$ is not exactly zero initially and the numerical equi-

valents of partial derivatives need not commute. Thus over a period of time the numerical solution of the evolution equations need not satisfy the constraints. For a fuller account of this problem see Dr. Smarr's lectures.

In what follows it will prove convenient to use a Newman-Penrose (NP) null tetrad $(l^a, n^a, m^a, \bar{m}^a)$ with the properties

$$l^a n_a = 1 = -m^a \bar{m}_a, \qquad \text{all other contractions vanish.}$$

Such a tetrad can be constructed from our orthonormal one as follows:

$$l = 2^{-\frac{1}{2}}(e^{(0)}+e^{(1)}), \quad n = 2^{-\frac{1}{2}}(e^{(0)}-e^{(1)}), \quad m = 2^{-\frac{1}{2}}(e^{(2)}+i\, e^{(3)}).$$

We replace $\underset{\sim}{E}, \underset{\sim}{B}$ by 3 complex scalars

$$\phi_0 = F_{ab} l^a m^b, \quad \phi_1 = \frac{1}{2} F_{ab}(l^a n^b + \bar{m}^a m^b), \quad \phi_2 = F_{ab} \bar{m}^a n^b,$$

and introduce the directional derivatives,

$$D = l^a \nabla_a, \quad \Delta = n^a \nabla_a, \quad \delta = m^a \nabla_a, \quad \bar{\delta} = \bar{m}^a \nabla_a .$$

Now in flat spacetime with Cartesian coordinates (t,x,y,z) the equations take the form:

$$\begin{aligned} E &\equiv D\phi_1 - \bar{\delta}\phi_0 = 0, \qquad & F &\equiv D\phi_2 - \bar{\delta}\phi_1 = 0, \\ F' &\equiv \Delta\phi_0 - \delta\phi_1 = 0, & E' &\equiv \Delta\phi_1 - \delta\phi_2 = 0 . \end{aligned} \tag{1.3}$$

With the notation established we can consider the following problem, the <u>characteristic initial value problem</u> (CIVP): given a 2-surface S_o, (say $t = t_o$, $x = x_o$,) and the 2 null 3-surfaces N_o, ($u \equiv t-x = t_o-x_o$), N'_o ($v \equiv t+x = t_o+x_o$) through it, (see fig. 1), pose suitable data on S_o, N_o, N'_o and solve for the fields in M, the future of $N_o \cup N'_o$.

The solution can be split into two parts. One can prescribe ϕ_0 arbitrarily on N_o, and ϕ_2 arbitrarily on N'_o. (Physically $\phi_0(\phi_2)$ represents the radiation crossing N_o (N'_o).) In addition ϕ_1 can be specified arbitrarily on S_o. Thus on S_o we must give ϕ_0, ϕ_1, ϕ_2, and ϕ_0 on N_o. Now $E = 0$ is an ordinary differential equation (ODE) giving $d\phi_1/dv$ along the generators u = const., y = const., z = const. of N_o, with initial data specified on S_o. Once this has been solved we know ϕ_1 on N_o. Next $F = 0$ is an ODE for ϕ_2 on N_o which we can solve. Similarly given ϕ_2 on N'_o we can solve in turn the ODE's $E' = 0$, $F' = 0$ for ϕ_1, ϕ_0 in N'_o. Thus we

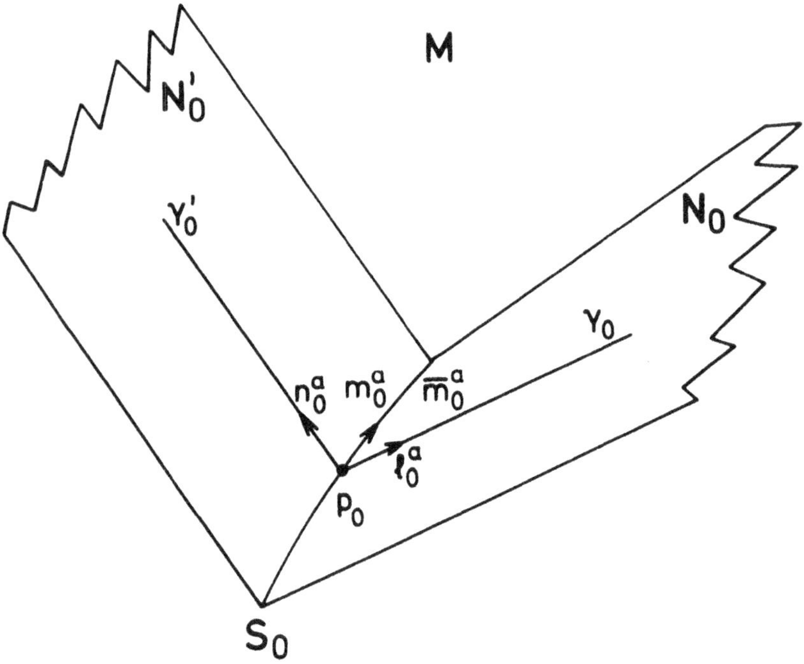

Figure 1

obtain ϕ_0, ϕ_1, ϕ_2 on $N_o \cup N_o'$.

It is straightforward to show that the equations

$$F' = 0, \quad E + E' = 0, \quad F = 0 \tag{1.4}$$

form a symmetric hyperbolic linear system for (ϕ_0, ϕ_1, ϕ_2). Because we know the solution on the characteristic surfaces N_o, N_o' we have a perfectly standard problem for which the existence and uniqueness of solutions is guaranteed. Although we may solve for the fields in M it is not a priori clear that we have solved Maxwell's equations in M. (We need to show $E = E' = 0$ in M, and not just $E + E' = 0$.) However it is straightforward to establish the identity

$$DE' - \Delta E \equiv \delta F - \bar{\delta}F' ,$$

which implies

$$(D + \Delta)E \equiv D(E + E') - \delta F + \bar{\delta}F' . \tag{1.5}$$

If we had solved the system (1.4) with perfect precision then

(1.5) would imply $(D+\Delta)E = 0$. But initially $E = 0$ and so $E = 0$ in M. Therefore we would have a solution of Maxwell's equations. In practice we know only that the right hand side of (1.5) is $O(\varepsilon)$, where ε is the truncation error, and that $E = O(\varepsilon)$ initially. Thus provided we integrate only over a finite range of u,v we are guaranteed that $E = O(\varepsilon)$ in M, so that we have a numerically satisfactory solution of Maxwell's equations in M.

For numerical purposes the system (1.4) has the highly desirable feature of being canonical hyperbolic; each of the equations is an equation in a hypersurface, i.e. it involves partial derivatives in at most 3 directions rather than 4. Thus if we use some explicit integration algorithm, e.g. leapfrog, we only have to handle problems in two space and one time variable. If we assume axisymmetry then we have only to handle hyperbolic problems in one space and one time variable, (which can be reduced to ODE problems.) With an additional symmetry we have only to handle ODE problems.

There is no intrinsic difficulty in including sources in the CIVP approach, and the use of curvilinear coordinates presents no problems. Serious problems arise however if we do not restrict S_o to be flat. For unless N_o is a light cone or a null hyperplane it will develop caustics, i.e. 2-surfaces where the hypersurface is no longer smooth. This topic will be discussed in some detail in chapter 3, but we point out here two of the problems it causes. If we match our coordinate/tetrad system closely to the geometry of the null hypersurfaces it will break down at a caustic, and the equations and their solutions become singular. Our numerical techniques need to be robust enough to handle this. The second problem occurs on the initial hypersurfaces. Beyond caustics on N_o or N'_o there will be regions of the hypersurface which are timelike related to S_o i.e. there exists a smooth timelike curve from points in these regions back to S_o. Thus these regions lie inside M. Clearly we cannot pose free data in these regions, since the solution there will be determined subsequently. A solution algorithm must take account of this.

2 EINSTEIN'S EQUATIONS

The first step in setting up the CIVP for general relativity is the introduction of a coordinate/tetrad system. As always in relativity this is not a trivial procedure; an infelicitous choice can greatly complicate a given problem. Here we describe a fairly natural choice which has proved extremely useful. A frame-independent treatment has been given in (6).

We start by choosing a spacelike 2-surface S_o. On S_o we set up a coordinate system x^A, $(A = 2,3)$. Next we construct one of

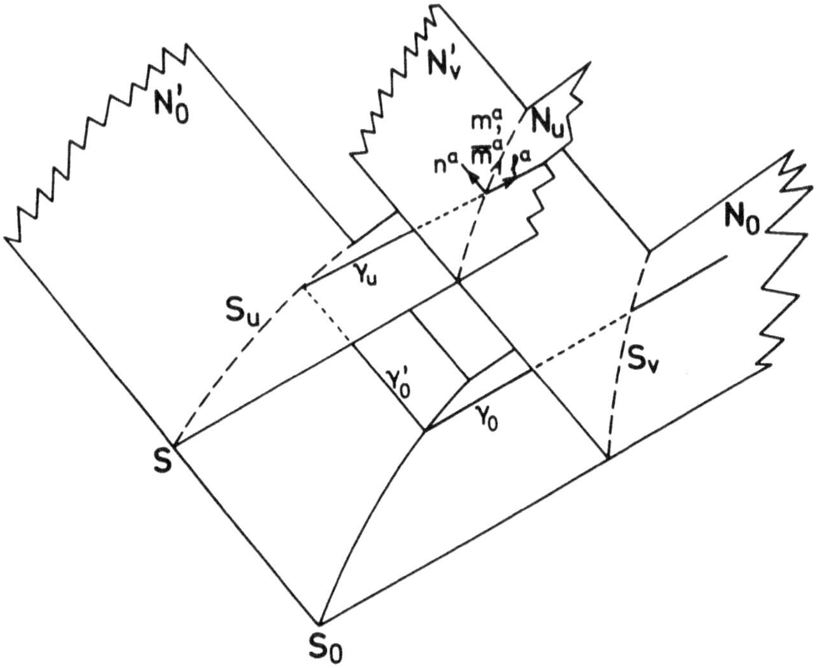

Figure 2

the null hypersurfaces N_o containing S_o. Let v be a (not necessarily affine) parameter along the generators γ_o of N_o. We label the 2-surfaces of constant v as S_v, fig. 2. If $l^a_o = dx^a/dv$ is the tangent vector to γ_o we can regard S_v as the result of moving S_o along the "lapse vector" l^a_o. (However, the "lapse function" $(l_a l^a)^{\frac{1}{2}}$ is zero!) Let N'_o be the other null hypersurface through S_o. We perform the same construction calling the parameter u, and the 2-surfaces of constant u S'_u. Let N_u be the other null hypersurface (besides N'_o) through S'_u. Let N'_v be the other null hypersurface through S_v, fig. 2. We define scalar functions u,v in spacetime M via u = constant on N_u, v = constant on N'_v. We next extend the definition of x^A from S_o to N'_o by requiring x^A = constant along the generators γ'_o and from N'_o to M by x^A = constant along the generators γ_u of N_u, fig. 2. Our coordinates are thus (u,v,x^2,x^3).

Our tetrad is chosen as follows. We first set

$$l_a = \nabla_a u , \qquad n_a = Q^{-1}(x^b)\nabla_a v \qquad \text{in M} \tag{2.1}$$

where Q will be chosen to set $l_a n^a = 1$. $m^a, \bar{m}^a$ are chosen on S_o

so as to "lie in S_o", (i.e. they span the tangent space of S_o). They are propagated onto N'_o by parallel propagation followed by projection so that they always lie in a S'_u. They are then propagated into M by a similar rule, so that they always lie in the 2-surface of intersection of a N_u and a N'_v, fig. 2. This means (6) that,

$$l^a = Q\partial/\partial v, \quad n^a = \partial/\partial u + C^A\partial/\partial x^A, \quad m^a = P^A\partial/\partial x^A, \tag{2.2}$$

where Q, C^A are real functions, P^A are complex, but are equivalent to 3 real functions. These 6 real functions are equivalent to the metric,

$$ds^2 = -|h_{AB}C^AP^B|^2\, du^2 + 2Q^{-1}dudv - 2h_{AB}C^A dudx^A - $$

$$- 2h_{C(A}h_{B)D}P^C\bar{P}^D dx^A dx^B , \tag{2.3}$$

where h_{AB} is the inverse of the positive definite matrix $h^{AB} = -2P^{(A}\bar{P}^{B)}$. They are our first 6 dependent variables. The connection components form the next set of dependent variables. Because of the way the tetrad was constructed certain simplifications occur. Using NP notation (4,9) the following relations hold:

$$\kappa = \nu = \varepsilon = 0, \quad \rho = \bar{\rho} \quad \mu = \bar{\mu} \quad \tau = \bar{\alpha} + \beta \quad \text{in M}$$
$$C^A = 0, \quad \gamma = \bar{\gamma} \quad \text{on } N'_o \tag{2.4}$$

The next 5 variables are the conformal factor Ω, and its first derivatives $s_0 = D\Omega$, $s_2 = \Delta\Omega$, $s_1 = \delta\Omega$. (D, Δ, δ were defined in chapter 1; s_1 is complex.) The Ricci tensor has 10 independent components. Our choice of conformal gauge ensures that the trace, the Ricci scalar vanishes. In NP notation the remaining 9 components appear as 3 real scalars Φ_{00}, Φ_{11}, Φ_{22} and 3 complex scalars Φ_{01}, Φ_{02}, Φ_{12}. The Weyl tensor has 10 independent components which can be written as 5 complex scalars Ψ_n, n = 0,1,...,4. In fact the Ψ_n actually vanish on $\mathcal{I}$, as does Ω, and it is more convenient to use $\phi_n = \Psi_n/\Omega$ which is finite and non-zero there. Thus we have 44 real dependent variables.

Of course in practice we do not construct the coordinate/frame system first, and then the connection and curvature tensors; the whole integration process proceeds as follows. We first choose a 2-surface S_o. Certain data is specified on S_o from which all the unknowns can be determined by differentiation and algebra. The details are described in appendix A. Next Q, Ψ_0, and Φ_{00} have to be specified on N_o. The choice of Q is the freedom available in the v-coordinate. Ψ_0 is the radiation and Φ_{00} the matter crossing N_o. They are the natural initial data for a coupled system of ordinary differential equations in N_o to determine the unknowns there. The details are described in appendix B. Then in an analogous manner we prescribe Q, Ψ_4, Φ_{22} on N'_o and solve an initial value problem for a system of ODE's to determine the unknowns in N'_o. The details are described in appendix B. A certain subset of the Einstein equations forms a quasilinear symmetric hyperbolic system for our unknowns in M, which we have to solve given the solution on $N_o \cup N'_o$. For the details see appendix C. The complement of this subset is a set of constraint equations on our unknowns. Only if the constraint equations are satisfied do we have a solution of Einstein's equations. Suppose we write the constraint equations in the form $\mathcal{C} = 0$. $\mathcal{C}$ is a rahter large array of quantities which vanish for a solution of Einstein's equations. Now by some rather lengthy manipulation of various identities one can establish

$$\mathcal{D}\mathcal{C} \equiv \mathcal{A}\mathcal{C} \tag{2.5}$$

Here $\mathcal{D}$ is a linear symmetric hyperbolic operator, and $\mathcal{A}$ is a matrix; all of the coefficients are known once we have solved for our unknowns in M. (The details can be found in (6).) If we had algorithms of perfect accuracy we would know that $\mathcal{C} = 0$ on $N_o \cup N'_o$ since we have already solved the equations there. Then by uniqueness $\mathcal{C} = 0$ in M, so that the constraints are automatically satisfied. In practice we know only that $\mathcal{C} = O(\varepsilon)$ on $N_o \cup N'_o$ where ε is the truncation error. However, provided we integrate the equations over only a finite range of u,v, (as in a compactified spacetime), and provided our unknowns remain regular, we are assured that $\mathcal{C} = O(\varepsilon)$ in M. This guarantee becomes invalid at and beyond caustics and singularities.

The quasilinear system of evolution equations that we use is canonical hyperbolic just as in the case of Maxwell's equations. Thus although we have a very large number of dependent variables we effectively have one less independent variable than expected. This is a significant simplification, especially for the numerical treatment of singularities.

3 CAUSTICS AND SINGULARITIES

Unfortunately the algorithms described in chapters 1, 2 only work smoothly for highly symmetrical problems. This is because the non-linearity of Einstein's equations produces (mathematical) singularities in the solutions. Consider for example the only non-linear equations on N_o in the vacuum case,

$$D\rho = \rho^2 + \sigma\bar{\sigma}, \qquad D\sigma = 2\rho\sigma + \Psi_0 , \tag{3.1}$$

where $D = d/dv$, and ρ is real. If ρ becomes positive and bounded away from zero it is easy to see that $\rho \to \infty$ in a finite v-time. Physically this singularity means that neighbouring generators of N_o have crossed, so that there is a local breakdown of our coördinate system and tetrad, i.e. a caustic. This caustic may also be a spacetime singularity.

The definition and classification of spacetime singularities is a complicated business; for a recent review see e.g. (10). Any numerically evolved spacetime is almost certain to be globally hyperbolic, and it appears from the results quoted in (10), (see especially §5.4.1), that in this case any spacetime singularity encountered by a numerical relativist will be a C^{0-} curvature singularity. A subset of these, the scalar polynomial (sp) curvature singularities can be characterised by the property that the components of the curvature tensor with respect to every regular frame become infinite. The complement of this subset is thought to consist of unstable singularities, which would probably not occur in a numerical evolution. We therefore restrict consideration for the time being to sp curvature singularities.

At a caustic the tetrad components of the curvature are likely to become infinite. As we shall see by rescaling the tetrad we can make it regular at the caustic. If the curvature components then become finite we can rule out the possibility of a sp curvature singularity. If they remain infinite we have a more difficult problem. Either we have to demonstrate that they are infinite in all regular frames, perhaps by considering curvature invariants, or we have to find a regular frame in which the curvature is finite.

Suppose now that we encounter a caustic which is not a curvature singularity. We would want to continue the numerical evolution further. To do this we need first a classification of caustics. Clearly this is a hopeless task in general. But here we are only interested in the local properties of stable caustics. A caustic on N_o will be said to be stable if it is stable under small perturbations of S_o and of the spacetime metric. For a more precise definition see (7). Since we are only interested in local properties at regular points in spacetime it is probably sufficient to consider stable caustics in Minkowski spacetime. A detailed study of such caustics has been carried out recently (7). If we restrict consideration to axisymmetric configurations there are precisely two possibilities.

The simplest caustic is the A2 or cusp line. To see an example consider Minkowski spacetime with coordinates $x^i = (t,x,y,z)$. In a neighbourhood of the origin we introduce new coordinates $a^i = (a,b,c,d)$ where

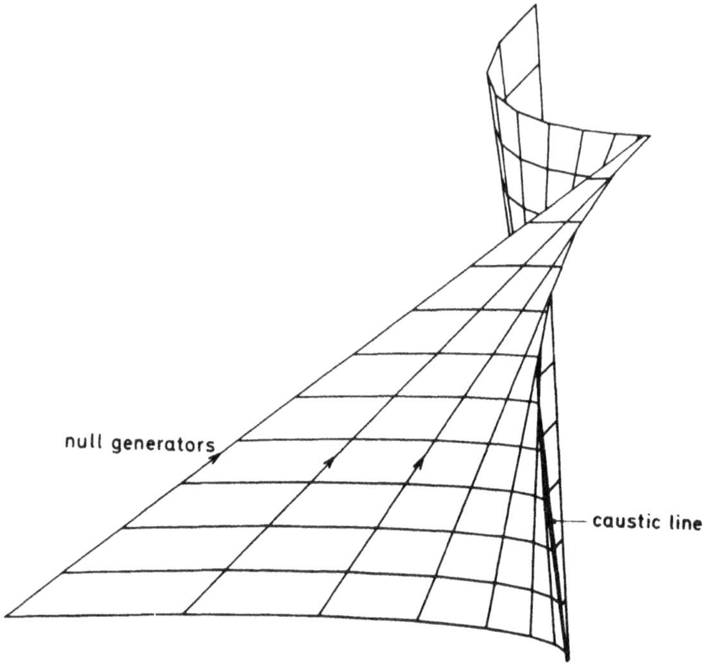

Figure 3

$$t = -2c^3+a+\frac{1}{2}b, \quad x = 3c^2-ac+\frac{1}{2}bc, \quad y = f(a-\frac{1}{2}b), \quad z=d \qquad (3.2)$$

and $f = (1-c^2)^{\frac{1}{2}}$. The Jacobian $|\partial x^i/\partial a^j|$ vanishes whenever $F = 0$, where $F \equiv 6cf^2 - a + \frac{1}{2}b$. We also introduce a NP tetrad which has components with respect to x^i-coordinates,

$$l^i = (1,-c,f,0), \; n^i = \frac{1}{2}(1,c,-f,0), \; m^i = 2^{-\frac{1}{2}}(0,f,c,i). \qquad (3.3)$$

Clearly this tetrad is perfectly regular in a neighbourhood of the origin. However, its components with respect to a^i-coordinates are

$$l^i=(1,0,0,0), \; n^i=(0,1,0,0), \; m^i=2^{-\frac{1}{2}}(6c^2f/F,0,f/F,i). \qquad (3.4)$$

This tetrad appears to be singular at $F = 0$, because the coordinate system breaks down there. Next consider the hypersurface Σ defined by $b = 0$. This hypersurface is null, l^i is tangent to the generators, defined by c=constant, d=constant and a is an affine parameter along the generators. $T(\Sigma)$ is spanned by $l, m, \bar{m}$. Thus the a^i-coordinates and frame are very similar to those introduced in chapter 2. To see what happens at $F = 0$, consider Fig. 3 which shows Σ embedded in Minkowski spacetime. There is a cusp at $F = 0$ which is intersected by every generator in a neighbourhood of the origin. This is clearly singular as can be seen by computing

$$\rho = \sigma = \frac{1}{2}F^{-1} \tag{3.5}$$

Even though our coordinate system has broken down at the caustic it is immediately obvious how to continue it through $F = 0$ since $\partial F/\partial a$, $\partial F/\partial b$, $\partial F/\partial c$ are all non-zero in a neighbourhood of the origin. Next consider the proper distance ds between the point $a = 0$, $c = c_o(1+\varepsilon)$ on S_o and the point $a = a_o$, $c = c_o$ on Σ. Elementary algebra shows that

$$ds^2 = -6F_o c_o{}^3 \varepsilon^2 + O(\varepsilon^3) \qquad F_o = F(c_o, a_o)$$

Now for sufficiently small a_o F_o has the same sign as c_o. Thus if $\varepsilon \neq 0$ $ds^2 < 0$ corresponding to a spacelike separation when the points lie on different generators. If $\varepsilon = 0$ it is trivial to show that $ds^2 = 0$ corresponding to a null separation. However, if a_o is increased until F_o changes sign then ds^2 changes sign also.

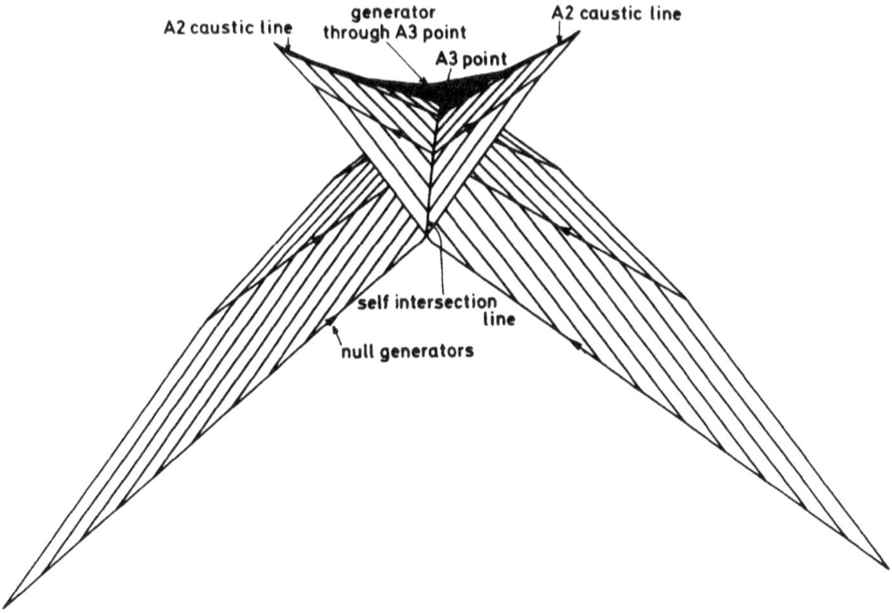

Figure 4

Thus points lying beyond the caustic, i.e. $cF < 0$ are timelike related to points on S_0. If such a caustic occurs on an initial null surface we cannot give constraint-free initial data at points beyond the caustic.

It should be noted that every other A2 caustic in Minkowski spacetime is locally diffeomorphic to this one. Thus we have treated the general case.

The only other stable caustic that can occur in an axisymmetric configuration in Minkowski spacetime is an A3 point which occurs when two A2 cusp lines touch, Fig.4. To construct an example we introduce new coordinates $p^i = (p,q,r,s)$, where

$$t = p+\tfrac{1}{2}q, \quad x = r(1-g)-\varepsilon gr(p-\tfrac{1}{2}q), \quad y = \varepsilon(gr^2+g-1)+g(p-\tfrac{1}{2}q), \quad z=s, \tag{3.6}$$

and $g = (1+r^2)^{-\frac{1}{2}}$, $\varepsilon = \pm 1$. The Jacobian vanishes whenever $G = 0$, where $G \equiv \varepsilon(p-\tfrac{1}{2}q) + 1 - 1/g^3$. We also introduce a NP tetrad which has components with respect to the x^i-coordinates,

$$l^i = (1,-\varepsilon gr,g,0), \quad n^i = \tfrac{1}{2}(1,\varepsilon gr,-g,0), \quad m^i = 2^{-\frac{1}{2}}(0,g,\varepsilon gr,i) \tag{3.7}$$

Clearly this tetrad is perfectly regular in a neighbourhood of the origin. However, its components with respect to p^i-coordinates are

$$l^i = (1,0,0,0), \quad n^i = (0,1,\tfrac{1}{2}(\varepsilon-1)r/G,0), \quad m^i=2^{-\frac{1}{2}}(0,0,-\varepsilon/g^2G,i). \tag{3.8}$$

This tetrad appears to be singular at $G = 0$, because the coordinate system breaks down there. Next consider the hypersurface Σ defined by $q = 0$. This hypersurface is null, l^i is tangent to the generators, defined by r = constant, s = constant, and p is an affine parameter along the generators. $T(\Sigma)$ is spanned by $l,m,\bar{m}$. To see what happens at $G = 0$, consider Fig.4 which shows Σ embedded in Minkowski spacetime. If $r \gtrless 0$ $G = 0$ corresponds to one of two A2 cusp lines, which touch at $r = 0$. The qualitatively new feature is a self-intersection line which touches the two cusp lines at their point of contact. This line corresponds to the intersection of widely separated generators. (A caustic corresponds to the intersection of neighbouring generators.) Note that

$$\rho = \sigma = -\tfrac{1}{2}/G \tag{3.9}$$

so that the connection components become singular at the caustic. Away from the A3 point $r = 0$ we have the A2 case already discussed. At the A3 point $\partial G/\partial r = 0, \partial G/\partial p, \partial G/\partial q, \partial^2 G/\partial r^2 \neq 0$ and so it is straightforward to continue the solution beyond the caustic.

The self-intersection line occurs at $H = 0$ where $H \equiv p+1-1/g$. Thus moving along a generator from S_0, $p = q = 0$, $r \neq 0$, it is

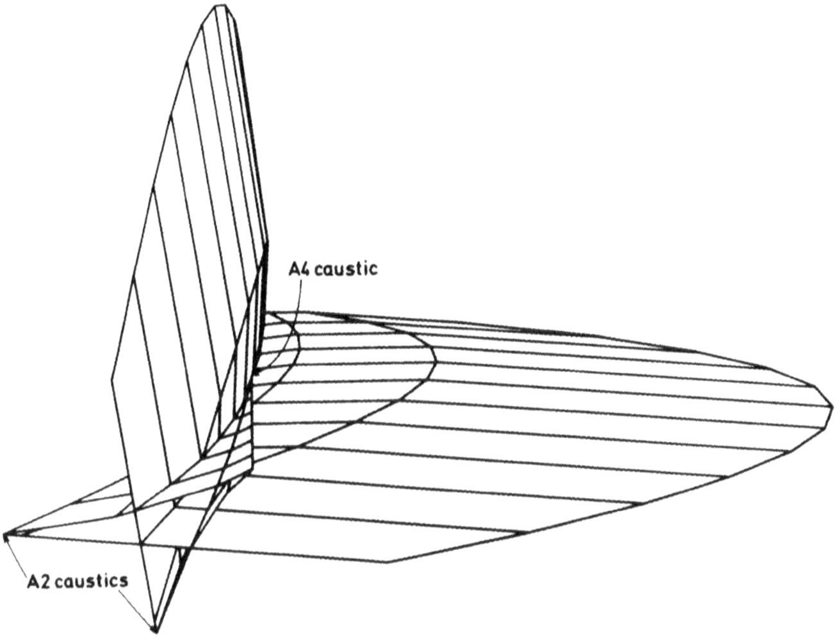

Figure 5

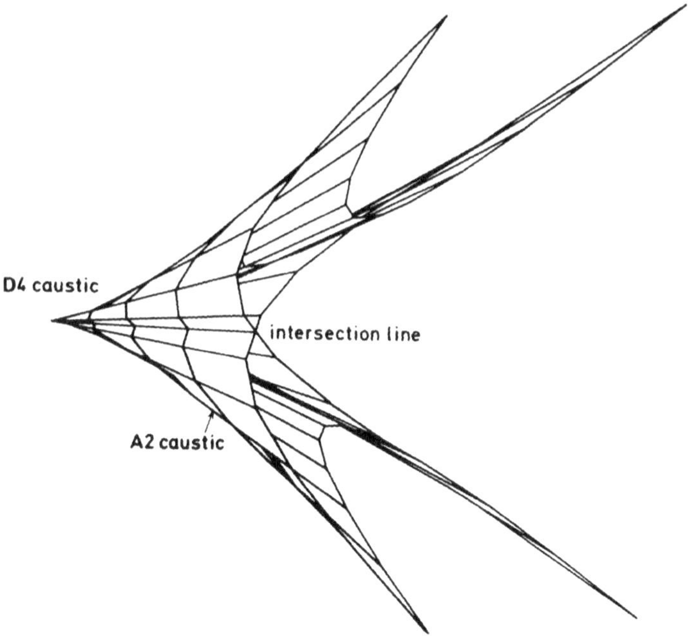

Figure 6

always encountered before the caustic. It is easy to see that points beyond this line are timelike related to S_o. Thus free data cannot be given beyond this line. It should be noted that the tetrad is perfectly regular there, except of course at $r = 0$. The location of the line is discussed in (7) and a practical algorithm will be published elsewhere. It should be noted that every other A3 caustic in Minkowski spacetime is locally diffeomorphic to this one. Thus we have treated the general case.

If the restriction to axisymmetry is removed caustics can become considerably more complicated. An A3 caustic can encounter an isolated A2 cusp line. For an instant three cusp lines and three lines of self-intersection touch at an A4 point, Fig. 5. An A3 caustic can touch a pair of touching cusp lines to form a $D4^+$ caustic, Fig. 6. For these and the one remaining case, the reader is referred to the discussion in (7).

4. NUMERICAL METHODS FOR QUASILINEAR ORDINARY DIFFERENTIAL EQUATIONS

As we have already seen singularities are a consequence of the non-linearity of Einstein's equations. Any robust algorithm for the numerical solution of Einstein's equations must be able to locate and handle these singularities successfully. In this chapter we discuss methods for the solution of ordinary differential equations which have singular behaviour analogous to that of Einstein's equations. The next chapter extends the treatment to hyperbolic systems.

We consider first a simple model for equation (3.1) viz.

<u>Problem 1</u>

$$y'(t) = 1 + y^2, \quad y(0) = 1, \tag{4.1}$$

with solution $y(t) = \tan(t+\frac{1}{4}\pi)$. This is a special case of

$$y'(t) = f(t,y), \quad y(0) = y_o \,. \tag{4.2}$$

We are interested in computing the evolution using a fixed step length h and some low order explicit method. (We are interested ultimately in the generalisation to systems of hyperbolic partial differential equations.) We establish the notation,

$$t_n = nh, \quad y_n = y(t_n), \quad f_n = f(t_n,y_n), \quad r_n = hf_n/y_n.$$

The simplest method to derive such algorithms is to replace $y(t)$ by an <u>approximant</u>, i.e. a function $Y(t)$ of given functional form, but containing free parameters. These parameters are fixed locally by known data, and then Y(t) is used to predict new values of Y. To see how this works consider the linear approximant $Y(t) = at + bt$, containing two free parameters a,b. Introduce $Y_n = Y(t_n)$, $F_n = f(t_n,Y_n)$. We then fix a,b by requiring $R_n = hF_n/Y_n$, $Y_n = y_n$, $Y'(t_n) = F_n$, $R_n = hF_n/Y_n$, finding $b = F_n$, $a = Y_n - bt_n$. Then

$$Y_{n+1} = a + b(t_n+h) = Y_n + hF_n , \tag{4.3}$$

an algorithm usually known as "Euler's method". Assuming that Y_n, F_n are accurately known the local error is $\Delta \equiv y_{n+1} - Y_{n+1} = O(h^2)$. Similarly the quadratic interpolant $Y(t)=a+bt+ct^2$ produces the "leap-frog" algorithm

$$Y_{n+1} = Y_{n-1} + 2hF_n , \tag{4.4}$$

with local error $\Delta = O(h^3)$.

The following table shows the result of applying Euler's and the leapfrog method to problem 1. The error is defined to be $y_{computed}/y_{exact} - 1$, and 1.23e-2 = 1.23×10^{-2} etc. For completeness we have also indicated the results for Runge-Kutta methods of second and fourth orders. The singularity occurs at t = 0.7854.

Error

t	Euler	Leap-frog	RK(2)	RK(4)	Padé(0,1)	Padé(1,1)
0.75	-4.1s-1	-6.1e-2	-3.0e-2	-8.0e-5	1.1e-1	-7.0e-4
0.76	-5.0e-1	-1.1e-1	-5.4e-2	-2.8e-4	1.6e-1	-1.0e-3
0.77	-6.4e-1	-2.1e-1	-1.2e-2	-1.7e-3	2.9e-1	-1.6e-3
0.78	-8.4e-1	-5.3e-1	-3.8e-1	-3.9e-2	1.8e0	-4.8e-3
0.79	-1.2e0	-2.0e0	-3.2e0	-2.6e2	-4.3e-1	5.7e-3
0.80	-1.8e0	-1.5e1	-5.9e2	-7.4e39	-1.9e-1	1.8e-3

It should be noted that while conventional methods give good results away from the singularity they are hopeless close to it. Furthermore, increasing the order of the method only gives worse results at the singularity. The reason is that no polynomial approximant can mimic the pole at $t = \frac{1}{4}\pi$. However this remark immediately suggests the approximant $Y(t) = a/(t+c)$, which leads to the non-linear algorithm, the Padé (0,1) algorithm,

$$Y_{n+1} = Y_n^{\,2}/(Y_n - hF_n) = Y_n/(1 - R_n) , \tag{4.5}$$

which appears to have been stated first, albeit implicitly, by

Lambert & Shaw (11). If $y_n \neq 0$ then the local error $\Delta = O(h^2)$, but if $y_n = 0$ then $\Delta = O(h)$ so that the method fails to be consistent. (To understand this, note that the approximant is either never zero or always zero.) If R_n is small then (4.5) reduces to the Euler method. We have also considered the approximant $Y(t) = (a+bt)/(c+dt)$ which leads to the Padé (1.1) algorithm,

$$Y_{n+1} = Y_{n-1} + (Y_n - Y_{n-1})^2/(Y_n - Y_{n-1} - \tfrac{1}{2} hF_n). \tag{4.6}$$

If $f_n \neq 0$ the Pade (1.1) algorithm has local error $\Delta = O(h^3)$ but if $f_n \neq 0$ it is inconsistent. (The approximant has not turning points.) If R_n is small, it reduces to leap-frog. The results of applying these two non-linear algorithms to the non-linear problem 1 are also shown in the table above. The Padé (0,1) method compares very favourably with Euler, while the Padé (1.1) method is clearly superior to leap-frog.

The reason why the above algorithms work so well on problem 1 is that near the singularity the exact solution has asymptotically the same form as the approximant. They work less well on

Problem 2

$$y'(t) = \frac{3}{2}(1 + y^{4/3})y^{1/3}, \quad y(0) = 1, \tag{4.7}$$

with solution $y(t) = \tan^{3/2}(t + \frac{1}{4}\pi)$. Lambert & Shaw (11) considered the approximant

$$Y(t) = \sum_{j=0}^{L} a_j t^j + b|c+t|^N, \quad N \neq 1,2,\ldots,L.$$

However they concluded that such a method required either

a) knowledge of $\partial f(t,y)$: y, $\partial^2 f/\partial y^2, \ldots$

or b) the solution of transcendental equations for c,N at each step.

To see this consider the simplest case $Y(t) = a/|b+t|^\lambda$. Then the algorithm is

$$Y_{n+1} = Y_n / |1 - R_n/\lambda|^\lambda . \tag{4.8a}$$

The approximant contains 3 free parameters and knowledge of y_n, f_n fixes only two of these. Lambert & Shaw assumed that y_{n-1} was also known and solved the transcendental equation

$$Y_n/Y_{n-1} = |1 + R_n/\lambda|^{\lambda} \tag{4.8b}$$

for λ at each step. However, if y_{n-1} is known then so is f_{n-1}. It then follows that

$$\lambda = R_{n-1}R_n/(R_n - R_{n-1}). \tag{4.8c}$$

Both the Lambert & Shaw algorithm (4.8a,b) and the generalised Padé (0,1) algorithm (4.8a,c) have local error $\Delta = O(h^3)$ provided $y_n, y_{n-1} \neq 0$ and are inconsistent otherwise.

Our approach can be generalised to a third order method which however requires the determination of the largest real root of a cubic equation at each stage. We use the approximant

$$Y(t) = (a + bt)/|c + t|^{\lambda} .$$

The generalised Padé (1.1) algorithm is

$$Y_{n+1} = Y_n(1 + A)/|1 + B|^{\lambda} \tag{4.9a}$$

where A is the largest real root of

$$2\alpha A^3 + 2(\gamma - 2\delta)A^2 + (\beta - 3\gamma)A + \gamma = 0 , \tag{4.9b}$$

$$B = 1 - (1-A)(A - R_n)/(A - R_{n-1}(1-A)) , \tag{4.9c}$$

$$\lambda = (A - R_n)/B , \tag{4.9d}$$

and

$$\alpha = 2R_{n-2} - R_{n-1} ,$$

$$\beta = R_{n-2} - 2R_{n-1} + R_n ,$$

$$\gamma = R_{n-1}(R_n + R_{n-2}) - 2R_nR_{n-2} ,$$

$$\delta = R_{n-2} - R_{n-1} .$$

Before comparing these more elaborate methods on problem 2 some comments on stability are necessary. The Padé methods should not be applied near zeros or turning points of the dependent variable. In constructing a practical algorithm this must be taken into account; one would normally use a conventional method, and only switch to a Padé method near the singularity. For this reason the usual stability criteria are irrelevant, although all of our methods are in fact A-stable. In practice the Padé algorithms will be presented with noisy rather than exact initial data.

This causes some algorithms, (not those presented here) to become unstable. Thus a more realistic method for comparing the algorithms is to integrate the equations using a conventional method and then to switch in the Padé method. In the following two tables problem 2 was integrated for $0 \le t \le 0.66$ using leapfrog (third order Runge-Kutta for the generalised Padé (1,1) method) and a step length of $h=10^{-3}$. The singularity occurs at $t_s = \frac{1}{4}\pi$ with a λ of 1.5. As before the error is $y_{computed}/y_{exact} - 1$.

	Padé (1,1)		Generalised Padé (0,1)		
t	error	$t_s-\frac{1}{4}\pi$	error	λ -1.5	$t_s-\frac{1}{4}\pi$
0.75	-3.2e-4	-7.2e-3	-4.1e-4	4.1e-3	7.7e-5
0.76	-3.8e-4	-5.2e-3	-4.1e-4	2.2e-3	3.6e-5
0.77	-3.6e-4	-3.3e-3	-9.6e-4	8.6e-4	1.6e-5
0.78	2.4e-3	-1.2e-3	-2.74-3	1.5e-4	1.0e-5
0.785	2.8e0	-3.5e-4	-3.6e-2	1.5e-5	9.9e-6

	Lambert & Shaw			Generalised Padé (1,1)		
t	error	λ-1.5	$t_s-\frac{1}{4}\pi$	error	λ-1.5	$t_s-\frac{1}{4}\pi$
0.75	-4.1e-4	2.7e-2	6.3e-4	-4.1e-7	-4.2e-3	-3.5e-5
0.76	-5.8e-4	2.3e-2	4.0e-4	-5.4e-7	-2.2e-3	-1.4e-5
0.77	-9.6e-4	1.8e-2	2.1e-4	-8.2e-7	-9.1e-4	-3.5e-6
0.78	-2.8e-3	1.3e-2	6.4e-5	-1.9e-6	-1.6e-4	-2.6e-7
0.785	-2.9e-2	-7.6e-3	3.4e-6	-7.8e-6	-1.9e-5	-5.1e-9

The Padé (1,1) and generalised Padé (0,1) methods cost about the same. Close to the singularity the latter is superior, and it delivers highly accurate information about the singularity. Although Lambert & Shaw's method is second order its cost is about the same as the third order Generalised Padé (1,1) method. While Lambert & Shaw's algorithm delivers results of comparable accuracy to the generalised Padé (0,1) method one can for the same cost improve the accuracy by an order of magnitude. Another interesting feature is apparent from the tables. Although the accuracy in the dependent variable drops as one approaches the singularity, the accuracy in λ, t_s actually improves!

In generalising these methods to coupled systems of equations two points need to be taken into account. The naive approach would be to apply a Padé method componentwise. Then for each dependent variable one can calculate a t_s. Because of various errors one will obtain a slightly different t_s for each component. Normally, however, one expects singular behaviour from the dependent variables at a common value of the independent variable. This leads to a considerable simplification. Suppose one labels the variables $y^{(1)},\ldots,y^{(N)}$. One first uses one of the above methods for one equation, say that for $y^{(1)}$. Then one may set $c^{(i)} = c^{(1)}$ for $i > 1$, which considerably simplifies the algebra. As an example consider the most complicated algorithm so far introduced, the generalised Padé (1,1) algorithm, and suppose it has been used to determine $A^{(1)}, B^{(1)}, \lambda^{(1)}$. Then for the remaining components we may use the much simpler algorithm

$$Y^{(i)}_{n+1} = Y^{(i)}_{n}(1 + A^{(i)}) \, / \, |1 + B|^{\lambda(i)} , \qquad (4.10a)$$

where $B = B^{(1)}$ is known, $A^{(i)}$ is the larger root of

$$A + (R^{(i)}_{n+1}(1-B) - R^{(i)}_{n} - B)A + R^{(i)}_{n} - R^{(i)}_{n-1}(1-B) = 0 , \qquad (4.10b)$$

and

$$\lambda^{(i)} = (A^{(i)} - R^{(i)}_{n})/B .$$

Similar simplifications apply to the lower order algorithms.

The second point is that one would never expect to compute $\lambda^{(i)} = 0$, because of rounding errors, etc. Thus at the singularity each $Y^{(i)} \to 0$ or $\to \infty$. One should monitor the possibility that $\lambda^{(i)}$ can become small, and in such an event integrate this variable using a conventional method. A numerical example is given in (8).

5. NUMERICAL METHODS FOR QUASILINEAR HYPERBOLIC SYSTEMS IN ONE SPACE AND ONE TIME VARIABLE

We consider first a single quasilinear equation

$$y_t + A(t,x,y)y_x = B(t,x,y) \qquad (5.1)$$

together with Cauchy data $y(0,x) = y_o(x)$ given for a suitable domain. Suppose that y becomes singular on the curve $t = s(x)$ and that an approximant $Y(t,x) = a/|t - s(x)|^{\lambda}$ is appropriate. Unless $s(x)$ is known we must approximate it. Unfortunately we have been unable to find a straightforward generalisation of the methods of chapter 4 which is stable in the sense of chapter 4.

We can however generalise Hartree's method (12). Suppose that y is known at regularly spaced x-values for $t = t_o$ and $t = t_1 = t_o - h$, and we want to find $y(t_o+h,x)$. We assume provisionally that A is y-independent. We can then integrate the characteristic equation $dx/dt = A$ back from the new point until it intersects the previous time levels at $(t_o,x_o),(t_1,x_1)$ by standard methods. Next we have to find $Y(t_o,x_o)$. Since we know y only on a grid of x-values at $t = t_o$, this involves interpolation. However, the standard interpolation techniques cannot be used here! If λ were known then $|Y|^{-1/\lambda}$ would be regular along

$t=t_0$ and so a standard interpolation formula could be used on it. (We can always arrange matters locally to set $Y > 0$.) In a similar manner we find $Y(t_1,x_1)$. Along the characteristic $x = x(t)$, (5.1) reduces to the ordinary differential equation

$$dy/dt = B(t,x(t),y(t,x(t)), \tag{5.2}$$

and we know y at $t = t_0$, $t = t_1$. This is precisely the problem we have solved in chapter 4, and a 3-level algorithm such as generalised Padé (0,1) may be used. (If y-values at 3 time levels can be carried then generalised Padé (1,1) could also be used.) This automatically generates a λ-value for use in interpolation at the next time level. Provided A is y-independent this procedure is obviously explicit. As in Hartree's original approach we have to solve the problem by iteration if A depends on y.

As an example we have considered

<u>Problem 3</u>

$$y_t - \tfrac{1}{3}y_x/(1+x)^2 = \tfrac{5}{3}\, y^{\frac{5}{3}}, \quad y(0,x)=(1+(1+x)^3/9)^{-\frac{3}{2}} \quad 0 \le x \le 1,$$

together with exact boundary conditions at $x = 1$. (The exact solution is

$$y = \text{sign}(d)/|d|^{3/9} \quad \text{where } d \equiv 1 - t + (1+x)^3/9 .$$

Details of the numerical integration are duscussed in (8). Using a steplength in both space and time of $h = 5 \times 10^{-3}$, the new Padé-Hartree method was employed whenever the algorithm thought it was within 100 steps of the singularity, and leapfrog was employed elsewhere. Near the singularity λ and the position were found to an accuracy 10^{-6}, and the relative error in the solution was $\sim 10^{-3}$. Away from the singularity this error improved and was only $100h^3$ after 500 steps.

There are no theoretical difficulties in extending the Padé-Hartree approach to symmetric hyperbolic systems of equations. One has to construct the system of characteristic lines through each point and integrate an ordinary differential equation along each characteristic. As in the case of systems of ordinary differential equations one should assume a common singularity for all of the dependent variables, and monitor the possibility of zeros or stationary points.

For large systems of equations the method of characteristics can become very laborious. For the special case of constrained canonical hyperbolic systems (defined below), much simpler methods can be constructed. Consider the system

$$y_t^{(i)} = B^{(i)}, \qquad 1 \leq i \leq I, \qquad (5.3a)$$

$$z_t^{(j)} + \sum_{i=1}^{I} C^{(ij)} y_x^{(j)} + \sum_{k=1}^{J} A^{(jk)} z_x^{(k)} = D^{(j)} \qquad 1 \leq j \leq J, \qquad (5.3b)$$

together with the constraint

$$\sum_{i=1}^{I} E^{(i)} y_x^{(i)} = F, \qquad (5.3c)$$

and suitable Cauchy data. Here the coefficients are functions of $t, x, y^{(i)}, z^{(j)}$, $A^{(jk)}$ is lower triangular, i.e. $A^{(jk)} \doteq 0$ for $k>j$, the Cauchy data satisfies (5.3c), and (5.3c) is compatible with (5.3a,b) i.e. if it is satisfied initially then it is satisfied for all t,x as a consequence of (5.3a,b). Such a system will be called constrained canonical hyperbolic. An example of such a system is Einstein's field equations in the CIVP formalism.

For simplicity we exhibit only the generalised Padé (0,1) algorithm. We use the approximants

$$Y^{(i)}(t,x) = Y_o^{(i)} |t-s(x)|^{-\lambda(i)}, \quad Z^{(j)} = Z_o^{(j)} |t-s(x)|^{-\mu(j)}. \qquad (5.4)$$

It is clear that we may evolve the system (5.3a) explicitly through one time step as if it were a system of ordinary differential equations using the generalised Padé (0,1) method. Thus we may regard the $\lambda^{(i)}$, $Y_o^{(i)}$ and $s(x)$ as known at each point on our computational grid. Now (5.3b) implies

$$\mu^{(j)} Z^{(j)} = \left(\sum_{i=1}^{I} C^{(ij)} \lambda^{(i)} Y^{(i)} - \sum_{k=1}^{J} A^{(jk)} \mu^{(k)} Z^{(k)}\right) s'(x) - D^{(j)}(t-s(x)). \qquad (5.5)$$

If $s'(x)$ were known this would be an explicit system of equations for the $\mu^{(j)}$. (It is here that the restriction that $A^{(jk)}$ be lower diagonal has been used. This condition could be relaxed at the cost of solving an algebraic linear system of equations.) A naive approach would be to obtain $s'(x)$ by finite-differencing $s(x)$ already known at the grid points. However, this leads to methods which are unstable in the sense of chapter 4. It is essential to use the constraint (5.3c) in the form

$$\left(\sum_{i=1}^{I} E^{(i)} \lambda^{(i)} Y^{(i)}\right) s'(x) = F(t-s(x)), \qquad (5.6)$$

which delivers $s'(x)$ to sufficient accuracy to guarantee stability. Thus all of the parameters in the approximants can be obtained explicitly and we can extrapolate forwards in time. Notice that we have to carry the y-variables at 2 time levels and the z-variables at only one.

As an example we have considered

Problem 4

$$y_t^{(1)} = 2y^{(2)} ,$$

$$y_t^{(2)} = \tfrac{1}{2}z ,$$

$$z_t + y_x^{(2)}/(1+x) = (6y^{(1)}-1)y^{(1)}y^{(2)}/ 2 ,$$

together with the constraint

$$y^{(1)}y_x^{(1)} + 2y_x^{(2)} = -(y^{(1)})^3(1+x)/2 .$$

Setting $s(x) \equiv 1 + \frac{1}{4}(1+x)^2$ we imposed initial data

$$y^{(1)}(0,x) = 2/s(x),\ y^{(2)}(0,x) = 1/s^2(x),\ z(0,x) = 4/s^3(x),$$

for $0 \leq x \leq 1$.

The exact solution is

$$y^{(1)} = 2/d(t,x),\ y^{(2)} = 1/d^2(t,x),\ z = 4/d^3(t,x),\ d \equiv -t+s(x).$$

The Padé algorithm was switched in when the algorithm thought it was within 100 steps of the singularity; otherwise leapfrog was used. Using a steplength in space and time $h = 10^{-2}$ the position of the singularity could be located to about 1 part in 10^5, the order of the singularity, $(\lambda^{(1)}, \lambda^{(2)}, \mu)$ to 1 part in 10^9, and the relative errors in the variables were $O(h^2)$. However, once the singularity had been passed and leapfrog switched on again the well-known instability that can occur with non-linear equations occurred. Further details can be found in (8).

We have not constructed Padé algorithms for the case of more than one space variable, because at present we do not need them. However, we anticipate no great difficulty in doing so.

6. SOME NUMERICAL SOLUTIONS OF EINSTEIN'S EQUATIONS

The goals of numerical relativity include the investigation of singularities and horizons and the generation and propagation of gravitational waves. As a first example we consider the numerical evolution of the Schwarzschild spacetime.

Let (t,r,θ,ϕ) be the standard Schwarzschild coordinates. We first introduce the standard retarded and advanced times,

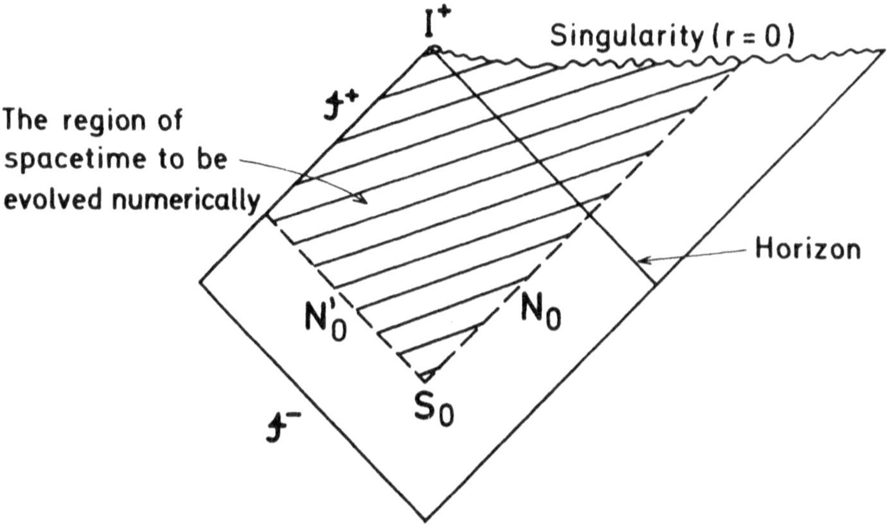

Figure 7

$$U = t-r-2M \ln |r/2M-1|, \quad V = t+r+2M \ln|r/2M-1| .$$

We now define new null coordinates $u = u(U)$, $v = v(V)$. Then the usual "physical" metric can be written

$$d\tilde{s}^2 = (fab)dudv - r^2 d\Sigma^2, \qquad d\Sigma^2 = d\theta^2 + \sin^2\theta d\phi^2 ,$$

where $f = 1-2M/r$, $a(u) = dU/du$ and $b(v) = dV/dv$. Now we make a conformal compactification with conformal factor $\Omega = h(u,v)/r$. The "unphysical" metric becomes

$$ds^2 = \Omega^2 d\tilde{s}^2 = (fab)(h/r)^2 dudv - h^2 d\Sigma^2 . \qquad (6.1)$$

$h(u,v)$ is not an arbitrary function because we have to choose a conformal gauge in which the unphysical Ricci scalar vanishes. This means that h has to satisfy a scalar wave equation

$$h_{uv} + \tfrac{1}{2}(fabM/r^3)h = 0 . \qquad (6.2)$$

The following choice of null coordinates was made:

$$u = -4M/U , \qquad v = \tan^{-1}(-\tfrac{1}{2}e^{-\frac{1}{4}V/M}/M) . \qquad (6.3)$$

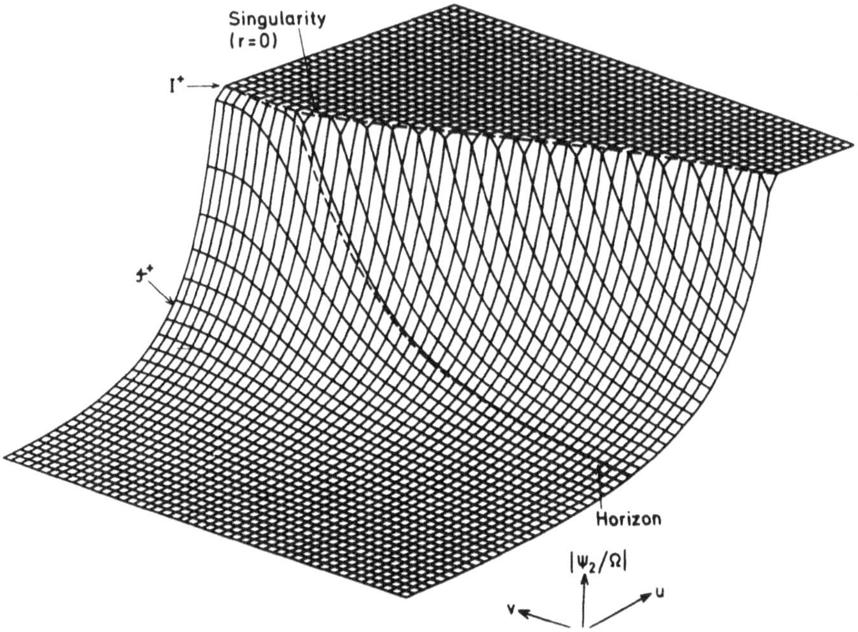

Figure 8

Although this results in a rather unconventional view of Schwarzschild, fig. 7, it was chosen so that the extremely irregular point I^+ was the last one to be dealt with by the computer programme. Specifying the unphysical Ricci tensor components Φ_{00} on N_o, (u = -1), and Φ_{22} on N_o' (v = -1), is equivalent to specifying h(u,v) on N_o, N_o'. Our choice was

$$h(-1,v) = (r(-1,v)/r(-1,-1))^{3/2}, \quad h(u,-1) = 1 , \qquad (6.4)$$

where r(u,v) is the usual Schwarzschild coordinate. Notice that although h may be chosen to be a function of r on the initial surfaces, such a choice is inadvisable within the spacetime. For setting h(u,v) = H(r) in equation (6.2) implies that h becomes singular on the horizon. Our choice of coordinates and conformal factor has been made to ensure that the horizon is regular. The remaining data that have to be given on N_o, N_o' are the Weyl tensor components Ψ_0, Ψ_4 (both zero), and the metric at S_o. Thus we do not give explicitly the position of the horizon or singularity. Further, this is a genuine 2-variable problem.

Using a step length of 1.1×10^{-2} in the u- and v-directions we have evolved the Schwarzschild spacetime numerically. Fig.8 shows $|\Psi_2(u,v)/\Omega|$. One can see that the complete shaded region of figure 7 has been obtained. The singularity was correctly

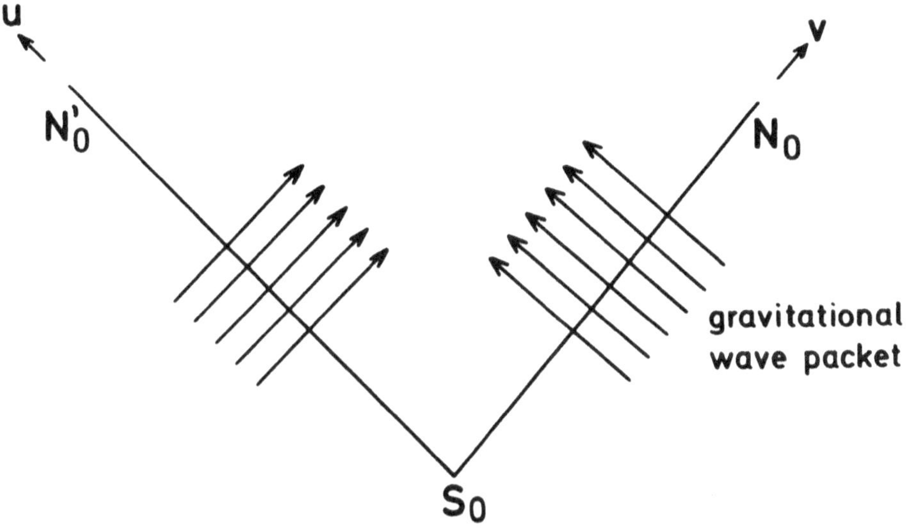

Figure 9

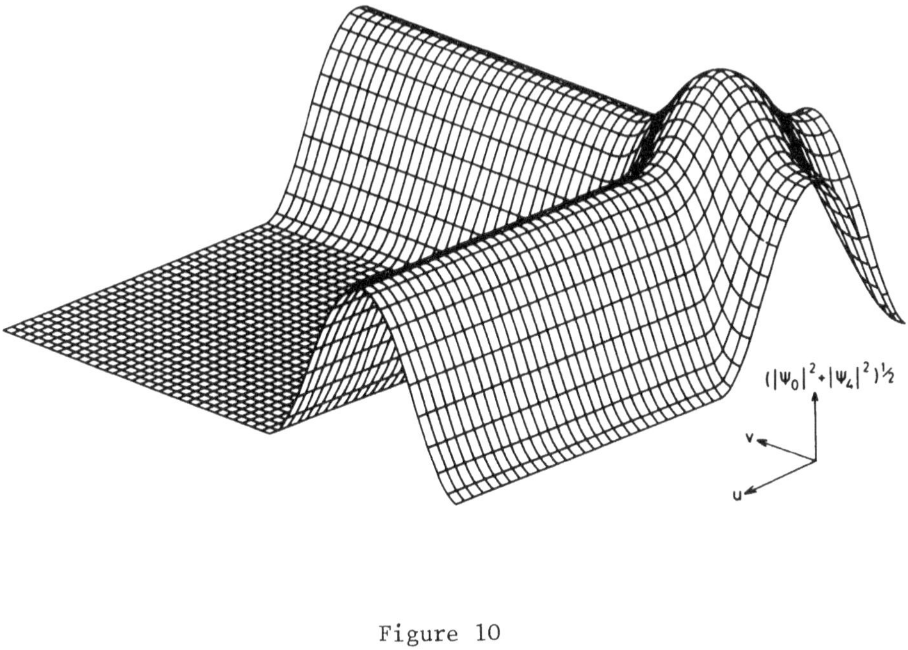

Figure 10

located to about one part in 10^7. Away from the singularity the errors in the dependent variables were typically $\lesssim 10^{-5}$, but close to the singularity this degraded to 10^{-2}.

The problem which is currently under investigation is the following. Consider two colliding gravitational wave packets in a vacuum spacetime, Figure 9. What happens? To analyse this problem we start from Minkowski spacetime with metric

$$ds^2 = dt^2 - dx^2 - dy^2 - dz^2.$$

Introducing null coordinates $u = t-z$, $v = t+z$ we have

$$ds^2 = 2dudv - dx^2 - dy^2 .$$

We next consider a situation where spacetime is Minkowski for $u < 0$, $v < 0$, and S_o is $u = 0$, $v = 0$. Now suppose a single gravitational wave packet crosses N_0. We specify this by setting $\Psi_4 = 0$, $\Psi_0 = \Psi_0(v,x,y)$ and evolving our equations. A very important case (at least theoretically), is that of plane waves in which there is no x,y-dependence, $\Psi_0 = \Psi_0(v)$. The wave simply propagates unaltered. Next consider two colliding plane waves within linearised gravity, (i.e. we linearise the Einstein field equations about the Minkowski background). The waves propagate unaltered until the interaction region is reached and after leaving this region they resume their original shape. Figure 10 shows $(|\Psi_0|^2 + |\Psi_4|^2)^{\frac{1}{2}}$ in a typical such configuration. The unaltered form of the wavefront can clearly be seen.

In the full non-linear theory however things are very different. To see this consider the pair of non-linear equations on N_o,

$$D\rho = \rho^2 + \sigma\bar{\sigma} , \quad D\sigma = 2\rho\sigma + \Psi_0, \quad D = d/dv. \tag{6.5}$$

Before the waves arrive $\rho = \sigma = 0$. Now suppose Ψ_0, the free datum, becomes non-zero only for $0 \leq v \leq v_0$. In linearised theory the quadratic terms in (6.5) do not contribute, and so for $v > v_o$,

$$\rho = 0, \qquad \sigma = \sigma_o = \int_0^{v_0} \Psi_0(v)dv . \tag{6.6}$$

However, if non-linearities are taken into account, then $\rho,\sigma \to \infty$ in a finite time. This can be seen most easily by considering the case of impulsive plane waves,

$$\Psi_0 = a\delta(v) , \qquad \Psi_4 = b\delta(u) , \tag{6.7}$$

where a,b are constants. Then at $v = v_o = 0+$, $\rho = 0$, $\sigma = a$, and for $v > 0$,

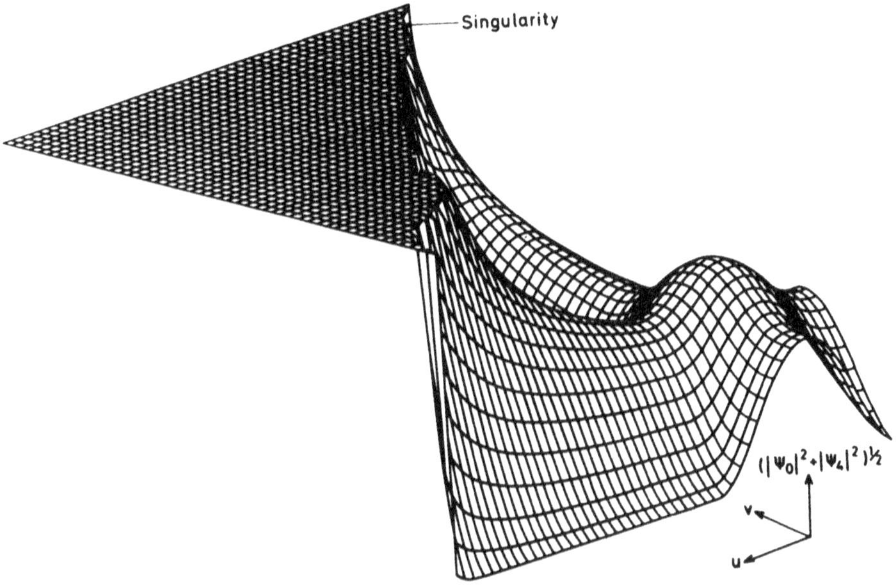

Figure 11

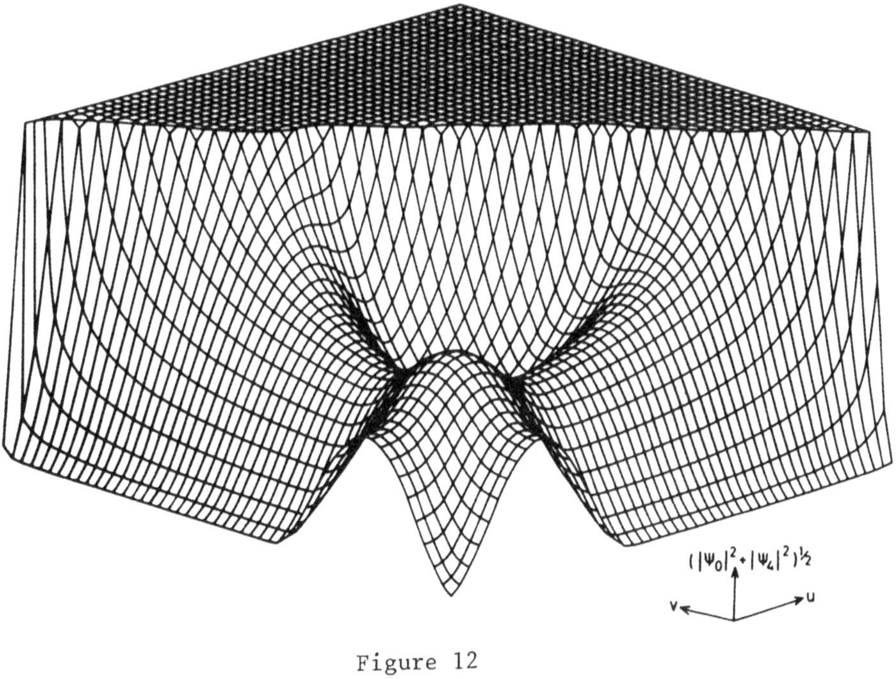

Figure 12

$$\rho = a^2v/(1 - a^2v^2), \qquad \sigma = a/(1 - a^2v^2), \tag{6.8}$$

with a singularity at $v = a^{-1}$. For a single impulsive plane wave this is an A2 caustic, which can be removed by changing to a different coordinate/tetrad system. For a pair of colliding impulsive plane waves $v = a^{-1}$ gives the intersection of N_0 with a spacetime singularity. The exact solution has been given by Szekeres (13) and by Khan & Penrose (14). We have integrated this spacetime numerically using a steplength $h = 7 \times 10^{-3}$. The relative errors were 0.1% away from the singularity, whose position and order was obtained to within 1%. (This was done using the first version of the code. Using a smaller steplength and the latest version should improve the accuracy considerably.)

It is sometimes claimed that this singularity is "unphysical" and would disappear if the wavefronts were smooth and/or the restriction to plane symmetry was dropped. It is very easy to consider smooth wavefronts. We considered the spacetime which was Minkowskian at S_0 and had the following wavefronts crossing N_0N_0':

$$\Psi_0 = af(v), \qquad \Psi_4 = bf(u), \qquad f(x) = h(x)h(1-x)x^2(1-x)^2 ,$$

where h(x) is the Heaviside step function. This was the data used to produce Fig. 10 from linearised theory. The equivalent figure for the full theory is shown as Fig. 11 and from a different angle as Fig. 12. Since there is no known exact solution we have no independent check on the accuracy.

Next we consider a single impulsive wave propagating into Minkowski spacetime, i.e. the data on N_0, N'_0 is

$$\Psi_0 = f(x,y)\delta(v), \qquad \Psi_4(u,x,y) = 0 \tag{6.9}$$

Here spacetime is Minkowskian for $v = t+z < 0$, t,x,y,z, are Cartesian coordinates in this region, and are continued according to the prescription of chapter 2. Somewhat surprisingly it turns out that generic impulsive waves have spacetime singularities behind them, $(v > 0)$. To make this statement more precise and outline the proof consider equation B1 from appendix B,

$$(D + 4\rho)\Psi_1 = (\bar{\delta} + 4\alpha - \pi)\Psi_0 . \tag{6.10}$$

For $v < 0$, spacetime is Minkowski and $\Psi_1 = 0$. Using (6.9) and integrating we find that at $v = 0_+$,

$$\Psi_1 = \overset{0}{\Psi}_1 = (\bar{\delta} + 4\alpha - \pi)f , \tag{6.11}$$

where the quantities in the bracket on the right take on their Minkowski values. A <u>generic impulsive wave</u> is one for which $\overset{0}{\Psi}_1 \neq 0$. In Cartesian coordinates with the frame of chapter 2

$\alpha = \pi = 0$ at $v = 0$. Thus in this system all waves are generic except those for which f = constant. The latter case is precisely plane waves. The equations to be solved on N_0 are ordinary differential equations and, assuming there is one spacelike Killing vector (a symmetry), they can be integrated analytically. It is tedious but straightforward to show that at the caustic $C^{abcd}C_{abcd}$, a curvature invariant becomes infinite. The restriction to impulsive waves also appears to be unnecessary, (but of course the equations cannot be integrated exactly in the general case); what matters for a singularity is that if Ψ_0 is non-zero only for $0 < v < v_0$ then $\Psi_1 \neq 0$ at $v = v_0$. The details and numerical examples will be published elsewhere.

If we replace x,y by plane polar coordinates $r = (x^2 + y^2)^{\frac{1}{2}}$, $\theta = \tan^{-1}(y/x)$ then a standard choice for m^a in the Minkowski region is $m^a = 2^{-\frac{1}{2}}(0,0,1,i/r)$. The data for a single impulsive wave is

$$\Psi_0 = f(r,\theta)\delta(v) \text{ on } N_0 , \qquad \Psi_4 = 0 \text{ on } N_0' . \tag{6.12}$$

Now consider the axisymmetric case $f = f(r)$. Then the only non-generic wave is given by $f(r) = \text{const.}/r^2$. This is of particular interest since it represents the limiting case of boosting a Schwarzschild black hole of mass m through velocity v in the negative z-direction, letting $v \to c$, holding $m/(1-v^2/c^2)^{\frac{1}{2}}$ constant, "a black hole moving at the speed of light". A problem of particular physical interest would be the collision of two such objects. Some analytic results already exist (15,16), and numerical investigations under way will be reported elsewhere.

APPENDIX A: THE CONSTRUCTION OF THE SOLUTION ON S_0

On S_0 we have to specify the following quantities as functions of x^A:

$$Q,\ P^A,\ \rho,\ \mu,\ \sigma,\ \lambda,\ \tau,\ \Omega,\ s_0,\ s_2,\ \phi_0,\ \phi_4,\ \Phi_{00},\ \Phi_{22} .$$

That is we have to specify the internal geometry (2-metric) of S_0, its extrinsic curvatures in each of N_0, N_0', local details of the conformal compactification, and the radiation and matter passing through S_0 transverse to N_0 and to N_0'.

Since the directional derivatives $\delta,\bar{\delta}$ lie on S_0 we can differentiate Ω, Q and obtain s_1, π from

$$\delta\Omega = s_1 , \qquad \bar{\delta} \ln Q = \bar{\tau} - \pi .$$

Another equation from (6) is

$$(\delta - \bar{\alpha} + \beta)\bar{P}^A = (\bar{\delta} - \alpha + \bar{\beta})P^A ,$$

and so by differentiating P^A we obtain $\bar{\alpha}-\beta$. But $\tau = \bar{\alpha}+\beta$ is already known, and so we find α, β. Because Q has to be specified on N_o' we know ΔQ, and we may compute γ from

$$\Delta \ln Q = -2\gamma .$$

We next form $F \equiv (s_0 s_2 - s_1\bar{s}_1)/\Omega$ which is finite even on $\mathcal{J}$. Now a batch of equations from (6) is

$$\delta s_0 = \tau s_0 - \rho s_1 - \sigma\bar{s}_1 - \Omega\Phi_{01} ,$$

$$\delta s_1 = \bar{\lambda} s_0 - (\bar{\alpha}-\beta)s_1 - \sigma s_2 - \Omega\Phi_{02} ,$$

$$\bar{\delta} s_1 = \rho s_0 + (\bar{\alpha}-\beta)s_1 - \rho s_2 - F - \Omega\Phi_{11} ,$$

$$\delta s_2 = -\mu s_1 + \overline{\lambda s_1} - 2\gamma s_2 - \Omega\Phi_{12} .$$

Thus by differentiating s_0, s_1, s_2 we obtain Φ_{11}, Φ_{01},Φ_{02},Φ_{12}. Another batch of equations is

$$\delta\rho - \bar{\delta}\rho = \tau\rho + (\bar{\beta} - 3\alpha)\sigma - \Omega\phi_1,$$

$$\delta\lambda - \bar{\delta}\mu = \tau\mu + (\bar{\alpha} - 3\beta)\lambda - \Omega\phi_3,$$

$$\delta\alpha - \bar{\delta}\beta = \rho\mu - \lambda\sigma - 2\alpha\beta + \alpha\bar{\alpha} + \beta\bar{\beta} + \Phi_{11} - \Omega\phi_2 .$$

Thus further differentiation produces ϕ_1, ϕ_2, ϕ_3. Although numerical differentiation could be ill-conditioned one might reasonably expect to be given analytic data on S_o.

APPENDIX B: THE CONSTRUCTION OF THE SOLUTION ON N_o and N_o'

On N_o we have to specify the data Q (which corresponds to the choice of v on N_o), Φ_{00} and Φ_0 (which give the amount of matter and radiation crossing N_o). Thus $D = Q\partial/\partial v$ is a known operator. Now the equations $D_1 = 0$, $E_1 = 0$, where

$$D_1 \equiv D\Omega - s_0 , \qquad E_1 \equiv Ds_0 + \Omega\Phi_{00}$$

are coupled linear ODE's for Ω, s_0. Since Ω, s_0 have already been given at S_o, i.e. v = 0, we have to solve a standard initial value problem. In a similar manner $N_1 = 0$, $N_2 = 0$, where

$$N_1 \equiv D\rho - \rho^2 - \sigma\bar{\sigma} - \Phi_{00} , \qquad N_2 \equiv D\sigma - 2\rho\sigma - \Psi_0,$$

constitute a standard initial value problem for ρ,σ. Once this has been solved we move on to $F^A_4 = 0$, where

$$F^A_4 \equiv DP^A - \rho P^A - \sigma\bar{P}^A .$$

This, like all the remaining problems in this section is a linear initial value problem since ρ, σ are already known. Once P^A has been determined we know $\delta = P^A\partial/\partial x^A$, and so we may compute $\delta \ln Q \equiv \alpha+\beta-\pi$, which allows us to write π as a linear combination of α, β and a known quantity. We use this result to eliminate π from the equations

$$AN_1 = 0, \quad AN_2 = 0, \quad B_1 = 0, \quad C_1 = 0, \quad E_3 = 0 ,$$

where

$$AN_1 \equiv D\alpha - \rho(\alpha+\pi) - \beta\bar{\sigma} - \bar{\Phi}_{01},$$

$$AN_2 \equiv D\beta - \sigma(\alpha+\pi) - \rho\beta - \Omega\phi_1 ,$$

$$B_1 \equiv D\phi_1 - \bar{\delta}\phi_0 + (4\alpha-\pi)\phi_0 - 4\rho\phi_1,$$

$$C_1 \equiv D\Phi_{01} - \delta\Phi_{00} + (2\alpha+2\beta-\pi)\Phi_{00} - 2\sigma\Phi_{01} + 2\sigma\bar{\Phi}_{01} + \bar{s}_1\phi_0 - s_0\phi_1 ,$$

$$E_3 \equiv Ds_1 - \bar{\pi}s_0 + \Omega\Phi_{01} .$$

Note that the δ-terms in $C_1 = 0$, $B_1 = 0$ are derivatives of quantities which have already been determined. Thus we have a linear system of ODE's for α, β, ϕ_1, Φ_{01}, s_1 together with initial values at S_o. Once these have been solved we also know π. Next we move on to the system of ODE's

$$N_4 = 0, \quad N_5 = 0, \quad B_2 = 0, \quad C_2 = 0, \quad C_3 = 0, \quad E_2 = 0 ,$$

$$N_4 \equiv D\mu - \delta\pi - \rho\mu - \sigma\lambda - \pi(\bar{\pi}-\bar{\alpha}+\beta) - \Omega\phi_2 ,$$

$$N_5 \equiv D\lambda - \bar{\delta}\pi - \rho\lambda - \bar{\sigma}\mu - \pi(\pi+\alpha-\bar{\beta}) - \bar{\Phi}_{02} ,$$

$$B_2 \equiv D\phi_2 - \bar{\delta}\pi_1 + \lambda\phi_0 + 2(\alpha-\pi)\phi_1 - 3\rho\phi_2 ,$$

$$C_2 \equiv D\Phi_{02} - \delta\Phi_{01} + \bar{\lambda}\Phi_{00} + 2(\beta-\bar{\pi})\Phi_{01} - \rho\Phi_{02} - 2\sigma\Phi_{11} + s_2\phi_0 - s_1\phi_1 ,$$

$$C_3 \equiv D\Phi_{11} - \delta\bar{\Phi}_{01} + \mu\Phi_{00} - \pi\Phi_{01} + (2\bar{\alpha}-\bar{\pi})\Phi_{01} - \sigma\Phi_{02} - 2\rho\Phi_{11} + \bar{s}_1\phi_1 - s_0\phi_2,$$

$$E_2 \equiv Ds_2 - \pi s_1 - \overline{\pi s_1} - F + \Omega\Phi_{11} . \quad (F \equiv (s_0 s_2 - s_1\bar{s}_1)/\Omega)$$

These form a coupled linear system for μ, λ, ϕ_2, Φ_{02}, Φ_{11}, s_2, which again have to be solved for initial values given on S_o. Finally we have to solve the system

$$B_3 = 0, \quad B_4 = 0, \quad C_4 = 0, \quad C_5 = 0, \quad F^A_3 = 0, \quad AN_3 = 0 ,$$

where

$$B_3 \equiv D\phi_3 - \bar{\delta}\phi_2 + 2\lambda\phi_1 - 3\pi\phi_2 - 2\rho\phi_3 \quad ,$$

$$B_4 \equiv D\phi_4 - \bar{\delta}\phi_3 + 3\lambda\phi_2 - 2(\alpha+2\pi)\phi_3 - \rho\phi_4 \; ,$$

$$C_4 \equiv D\Phi_{12}-\delta\Phi_{11}+\mu\Phi_{01}+\overline{\lambda\Phi}_{01}-\pi\Phi_{02}-2\bar{\pi}\Phi_{11}-\rho\Phi_{12}-\sigma\bar{\Phi}_{12}+s_2\phi_1-s_1\phi_2 \; ,$$

$$C_5 \equiv D\Phi_{22}-\delta\bar{\Phi}_{12}+\overline{\lambda\Phi}_{02}+2\mu\Phi_{11} - 2\pi\Phi_{12}-2(\beta+\bar{\pi})\bar{\Phi}_{12}-\rho\Phi_{22}+s_2\phi_2-s_1\phi_3,$$

$$F^A_3 \equiv DC^A - (\bar{\tau}+\pi)P^A - (\tau+\bar{\pi})\bar{P}^A,$$

$$AN_3 \equiv D\gamma - (\tau+\bar{\pi})\alpha - (\bar{\tau}+\pi)\beta - \tau\pi-\Omega\phi_2- \Phi_{11} \; .$$

This completes the determination of the solution on N_o.

The treatment of N'_o is very similar. We require as initial data that Q, Φ_{22}, Ψ_4 be specified on N'_o. Since $C^A = 0$ on N'_o, $\Delta = \partial/\partial u$ there. Further $\gamma = -\frac{1}{2}\Delta \ln Q$, and so γ is real. (However, for convenience later we do not make use of this fact when writing out the equations on N'_o.) Now the equations $D'_1 = 0$, $E'_1 = 0$, where

$$D'_1 \equiv \Delta\Omega - s_2 \; , \qquad E'_1 \equiv \Delta s_2 + (\gamma+\bar{\gamma})s_2 + \Omega\Phi_{22},$$

are coupled linear ODE's for Ω, s_2 which can be solved given initial values on S_o, $u = 0$. Next $N'_1 = 0$, $N'_2 = 0$, where

$$N'_1 \equiv \Delta\mu+\mu(\mu+\gamma+\bar{\gamma})+\lambda\bar{\lambda}+\Phi_{22} \; , \qquad N'_2 \equiv \Delta\lambda+\lambda(2\mu+3\gamma-\bar{\gamma}) + \Omega\phi_4$$

are coupled non-linear ODE's for μ, λ on N'_o. All subsequent equations on N'_o are linear ODE systems.

Next we solve $F^A_6 = 0$, where

$$F^A_6 \equiv \Delta P^A + (\mu-\gamma+\bar{\gamma})\, P^A + \overline{\lambda P}^A \; ,$$

for P^A so that $\delta = P^A\partial/\partial x^A$ is known. Once we have δ we can solve

$$AN'_1 = 0, \quad AN'_2 = 0, \quad B'_1 = 0, \quad C'_1 = 0, \quad E'_3 = 0,$$

where

$$AN'_1 \equiv \Delta\alpha-\bar{\delta}\gamma+(\nu-\bar{\gamma})\alpha+(\bar{\tau}-\bar{\beta})\;\gamma(\tau+\beta)\;\lambda+\Omega\phi_3 \; ,$$

$$AN'_2 \equiv \Delta\beta-\delta\gamma+\mu\tau-\beta(\gamma-\bar{\gamma}-\mu)+\alpha\bar{\lambda}+\Phi_{12},$$

$$B'_1 \equiv \Delta\phi_3-\delta\phi_4+2(\gamma+2\mu)\phi_3+(\tau-4\beta)\phi_4 \; ,$$

$$C'_1 \equiv \Delta\Phi_{12}-\delta\Phi_{22}+2(\mu+\bar{\gamma})\Phi_{12}+2\overline{\lambda\Phi}_{12}-\tau\Phi_{22}+\bar{s}_1\bar{\phi}_4-s_2\bar{\phi}_3 \; ,$$

$$E_3 \equiv \Delta s_1-(\gamma-\bar{\gamma})s_1+\tau s_2+\Omega\Phi_{12}$$

for α, β, ϕ_3, Φ_{12}, s_1. We next solve $N_3' = 0$ for π, where

$$N_3' \equiv \Delta\pi + (\pi+\bar{\tau})\mu + (\bar{\pi}+\tau)\lambda + (\gamma-\bar{\gamma})\pi + \Omega\phi_3 + \bar{\Phi}_{12} .$$

The next set of equations is

$$B_2 = 0, \quad C_2' = 0, \quad C_3' = 0, \quad N_4' = 0, \quad N_5' = 0, \quad E_2' = 0 ,$$

where

$$B_2' \equiv \Delta\phi_2 - \delta\phi_3 + 3\mu\phi_2 - 2(\beta-\tau)\phi_3 - \sigma\phi_4 ,$$

$$C_2' \equiv \Delta\Phi_{02} - \delta\Phi_{12} + (2\gamma-2\bar{\gamma}+\mu)\Phi_{02} + 2\bar{\lambda}\Phi_{11} + 2(\tau-\bar{\alpha})\Phi_{12} - \sigma\Phi_{22} + s_0\bar{\phi}_4 - s_1\bar{\phi}_3 ,$$

$$C_3' \equiv \Delta\Phi_{11} \; \bar{\delta}\Phi_{12} + \lambda\Phi_{02} + 2\mu\Phi_{11} + (\bar{\tau}-2\bar{\beta})\Phi_{12} + \tau\bar{\Phi}_{12} - \rho\Phi_{22} + s_1\phi_3 - s_2\phi_2 ,$$

$$N_4' \equiv \Delta\rho - \bar{\delta}\tau - (\gamma+\bar{\gamma}-\mu)\rho + \sigma\lambda - (\bar{\beta}-\alpha-\bar{\tau})\tau + \Omega\phi_2 ,$$

$$N_5' \equiv \Delta\sigma - \delta\tau + (\mu-3\gamma+\bar{\gamma})\sigma + \bar{\lambda}\rho + (\tau+\beta-\bar{\alpha})\tau + \Phi_{02} ,$$

$$E_2' \equiv \Delta s_0 - (\gamma+\bar{\gamma})s_0 + \bar{\tau}s_1 + \tau\bar{s}_1 - F + \Omega\Phi_{11} ,$$

which have to be solved for ϕ_2, Φ_{02}, Φ_{11}, ρ, σ, s_0. The final set of equations is

$$B_3' = 0, \quad B_4' = 0, \quad C_4' = 0, \quad C_5' = 0 ,$$

where

$$B_3' \equiv \Delta\phi_1 - \delta\phi_2 - 2(\gamma-\mu)\phi_1 + 3\tau\phi_2 - 2\sigma\phi_3 ,$$

$$B_4' \equiv \Delta\phi_0 - \delta\phi_1 - (4\gamma-\mu)\phi_0 + 2(2\tau+\beta)\phi_1 - 3\sigma\phi_2 ,$$

$$C_4' \equiv \Delta\Phi_{01} - \delta\Phi_{11} + (\mu-2\gamma)\Phi_{01} + \bar{\lambda}\bar{\Phi}_{01} + \bar{\tau}\Phi_{02} + 2\tau\Phi_{11} - \rho\Phi_{12} - \sigma\bar{\Phi}_{12} + s_0\bar{\phi}_3 - s_1\bar{\phi}_2 ,$$

$$C_5' \equiv \Delta\Phi_{00} - \bar{\delta}\Phi_{01} + (\mu-2\gamma-2\bar{\gamma})\Phi_{00} + 2(\alpha+\bar{\tau})\Phi_{01} + 2\tau\bar{\Phi}_{01} - \bar{\sigma}\Phi_{02} - 2\rho\Phi_{11} + s_0\phi_2 - \bar{s}_1\phi_1 ,$$

which have to be solved for ϕ_1, ϕ_0, Φ_{01}, Φ_{00}. This completes the solution on N_0'.

APPENDIX C: THE CONSTRUCTION OF THE SOLUTION IN M

It is shown in (6) that the following set of equations,

$(F_5,\ F_3^A,\ F_4^A,\ N_1,\ N_1',\ N_2,\ N_2',\ N_3',\ AN_3,\ AN_2,\ AN_1,$

$B_4',\ B_1+B_3',\ B_2+B_2',\ B_3+B_1',\ B_4,\ 2C_5',\ 2C_1+2C_4'+C_6,$

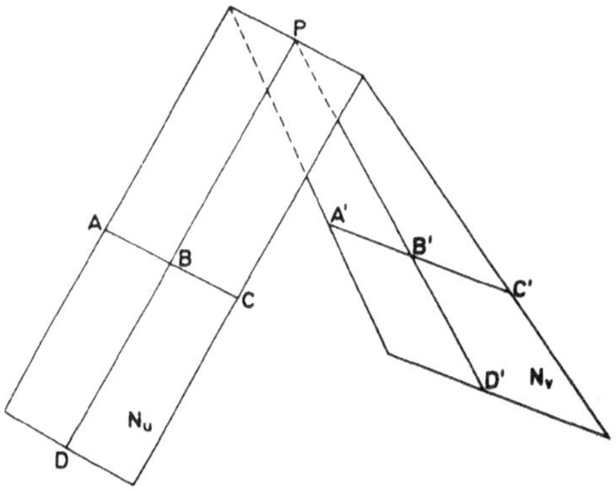

Figure 13

$$C_2+C'_2,\ C_3+C'_3,\ 2C'_1+2C_4+C_6,\ 2C_5,\ D_1,\ E_1,\ E_2,\ E_3) = 0,$$

form a quasilinear symmetric hyperbolic system of equations for our unknowns in M. Here

$$C_6 \equiv \delta\Phi_{11}-\bar{\delta}\Phi_{02}+\mu\Phi_{01}-\overline{\lambda\Phi}_{01}+2(\bar{\beta}-\alpha)\Phi_{02}-\rho\Phi_{12}+\sigma\bar{\Phi}_{12}-s_2\phi_1+s_1(\phi_2+\bar{\phi}_2)-s_0\bar{\phi}_3\ ,$$

and all of the other quantities were defined in appendix B. As initial values we have the solution on $N_o \cup N'_o$.

The system is canonical hyperbolic. Each equation involves one of the directional derivatives D, Δ or $D+\Delta$, and 0,1 or 2 "side derivative" terms in $\delta, \bar{\delta}$, so that in general no more than 3 directional derivatives occur in each equation. If an explicit finite differencing scheme is used then the $D+\Delta$ derivatives can be eliminated. Consider e.g. $B_1 + B'_3 = 0$. We use $B_1 = 0$ to obtain one estimate of ϕ_1 at a new grid point, and $B'_3 = 0$ to obtain a second estimate. We then use the average of the two estimates. To see in more detail how this works, consider an explicit 3 level method and the computational grid of Fig.13. We want to use known values of the solution at A,B,C,D,A',B',C', D' to estimate a new value at P. We use values along the line DBP to integrate

$$(F_3^A,\ F_4^A,\ N_1,\ N_2,\ AN_3,\ AN_2,\ AN_1,\ D_1,\ E_1,\ E_2,\ E_3) = 0\ ,$$

for C^A, P^A, ρ, σ, γ, β, α, Ω, s_0, s_2, s_1. We treat these 19 real equations effectively as if they were ordinary differential equations. We use values along the line D'B'P to integrate

$$(F_5,\ N_1',\ N_2',\ N_3') = 0\ ,$$

for Q, μ, λ, π, as if they were 6 ordinary differential equations. Next we integrate the equations

$$(B_1,\ B_2,\ B_3,\ B_4,\ C_1,\ C_2,\ C_3,\ C_4,\ C_5) = 0$$

in the surface DABCP for ϕ_1, ϕ_2, ϕ_3, ϕ_4, Φ_{01}, Φ_{02}, Φ_{11}, Φ_{12}, Φ_{22}. In the general case (no symmetries), we treat these as hyperbolic equations in one time and two space variables. At present the finite size of computers restricts us to problems with at least one spacelike Killing vector. Thus we effectively integrate equations in one time and one space variable. Next we integrate

$$(B_1',\ B_2',\ B_3',\ B_4',\ C_1',\ C_2',\ C_3',\ C_4',\ C_5') = 0$$

in the surface D'A'B'C'P for ϕ_3, ϕ_2, ϕ_1, ϕ_0, Φ_{12}, Φ_{02}, Φ_{11}, Φ_{01}, Φ_{00}. Finally we average the values of ϕ_1, ϕ_2, ϕ_3, Φ_{01}, Φ_{02}, Φ_{11}, Φ_{12} at P.

Although this algorithm clearly uses a great deal of computer store the reduction in the number of effective independent variables makes it extremely fast.

REFERENCES

1. Walker, M.: 1982, "On the positivity of total gravitational energy at retarded times", preprint MPA 19, Max-Planck-Institute for Astrophysics.

2. Penrose, R.: 1968, "Battelle Rencontres", (ed. DeWitt, C.M., and Wheeler, J.A., Benjamin), pp.121-235.

3. Schmidt, B.G.: 1979, "Isolated Gravitating Systems in General Relativity", (ed. Ehlers, J., Soc. Italiana di Fisica), pp.11-49.

4. Newman, E.T. and Tod, K.P.: 1980, "General Relativity and Gravitation", (ed. Held, A., Plenum), pp.1-36.

5. Friedrich, H.: 1981, Proc.Roy.Soc.Lond. A375, pp.169-184.

6. Stewart, J.M. and Friedrich, H.: 1982, Proc.Roy.Soc.Lond. (in press).

7. Friedrich, H. and Stewart, J.M.: 1982, Proc.Roy.Soc.Lond. (in press).

8. Corkill, R. and Stewart, J.M.: 1982, "Numerical Relativity II" preprint MPA 32 , Max-Planck-Institute for Astrophysics.

9. Newman, E.T. and Penrose, R.: 1962, J.Math.Phys. 3, pp.566-578.

10. Tipler, F.J., Clarke, C.J.S. and Ellis, G.F.R.: 1980, "General Relativity and Gravitation", (ed. Held,A., Plenum), pp.97-206.

11. Lambert, J.D. and Shaw, B.: 1966, Mathematics of Computation 20, pp.11-20.

12. Hartree, D.R.: 1958, "Numerical Analysis", (2nd ed., Oxford).

13. Szekeres, P.: 1972, J.Math.Phys. 13, pp.286-294.

14. Khan, X. and Penrose, R.: 1971, Nature 229, pp.185-186.

15. Curtis, G.E.: 1978, General Relativity & Gravitation 9, pp.999-1008.

16. D'Eath, P.D.: 1978, Phys.Rev.D18, pp.990-1019.

PARTICIPANTS

DR. ROBERTO BEDOGNI
OSSERVATORIO ASTRONOMICO
UNIVERSITARIO
VIA ZAMBONI, 33
CASELLA POSTALE 596
40100 BOLOGNA
I T A L Y

DR. ERICH BETTWIESER
UNIVERSITAETS-STERNWARTE
GOETTINGEN
GEISMARLANDSTR. 11
D-3400 GOETTINGEN
FED. REP. OF G E R M A N Y

DR. WOLFGANG BRINKMANN
MAX-PLANCK-INSTITUT FUER
EXTRATERRESTRISCHE PHYSIK
D-8046 GARCHING BEI MUENCHEN
FED. REP. OF G E R M A N Y

DR. J. ROBERT BUCHLER
UNIVERSITY OF FLORIDA
DEPARTMENT OF PHYSICS
215 WILLIAMSON HALL
GAINESVILLE, FL 32611
U S A

DR. JOHN I. CASTOR, L-18
LAWRENCE LIVERMORE NATIONAL LABORATORY
P.O. BOX 808
LIVERMORE, CA 94550
U S A

DR. ERNST DORFI
MAX-PLANCK-INSTITUT
FUER KERNPHYSIK
POSTFACH 103980
SAUPFERCHECKWEG 1
D-6900 HEIDELBERG 1
FED. REP. OF G E R M A N Y

DR. JAMES W. EASTWOOD
UKAEA CULHAM LABORATORY
ABINGDON, OXON OX14 3DB
E N G L A N D

DR. H. FALK
UNIVERSITY OF WESTERN ONTARIO
DEPARTMENT OF ASTRONOMY
LONDON, ON N6A 3K7
C A N A D A

DR. BRUCE A. FRYXELL
UNIVERSITY OF CALIFORNIA
SANTA CRUZ
LICK OBSERVATORY
SANTA CRUZ, CA 95064
U S A

DR. D.F. GRIFFITHS
DEPT. OF MATHEMATICS
UNIV. OF DUNDEE
DUNDEE, DD1 4HN
S C O T L A N D

DR. JOHN HAWLEY
UNIVERSITY OF ILLINOIS
DEPARTMENT OF ASTRONOMY
341 ASTRONOMY BUILDING
1011 WEST SPRINGFIELD
URBANA, IL 61801
U S A

DR. WOLFGANG HILLEBRANDT
MAX-PLANCK-INSTITUT FUER
ASTROPHYSIK
KARL-SCHWARZSCHILD-STR. 1
D-8046 GARCHING
FED. REP. OF G E R M A N Y

DR. RICHARD I. KLEIN, L-18
LAWRENCE LIVERMORE NATIONAL LABORATORY
P.O. BOX 808
LIVERMORE, CA 94550
U S A

DR. RICHARD ISAACSON
DEPARTMENT OF PHYSICS AND ASTRONOMY
UNIVERSITY OF PITTSBURGH
PITTSBURGH, PA 15260
U S A

DR. JACQUES LEORAT
OBS. DE MEUDON
PLACE JULES JANSSON
F-92190 MEUDON
F R A N C E

DR. PASQUALE LONDRILLO
OSSERVATORIO ASTRONOMICO
UNIVERSITARIO
VIA ZAMBONI, 33
CASELLA POSTALE 596
40100 BOLOGNA
I T A L Y

DR. PHILIP MARCUS
MIT
DEPARTMENT OF MATHEMATICS
CODE 2-337
CAMBRIDGE, MA 02139
U S A

DR. J. MEYER-TER-VEHN
MAX-PLANCK-INSTITUT
FUER QUANTENOPTIK
D-8046 GARCHING BEI MUENCHEN
FED. REP. OF G E R M A N Y

DR. BARBARA W. MIHALAS
SACRAMENTO PEAK NATIONAL OBSERVATORY
SUNSPOT, N.M. 88349
U S A

DR. DIMITRI MIHALAS
LOS ALAMOS NATIONAL LABORATORY
P.O. BOX 1663
LOS ALAMOS, N.M. 87545
U S A

DR. P.J. MOREL
UNIVERSITE DE NICE
OBSERVATOIRE
B.P. 252 06007 NICE CEDEX
F R A N C E

DR. A. NESIS
KIEPENHEUER-INSTITUT
FUER SONNENPHYSIK
SCHOENECKSTR. 6
D-7800 FREIBURG IM BREISGAU
FED. REP. OF G E R M A N Y

DR. MICHAEL J. NEWMAN, X-2, MS-220
LOS ALAMOS NATIONAL LABORATORY
P.O. BOX 1663
LOS ALAMOS, N.M. 87545
U S A

DR. MICHAEL L. NORMAN
MAX-PLANCK-INSTITUT
FUER ASTROPHYSIK
KARL-SCHWARZSCHILD-STR. 1
D-8046 GARCHING BEI MUENCHEN
FED. REP. OF G E R M A N Y

DR. GORDON L. OLSON, X-7, MS-625
LOS ALAMOS NATIONAL LABORATORY
P.O. BOX 1663
LOS ALAMOS, N.M. 87545
U S A

DR. ANNICK POUQUET
UNIVERSITE DE NICE
OBSERVATOIRE
B.P. 252
F-06007 NICE CEDEX
F R A N C E

DR. JEAN-PIERRE POYET
OBSERVATOIRE DE L'UNIVERSITE
DE TOULOUSE
1 AVENUE CAMILLE-FLAMMARION
F-31500 TOULOUSE
F R A N C E

DR. REINHARD SCHLICKEISER
MAX-PLANCK-INSTITUT
FUER RADIOASTRONOMIE
AUF DEM HUEGEL 69
D-5300 BONN 1
FED. REP. OF G E R M A N Y

DR. RODNEY SCHULTZ, X-2, MS-220
LOS ALAMOS NATIONAL LABORATORY
P.O. BOX 1663
LOS ALAMOS, N.M. 87545
U S A

DR. LARRY L. SMARR
UNIVERSITY OF ILLINOIS
DEPARTMENT OF ASTRONOMY
341 ASTRONOMY BUILDING
1011 WEST SPRINGFIELD
URBANA, IL 61801
U S A

DR. ROBERT F. STEIN
MICHIGAN STATE UNIVERSITY
DEPARTMENT OF ASTRONOMY AND
ASTROPHYSICS
EAST LANSING, MI 48824
U S A

DR. JOHN M. STEWART
UNIVERSITY OF CAMBRIDGE
DEPT. OF APPLIED MATHEMATICS
AND THEORETICAL PHYSICS
SILVER STREET
CAMBRIDGE CB3 9EW
E N G L A N D

DR. WERNER M. TSCHARNUTER
INSTITUT FUER ASTRONOMIE
DER UNIVERSITAET WIEN
TUERKENSCHANZSTR. 17
A-1180 WIEN
A U S T R I A

DR. DAVID L. TUBBS, X-2, MS-220
LOS ALAMOS NATIONAL LABORATORY
P.O. BOX 1663
LOS ALAMOS, N.M. 87545
U S A

DR. GEERT DICK VAN ALBADA
KAPTEYN LABORATORIUM
DER RIJKSUNIVERSITEIT TE GRONINGEN
NETTELBOSJE 2
POSTBUS 800
9700 AV GRONINGEN
THE N E T H E R L A N D S

DR. ROBERT P. WEAVER, X-7, MS-625
LOS ALAMOS NATIONAL LABORATORY
P.O. BOX 1663
LOS ALAMOS, N.M. 87545
U S A

DR. GARY L. WELTER
ASTRONOMISCHE INSTITUTE
DER UNIVERSITAET BONN
SONDERFORSCHUNGSBEREICH 131
AUF DEM HUEGEL 71
D-5300 BONN 1
FED. REP. OF G E R M A N Y

DR. JAMES R. WILSON, L-35
LAWRENCE LIVERMORE NATIONAL LABORATORY
P.O. BOX 808
LIVERMORE, CA 94550
U S A

DR. KARL-HEINZ A. WINKLER
MAX-PLANCK-INSTITUT
FUER ASTROPHYSIK
KARL-SCHWARZSCHILD-STR. 1
D-8046 GARCHING BEI MUENCHEN
FED. REP. OF G E R M A N Y

DR. PAUL R. WOODWARD, L-71
LAWRENCE LIVERMORE NATIONAL LABORATORY
P.O. BOX 808
LIVERMORE, CA 94550
U S A

DR. HAROLD W. YORKE
UNIVERSITAETS-STERNWARTE
GOETTINGEN
GEISMARLANDSTR. 11
D-3400 GOETTINGEN
FED. REP. OF G E R M A N Y

INDEX

GPSR Compliance

The European Union's (EU) General Product Safety Regulation (GPSR) is a set of rules that requires consumer products to be safe and our obligations to ensure this.

If you have any concerns about our products, you can contact us on ProductSafety@springernature.com

In case Publisher is established outside the EU, the EU authorized representative is:

Springer Nature Customer Service Center GmbH
Europaplatz 3
69115 Heidelberg, Germany

Batch number: 10761697

Printed by Printforce, the Netherlands